S0-BPI-906

PROGRESS IN CLINICAL AND BIOLOGICAL RESEARCH

RECENT TITLES

Vol 290: **Enzymology and Molecular Biology of Carbonyl Metabolism 2: Aldehyde Dehydrogenase, Alcohol Dehydrogenase, and Aldo-Keto Reductase,** Henry Weiner, T. Geoffrey Flynn, *Editors*

Vol 291: **QSAR: Quantitative Structure-Activity Relationships in Drug Design,** J.L. Fauchère, *Editor*

Vol 292: **Biological and Synthetic Membranes,** D. Allan Butterfield, *Editor*

Vol 293: **Advances in Cancer Control: Innovations and Research,** Paul N. Anderson, Paul F. Engstrom, Lee E. Mortenson, *Editors*

Vol 294: **Development of Preimplantation Embryos and Their Environment,** Koji Yoshinaga, Takahide Mori, *Editors*

Vol 295: **Cells and Tissues: A Three-Dimensional Approach by Modern Techniques in Microscopy,** Pietro M. Motta, *Editor*

Vol 296: **Developments in Ultrastructure of Reproduction,** Pietro M. Motta, *Editor*

Vol 297: **Biochemistry of the Acute Allergic Reactions: Fifth International Symposium,** Bruce U. Wintroub, Alfred I. Tauber, Arlene Stolper Simon, *Editors*

Vol 298: **Skin Carcinogenesis: Mechanisms and Human Relevance,** Thomas J. Slaga, Andre J.P. Klein-Szanto, R.K. Boutwell, Donald E. Stevenson, Hugh L. Spitzer, Bob D'Motto, *Editors*

Vol 299: **Perspectives in Shock Research: Metabolism, Immunology, Mediators, and Models,** John C. Passmore, *Editor*, Sherwood M. Reichard, David G. Reynolds, Daniel L. Traber, *Co-Editors*

Vol 300: **Alpha$_1$-Acid Glycoprotein: Genetics, Biochemistry, Physiological Functions, and Pharmacology,** Pierre Baumann, Chin Bin Eap, Walter E. Müller, Jean-Paul Tillement, *Editors*

Vol 301: **Prostaglandins in Clinical Research: Cardiovascular System,** Karsten Schrör, Helmut Sinzinger, *Editors*

Vol 302: **Sperm Measures and Reproductive Success: Institute for Health Policy Analysis Forum on Science, Health, and Environmental Risk Assessment,** Edward J. Burger, Jr., Robert G. Tardiff, Anthony R. Scialli, Harold Zenick, *Editors*

Vol 303: **Therapeutic Progress in Urological Cancers,** Gerald P. Murphy, Saad Khoury, *Editors*

Vol 304: **The Maillard Reaction in Aging, Diabetes, and Nutrition,** John W. Baynes, Vincent M. Monnier, *Editors*

Vol 305: **Genetics of Kidney Disorders,** Christos S. Bartsocas, *Editor*

Vol 306: **Genetics of Neuromuscular Disorders,** Christos S. Bartsocas, *Editor*

Vol 307: **Recent Advances in Avian Immunology Research,** Balbir S. Bhogal, Guus Koch, *Editors*

Vol 308: **Second Vienna Shock Forum,** Günther Schlag, Heinz Redl, *Editors*

Vol 309: **Advances and Controversies in Thalassemia Therapy: Bone Marrow Transplantation and Other Approaches,** C. Dean Buckner, Robert Peter Gale, Guido Lucarelli, *Editors*

Vol 310: **EORTC Genitourinary Group Monograph 6: BCG in Superficial Bladder Cancer,** Frans M.J. Debruyne, Louis Denis, Ad P.M. van der Meijden, *Editors*

Vol 311: **Molecular and Cytogenetic Studies of Non-Disjunction,** Terry J. Hassold, Charles J. Epstein, *Editors*

Vol 312: **The Ocular Effects of Prostaglandins and Other Eicosanoids,** Laszlo Z. Bito, Johan Stjernschantz, *Editors*

Vol 313: **Malaria and the Red Cell 2,** John W. Eaton, Steven R. Meshnick, George J. Brewer, *Editors*

Vol 314: **Inherited and Environmentally Induced Retinal Degenerations,** Matthew M. LaVail, Robert E. Anderson, Joe G. Hollyfield, *Editors*

Vol 315: **Muscle Energetics,** Richard J. Paul, Gijs Elzinga, Kazuhiro Yamada, *Editors*

Please contact publisher for information about previous titles in this series.

INHERITED AND ENVIRONMENTALLY INDUCED RETINAL DEGENERATIONS

INHERITED AND ENVIRONMENTALLY INDUCED RETINAL DEGENERATIONS

Proceedings of the International Symposium on Retinal Degenerations, Held in San Francisco, California, September 2 and 3, 1988

Editors

Matthew M. LaVail
Departments of Anatomy and Ophthalmology
University of California
San Francisco

Robert E. Anderson
Cullen Eye Institute
Baylor College of Medicine
Houston, Texas

Joe G. Hollyfield
Cullen Eye Institute
Baylor College of Medicine
Houston, Texas

ALAN R. LISS, INC. • NEW YORK

Address all Inquiries to the Publisher
Alan R. Liss, Inc., 41 East 11th Street, New York, NY 10003

Printed in the United States of America

The publication of this volume was facilitated by the authors and editors who submitted the text in a form suitable for direct reproduction without subsequent editing or proofreading by the publisher.

Figures 2, 5, and 6 from the chapter by Nir and Papermaster were reproduced from Nir I et al: Opsin Distribution and Synthesis in Degenerating Photoreceptors of rd Mutant Mice, Journal of Experimental Eye Research, in press, with permission of Academic Press.

Library of Congress Cataloging-in-Publication Data

International Symposium on Retinal Degenerations (1988 : San Francisco, Calif.)
Inherited and environmentally induced retinal degenerations.

(Progress in clinical and biological research ; v. 314)
"Satellite meeting to the VIII International Congress of Eye Research"--Pref.
Includes index.
1. Retinal degeneration--Congresses. 2. Retinitis pigmentosa--Congresses. I. LaVail, Matthew M. II. Anderson, Robert E. III. Hollyfield, Joe G. IV. International Congress of Eye Research (8th : 1988 : San Francisco, Calif.) V. Title. VI. Series. [DNLM: 1. Light--adverse effects--congresses. 2. Photoreceptors--pathology--congresses. 3. Pigment Epithelium of Eye--pathology--congresses. 4. Pigment Epithelium of Eye--transplantation--congresses. 5. Retinal Degeneration--congresses. 6. Retinitis Pigmentosa--congresses. W1 PR668E v. 134 / WW 270 I625i 1988]
RE661.D3I57 1989 617.7'3 89-8250
ISBN 0-8451-5164-9

Dedicated to

Alan M. Laties, M.D.

Professor of Ophthalmology, University of Pennsylvania School of Medicine and Chairman of the Scientific Advisory Board of the Retinitis Pigmentosa Foundation Fighting Blindness since its inception—for his untiring efforts and determination in seeking the means to prevent or cure retinitis pigmentosa and allied retinal degenerations.

Contents

Contributors

G.M. Acland, School of Veterinary Medicine, University of Pennsylvania, Philadelphia, PA 19104 **[427]**

G. Adam, Molecular Genetics Unit, University of Newcastle-upon-Tyne, England **[83]**

Ruben Adler, Retinal Degenerations Research Center, Department of Ophthalmology, Johns Hopkins University School of Medicine, Baltimore, MD 21205 **[169]**

G.D. Aguirre, School of Veterinary Medicine, University of Pennsylvania, Philadelphia, PA 19104 **[427]**

Sahar Al-Mahdawi, Department of Pathology, Institute of Ophthalmology, London WC1H 9QS, England **[183]**

R.A. Alvarez, Cullen Eye Institute, Baylor College of Medicine, Houston, TX 77030 **[427]**

R.E. Anderson, Cullen Eye Institute, Baylor College of Medicine, Houston, TX 77030 **[427, 513]**

Björn-Erik Andersson, Department of Ophthalmology, University of Linköping, S-58185 Linköping, Sweden **[539]**

Mary F. Barbe, Department of Anatomy, Medical College of Pennsylvania, Philadelphia, PA 19104 **[523]**

David W. Batey, Division of Nutritional Sciences, School of Public Health, University of California, Los Angeles, CA 90024 **[331]**

Nicolas G. Bazan, Louisiana State University Eye Center, New Orleans, LA 70112 **[191]**

D. Beezley, Department of Physiology, Colorado State University, Fort Collins, CO 80523 **[555]**

Joanna Belin, Department of Chemical Pathology, University College and Middlesex School of Medicine, London W1P 6DB, England **[57]**

S.S. Bhattacharya, Molecular Genetics Unit, University of Newcastle-upon-Tyne, England **[83]**

Janet C. Blanks, Department of Ophthalmology, University of Southern California School of Medicine, Los Angeles, CA 90033 **[217]**

François-Xavier Borruat, Hôpital Ophtalmique, Université de Lausanne, 1004 Lausanne, Switzerland **[3]**

D.J. Bower, MRC Human Genetics Unit, Edinburgh, Scotland **[83]**

M. Brittis, Department of Ophthalmology, Columbia University, New York, NY 10032 **[659]**

Ann H. Bunt-Milam, Department of Ophthalmology, University of Washington, Seattle, WA 98195 **[19]**

Michael Burroughs, Sensory Sciences Center, Graduate School of Biomedical Sciences, University of Texas Health Science Center, Houston, TX 77030 **[291]**

Ruth B. Caldwell, Department of Anatomy, Medical College of Georgia, Augusta, GA 30912 **[393, 409]**

Rafael Cantera, Laboratory of Molecular Neuroanatomy, Department of Zoology, University of Lund, S-22362 Lund, Sweden **[275]**

Louvenia Carter-Dawson, Sensory Sciences Center, Graduate School of Biomedical Sciences, University of Texas Health Science Center, Houston, TX 77030 **[291]**

M. Caslake, Department of Pathological Biochemistry, University of Glasgow, Royal Infirmary, Glasgow G4 0SF, Scotland **[39]**

Gerald J. Chader, Laboratory of Retinal Cell and Molecular Biology, National Eye Institute, National Institutes of Health, Bethesda, MD 20892 **[275]**

Michael H. Chaitin, Bascom Palmer Eye Institute and the Department of Anatomy and Cell Biology, University of Miami School of Medicine, Miami, FL 33136 **[265]**

Robert J. Collier, Department of Ophthalmology, School of Medicine and Dentistry, University of Rochester, Rochester, NY 14642 **[569]**

C.A. Converse, Department of Pharmacy, University of Strathclyde, Glasgow G1 1XW, Scotland **[39]**

John R. Cotter, Department of Ophthalmology, University of Kansas Medical Center, Kansas City, KS 66103 **[357]**

G. Cremer-Bartels, Eye Hospital of the Westfaelian Wilhelms-University Münster, 4400 Münster, Federal Republic of Germany **[77]**

Jess Cunnick, Department of Biochemistry, Kansas State University, Manhattan, KS 66506 **[441]**

Michael Danciger, Department of Biology, Loyola Marymount University, Los Angeles, CA 90045 **[143]**

Coca del Cerro, Departments of Neurobiology and Anatomy, University of Rochester Medical School, Rochester, NY 14642 **[673]**

Manuel del Cerro, Departments of Neurobiology and Anatomy, University of Rochester Medical School, Rochester, NY 14642 **[673]**

A. Margreet de Leeuw, Department of Ophthalmology, University of Washington, Seattle, WA 98195 **[19]**

L.A. Donoso, Wills Eye Hospital, Philadelphia, PA 19107 **[493]**

Yusuf K. Durlu, Department of Ophthalmology, Tohoku University School of Medicine, Sendai, Japan **[585]**

Curtis D. Eckhert, Division of Nutritional Sciences, School of Public Health, University of California, Los Angeles, CA 90024 **[331]**

Graig E. Eldred, Department of Ophthalmology, University of Missouri School of Medicine, Columbia, MO 65212 **[113]**

E. El-Hifnawi, Department of Anatomy, Medical University of Lübeck, D-2400 Lübeck, Federal Republic of Germany **[343]**

C. Fahlman, School of Veterinary Medicine, University of Pennsylvania, Philadelphia, PA 19104 **[427]**

Debora B. Farber, Jules Stein Eye Institute, UCLA School of Medicine, Los Angeles, CA 90024 **[143]**

Lynette Feeney-Burns, Mason Institute of Ophthalmology, University of Missouri, Columbia, MO 65212 **[601]**

Malinda E.C. Fitzgerald, Department of Anatomy and Neurobiology, The University of Tennessee, Memphis, TN 38163 **[409]**

Chun-Lan Gao, Mason Institute of Ophthalmology, University of Missouri, Columbia, MO 65212 **[601]**

H. Gerding, Eye Hospital of the Westfaelian Wilhelms-University Münster, 4400 Münster, Federal Republic of Germany **[77]**

W. Gerlach, Eye Hospital of the Westfaelian Wilhelms-University Münster, 4400 Münster, Federal Republic of Germany **[77]**

P. Gouras, Department of Ophthalmology, Columbia University, New York, NY 10032 **[659]**

David J. Gower, Section of Neurosurgery, University of Oklahoma, Oklahoma City, OK 73126 **[523]**

W. Gremm, Eye Hospital of the Westfaelian Wilhelms-University Münster, 4400 Münster, Federal Republic of Germany **[77]**

Donald A. Grover, Department of Ophthalmology, University of Rochester Medical School, Rochester, NY 14642 **[673]**

P. Guntermann, Eye Hospital of the Westfaelian Wilhelms-University Münster, 4400 Münster, Federal Republic of Germany **[77]**

Gregory S. Hageman, Department of Ophthalmology, Bethesda Eye Institute, St. Louis University, St. Louis, MO 63110 **[217]**

L. Hanneken, Eye Hospital of the Westfaelian Wilhelms-University Münster, 4400 Münster, Federal Republic of Germany **[77]**

David R. Hardten, Department of Ophthalmology, University of Kansas Medical Center, Kansas City, KS 66103 **[377]**

Lloyd A. Horrocks, Department of Physiological Chemistry, Ohio State University College of Medicine, Columbus, OH 43205 **[49]**

Yoshihiro Hotta, Laboratory of Mechanisms of Ocular Diseases, National Eye Institute, NIH, Bethesda, MD 20892 **[99]**

Mei-Hui Hsu, Division of Nutritional Sciences, School of Public Health, University of California, Los Angeles, CA 90024 **[331]**

Stephen E. Hughes, Central Institute for the Deaf, St. Louis, MO 63110 **[687]**

L. Huq, Department of Pharmacy, University of Strathclyde, Glasgow G1 1XW, Scotland **[39]**

Ali A. Hussain, Comparative Ophthalmology Unit, The Animal Health Trust, Kennett CB8 7PN, England **[183]**

George Inana, Laboratory of Mechanisms of Ocular Diseases, National Eye Institute, NIH, Bethesda, MD 20892 **[99]**

C.F.I. Inglehearn, MRC Human Genetics Unit, Edinburgh, Scotland **[83]**

Margaret J. Irons, Division of Anatomy and Experimental Morphology, Department of Biomedical Sciences, Faculty of Health Sciences, McMaster University, Hamilton, Ontario, Canada L8N 3Z5 **[301]**

Sei-ichi Ishiguro, Department of Ophthalmology, Tohoku University School of Medicine, Sendai, Japan **[585]**

Samuel G. Jacobson, Department of Ophthalmology, Bascom Palmer Eye Institute, University of Miami School of Medicine, Miami, FL 33136 **[3]**

Harry Jansen, Department of Anatomy, Faculty of Medicine, Erasmus University, 3000 DR Rotterdam, The Netherlands **[233]**

M. Jay, Moorfields Eye Hospital, London, England **[83]**

Lincoln V. Johnson, Department of Anatomy and Cell Biology, University of Southern California School of Medicine, Los Angeles, CA 90033 **[217]**

W.A. Keegan, Department of Pharmacy, University of Strathclyde, Glasgow G1 1XW, Scotland **[39]**

Nancy G. Kennaway, Department of Medical Genetics, The Oregon Health Sciences University, Portland, OR 97201 **[99]**

H. Kjeldbye, Department of Ophthalmology, Columbia University, New York, NY 10032 **[659]**

Christine A. Kozak, National Institute of Allergy and Infectious Diseases, NIH, Bethesda, MD 20891 **[143]**

V. Kramer, Eye Hospital of the Westfaelian Wilhelms-University Münster, 4400 Münster, Federal Republic of Germany **[77]**

K. Krause, Eye Hospital of the Westfaelian Wilhelms-University Münster, 4400 Münster, Federal Republic of Germany **[77]**

W. Kühnel, Department of Anatomy, Medical University of Lübeck, D-2400 Lübeck, Federal Republic of Germany **[343]**

Matthew M. LaVail, Departments of Anatomy and Ophthalmology, Beckman Vision Center, University of California, San Francisco, CA 94143 **[315, 513, 577]**

Eliot Lazar, Departments of Neurobiology and Anatomy, University of Rochester Medical School, Rochester, NY 14642 **[673]**

Rehwa H. Lee, Jules Stein Eye Institute, UCLA School of Medicine, Los Angeles, CA 90024 **[155]**

Lawrence E. Leguire, Department of Ophthalmology, Ohio State University College of Medicine, Columbus, OH 43205 **[49]**

L. Li, Department of Anatomy, Bowman Gray School of Medicine, Wake Forest University, Winston Salem, NC 27103 **[645]**

S. Lindsay, Molecular Genetics Unit, University of Newcastle-upon-Tyne, England **[83]**

Richard N. Lolley, Laboratory of Developmental Neurology, VA Medical Center, Sepulveda, CA 91343 **[155]**

R. Lopez, Department of Ophthalmology, Columbia University, New York, NY 10032 **[659]**

Michael T. Matthes, Departments of Anatomy and Ophthalmology, Beckman Vision Center, University of California, San Francisco, CA 94143 **[315]**

M.B. Maude, Cullen Eye Institute, Baylor College of Medicine, Houston, TX 77030 **[427]**

T. McLachlan, Department of Pharmacy, University of Strathclyde, Glasgow G1 1XW, Scotland **[39]**

M. Morita, Department of Anatomy and Neurobiology, Colorado State University, Fort Collins, CO 80523 **[555]**

Takashi Muramatsu, Department of Biochemistry, Kagoshima University Faculty of Medicine, Kagoshima-shi 890, Japan **[69, 577]**

Muna I. Naash, Cullen Eye Institute and Department of Biochemistry, Baylor College of Medicine, Houston, TX 77030 **[513]**

Akira Nakajima, Department of Ophthalmology, Juntendo University School of Medicine, Tokyo, Japan **[99]**

Kristina Narfström, Department of Ophthalmology, University of Linköping, S-58185 Linköping, Sweden **[275]**

Martha Neuringer, Oregon Regional Primate Research Center, Beaverton, OR 97005 **[601]**

Sven Erik G. Nilsson, Department of Ophthalmology, University of Linköping, S-58185 Linköping, Sweden **[275, 539]**

Izhak Nir, Department of Pathology, University of Texas Health Science Center, San Antonio, TX 78284 **[251]**

Werner K. Noell, Department of Ophthalmology, University of Kansas Medical Center, Kansas City, KS 66103 **[357, 377]**

Mary F. Notter, Departments of Neurobiology and Anatomy, University of Rochester Medical School, Rochester, NY 14642 **[673]**

P.J. O'Brien, Laboratory of Retinal Cell and Molecular Biology, National Eye Institute, NIH, Bethesda, MD 20892 **[427]**

Norio Ohba, Department of Ophthalmology, Kagoshima University Faculty of Medicine, Kagoshima-shi 890, Japan **[69, 577]**

John Olchowka, Departments of Neurobiology and Anatomy, University of Rochester School of Medicine, Rochester, NY 14642 **[673]**

D.T. Organisciak, Department of Biochemistry, Wright State University, Dayton, OH 45435 **[493]**

C.J. Packard, Department of Pathological Biochemistry, University of Glasgow, Royal Infirmary, Glasgow G4 0SF, Scotland **[39]**

David S. Papermaster, Department of Pathology, University of Texas Health Science Center, San Antonio, TX 78284 **[251]**

S.S. Papiha, Molecular Genetics Unit, University of Newcastle-upon-Tyne, England **[83]**

E.L. Pautler, Department of Physiology, Colorado State University, Fort Collins, CO 80523 **[555]**

John S. Penn, Cullen Eye Institute, Baylor College of Medicine, Houston, TX 77030 **[623]**

E. Bradly Pewitt, Department of Ophthalmology, University of Kansas Medical Center, Kansas City, KS 66103 **[357]**

Luis Politi, Retinal Degenerations Research Center, Department of Ophthalmology, Johns Hopkins University School of Medicine, Baltimore, MD 21205 **[169]**

Philip J. Polkinghorne, Department of Clinical Ophthalmology, Institute of Ophthalmology, London WC1H 9QS, England **[57]**

Luke Qi Jiang, Departments of Neurobiology and Anatomy, University of Rochester Medical School, Rochester, NY 14642 **[673]**

Luo Qingli, Department of Ophthalmology, University of Washington, Seattle, WA 98195 **[19]**

D.S. Reeves, Department of Biochemistry, Wright State University, Dayton, OH 45435 **[493]**

Rouel S. Roque, Department of Anatomy, Medical College of Georgia, Augusta, GA 30912 **[393]**

Munefumi Sameshima, Department of Ophthalmology, Kagoshima University Faculty of Medicine, Kagoshima-shi 890, Japan **[577]**

Somes Sanyal, Department of Anatomy, Faculty of Medicine, Erasmus University, 3000 DR Rotterdam, The Netherlands **[233, 275]**

Randall Sapp, Division of Biological Sciences, University of Missouri, Columbia, MO 65211 **[467]**

P.G. Sealey, MRC Human Genetics Unit, Edinburgh, Scotland **[83]**

D. Seibt, Eye Hospital of the Westfaelian Wilhelms-University Münster, 4400 Münster, Federal Republic of Germany **[77]**

Susan L. Semple-Rowland, Department of Ophthalmology, University of Florida College of Medicine, Gainesville, FL 32610 **[455]**

J. Series, Department of Pathological Biochemistry, University of Glasgow, Royal Infirmary, Glasgow G4 0SF, Scotland **[39]**

Barry T. Shannon, Department of Pediatrics, Ohio State University College of Medicine, Columbus, OH 43205 **[49]**

H.J. Sheedlo, Department of Anatomy, Bowman Gray School of Medicine, Wake Forest University, Winston-Salem, NC 27103 **[645]**

J. Shepherd, Department of Pathological Biochemistry, University of Glasgow, Royal Infirmary, Glasgow G4 0SF, Scotland **[39]**

Takashi Shiono, Department of Ophthalmology, Tohoku University School of Medicine, Sendai, Miyagi 980, Japan **[99]**

Martin S. Silverman, Central Institute for the Deaf and the Department of Anatomy and Neurobiology, Washington University, St. Louis, MO 63110 **[687]**

Susan M. Slapnick, Division of Neurosurgery, University of Wisconsin, Madison, WI 53706 **[393, 409]**

Anthony D. Smith, Department of Chemical Pathology, University College and Middlesex School of Medicine, London W1P 6DB, England **[57]**

William S. Stark, Division of Biological Sciences, University of Missouri, Columbia, MO 65211 **[467]**

B. Sullivan, Department of Ophthalmology, Columbia University, New York, NY 10032 **[659]**

Barbro Swenson, Department of Ophthalmology, University of Linköping, S-58185 Linköping, Sweden **[539]**

Dolores Takemoto, Department of Biochemistry, Kansas State University, Manhattan, KS 66506 **[441]**

Katsuhida Takumi, Department of Ophthalmology, Kagoshima University Faculty of Medicine, Kagoshima-shi 890, Japan **[577]**

Makoto Tamai, Department of Ophthalmology, Tohoku University of Medicine, Sendai, Japan **[585]**

Ola Textorius, Department of Ophthalmology, University of Linköping, S-58185 Linköping, Sweden **[539]**

Lisa A. Thum, Cullen Eye Institute, Baylor College of Medicine, Houston, TX 77030 **[623]**

J.E. Turner, Department of Anatomy, Bowman Gray School of Medicine, Wake Forest University, Winston Salem, NC 27103 **[645]**

Michael Tytell, Department of Anatomy, Bowman Gray School of Medicine, Wake Forest University, Winston-Salem, NC 27103 **[523]**

Fumiyuki Uehara, Department of Ophthalmology, Kagoshima University Faculty of Medicine, Kagoshima-shi 890, Japan **[69, 577]**

Robert J. Ulshafer, Department of Ophthalmology, University of Florida, Gainesville, FL 32610 **[131]**

Kazuhiko Unoki, Department of Ophthalmology, Kagoshima University Faculty of Medicine, Kagoshima-shi 890, Japan **[69, 577]**

Theo van Veen, Laboratory of Molecular Neuroanatomy, Department of Zoology, University of Lund, S-22362 Lund, Sweden **[275]**

Mary J. Voaden, Department of Visual Science, Institute of Ophthalmology, London WC1H 9QS, England **[57, 183]**

H-M. Wang, Department of Biochemistry, Wright State University, Dayton, OH 45435 **[493]**

Richard G. Weleber, Department of Ophthalmology, The Oregon Health Sciences University, Portland, OR 97201 **[99]**

M.G. Wetzel, Laboratory of Retinal Cell and Molecular Biolgoy, National Eye Institute, NIH, Bethesda, MD 20892 **[427]**

Stanley J. Wiegand, Departments of Neurobiology and Anatomy, University of Rochester Medical School, Rochester, NY 14642 **[673]**

Barbara Wiggert, Laboratory of Retinal Cell and Molecular Biology, National Eye Institute, NIH, Bethesda, MD 20892 **[275]**

Lowell L. Williams, Department of Pediatrics, Ohio State University College of Medicine, Columbus, OH 43205 **[49]**

Nicholas J. Willmott, Department of Visual Science, Institute of Ophthalmology, London WC1H 9QS, England **[183]**

A. Xie, Department of Biology, Beijing Normal University, Beijing, People's Republic of China **[493]**

Douglas Yasumura, Departments of Anatomy and Ophthalmology, University of California School of Medicine, San Francisco, CA 94143 **[577]**

Seymour Zigman, Department of Ophthalmology, School of Medicine and Dentistry, University of Rochester, Rochester, NY 14642 **[569]**

Carmelann Zintz, Laboratory of Mechanisms of Ocular Diseases, National Eye Institute, NIH, Bethesda, MD 20892 **[99]**

Preface

The fields of inherited and environmentally induced retinal degenerations have grown enormously in the past two decades. As a consequence, there has been a need for both increased interaction among scientists in these fields and rapid dissemination of research findings to basic and clinical vision scientists. One forum for such exchanges has been the international symposia organized in conjunction with the VI and VII International Congresses of Eye Research in 1984 and 1986, respectively. Soon after each of these symposia took place, proceedings volumes were published. The increased interest in these symposia and the timely publication of their contents stimulated us to continue this biannual event. Thus, the *International Symposium on Retinal Degenerations* was held in San Francisco, California, September 2–3, 1988, as a satellite meeting of the VIII International Congress of Eye Research. As in the past, a number of chapters in this volume have been contributed by individuals who could not attend the symposium, thus providing a broad sampling of contemporary research on inherited and environmentally induced retinal degenerations.

It is our pleasure to recognize the agencies that have provided most of the funding for the studies presented here. These are the National Eye Institute, Research to Prevent Blindness, and the Retinitis Pigmentosa Foundation Fighting Blindness. We are particularly indebted to the Retinitis Pigmentosa Foundation Fighting Blindness, which provided travel funds for scientists from several countries so that they could participate in the *International Symposium on Retinal Degenerations,* paid for all of the local costs of the symposium, and financed the publication of this volume.

We wish to thank Dr. David Maurice and Susan Marsh, the Secretariat and Conference Coordinator, respectively, of the VIII International Congress of Eye Research, for their cooperation and assistance in coordinating the two meetings. We also thank Ms. Gloria Riggs for her administrative and secretarial assistance in organizing the symposium and this volume.

Matthew M. LaVail
Robert E. Anderson
Joe G. Hollyfield

I. RETINITIS PIGMENTOSA AND OTHER HUMAN RETINAL DEGENERATIONS

Inherited and induced retinal degenerations are often most conveniently studied in animal models. Indeed, the ready availability of tissues of defined age and genetic background in animal models have made them extremely important for the exploration of photoreceptor and retinal pigment epithelial cell degeneration mechanisms. However, experimental observations on animal models ultimately must be compared with those of human diseases, so it is of paramount importance to study critically the disorders in human patients and human donor tissues.

The eleven papers in this section consider various aspects of human retinal degenerations, most of them with one or more forms of retinitis pigmentosa. The first two papers describe advances in the characterization of retinitis pigmentosa; quantification of visual function and histopathological analysis of donor eye tissues. The next five papers present investigations on blood obtained from patients with retinitis pigmentosa, with findings on lipid and fatty acid metabolism, glycosyltransferase in lymphocytes, and monapterin concentration in serum and blood cells. Two papers follow on current molecular genetics studies of X-linked retinitis pigmentosa and gyrate atrophy. The final two papers consider the possible role of lipofuscin and zinc content of melanosomes in age-related maculopathies.

Inherited and Environmentally Induced
Retinal Degenerations, pages 3–17

ADVANCED RETINITIS PIGMENTOSA: QUANTIFYING VISUAL FUNCTION

François-Xavier Borruat and Samuel G. Jacobson

Hôpital Ophtalmique, Université de Lausanne, 1004 Lausanne, Switzerland (F.-X.B.) and
Department of Ophthalmology, Bascom Palmer Eye Institute, University of Miami School of Medicine, Miami, FL 33136, USA (S.G.J.)

INTRODUCTION

Retinitis pigmentosa (RP) is a group of inherited retinal degenerations in which there is progressive loss of peripheral vision and later central vision. Guidelines for a thorough evaluation of patients with RP have been proposed (Marmor et al, 1983) and four tests of visual function have been strongly recommended - visual acuity, kinetic light-adapted perimetry, dark-adapted thresholds and full field electroretinography. In RP patients with peripheral vision, the four recommended tests can provide useful information about the level of disability and even the functional subtype of RP. However, in patients with advanced disease (i.e. those who retain only a central island of vision), some of these tests are far less informative and accurate.

In the present study, we describe test strategies that have been designed specifically to measure the visual function of the more central retinal region of patients with advanced retinal degeneration. The results we obtained with these methods in a number of advanced RP patients show that different patterns of visual dysfunction can be detected even in these small residual central islands.

METHODS

The patients whose test results are described in this study have typical RP, are at an advanced stage of the disease (i.e. have a visual field diameter no greater than

40°, as measured with a Goldmann kinetic perimeter using the V-4e test target), and have a non-detectable computer-averaged electroretinogram.

Fundus Perimetry

Fundus perimetry is a technique that permits measurement of the visual field while viewing the fundus. With a modified fundus photoperimeter (Canon CPP-1, 45° field; Parel et al, 1986), light-adapted kinetic perimetry was performed using white targets (Goldmann sizes V, 103 min diameter, and I, 6.5 min; 318 cd/m^2) on a white background light (10 cd/m^2). The patient's fixation was monitored continuously and the test target was moved from non-seeing to seeing areas. The fundus with superimposed visual field was photographed at the end of the test.

Automated Static Perimetry and Dark Adaptometry

The modified automated perimeter, testing techniques and methods of data analysis have been previously described (Jacobson et al, 1986a; Apáthy et al, 1987). Among the test strategies used in this study were: 1) a central test with 38 loci covering 48° of central field; 2) a peripheral test with 51 loci on a 12° grid covering visual field beyond the central test; 3) a profile test with 31 loci at 2° spacing extending 30° nasal and temporal to the fovea along the horizontal meridian; and 4) a macula test with 17 loci on a 2° grid covering a square central field, 6° on a side. Tests were performed either in the light-adapted state with a white target (103 min diameter) on a white background light (10 cd/m^2), or dark-adapted with two monochromatic stimuli of different wavelengths (λ_{max}= 500 nm and 650 nm, 103 min diameter).

In a typical testing session, the extent of the visual field was first determined with light-adapted suprathreshold static screening (white target, 3183 cd/m^2) using the central and peripheral tests. All further testing was performed with threshold strategy. Principles of two-color dark-adapted static perimetry have been published (Zeavin and Wald, 1956; Massof and Finkelstein, 1979; Jacobson and Apáthy, 1988). Whether rods and/or cones were mediating vision at each test locus was determined from the sensitivity difference between the 500 and 650 nm stimuli. Sensitivity losses for rods and cones were calculated for each locus by comparison to normal mean values (Apáthy et al, 1987).

Dark adaptometry was performed with the modified automated perimeter (Jacobson et al, 1986a) and always followed dark-adapted perimetry. The test locus for adaptometry was selected from amongst those examined with perimetry. Following a bleach, the patient was tested alternately with the 500 and 650 nm stimuli until sensitivity recovered to the levels previously determined during perimetry.

Dark Adapted Visual Evoked Cortical Potentials (VECP)

The technique of two-color dark-adapted flash VECPs in advanced RP patients has been described (Jacobson et al, 1985). Briefly, VECPs were recorded between a midline scalp electrode located 5 cm above the inion and a midfrontal electrode in response to monocularly-delivered ganzfeld flashes of light. The dark-adapted patient was tested with blue and red light flashes, scotopically matched for ERG b-wave amplitude in normal subjects (Jacobson et al, 1985). When a VECP was recordable in response to the suprathreshold blue stimulus, an intensity series with blue light was performed to determine the threshold.

RESULTS

Figure 1 shows representative examples of fundus perimetric test results on 4 patients with advanced RP. Patient A, a 56 year old man with simplex RP and visual acuity of 0.8 (20/25), is an example of good correspondence between fundus structure and function. There is a central retinal area with preserved retinal pigmented epithelium (RPE) and the visual field isopter with the size V target corresponds well with this area of preserved RPE; with the size I target, the visual field is smaller. In contrast, Patient B, a 39 year old woman with multiplex RP and visual acuity of 0.25 (20/80), shows a lack of correspondence between fundus appearance and function. Preserved RPE appears to extend to the vessel arcades but the measured visual field is much smaller.

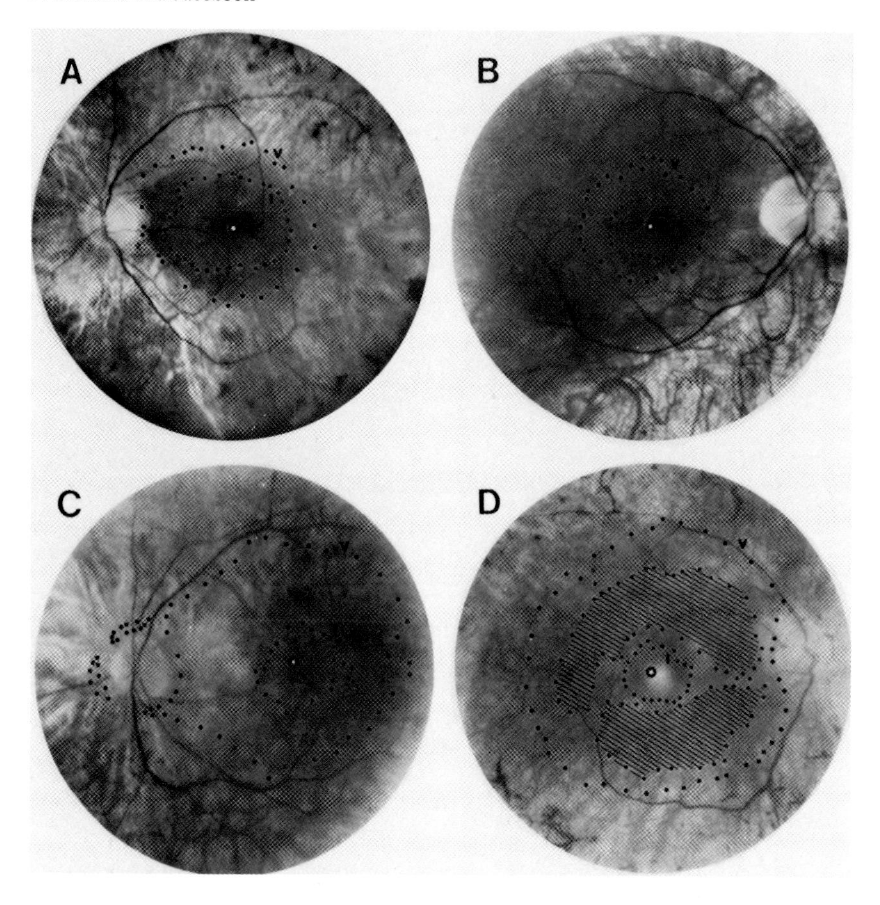

Figure 1. Kinetic fundus perimetry in 4 advanced RP patients (**A-D**). Isopters for the V and I size targets are labelled. Hatched areas in D represent scotomas to the V target. A white star is the patient's locus of fixation.

Patients C and D have subtle perimetric findings that were not detected on Goldmann kinetic fields. Patient C, a 40 year old man with multiplex RP and visual acuity of 0.05 (20/400), shows peri-papillary preservation of function. The visual field (size V) extends to the vessel arcades and there are preserved islands of vision around the optic disc, detected only with the larger target. Patient D, a 55 year old woman with multiplex RP and visual acuity of 0.7

(20/60), has a visual field isopter (V target) which is comparable to that of Patient C. Within the central field, however, there are perifoveal scotomas and these likely account for the patient's complaints of difficulty reading.

Figure 2 shows the static perimetric test strategies and representative results in two patients with advanced RP. In Figure 2A, the entire field of vision was screened with suprathreshold static stimuli, light-adapted; both patients show no responses on the 'peripheral test'. P1, a 44 year old woman with multiplex RP, detected many stimuli in the inner octagonal region which represents the 'central test'. P2, a 29 year old woman with simplex RP, has a far smaller visual field and detected stimuli only at a few central loci.

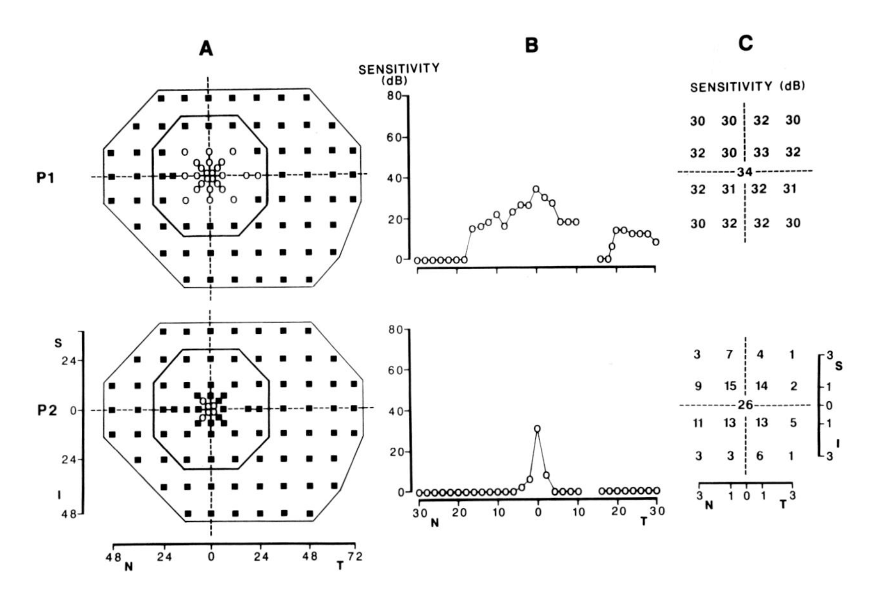

Figure 2. Light-adapted static perimetric test results from the right eye of 2 advanced RP patients (P1,P2). **A.** Suprathreshold screening results for both the 'central test' (inner octagonal region) and 'peripheral test' (remaining delineated area); open circles represent detected stimuli, and filled squares, no detection. **B.** Threshold results for the 'profile test'. **C.** Threshold results for the 'macula test'. X and Y axes are visual field eccentricities in degrees; N,T,S,I: nasal, temporal, superior, inferior.

In Figures 2B and 2C the results of threshold testing further define the functional differences in these two patients. With the 'profile test' (Fig. 2B), it is evident that P1 retains function temporal to the physiologic blind spot but not at nasal loci of comparable eccentricity, another example of peri-papillary preservation of function. Both 'profile test' (Fig. 2B) and 'macula test' (Fig. 2C) results indicate good sensitivity within the tunnel of vision in P1 (only about 5-8 decibels, dB, of sensitivity loss), but there is markedly decreased sensitivity except at the foveal locus in P2. Although the maximum extent of the visual field and the level of function within it can be determined with these light-adapted tests, they do not provide specific information about rod and cone dysfunction. Such information can be gained from dark-adapted perimetry.

Figure 3 shows results of two-color dark-adapted static perimetry and adaptometry for a normal subject. In the upper part of Figure 3A, sensitivities for 500 and 650 nm stimuli are plotted against eccentricity. The graph below displays the sensitivity difference and the photoreceptor mediation (along the horizontal axis) at each locus. In the normal subject, rods (R) mediate detection at threshold across the retina except at the foveal locus and 1 or 2 parafoveal loci, which show mixed (M) rod and cone mediation. Because of the relatively large size of the test target, pure cone function is usually not demonstrable at the foveal locus in normal subjects.

In Figure 3B are normal results of the macula test for the two colors (upper part), sensitivity differences (lower left) and the photoreceptor mediation (lower right) at these 17 loci. In Figure 3C, to demonstrate the principles of two-color dark-adapted testing, dark adaptometry results using the two colors are shown for 4 loci representing different eccentricities within the macula test. At greater eccentricities from the fovea where rod sensitivity is higher, the usual biphasic dark adaptation curve is very evident (lower two graphs). At the first plateau, sensitivity differences between colors are all indicative of cone mediation, i.e. $\leqslant$12 dB. After several minutes there is a rod-cone break followed by an increase in rod sensitivity; sensitivity differences also increase, progressing through mixed (13-27 dB) to exclusively rod ($\geqslant$28 dB) mediation (Jacobson et al, 1986a; Apáthy et al, 1987).

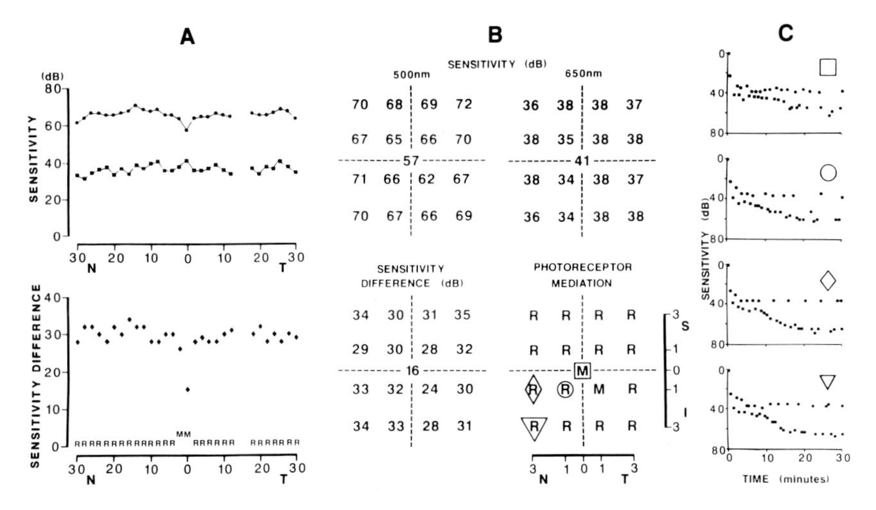

Figure 3. Two-color dark-adapted static perimetry and dark adaptometry in a normal subject. **A.** Profile test results. Upper: sensitivities for 500 nm (upper data points) and 650 nm (lower data points) stimuli. Lower: sensitivity differences (500-650 nm) and photoreceptor mediations (letters along X axis). **B.** Macula test results. Upper: sensitivities for 500 and 650 nm stimuli. Lower: sensitivity differences (left) and mediations (right). **C.** Dark adaptometry results for 4 different loci from the macula test (denoted by symbols in Fig. 3B, lower right). X axis is time post-bleach; Y axis is sensitivity in dB. In each graph, the upper data points are the recovery of sensitivity to the 650 nm stimulus and the lower ones, the 500 nm stimulus. R, rods; M, mixed rods and cones.

Patients with advanced RP can show remarkable differences in the patterns of rod and cone dysfunction within their residual central island of vision. To exemplify these functional differences, Figure 4 displays results of two-color dark-adapted perimetry in 3 patients. P3, a 52 year old man with multiplex RP and visual acuity of 0.5 (20/40), has detectable rod function throughout most of his remaining central 20° of visual field. On the macula test, rod sensitivity was 10-20 dB lower than normal and cone sensitivity loss about 4 to 9 dB. One locus of cone function was detected at the edge of the visual field (Fig. 4A, upper). P4 is a 63 year old woman with multiplex RP, a visual acuity of 0.2 (20/100) and a slightly larger extent of visual field than P3 on the horizontal profile test (Fig. 4A, middle). A pattern of mixed and cone mediated loci was found and, on the macula test, rod sensitivity, when measurable, was decreased by 18 to 26 dB; cone sensitivity loss ranged from 11 to 15 dB. P5 is a 38 year old woman with autosomal recessive RP and visual acuity of 0.2 (20/100). Although younger than P3 and P4, P5 has the more severe rod and cone dysfunction. There is only cone function detectable and cone sensitivity is decreased by 20 to 30 dB across the macula test area. Rod sensitivity loss can only be inferred but must be at least greater than 40 dB.

Further examples of the diversity of functional findings in advanced RP patients are shown in Figure 5.

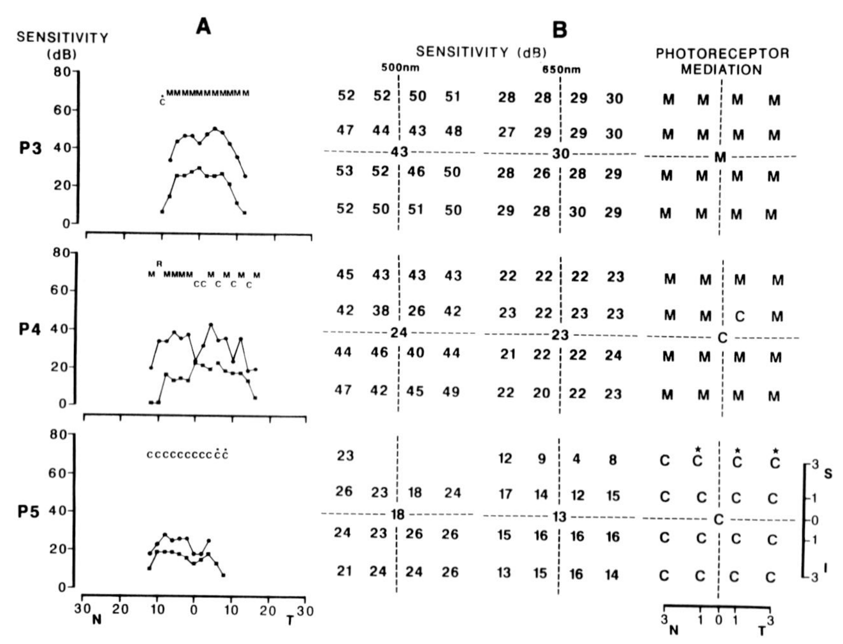

Figure 4. Two-color dark-adapted perimetry in 3 advanced RP patients (P3,P4,P5). **A.** Profile test results. Photoreceptor mediations are shown above the data. **B.** Macula test results. (*) over mediation indicates that one of the two stimulus colors was not detected (Apáthy et al, 1987). R, rods; C, cones; M, mixed rods and cones.

P6-8 were tested not only with dark-adapted perimetry but also with two-color dark adaptometry. P6, a 25 year old woman with simplex RP and 1.0 (20/20) acuity, shows nearly normal rod function in the central 10° (pattern similar to P3, Fig. 4). Dark adaptometry was performed at a locus with almost normal rod sensitivity (3° nasal, 3° inferior field). The cone plateau was reached relatively rapidly and cone sensitivity was in the normal range but rod adaptation appears slightly slower than normal (Fig. 3C, lower). P7, a 43 year old man having autosomal dominant RP with reduced penetrance and a visual acuity of 0.3 (20/60), shows a photoreceptor pattern with mainly mixed and cone loci and only a few rod loci (similar to P4, Fig. 4). The foveal locus was tested with dark adaptometry and showed only cone function. The cone plateau was reached later than in P6 and in normals (Fig. 3C, upper) and cone sensitivity was decreased by about 8 dB. P8 is a 53 year old man with multiplex RP; his visual acuity is 0.7 (20/30). His photoreceptor pattern shows only cone function, but in

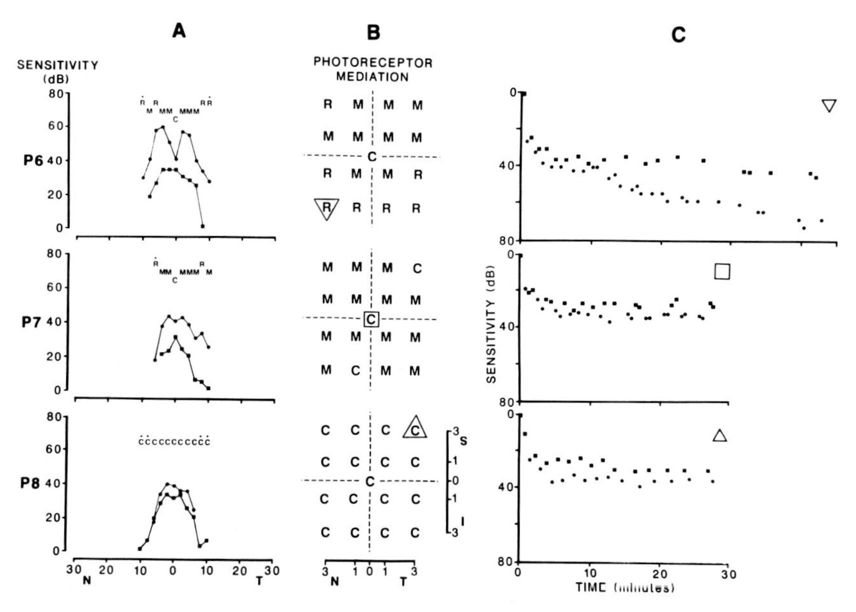

Figure 5. Two-color dark-adapted perimetry and adaptometry in 3 other advanced RP patients (P6-P8). **A.** Profile test results. **B.** Macula test results. **C.** Dark adaptometry results at the loci denoted by the symbols on the macula test in B.

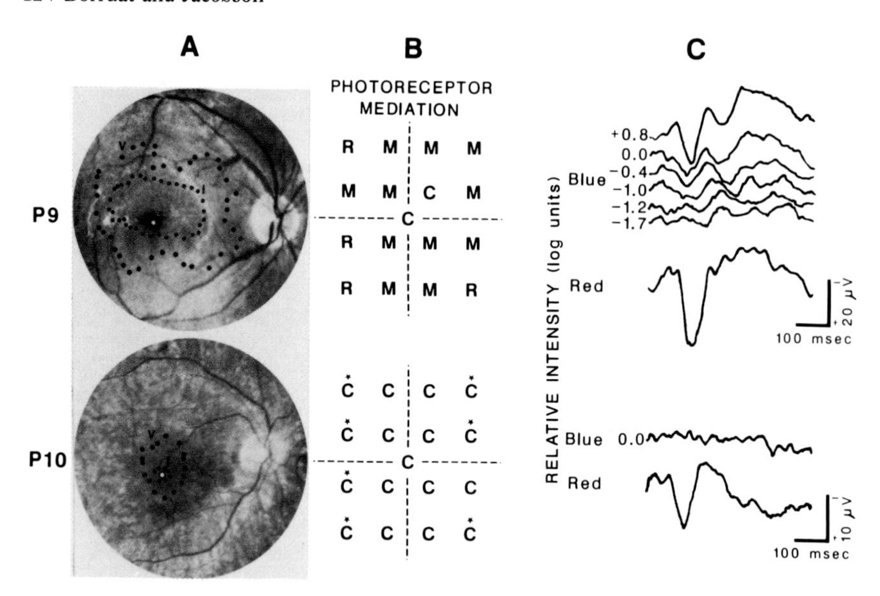

Figure 6. Fundus perimetry (A), two-color dark-adapted perimetry (B) and two-color dark-adapted VECPs (C) in 2 advanced RP patients (P9,P10). For the VECPs, the "Blue 0.0" stimulus is scotopically matched for the ERG with the "Red" stimulus. An intensity series with blue light flashes is shown for P9.

contrast to P5 (Fig. 4), there is excellent cone sensitivity within the residual central island. Dark adaptometry confirms the good cone function and no measurable rod adaptation.

Results of fundus perimetry, dark-adapted static perimetry and visual evoked cortical potentials are shown for two advanced RP patients in Figure 6. P9 is the same patient as P6 but the test results are one year later than those in Figure 5. With fundus perimetry the visual field does not conform to any obvious retinal boundary (Fig. 6A, upper). In the macula test (Fig. 6B, upper) there are more mixed and cone loci than one year before, indicating worsening rod (and cone) function between test dates. The VECP confirms the presence of substantial rod function centrally: there are recordable waveforms for intensities of blue light known to evoke only rod retinal responses. With the red light, there is also a waveform which is considered to be a mixed response of cones and rods. P10, a 61 year old woman with multiplex RP, has a far smaller island on fundus perimetry than P9 and only cone function is detectable therein. The blue light VECP is flat but there is a substantial signal to red flashes of light, again providing an electrophysiological confirmation of the psychophysical findings.

DISCUSSION

Progression of RP to the advanced stage, in which only a central island of vision remains, is usually thought to occur only late in life but considerable numbers of RP patients in their third and fourth decades of life progress to this stage. When applied to this large group of advanced RP patients, the recommended visual function tests such as the full field electroretinogram and the Goldmann kinetic field do little more than confirm the severity of the peripheral visual loss. We have sought further ways to quantify the residual function in advanced RP for the following purposes: 1) to attempt functional subtyping in these patients such as has been possible in less advanced RP; 2) to determine the patterns of central visual loss in RP as a step toward understanding the underlying mechanisms; 3) to provide means to determine the complete natural history of RP and to monitor advanced patients in future treatment trials; and, 4) to permit an increase in accuracy of clinico-pathological correlation and an enhancement of scientific value of postmortem retinal tissue from donors with advanced disease.

Two subtypes of autosomal dominant and simplex/ multiplex RP have been defined by two-color dark-adapted static perimetric testing of peripheral visual function (Massof and Finkelstein, 1979). In the 'regionalized' form of RP, rod and cone function are lost together; mid-peripheral dysfunction can become severe while central and peripheral islands still retain substantial rod and cone function. In the 'diffuse' form, there is severe rod dysfunction uniformly across the visual field from an early age and later, there is loss of cone function. These subtypes were recently shown in autosomal dominant RP to have different disease mechanisms (Kemp et al, 1988a).

In autosomal dominant RP where family members with preserved peripheral vision can be tested, the definition of psychophysical subtype in relatives with advanced disease is not problematic (Massof and Finkelstein, 1981; Lyness et al, 1983). Subtyping simplex RP patients with advanced disease, however, can be difficult. Patients with only cone function measurable in the residual central island (e.g. P5,P8, and P10) could have either the diffuse form of RP or an end stage of regionalized RP. For example, serial testing on P6/P9 (Figs. 5 and 6) showed loci on the macula test that progressed from R to M and M to C, suggesting further loss of rod function over a period of one year. With progression it is possible that all loci would be cone-mediated, making this pattern indistinguishable from a diffuse-type central island. Those patients with substantial rod function (P3,P4,P6/P9 and P7) in the central island must exemplify, however, the regionalized subtype.

The demonstration of slowed adaptation in some RP patients (Alexander and Fishman, 1984) and in vitamin A deficiency (Kemp et al, 1988b) has rekindled interest in adaptometry measurements in RP. It has been suggested that patients with slowed rod adaptation may represent a further subtype of regionalized RP (Ernst et al, 1987). As we showed in Figure 5, two-color dark adaptometry can be performed within the central few degrees of visual field in advanced RP patients. Those advanced patients identified by dark-adapted perimetry to have considerable rod function in their central islands therefore become candidates for adaptometry to determine if the variant of regionalized RP with slowed rod adaptation can be detected in patients at advanced stages of the disease and in this central retinal region.

Our application of fundus kinetic perimetry to advanced RP produced some interesting findings. As has been reported for gyrate atrophy (Enoch et al, 1984), retinal boundaries in some RP patients corresponded well to the visual field extent (Fig. 1A). In other patients (Figs. 1B and 6), however, there was no close correspondence. These observations need tc be extended to determine whether such differences in fundus perimetric findings represent different stages of the same disease or possibly different underlying disease mechanisms.

Fundus perimetry also revealed two other details in the central function of advanced RP patients that, to our knowledge, have not been previously described. First, small areas of function were detected around the optic nerve head but function was not measurable in the temporal retina at the same eccentricities (Fig. 1C). A similar finding was evident in another patient (P1) with the profile test (Fig. 2B). Second, pericentral scotomas were detected within the isopter defined by the V target (Fig. 1D). Such scotomas could account for otherwise unexplained visual complaints (especially with reading) in some RP patients. The mechanisms underlying peri-papillary preservation of function and the regional variations of dysfunction within the central retina are as yet unknown.

A frequently asked question by advanced RP patients is: "How long will I keep my remaining vision?" The natural history of central retinal function in advanced RP patients has not been studied. Accurate means to assess macular function, like the testing techniques described herein, may permit monitoring these patients for the purpose of natural history studies and even future treatment trials. Such methods, for example, would be particularly suited to determine the efficacy of recently proposed treatments for macular edema in RP (Newsome and Blacharski, 1987; Cox et al, 1988). Other non-invasive techniques to measure localized retinal function such as the focal electroretinogram and fundus reflectometry (reviewed in Jacobson et al, 1986b) may also be of value for these purposes.

One of the more important ways to understand the cellular mechanisms of RP is through the study of postmortem RP eye donor tissue and much effort has been expended to facilitate the acquisition of such tissue. The majority of donor eyes come from patients with advanced stages of RP and these are the patients whose residual visual function is usually quantified only with a measure of visual acuity and possibly a Goldmann kinetic field. As evidenced by the findings in many patients in this study (e.g. P3,P4,P6,P7, and P10), substantial rod and cone function can be retained centrally after extensive peripheral visual loss has occurred. Assuming that there is good morphological correlation to the psychophysical findings, there are advanced RP patients who have in their central retina substantial numbers of functioning rods and cones which may be amenable to cell biological, biochemical and possibly molecular biological studies.

REFERENCES

Alexander KR, Fishman GA (1984). Prolonged rod dark adaptation in retinitis pigmentosa. Br J Ophthalmol 68:561-569.

Apáthy PP, Jacobson SG, Nghiem-Phu L, Knighton RW, Parel J-M (1987). Computer-aided analysis in automated dark-adapted static perimetry. In: Greve EL, Heijl A (eds): "Seventh international visual field symposium", Dordrecht: Martinus Nijhoff/Dr W Junk, pp 277-284.

Cox SN, Hay E, Bird AC (1988). Treatment of chronic macular edema with acetazolamide. Arch Ophthalmol 106:1190-1195.

Enoch JM, O'Donnell J, Williams RA, Essock EA (1984). Retinal boundaries and visual function in gyrate atrophy. Arch Ophthalmol 102:1314-16.

Ernst W, Kemp CM, Moore AT (1987). Abnormal rod dark adaptation in autosomal dominant retinitis pigmentosa. Invest Ophthalmol Vis Sci Suppl 28:236.

Jacobson SG, Knighton RW, Levene RM (1985). Dark- and light-adapted visual evoked cortical potentials in retinitis pigmentosa. Doc Ophthalmol 60:189-196.

Jacobson SG, Voigt WJ, Parel JM, Apáthy PP, Nghiem-Phu L, Myers SJ, Patella VM (1986a) Automated light- and dark-adapted perimetry for evaluating retinitis pigmentosa. Ophthalmol 93:1604-1611.

Jacobson SG, Parel J-M, Knighton RW (1986b) Hereditary retinal degeneration: advances in methods of testing visual function. In: Smith JL (ed): "Neuro-ophthalmology now", New York: Field, Rich & Associates, pp 93-108.

Jacobson SG, Apáthy, PP (1988). Automated rod and cone perimetry in retinitis pigmentosa. In: Smith JL, Katz RS (eds): "Neuro-ophthalmology enters the nineties", Hialeah: Dutton press, pp 35-47.

Kemp CM, Jacobson SG, Faulkner DJ (1988a) Two types of visual dysfunction in autosomal dominant retinitis pigmentosa. Invest Ophthalmol Vis Sci 29:1235-1241.

Kemp CM, Jacobson SG, Faulkner DJ, Walt RW (1988b) Visual function and rhodopsin levels in humans with vitamin A deficiency. Exp Eye Res 46:185-187.

Lyness AL, Ernst W, Quinlan MP, Clover GM, Arden GB, Carter RM, Bird AC, Parker JA (1985) A clinical, psychophysical, and electroretinographical survey of patients with autosomal dominant retinitis pigmentosa. Br J Ophthalmol 69:326-339.

Marmor MF et al (1983) Retinitis pigmentosa: a symposium on terminology and methods of examination. Ophthalmol 90:126-131.

Massof RW, Finkelstein D (1979) Rod sensitivity relative to cone sensitivity in retinitis pigmentosa. Invest Ophthalmol Vis Sci 18:163-172.

Massof RW, Finkelstein D (1981) Two forms of autosomal dominant primary retinitis pigmentosa. Doc Ophthalmol 51:289-346.

Newsome DA, Blacharski PA (1987). Grid photocoagulation for macular edema in patients with retinitis pigmentosa. Am J Ophthalmol 103:161-166.

Parel J-M, Jacobson SG, Chiu MT, Borruat F-X, Denham D, Nose I (1986). Fundus perimetry in retinitis pigmentosa. Ophthalmol 93(8):125.

Zeavin BH, Wald G (1956). Rod and cone vision in retinitis pigmentosa. Am J Ophthalmol 42:253-268.

ACKNOWLEDGMENTS

This work was performed at the Retinitis Pigmentosa Center of Bascom Palmer Eye Institute and was supported in part by the National Retinitis Pigmentosa Foundation, Inc. (Baltimore, MD) and The Chatlos Foundation, Inc. (Longwood, FL). We thank Professor C. Gailloud for constant encouragement; Mr. P.P. Apáthy, Mr. Jean-Marie Parel and Dr. R.W. Knighton for critical advice and help; and Ms. Barbara French for assistance with artwork.

Inherited and Environmentally Induced
Retinal Degenerations, pages 19–38

OBSERVATIONS FROM THE FIRST YEAR OF OPERATION OF THE U.S. RETINITIS PIGMENTOSA HISTOPATHOLOGY LABORATORY

Ann H. Bunt-Milam, Luo Qingli, and
A. Margreet de Leeuw
Department of Ophthalmology RJ-10,
University of Washington, Seattle, WA
98195

The U.S. Retinitis Pigmentosa Histopathology Laboratory was established on July 1, 1987, supported by the Louis Berkowitz Family Foundation and the National Retinitis Pigmentosa Foundation. The purpose of this service laboratory is to provide light and electron microscopic pathology reports on donor eyes and retinas received through the Retina Donor Program of the Retinitis Pigmentosa Foundation. In our first year, we have received 19 donor eyes from patients of different ages diagnosed as having retinitis pigmentosa (RP) or senile macular degeneration, or as carriers of RP. In addition, eyes matched for age and *post mortem* intervals have been studied as controls. Detailed protocols have been developed for reporting qualitative and quantitative changes in each layer and cell type in the diseased retinas, with emphasis on documenting novel observations that may provoke new questions or avenues for research on the etiology/treatment of retinal diseases.

Our microscopic findings from each retina will be correlated with data obtained from the opposite donor retina in experiments conducted at Retinitis Pigmentosa Research Centers using different methodologies such as biochemistry, tissue culture or molecular biology. Our histopathologic observations on retinas from patients with different genetic forms of RP and varying ages will be compared in order to construct longitudinal histopathologic descriptions of the different forms of RP. This will provide a useful data base for future research.

A final function of the U.S. Retinitis Pigmentosa Histopathology Laboratory is to maintain a registry of diseased and normal retinal tissues that are available either as fixed tissue, embedded blocks, or microscopic sections. These will be distributed to qualified investigators who wish to apply special morphologic research techniques in their studies of human retinal dystrophies. Tissues are currently available from patients with RP and senile macular degeneration, from RP carriers, and from normal, age and *post mortem* matched controls.

As reviewed recently (Marshall and Heckenlively, 1988; Newsome, 1988; Pagon, 1988), histopathologic studies have led to the concept that the primary abnormality in RP is focal death of both rod and cone photoreceptors. The zone of initial loss of rods and cones is usually the mid peripheral retina. Rods are usually more severely affected than cones and both cell types show initial shortening and disorganization of the outer segment (OS) prior to photoreceptor death (Fig. 1). The foveal photoreceptors are usually decreased in number and size, commonly remaining as a monolayer of cone inner segments (IS) with attenuated or absent OS.

Secondary to death of photoreceptors, abnormalities frequently occur in the retinal pigment epithelium (RPE), including loss of melanin pigment, cellular atrophy or proliferation to form redundant layers of RPE, and finally migration of RPE cells into the inner retina. Ultimately, melanin laden cells encircle attenuated, hyalinized central retinal blood vessels, producing bone spicule pigmentation of the fundus.

Changes in the inner layers of the neurosensory retina are variable, ranging from total loss of interneurons and ganglion cells and replacement by glia to retention of an apparently normal complement of non-photoreceptor neurons. Studies of animal models of hereditary photoreceptor degeneration also suggest that the inner retinal neurons may remain normal in number and function (Eisenfeld, *et al.*, 1984) or may show degenerative changes (Grafstein *et al.*, 1972). The degree of pathology of inner retinal neurons, possibly reflecting transneuronal degeneration secondary to photoreceptor death, may be correlated with the severity and/or time of onset of photoreceptor death. Any degenerative changes in the inner retinal neurons in RP are extremely significant in view of ongoing research efforts to replace lost photoreceptors by transplantation, as described elsewhere in this volume. Obviously, if the inner retinal layers do not retain function after photoreceptor death due to RP, such photoreceptor transplants might not suffice to restore effective visual processing by the retina.

Other inconstant histopathologic findings in RP are cellular epiretinal membranes originating from the optic nerve head (Szamier, 1981), drusen of the optic nerve head (Puck *et al.*, 1985; Spencer, 1978), extensive and heterogeneous extracellular deposits in the region of Bruch's membrane, and focal loss of the choriocapillaris. Extensive literature citations on these topics are found in the three recent reviews cited above.

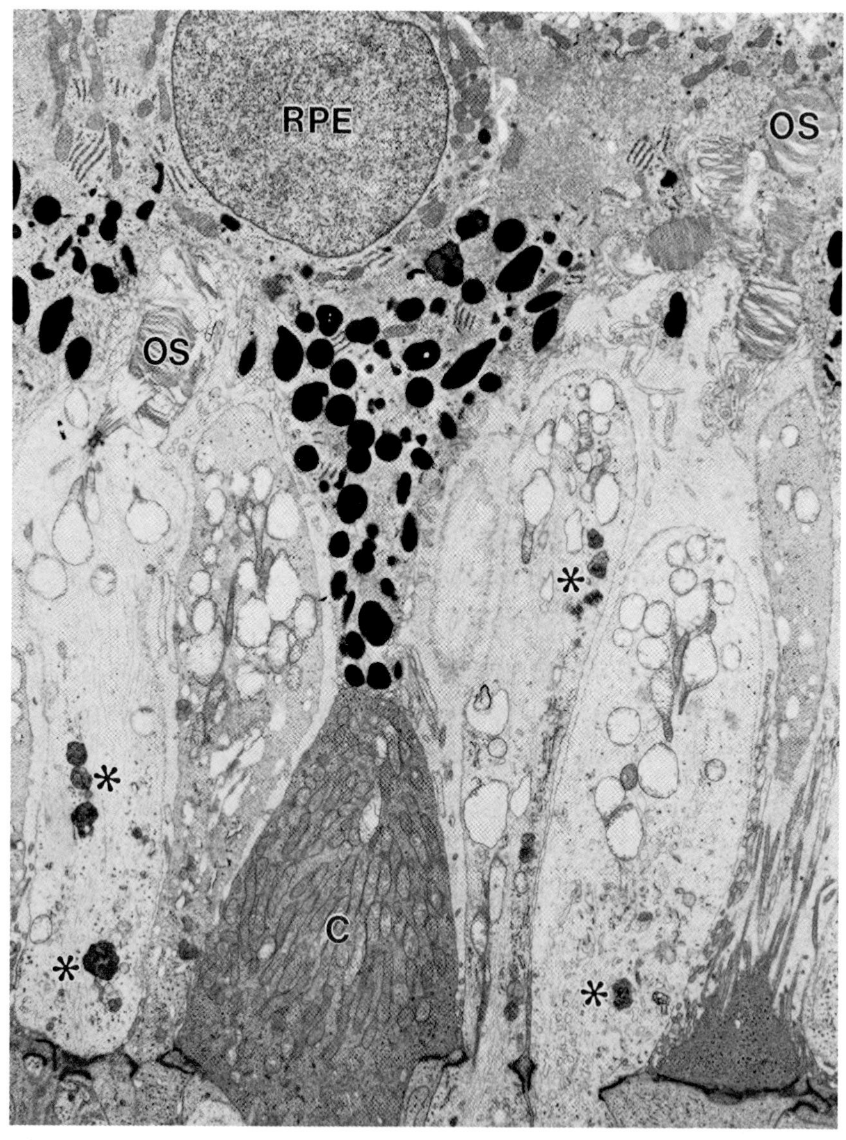

Fig. 1. The mid peripheral retina of a 31 year old man with autosomal recessive RP, as described by Bunt-Milam *et al.*, 1983. Pathologic rod changes include shortened OS, swollen mitochondria and dense IS granules (*). The cones (C) are invested by conical RPE sheaths. X20,000.

RESULTS

Based on the 10 RP eyes we have analyzed in detail this past year, we offer the following new observations and questions they have provoked. We hope that some of our queries may be amenable to laboratory experimentation and trigger new ideas about this poorly understood retinal disease.

Case #87-001.

Clinical: A 65 year old woman diagnosed as having "atypical" RP of unknown genetic type, with prominent bone spicule formation and loss of side vision for the past 21 years but no difficulty with dim vision when last examined 13 years prior to death. At this time, there was marked visual field constriction (within the 15° marker) and the ERG was completely extinguished in both eyes.

Gross Pathology: Heavy bone spicule pigmentation 360° from mid periphery to ora serrata, asteroid hyalosis.

Microscopic Pathology: OS were limited to stubby parafoveal rods and cones. The connecting cilia, periciliary ridges, and striated rootlets of these photoreceptors were morphologically normal (Fig. 2). A novel observation was the presence of one or more multivesicular bodies in some of the disorganized OS, which were not fused with their IS (Figs. 3 and 4). Cone IS in the fovea were greatly reduced in number. Photoreceptors were absent from the mid periphery, and the far periphery contained no rods and a single row of cone IS that lacked OS. The mid periphery showed melanin clumping in the RPE and loss of the choriocapillaris.

Comment: This case shows marked loss of rods and cones in the mid periphery, correlating well with the patient's constricted visual fields. Occurrence of normal appearing connecting cilia (also see Bunt-Milam *et al.*, 1983; Fliesler *et al.*, 1988) stands in contrast to the suggestions that RP might represent one expression of a generalized cilium defect (Arden and Fox, 1979; Finkelstein *et al.*, 1982; Hunter *et al.*, 1986). The finding of multivesicular bodies in the few remaining OS of the centralmost photoreceptors is a novel one, and raises the possibility of OS autodigestion. Multivesicular bodies are components of the lysosomal system in which endocytosed molecules and intracellular organelles are catabolized in a variety of cell types (Alberts *et al.*, 1983). An age and *post mortem* matched retina was processed identically; although the rod and cone OS appeared disorganized, they did not contain multivesicular bodies. Autophagosomes have been described in photoreceptor IS (Reme, 1977) and they are prominent in the IS of RP retinas (Szamier and Berson, 1977; Szamier *et al.*, 1979). To our knowledge, the present case is the first to show these lysosomal organelles in diseased OS.

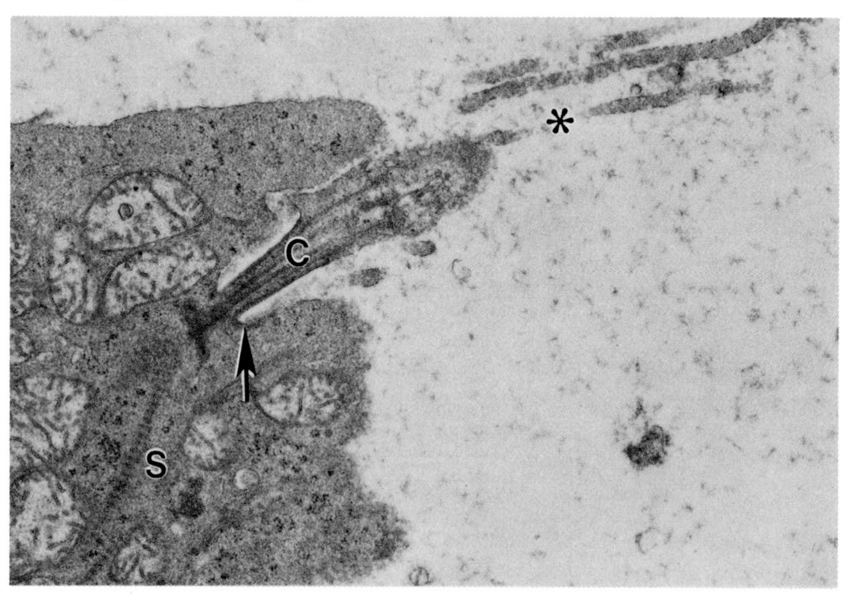

Fig.2. Case #87-001. Longitudinal section of macular photoreceptor. Note absence of OS and presence of morphologically normal connecting cilium (C), periciliary ridge (arrow), and striated rootlet (S). (*), calicyl processes. X15,000.

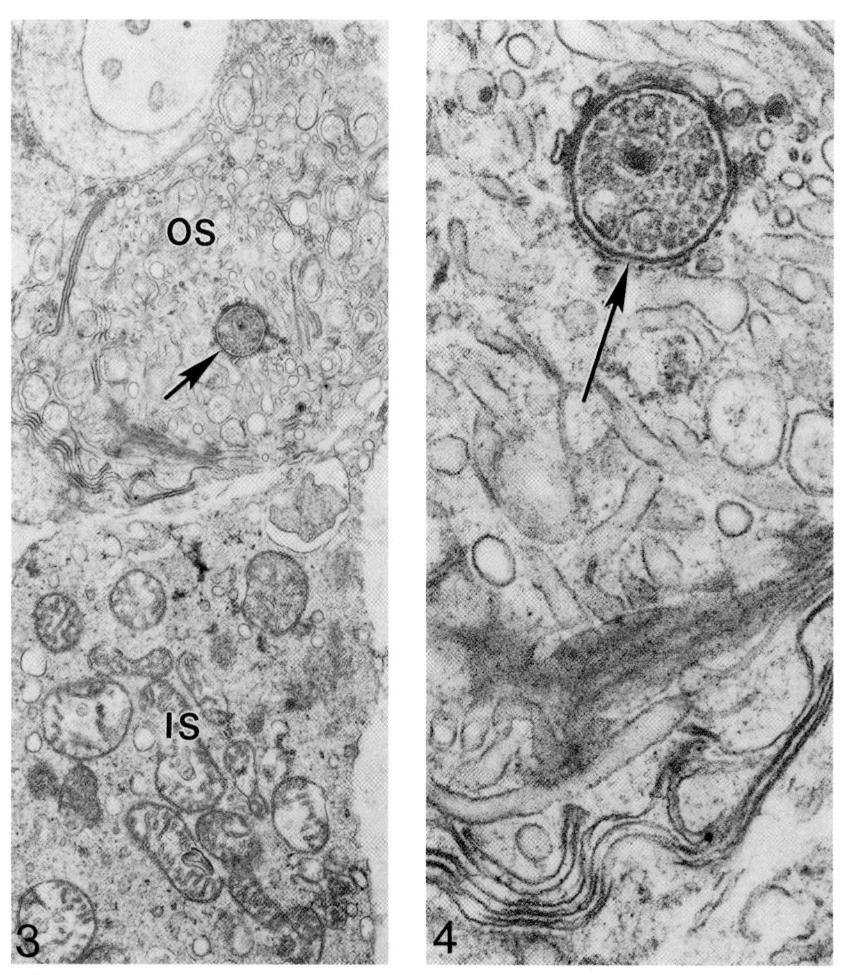

Fig. 3. Case #87-001. Macular photoreceptor. Note severe disorganization of OS membranes and presence of multivesicular body (arrow) within the OS, which is not fused with the cytoplasm of the IS. X15,000.

Fig. 4. Higher magnification of Fig. 3. Note fine structure of multivesicular body within the OS. The walls of the multivesicular body show characteristic electron dense plaque material and the interior is filled with fine vesicles. X50,000.

Case #87-002.

Clinical: A 53 year old man with a history of RP of unknown type.

Gross Pathology: Bone spicule pigmentation increases from the mid to far periphery, accompanied by prominent coalescing zones of depigmented RPE. The macula is thin and covered by an epiretinal membrane. The retinal blood vessels are very attenuated but not hyalinized.

Microscopic Pathology: The retina shows profound loss of photoreceptors in the mid and far periphery and retention of some photoreceptor inner segments and shortened OS in the centralmost retina. A thin, cellular membrane covers the macula. The regions of total photoreceptor loss show reactive gliosis of Müller cells, with hypertrophy of glial fibers and migration of very large Müller nuclei to the outer retina. Reactive changes of the RPE include duplication and migration of individual pigmented cells into the retina, with focal loss of RPE. In these regions, hypertrophied Müller cell processes fill the outer retina and the external limiting membrane abuts the few remaining zones of RPE (Fig. 5).

Comment: In some regions of this retina, photoreceptors are totally absent and replaced by hypertrophied Müller processes linked by *zonulae adherentes* junctions. These processes closely appose but do not form intercellular junctions with the RPE as judged by electron microscopy.

Several laboratories have recently reported some success in transplanting neonatal retinal photoreceptors into the subretinal space of rat retinas rendered experimentally free of photoreceptors by light damage or hereditary disease. If such transplants are to be attempted in the future in RP retinas where photoreceptors have degenerated, it will be necessary to determine whether a subretinal space is still existent (i.e. that reactive Müller glial cells have not formed intercellular junctions with the RPE, which might present a mechanical barrier to introduction of photoreceptor transplants). In this case of advanced RP, the subretinal space appears to remain unsealed and potentially patent for photoreceptor transplantation. However, the prominent external limiting membrane and hypertrophied Müller processes in the outer retina may represent another potential barrier to reestablishment of synaptic connections between transplanted photoreceptors and second order neurons of the host retina.

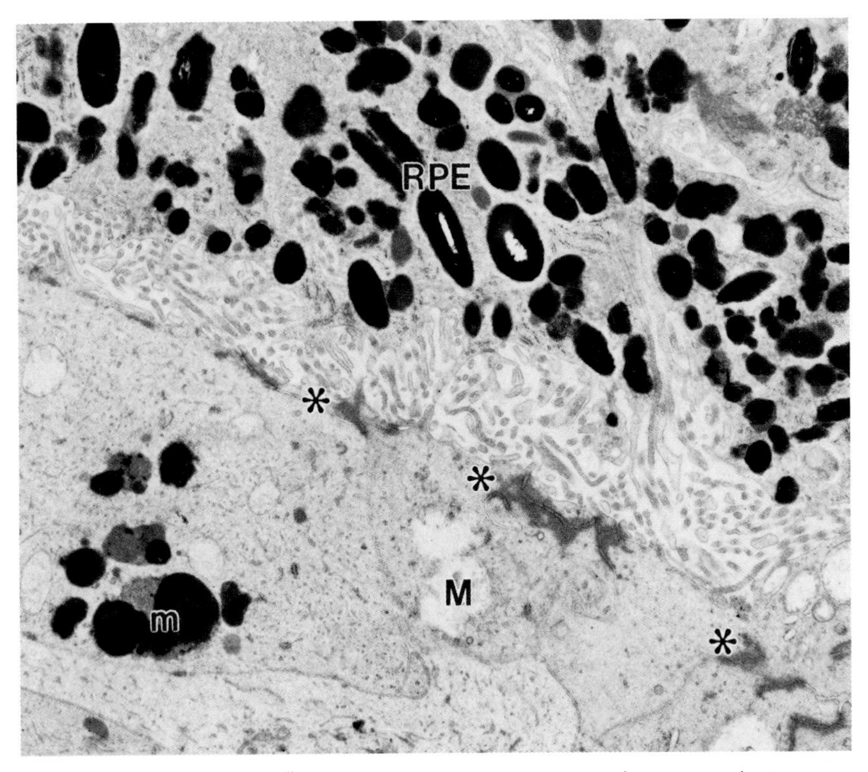

Fig. 5. Case #87-002. In the mid periphery, photoreceptors are replaced by hypertrophied Müller processes (M) that contain phagocytosed melanin (m). Intercellular junctions (*) are formed between Müller processes but are absent between the Müller cells and the RPE. X7,500.

Case #87-006.

Clinical: A 67 year old man with a history of RP, legally blind for the past 14 years.

Gross Pathology: Conspicuous bone spicule pigmentation from superior and inferior temporal vessels to the ora serrata. The macula is thin and there is a flat, uniformly black lesion 2 mm in diameter just superior to the fovea.

Microscopic Pathology: The black lesion is a choroidal nevus. The macula lacks photoreceptors except for a small group of IS in the floor of the fovea that lack OS. Photoreceptors are scant in the mid and far periphery and have only a few, very short OS. RPE cells have migrated into the inner retina and surround branches of the central retinal vessels in the mid and far periphery.

Comment: This represents an advanced case of RP with few remaining photoreceptors and classic reactive changes of the RPE.

Case #88-002.

Clinical: A 96 year old woman with a history of RP.

Gross Pathology: Prominent bone spicule pigmentation, most severe superiorly and temporally. Choroidal sclerosis is noted nasal to the optic nerve head.

Microscopic Pathology: This retina shows markedly reduced numbers of photoreceptors and only a few, very short OS are found in the parafovea. There is severe loss of inner retinal neurons and hyalinization of central retinal vessels. An interesting feature is the marked thickening of Bruch's membrane, which contains strata of different densities and multiple round deposits which contain nuclear sized inclusions (Fig. 6a and b). There is focal loss of the choriocapillaris over these round deposits. Electron microscopy of Bruch's membrane reveals these deposits as ring shaped formations, each approximately 10-15 μm in diameter, some with clear centers and others filled with granular debris. The walls of these rings contain calcified spherules with concentric laminations (Figs. 7 and 8).

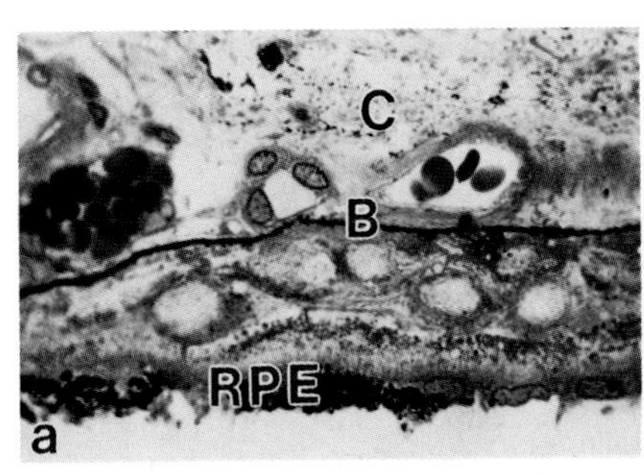

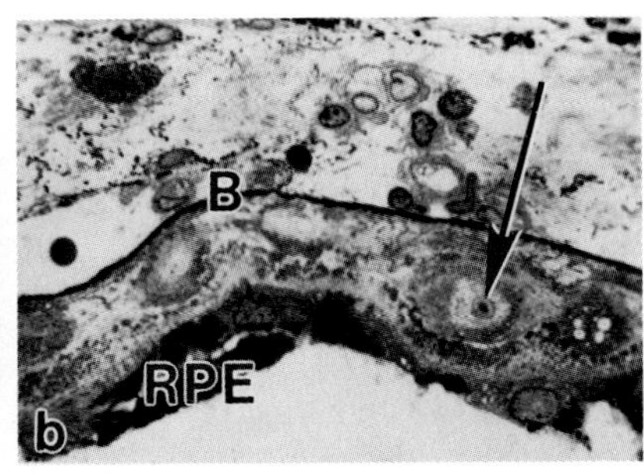

Fig. 6a and b. Case #88-002. Light micrographs of markedly thickened Bruch's membrane (B) overlying the RPE. The retina is artifactually detached. There is depigmentation of the choroid (C) and loss of choriocapillaris vessels. Large (10 - 15 µm) ring shaped deposits lie between Bruch's membrane and the RPE, some with a dense central inclusion (arrow). X350.

Comment: This retina is from our oldest RP donor to date. It is likely that the reactive changes in Bruch's membrane and the marked atrophy of the inner retinal layers reflect the advanced age of this patient. The significance of the unusual ring shaped deposits in Bruch's membrane is unknown.

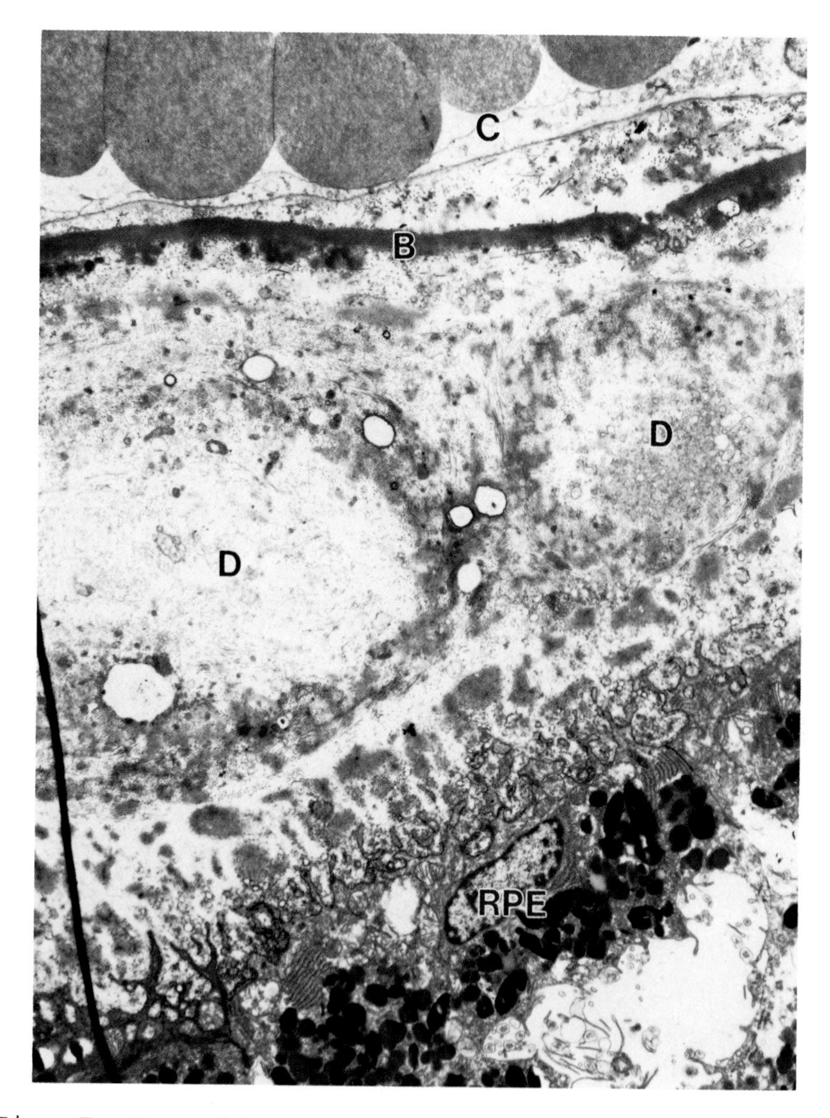

Fig. 7. Case #88-002. Electron micrograph of RPE, Bruch's membrane (B), ring shaped deposits (D), and choriocapillaris (C) which contains erythrocytes. X7,000.

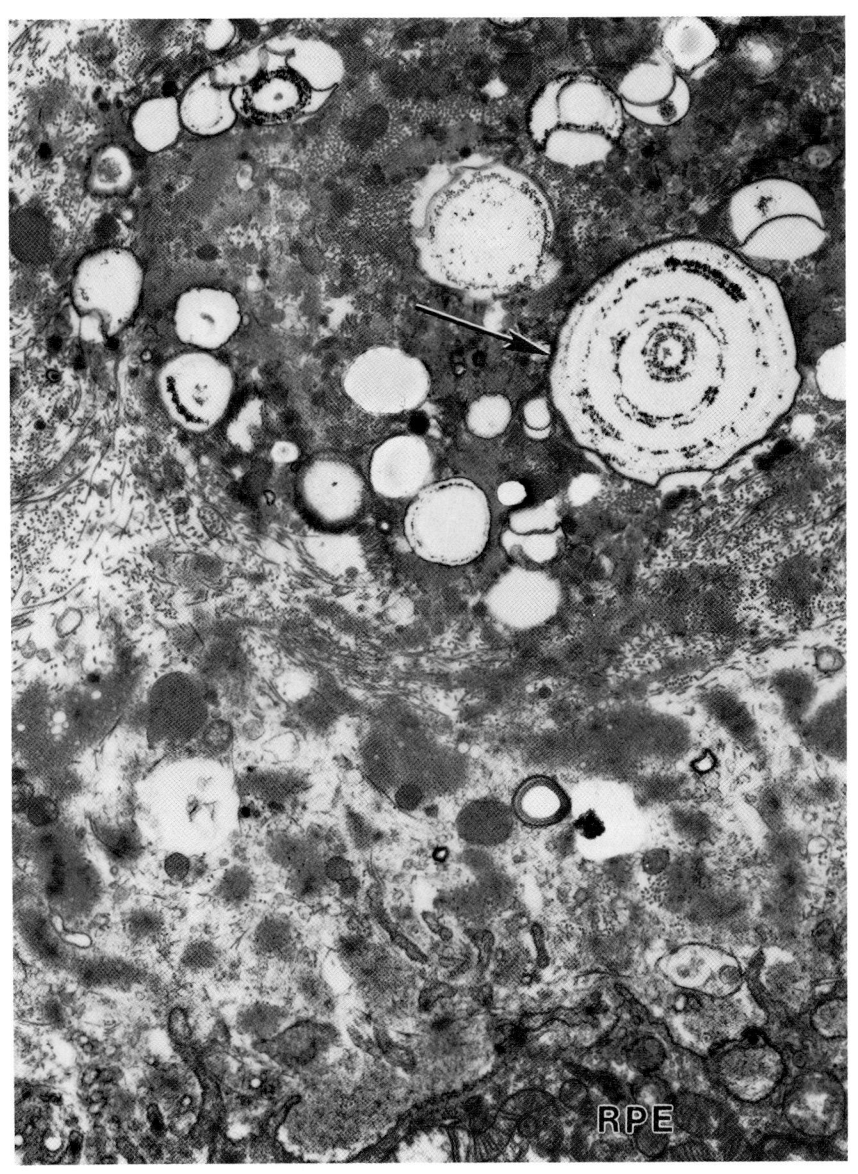

Fig. 8. Case #88-002. Higher magnification EM of region as shown in Fig. 8. Note abundant extracellular debris and calcified granules (arrow) between Bruch's membrane and RPE. X12,500.

Case #88-003.

Clinical: A 74 year old man with RP diagnosed at age 6.

Gross Pathology: Marked bone spicule pigmentation extending from mid periphery to ora serrata. Choroidal sclerosis in region of macula and encircling the optic nerve head.

Microscopic Pathology: No photoreceptors could be identified in this retina by light microscopy. The RPE shows reactive changes, most severe in the mid periphery, with bilayering and migration of RPE cells into the neurosensory retina, where they surround inner retinal blood vessels. By electron microscopy (Fig.9), the two layers of RPE are unusual in that their apical surfaces abut, enclosing small lumina encircled by microvilli. The basal RPE surfaces face Bruch's membrane and the external limiting membrane of the neurosensory retina. Granular deposits and debris are found in Bruch's membrane and also between the vitread layer of RPE and the adjacent gliotic retina.

Comment: This retina shows advanced RP with unusual reactive changes in the RPE. Bilayering of the RPE over regions of photoreceptor death is common, but in this case the bilayers are organized apex to apex, reminiscent of the two layers of epithelium that cover the ciliary body. The basal surface of the vitread layer of RPE is associated with extracellular deposits resembling the debris commonly found in Bruch's membrane. This suggests that the source of these extracellular deposits is the RPE and not solely the choriocapillaris.

Fig. 9.(Next Page). Case #88-003. A region of mid peripheral retina that lacks photoreceptors and has undergone reactive Müller cell gliosis. Note facing apical surfaces of the two layers of RPE, forming a small lumen (*) surrounded by microvilli. Extracellular granular debris (G) lies adjacent to the basal surfaces of both RPE layers. X10,000.

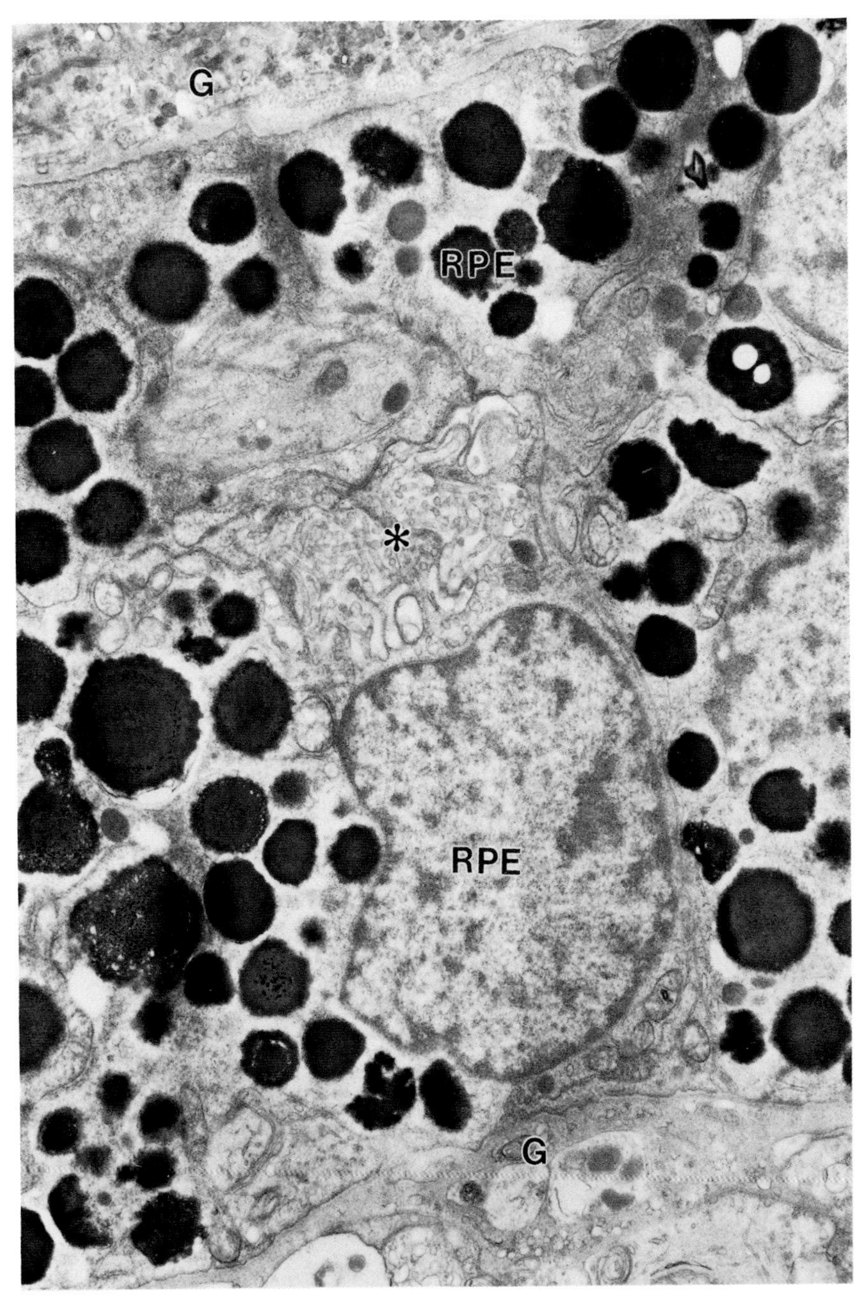

Fig. 9. See previous page for description.

Cases #88-005 and 88-006.

Clinical: A 71 year old male RP patient whose daughter also has RP. This pair of eyes was submitted frozen in normal saline with a *post mortem* interval of only 1.75 hours. The frozen eyes were slit at the pars plana and allowed to thaw and fix in buffered 4% paraformaldehyde for 8 hours. Retinas were processed for light and electron microscopic immunocytochemistry, including cryostat and LR-White sectioning.

Gross Pathology (OS and OD): Marked bone spicule pigmentation and dry, thin macular degeneration.

Microscopic Pathology: The tissue is very badly preserved. The photoreceptor layer is totally absent and melanin containing macrophages are found throughout the inner retina.

Comment: This case represents advanced RP with total loss of photoreceptors and very poor preservation of the tissue. The intent of these studies was to preserve the previously frozen retinas in a fixative compatible with immunocytochemistry studies. However, even though the retinas had been frozen promptly *post mortem*, since the saline did not contain a cryoprotectant such as 30% sucrose, the ice crystal damage was extensive and the morphology was far too poor for even light microscopy. Given the total absence of photoreceptors, further immunocytochemical studies were not attempted.

Cases #88-008 and 88-009.

Clinical: A 65 year old man with X-linked RP. He had a history of nyctalopia at age 6-8 years and his vision was severely compromised by age 40, with visual acuity reduced to only hand motion at 10 ft.

Gross Pathology (OS and OD): The retina shows heavy bone spicule formation from the nasal edge of the disc and just above and below the fovea to the ora serrata. There is clumping of melanin in the RPE, most pronounced temporally.

Microscopic Pathology: Both maculas show total loss of photoreceptors and RPE, with marked loss of the choriocapillaris and depigmentation of a thinned choroid. The neurons of the inner retina show marked cystic edema and are diminished in number by approximately one half. A preretinal membrane is present OD. Both peripheral retinas show absence of photoreceptors and focal depigmentation, proliferation, or loss of RPE. RPE cells encircle central retinal vessels and a few have migrated into the vitreous cavity OD.

Comment: This pair of retinas shows advanced degeneration of the photoreceptors and RPE, consistent with the patient's marked loss of vision for the previous 25 years. The inner retinal neurons are reduced in number by approximately one half and there is marked bone spicule pigmentation OU,characteristic of long standing RP.

Case #88-010.

Clinical: A 67 year old woman with RP of unknown type.

Gross Pathology: There is heavy bone spicule pigmentation 360° and numerous small drusen throughout, most pronounced inferiorly. The macula is thinned with black, granular pigmentation. The choroid shows some depigmentation in a tigroid pattern.

Microscopic Pathology: Photoreceptors and RPE cells are absent in the fovea, and some stubby inner segments with tiny OS are found in the parafovea. The inner retina shows cystic edema, thickening of the walls of central retinal vessels, and a contracted epiretinal membrane. Müller fibers are hypertrophied to fill the outer retina and extremely large Müller cell nuclei lie in the region formerly occupied by photoreceptors. The remainder of the retina shows numerous scattered drusen, loss of RPE and photoreceptors, Müller cell hypertrophy, and melanin containing cells surrounding central retinal vessels.

Comment: This represents a well preserved

specimen of advanced RP, consistent with the short *post mortem* interval of 2.5 hours. The only photoreceptors identified lie in the parafovea and have extremely attenuated IS and OS. Further ophthalmologic history is needed on this patient in order to correlate her changes in visual function with the striking histopathologic changes observed here.

DISCUSSION

This paper summarizes very briefly some of the studies conducted during the past year in the newly formed U.S. Retinitis Pigmentosa Histopathology Laboratory. Several new observations have been made regarding photoreceptors, the RPE, and Bruch's membrane. It is anticipated that the detailed analyses of each donor retina by light and electron microscopy will lead ultimately to improved understanding of the histopathologic time course of progression of the different genetic forms of RP. The histopathologic analyses will also facilitate correlation and interpretation of experimental results obtained from the opposite donor retinas. Goals for the coming year include obtaining more complete ophthalmologic histories for each donor, which will enable us to better correlate our morphologic observations with functional (visual) defects and the genetic form of the disease. In addition, we will continue to coordinate distribution of these valuable tissues to other research laboratories for analysis by specialized morphologic techniques, such as lectin cytochemistry. We also hope to obtain donor retinas processed in fixatives that are compatible with the new techniques of immunocytochemistry and *in situ* hybridization of nucleic acid probes. We welcome feedback on our initial studies and will be delighted to provide detailed copies of our protocols to interested investigators upon request.

ACKNOWLEDGEMENTS

This research was supported by the Louis Berkowitz Family Foundation and the National Retinitis Pigmentosa Foundation, and in part by NIH Research Grants EY0-1311 and -1730. The authors thank I. Klock, F. Dahlan, D. Possin, B. Clifton, R. Jones, and C. Stephens for technical assistance.

REFERENCES

Alberts B, Bray D, Lewis J, *et al.*(1983). "Molecular Biology of the Cell", New York, Garland Publishing Co, 1983. p 369.

Arden GB, Fox B (1979). Increased incidence of abnormal nasal cilia in patients with retinitis pigmentosa. Nature 279:534-536.

Bunt-Milam AH, Kalina RE, Pagon RA (1983). Clinical-ultrastructural study of a retinal dystrophy. Invest Ophthalmol Vis Sci 24:458-469.

Eisenfeld AJ, LaVail MM, LaVail J (1984). Assessment of possible transneuronal changes in the retina of rats with inherited retinal dystrophy: cell size, number, synapses, and axonal transport by retinal ganglion cells. J Comp Neurol 223:22-34.

Finkelstein D, Reissig M, Kashima H, *et al.*(1982). Nasal cilia in retinitis pigmentosa. Birth Defects 18:197-206.

Fliesler SJ, Chaitin MH, Jacobson SG (1988). X-linked retinitis pigmentosa (XLRP): light and electron microscopic analyses. 8th Int Cong Eye Res Abstracts, p 45.

Grafstein B, Murray M, Ingoglia N (1972). Protein synthesis and axonal transport in retinal ganglion cells of mice lacking visual receptors. Brain Res. 44:37-48.

Hunter DG, Fishman G, Mehta RS, *et al.*(1986). Abnormal sperm and photoreceptor axonemes in Usher's syndrome. Am J Ophthalmol 104:385-389.

Marshall J, Heckenlively JR (1988). Pathologic findings and putative mechanisms in retinitis pigmentosa. In: JR Heckenlively, (ed.) "Retinitis Pigmentosa", Philadelphia: JB Lippincott, pp 37-67.

Newsome DA (1988). Retinitis pigmentosa, Usher's syndrome, and other pigmentary retinopathies. In: DA Newsome, (ed): "Retinal Dystrophies and Degenerations", New York: Raven Press, pp 161-194.

Pagon RA (1988). Retinitis Pigmentosa: A Review. Survey of Ophthalmology, (In Press).

Puck A, Tso MOM, Fishman GA (1985). Drusen of the optic nerve associated with retinitis pigmentosa. Arch Ophthalmol 103: 231-234.

Reme CE (1977). Autophagy in visual cells and pigment epithelium. Invest Ophthalmol Vis Sci 16:807-815.

Spencer WH (1978). Drusen of the optic disc and aberrant axoplasmic transport. Ophthalmol 85:21-38.

Szamier RB (1981). Ultrastructure of the preretinal membrane in retinitis pigmentosa. Invest Ophthalmol Vis Sci 21:227-236.

Szamier RB, Berson EL (1977). Retinal ultrastructure in advanced retinitis pigmentosa. Invest Ophthalmol Vis Sci 16: 947-962.

Szamier RB, Berson EL., Klein R, *et al.* (1979). Sex-linked retinitis pigmentosa: ultrastructure of photoreceptors and pigment epithelium. Invest Ophthalmol Vis Sci 18:145-160.

Inherited and Environmentally Induced
Retinal Degenerations, pages 39–47

FURTHER EPIDEMIOLOGICAL STUDIES ON LIPID METABOLISM IN RETINITIS PIGMENTOSA

C.A. Converse, W.A. Keegan, L. Huq, J. Series*, M. Caslake*, T. McLachlan, C.J. Packard* and J. Shepherd*

Department of Pharmacy, University of Strathclyde and *Department of Pathological Biochemistry, University of Glasgow, Glasgow, Scotland.

Retinitis pigmentosa (RP) is a hereditary disease of the retina which appears to affect primarily the photoreceptor cells and retinal pigment epithelium. In the retina, photoreceptor outer segments are continuously being renewed. One constituent of the outer segments which has to be supplied for this renewal is docosahexaenoic acid (C22:6), since the outer segment discs are particularly rich in this polyunsaturated fatty acid (Anderson et al, 1974; Stone et al, 1979). Dietary studies on rats and rhesus monkeys have shown that animals deprived of C22:6 or its precursors may develop visual defects reminiscent of RP (Hands et al, 1965; Wheeler et al, 1975; Neuringer et al, 1984). Thus, it may be relevant to ask whether RP is itself caused by a defect in the supply of lipids to the retina.

Our approach to this question has been to measure levels of various lipids in the blood of RP patients, on the theory that a defect in lipid metabolism which affects the eye may also be measurable in the blood. We have found a significant tendency for hyperlipidemia in RP patients, as reflected in raised cholesterol and low-density lipoprotein (LDL) levels (Converse et al, 1983, 1985). However, some patients are hypolipidemic: these patients will be discussed in particular in this paper. We have also reported low levels of plasma C22:6 in X-linked and autosomal dominant RP (Converse et al, 1983, 1987a) and an increased prevalence of unusual isoforms of plasma apolipoprotein E in simplex patients (Converse et al, 1988).

We have investigated various factors which can influence plasma lipid levels in some detail. We find no differences between RP patients and their unaffected relatives in exercise, level of employment or diet, yet the RP patients have, on average, higher cholesterol and LDL levels (Converse et al, 1987b). Since hyperlipidemia may also be secondary to hypothyroidism, this paper also reports on studies on thyroid function indices in RP patients. It does not appear that hypothyroidism can account for the prevalence of hyperlipidemia in RP.

METHODS

For the LDL studies, fasting blood samples from 165 RP patients were collected in potassium EDTA tubes. Plasma was prepared and cholesterol, triglycerides, VLDL (very low-density lipoprotein), LDL and HDL (high-density lipoprotein) values determined by standard enzyme-based automated assay procedures, according to the Lipid Research Clinics Manual of Laboratory Operations. Results were compared to the Lipid Research Clinic (LRC) tables (US Public Health Service, 1980), and patients classified by percentile for their age and sex. Genealogical information was also collected to ascertain mode of inheritance.

The procedures for measuring LDL turnover have been described previously (Packard et al, 1983). In brief, on the morning of the first day 120 ml of fasting venous blood is collected from each patient and the LDL fraction prepared by rate zonal centrifugation. An aliquot of this fraction is labelled with ^{125}I and another with ^{131}I. The latter is then treated with 1,2-cyclohexanedione to block its interaction with cell membrane receptors. Both tracers are then injected back into the original donor. Blood samples are taken every day for two weeks, to monitor the disappearance of ^{125}I and ^{131}I from the plasma. The plasma lipoprotein levels and the composition of the LDL are also monitored.

^{125}I-LDL clearance measures total catabolism while the modified tracer provides an index of receptor-independent removal. The radioactivity present in blood samples taken over a two-week period is used to determine the removal rate of the LDL apoprotein. Plasma decay curves are constructed and used to calculate fractional metabolic rates by the procedure of Matthews (1957). Plasma LDL pool sizes are determined and this permits calculation of the amount of LDL catabolised each day (FCR x pool in mg) in mg apo-LDL per kg body weight per day. Under the steady state conditions of the study this value equals the synthetic rate of LDL apoprotein.

For thyroid function tests, serum or plasma samples were obtained from 119 RP patients and stored at -20°C until assay. They were assayed for thyroid-stimulating hormone (TSH) by the biochemistry laboratory at the Royal Infirmary. A few samples were also tested for auto-antibodies to thyroxine at the Western Infirmary, Glasgow. TSH values were compared to those of a normal Glasgow population which was under investigation in the Royal Infirmary at the same time.

RESULTS

The distribution of LDL levels in the RP population is shown in Figure 1. Compared to the LRC normal population, there are more patients with LDL levels in the lowest and highest percentiles for their age and sex. The tendency for

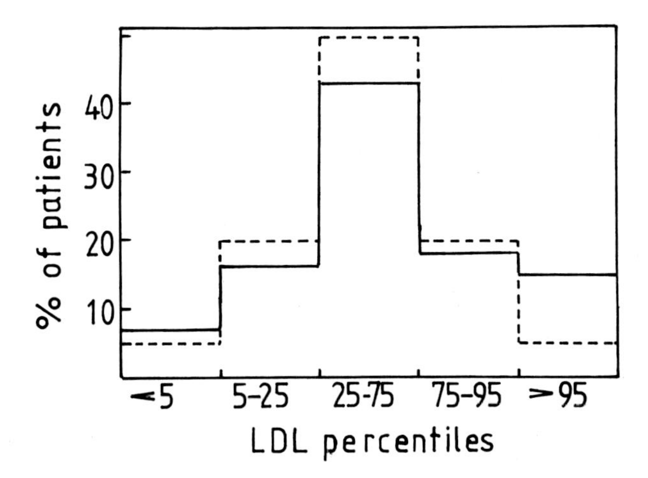

Figure 1: The distribution of LDL levels within the RP population. Solid line: Actual distribution of RP patients in each percentile (as determined from the LRC tables for each patient's age and sex). Dashed line: the expected distribution.

hyperlipidemia in RP has been described previously (Converse et al, 1983, 1985). The hypolipidemic population (LDL levels in the lowest 5% of the population) consisted of 13 individuals, from different families. Eleven of these people had autosomal dominant RP, as determined by their family histories. Since we were looking at all RP patients, not just one per family, it might be possible that autosomal dominant patients are over-represented in our population, and this could explain their prevalence in the hypo-LDL group as well. However, this was not the case: in the total population of 165 people, only 60 (36%) had autosomal dominant RP, while 73 (44%) had simplex or presumed autosomal recessive RP (ie, mixed multiplex), 31 (19%) were from X-linked families, and one was undetermined.

The statistical analysis of patients with the autosomal dominant type of RP compared to other types (Table 1) shows that hypolipidemia is significantly more common in the former group.

Table 1. Contingency table comparing the incidence of low LDL levels in autosomal dominant RP to that in other types of RP.

	Low LDL*	Normal or high LDL
AD RP	11	49
Other types	2	103

Chi-squared test: T=14.20, $p<.001$

*Low LDL is defined as LDL levels in the lowest 5% for each patient's age and sex.

We previously have descibed two autosomal dominant families with a tendency for low LDL levels (Converse et al, 1983, 1987a). Both these families have a late onset disease; some of the patients appear to have albipunctate deposits in the retina. Two of the patients, JR (77, male, pedigree 2378) and DH (63, male, pedigree 6043) volunteered to participate in a study of LDL turnover rates. When the fractional LDL clearance rate (FCR) per day was determined it was seen that DH had a normal FCR, but JR had slow clearance both of native and cyclohexanedione-modified LDL (FCR values (native): DH 0.435 pools/day, JR 0.248 pools/day). The amount of LDL synthesised per day was calculated to be 11.7 mg/dl for DH and 5.5 mg/dl for JR, the normal range being 10-15 mg/dl. Thus JR synthesised LDL at about half the normal rate.

For tests of thyroid function, 119 RP patients were compared to 500 controls. The latter population had the following distribution of TSH levels: <5 mU/l: 486 people; 5-10 mU/l: 7; 10-25: 5, >25: 2. For the purposes of comparison with the RP population, a cut-off of 10 mU/l was used, and anyone with over 10 mU/l in their blood was regarded as having hypothyroidism (when the thyroid is less active, there is less inhibition of TSH levels by thyroxine, and TSH levels rise). The results of this survey are shown in Table 2. It was found that there is a statistically significant increase in hypothyroidism in RP, although this is of borderline significance and does not involve a very large percentage of the RP population.

Table 2. Contingency table comparing hypothyroidism in RP and control populations.

	Hypothyroid	Normal
RP	5	114
Controls	7	493

Chi squared test: T=3.97, $p<.05$

DISCUSSION

Retinitis pigmentosa is not a single disease but a family of diseases which vary in their mode of inheritance, and as such may have a variety of etiologies. However, the common features of the various types of RP may mean that common biochemical pathways are affected, although the particular biochemical lesions in different types of RP ultimately may be traced to different genetic loci.

Two features of the outer retina which have been the focus of considerable research interest are the continual renewal of rod outer segments, and the requirement for a specific lipid, docosahexaenoic acid, as a constituent of outer segment membranes. It is possible in a slowly degenerative disease such as RP that even a small reduction in the supply of required lipids may affect the function and perhaps the renewal of outer segment discs, or of other membrane-mediated events in the retina. For this reason, it is particularly interesting to look at the plasma lipid transport system to see if there are any consistent defects in the lipids or their transport, bearing in mind that each type of RP should be considered separately if possible.

We have described hyperlipidemia in a number of RP patients. This hyperlipidemia is not the result of diet, exercise, employment, or any other aspects which we could identify of the lifestyle of the patients (Converse et al, 1987b). We now have looked at thyroid function, and although we do find an increase in hypothyroidism, from 1% in a control population to 4% in our RP patients, there is no apparent correlation of hypothyroidism with one type of RP, and the incidence of this disorder would appear to be too low to explain the more general hyperlipidemia.

Although hyperlipidemia is more common than hypolipidemia in RP, several aspects of the latter are of particular interest. We found that almost all the patients with very low levels of plasma LDL had autosomal dominant RP. This does not mean that most autosomal dominant patients have low LDL levels: the majority have average or high levels. LDL levels may vary even within the same family. Nevertheless the prevalence of this condition in autosomal dominant RP may indicate that there is some common biochemical defect in these patients which under certain circumstances (extrinsic or intrinsic) expresses as hypolipidemia.

It may be that low LDL is merely linked, in a genetic sense, to the gene for one of the types of autosomal dominant RP. Arguing against this possibility is the fact that lipids are known to be important for retinal function, and that in another disease, abetalipoproteinemia (Bassen-Kornzweig disease), there is no apolipoprotein B (apo B) and hence no LDL in the blood (apo B is the main protein in LDL) and a form of retinitis pigmentosa results (Herbert et al, 1983). The albipunctate appearance of the retina in some of our autosomal dominant patients resembles that in abetalipoproteinemia.

We have now examined the possibility that low LDL results from a defect in apo B, and causes defective lipid transport to the retina. One of the two patients tested had a low rate of LDL synthesis, reminiscent of that of a German RP patient (Vega et al, 1987). It may be that there is some defect in our patient's apo B. However, there are several problems with this interpretation: first, the patient is 77 and we have little information on LDL synthesis levels in older people, and second, one patient is hardly a statistically valid sample. The LDL turnover test is expensive and time-consuming and demands particularly committed patients. Thus we are now engaged in an alternative approach, using gene probes to look for apo B polymorphisms and associated functional variations (Demant et al, 1988) in RP.

Since the hyperlipidemia and hypolipidemia which we have described are general phenomena which could be the result of a variety of biochemical defects, it is important to try to get closer to the basic defect which is perhaps responsible for the pathogenesis of each type of RP. This means looking at specific pathways and enzymes or other proteins. The above study of apo B is an attempt to do this, and so are our studies of isoforms of apo E (Converse et al 1987a, 1988), following on the earlier studies of Jahn and his colleagues (1987). We have also described significantly decreased levels of C22:6 in X-linked RP and in one autosomal dominant family (Converse et al, 1983, 1987a). This finding has been confirmed in autosomal dominant RP and in Ushers syndrome patients in Louisiana (Anderson et al, 1987; Bazan et al, 1986). One large autosomal dominant RP does not show low C22:6 (Dehning and Garcia, 1987); this is not unexpected, since there are several types of autosomal dominant RP, which may not have the same biochemical characteristics. Low plasma C22:6 has also been described in an animal model of RP, the miniature poodle (Anderson et al, 1988). Thus there is considerable evidence that low C22:6, whether resulting from defects in lipid handling or fatty acid biosynthesis, is associated with RP; this is a fruitful field for further investigation.

We thank Ms. Mary-F. Collins for her assistance in obtaining patients, Dr. T. Demant for helpful discussions, and Professor W.S. Foulds and his staff at the Tennent Institute of Ophthalmology for their support. This research was funded by the National Retinitis Pigmentosa Foundation, the George Gund Foundation, the British Retinitis Pigmentosa Society and the W.H. Ross Foundation (Scotland) for the Study of Prevention of Blindness.

REFERENCES

Anderson RE, Benolken RM, Dudley PA, Landis DJ, Wheeler TG (1974). Polyunsaturated fatty acids of photoreceptor membranes. Exp Eye Res 18: 205-213.

Anderson RE, Maude MB, Alvarez RA, Acland GM, Aguirre GD (1988). Plasma levels of docosahexaenoic acid in miniature poodles with an inherited retinal degeneration. Invest Ophthalmol Vis Sci 29(Suppl): 169.

Anderson RE, Maude MB, Lewis RA, Newsome DA, Fishman GA (1987). Abnormal plasma levels of polyunsaturated fatty acid in autosomal dominant retinitis pigmentosa. Exp Eye Res 44: 155-159.

Bazan NG, Scott BL, Reddy TS, Pelias MZ (1986). Decreased content of docosahexaenoate and arachidonate in plasma phospholipids in Usher's syndrome. Biochem Biophys Res Comm 141:600-604.

Converse CA, Hammer HM, Packard CJ, Shepherd J (1983). Plasma lipid abnormalities in retinitis pigmentosa and related conditons. Trans Ophthalmol Soc UK 103: 508-512.

Converse CA, Huq L, McLachlan T, Bow AC, Alvarez E (1988). Apolipoprotein E isotypes in retinitis pigmentosa. Invest Ophthalmol Vis Sci 29(Suppl): 169.

Converse CA, McLachlan T, Bow AC, Packard CJ, Shepherd J (1987a). Lipid metabolism in retinitis pigmentosa. In

Hollyfield JG, LaVail MM, Anderson RE (eds): Degenerative Retinal Disorders: Clinical and Laboratory Investigations," New York: Alan R. Liss, pp. 93-101.
Converse CA, McLachlan T, Hammer HM, Packard CJ, Shepherd J (1985). Hyperlipidemia in retinitis pigmentosa. In LaVail MM, Hollyfield JG, Anderson RE (eds): Retinal Degeneration: Experimental and Clinical Studies," New York: Alan R. Liss, pp. 63-74.
Converse CA, McLachlan T, Packard CJ, Shepherd J (1987b). Lipid abnormalities in retinitis pigmentosa. In Zrenner E, Krastel H, Goebel HH (eds): Research in Retinitis Pigmentosa," Oxford: Pergamon Press, pp. 557-561.
Dehning DO, Garcia CA (1987). Lipid abnormalities in autosomal dominant retinitis pigmentosa. Invest Ophthalmol Vis Sci 28(Suppl): 346.
Demant T, Houlston RS, Caslake MJ, Series JJ, Shepherd J, Packard CJ (1988). Catabolic rate of low density lipoprotein is influenced by variation in the apolipoprotein B gene. J Clin Invest 82: 797-80.
Herbert PN, Assman G, Gotto AM Jr, Fredrickson DS (1983). Familial lipoprotein deficiency: Abetalipoproteinemia, hypobetalipoproteinemia and Tangier Disease. In Stanbury JB, Wyngaarden JB, Fredrickson DS, Goldstein JL, Brown MS (eds): The Metabolic Basis of Inherited Disease. New York: McGraw-Hill, pp. 589-621.
Jahn CE, Oette K, Esser A, v Bergmann K, Leiss O (1987). Increased prevalence of apolipoprotein E2 in patients with retinitis pigmentosa. Ophthalmic Res 19: 285-288.
Neuringer M, Connor WE, Van Petten C, Barstad L (1984). Dietary omega-3 fatty acid deficiency and visual loss in infant rhesus monkeys. J Clin Invest 73: 272-276.
Stone WL, Farnsworth CC, Dratz EA (1979). A reinvestigation of the fatty acid content of bovine, rat and frog retinal rod outer segments. Exp Eye Res 28: 387-397.
US Public Health Service, NIH (1980). "The Lipid Research Clinics Population Studies Data Book, Vol I: The Prevalence Study". Washington (NIH Publication 80-1527).
Vega GL, v Bergmann K, Grundy SM, Beltz W, Jahn CE, East C (1987). Increased catabolism of VLDL-apolipoprotein B and synthesis of bile acids in a case of hypobetalipoproteinemia. Metabolism 36: 262-269.
Wheeler TG, Benolken RM, Anderson RE (1975). Visual membranes: specificity of fatty acid precursors for the electrical response to illumination. Science 188: 1312-1314.

Inherited and Environmentally Induced
Retinal Degenerations, pages 49–56

SERUM FATTY ACID PROPORTIONS IN RETINITIS PIGMENTOSA MAY BE AFFECTED BY A NUMBER OF FACTORS

Lowell L. Williams, Lloyd A. Horrocks, Lawrence E. Leguire, and Barry T. Shannon

Departments of Pediatrics (L.L.W.,B.T.S.),Physiological Chemistry (L.A.H.), Ophthalmology(L.E.L.) Ohio State University College of Medicine, Columbus, Ohio 43205

Variations in serum fatty acid (FA) proportions of 67 Retinitis pigmentosa patients suggested a consideration of other factors which might affect these values. Serum FA,particularly arachidonic acid, may be altered during immunoregulatory substance formation in addition to genetic and dietary controls. Parallel immune system studies of these patients showed FA patterns consistent with increased prostaglandin-mediated (PG) lymphocyte suppression in 27% and possible block of PG immunoregulation in another 37%. While differences found between total serum FA of RP patients and age-matched normals included lower 18:2ω6 and higher 22:6ω3, the dominant and recessive RP patients showed similar proportion. However, lower than normal 22:6ω3 was found in 7 Usher RP.

INTRODUCTION

Polyunsaturated fatty acid proportions (FA) in serum are important in providing substrates for enzymes of visual function (Anderson et al.,1977; Birkle and Bazan,1986).Serum FA deficiencies have been proposed as possible contributing factors to disease in patients with the hereditary progressive retinal deterioration, Retinitis pigmentosa (RP) since lower than normal FA were found in patients with X-linked RP (Converse et al.,1985) and in an animal model (rd mice) of inherited visual cell degeneration (Bazan et al.,1984).Serum FA disproportions may be an epiphenomena or may vary in pattern with differing RP disease genetics (Converse et al.1987) but it is also possible that they may be depleted by other metabolic needs. For example, serum arachidonic acid (free and stored in phospholipids) serves as substrate for the formation of the important immunoregulatory substances, prostaglandins and leukotrienes (Birkle and Bazan,1986). Membrane stability is dependent upon FA composition (Hise et al.1986). Serum FA may also directly affect lymphocyte immune behavior (Traill and Wick,1984; Schreiner et al.,1988). In the present study, to determine a possible competitive role of immunoregulatory metabolic requirements on serum FA proportions, we compared interactions between serum FA and immunoregulation in RP patients.

METHODS

Sixty-seven subjects, in whom Retinitis pigmentosa had been confirmed by visual field, electroretinogram, and ophthalmologic examinations (Heckenlively,1988) volunteered participation in the study conducted with Human Subjects Research Guidelines. Comprised of 38 females and 29 males, ages ranged from 17 to 72, with an average of 47 years, suggesting a preponderately older group with long-standing RP (average % of life with RP symptoms = 43%). Dominant inheritance was found in 28 while 39 had recessive and simplex patterns. Members of 58 families were included. Usher's Syndrome with neurosensory hearing loss was found in 7. Fasting blood samples were examined both for FA and immunologic responses of lymphocytes and serum including anti-viral reactions. Age-matched normals and family members of RP patients were recruited for sampling.

The study was designed to evaluate possible relationships between serum FA and immunologic parameters measured in the same patient sample. Serum proportions of free and total esterified FA were measured by standard methods using chloroform:methanol (2:1) extraction, saponification, methylation, and identification of peaks with gas liquid chromatography (Williams et al.,1988). Dr. RE Anderson's laboratory (Houston TX)measured FA proportions of the phospholipid compartment from portions of RP and control samples (Anderson et al.,1977). Mitogenic responses of lymphocytes, modified by indomethacin to reveal prostaglandin suppression, (Goodwin et al.,1977), and numbers of peripheral T-cells showing activation were measured in the same samples and have been reported (Williams et al.,1988b). Serum anti-viral antibodies to rubella, rubeola, herpes simplex I and II, and varicella zoster viruses were measured by ELISA (Williams et al. 1988c). Statistical analyses between patients and controls by ANOVA used the IBM CRISP statistical program.

RESULTS

Serum Fatty Acids

In free FA, differences from normal FA in RP sera included higher proportions of 14:1, 16:1, and 20:1, but lower than normal proportions of 18:2ω6, 20:4ω6, and 20:5ω3 ($p<0.001$). The long-chain polyunsaturate, 22:6ω3, was slightly below normal. Dominant and recessive RP patients showed similar patterns. In total lipids from RP sera, FA proportions greater than normal included 16:0, 16:1, 18:2ω6, 20:2ω6, 22:4ω6, and 22:6ω3 ($p<0.001$). Average linoleic acid, 18:2ω6, was below normal in both dominant and recessive ($p<0.001$) but average arachidonic acid was normal. However, in the Usher Syndrome patients both 20:4ω6 and 22:6ω3 were lower than normals ($p<0.005$).

Although we found total lipid RP FA profiles different from normals, dominant and recessive RP groups were similar. However, calculation of the linoleic to oleic acid (L/O) ratio, an index possibly having clinical application after infectious mononucleosis (Williams et al.,1988a), showed a difference between RP heredity groups. Recessive RP had a L/O ratio of 1.20, falling at the lower range of normal(1.4 +0.2; mean +SD),while dominant RP had a L/O ratio of 1.0, lower than normal ($p<0.001$). This lower value suggested a more severe disease process (Williams et al.,1988a).

Serum Phospholipid Fatty Acids

Additional FA determinations of the phospholipid compartment of these samples (Cullen Eye Institute, R.E.A.), demonstrated differences between RP and control sera in the ratios of 22:6/16:0 and 20:5/16:0 and in 22:6 nmol/ml plasma and 22:6/ug P ($p<0.001$ using a two-tailed test with pairwise comparisons of age-adjusted means). In contrast, no difference appeared between the dominant and recessive RP groups except that the ratio of 20:5/16:0 showed change ($p<0.0186$). These data agreed with the FA results found in the Columbus laboratory. The average of all RP patients showed a higher 22:6ω3 proportion in total lipids than age-matched controls. In addition, 22:4ω6 was elevated in the phospholipid compartment of RP in comparison to controls, in agreement ($p<0.005$). Tests for heterogeneous regression (Houston results) were not significant for the ratios of 22:6 and 20:5 to 16:0 and for 22:6/ug P/ml plasma in the RP groups, but the higher proportion of 16:0 in RP may have obscured this relationship. However, there was some evidence for heterogeneous age regression for 22:6 nmol/ml plasma ($p<0.0133$) and marginal evidence for 22:6/ug P ($p<0.0743$) suggesting an accumulation of 22:6ω3 over time.

Relationship of Serum Arachidonic Acid to Immunologic Expression in RP

Unusual immunoregulation responses in these patients, reported elsewhere, included increased numbers of activated T-cells identified by expression of the Ta1 epitope (Williams et al.,1988b), and unusually high serum antibodies to rubella virus in the majority of the RP patients (Williams et al., 1988c). Correlations were found between serum FA in these RP patients and their hyperimmune responses:

Table 1. Immunoregulation Patterns Associated with Serum Total Arachidonic Acid Proportions "

	Low AA"	Medium AA	High AA	Normals (20)
AA Serum %	7%	7 - 10%	10%	7.5 - 9.5%
RP Patient Number	18	24	25	none
PG Effect on Lymphs'	all=18	none=18 equiv=6	none=19 equiv=6	none=15 equiv=5
% with Increased Activation+	56%	54%	52%	5%
High-anti-Rubella virus Antibodies@	64%	55%	58%	2%
Pattern	1	2	3	

(note : no age, sex or heredity patterns were represented in greater numbers in any of the RP groups separated by serum AA proportions.)

" AA = arachidonic acid proportions = % total FA extraction
' PG = prostaglandin effect was present when indomethacin, added *in vitro*, enhanced lymphocyte mitogen stimulation to demonstrate PG-mediated suppression (Goodwin et al,1977). This effect is seen largely in the low AA group.
Activated T-cells were identified by the fluorescent activation epitope,Ta1, using flow cytometry (Williams et al. 1988b). All RP groups show similar activation elevated over normal range.
+ The percent of RP patients with elevated numbers of activated cells in each category. The RP groups are similar.
@ High ELISA anti-rubella antibodies are serum titers greater than 0.415 (equivalent to HI titer > 1:256). RP groups show similar incidence, but all higher than normals.

Table 1 demonstrates three immune patterns in the RP population suggesting a spectrum of metabolic relationships. Pattern 1 - Low AA, found in 18 RP patients, includes low total arachidonic acid, decreased lymphocyte stimulation to the T-cell mitogen, phytohemagglutinin (PHA), and increased indomethacin enhancement of that response. An "inflammatory-type of reaction" of increased prostaglandin E2 production is suggested by this combination (Goodwin et al., 1977). In Pattern 2, normal proportions of arachidonic acid and PHA stimulations were found, but these were not consistent with the T-cell activation found in 56% of the patients. Pattern 3 - High AA, found in 25 RP patients, combined markedly elevated arachidonic acid proportions and high responses to PHA without an indomethacin effect. This unusual pattern could represent a block of PG formation (macrophage inhibition) which is inconsistent with elevated activated T-cell numbers also present in the samples. These three patterns then suggest a spectrum of immune and biochemical responses.

DISCUSSION

A genetic defect in lipid metabolism may cosegregate with expression of retinal degeneration in Retinitis pigmentosa (RP)(Converse et al.,1987a,b). Different proportions of FA measured in different RP populations may result from variable genetic contributions or different methodologies (Converse et al.,1987b). However, other environmental factors may play additional roles. Altered immunoregulation in RP, measured by several centers, may be one of these factors. Of particular significance to serum FA is the presence of an "inflammatory-type" of immune response demonstrated by increased lymphokine production (Dinarella and Mier,1987). In the reaction, serum arachidonic acid (AA) may be used as a precursor to form immunoregulatory agents, including prostaglandins (PG) and leukotrienes (Farrar and Humes,1985). A portion of the RP patients in this study appeared to have this response since low total serum AA appeared at the same time as increased PGE2-mediated suppression of lymphocyte responses and T-cell activation epitope increase (Table 1) (Goodwin et al.,1977).

However, unusually high proportions of AA in other RP patient sera, confirmed by two laboratories, showing little evidence of PG effects on lymphocytes, introduced the possibility that PG formation was blocked in those patients. Since some RP appeared to have normal levels of both of these measures despite other immunologic evidence of activation, a spectrum of immune reactions was proposed. AA proportions would be affected only at both extremes. In agreement with this concept, both hyperimmune (Galbraith and Fudenberg,1984; Hendricks and Fishman,1985) and hypoimmune (Detrick et al., 1985) responses have been found in RP. Similar extremes of immunoregulation are found in chronic infectious diseases in which the host and invading agent alternate strategies and dominance over time (Tishon et al;1988; Fauci,1988).

The altered immunoregulation in RP could be consistent with host reactions to an infectious agent (Galbraith and Fudenberg,1984). Data from these patients support this possibility. High serum anti-rubella antibodies measured in the RP patients of this study suggest a latent rubella virus infection (Williams et al., 1988c). It is possible that the rubella virus, known to have retinal tropism and toxicity (Waxham and Wolinsky,1984), may be present latently in some genetically-defective retinas in RP. It is known that secondary infections, common in genetic diseases in general, may cause more pathologic changes than the genetic defect itself, with cystic fibrosis being an example (Lloyd-Still et al., 1981). The possibility that a secondary, latent infection might be superimposed upon retinal degeneration in RP may elucidate some differences measured in FA as well as immunologic responses. Further, the immune privileged site of the eye may have obscured expected signs of rubella virus infection (Streilein and Wegmann,1987; Waxham and Wolinsky,1984).

Further understanding of FA proportions may also result from separation of disease entities within the spectrum of the RP syndrome (Heckenlively, 1988). Our 7 Usher's Syndrome patients expressed low serum total 20:4ω6 and 22:6ω3, as previously reported (Bazan et al.,1986) and showed a PG effect on lymphocyte behavior. In addition, although FA proportions were similar between 28 dominant and 39 recessive patients, a new metabolic index, the linoleic/oleic (L/O) ratio, was different between them, suggesting differing disease processes (Williams et al., 1988a). In RP, variable clinical manifestations are common. with some patients limiting retinal loss to marginal areas and retaining sight during long lifetimes, while others show a rapid deterioration to blindness. These characteristics suggest that there may be other, as yet undetected, infectious, metabolic or immunologic influences on serum FA which have limited our understanding of RP pathogenesis.

REFERENCES

Anderson RE,Benolken RM, Jackson MB, Maude MB (1977). The relationship between membrane fatty acids and the development of the rat retina. Adv Exper Med Biol 83:547-559.

Bazan NG, Reddy TS, Dobard P (1984). Metabolic and structural alterations in retinal membrane lipids in mice with inherited blindness. Suppl Invest Ophthal Vis Sci 25:114.

Birkle DL, Bazan NG (1986). The arachidonic cascade and phospholipid and docosahexaenoic acid metabolism in the retina. Prog Retinal Res 5:309-334.

Bazan NG, Scott BL, Reddy TS, Pelias MZ (1986). Decreased content of docosahexaenoate and arachidoneate in plasma phospholipids in Usher's Syndrome. Biochem Biophys Res Comm 141:600-604.

Converse CA, McLachlen T, Bow AC, Packard CJ, Shepherd J (1987a). Lipid metabolism in Retinitis pigmentosa In "Degenerative Retinal Disorders: Clinical and Laboratory Investigations". New York: Alan Liss, pp 93-101.

Converse CA, McLachlen T, Packard CJ, Shepherd J (1987b). Lipid abnormalities in Retinitis pigmentosa. Advan Biosci 62:557-561.

Detrick B, Newsome DA, Percopo CM, Hooks JJ (1985). Class II antigen expression and gamma interferon modulation of monocytes and retinal epithelial cells from patients with Retinitis pigmentosa. Clin Immunol Immunopath 36:102-111.

Dinarello CA, Mier JW (1987) . Current concepts: lymphokines. N Engl J Med 317:940-945.

Farrar WL, Humes JL (1985). The role of arachidonic acid metabolism in the activities of interleukin 1 and 2. J Im-

munol 135:1153-1159.
Galbraith GM, Fudenberg HH (1984). One subset of patients with Retinitis pigmentosa has immunologic defects. Clin Immunol Immunopath 31:254-260.
Fauci AS (1988). The Human Immunodeficiency Virus: infectivity and mechanisms of pathogenesis. Science 239:617-622.
Heckenlively JR (1988). In "Retinitis Pigmentosa", Philadelphia: JB Lippincott. Chaps 1-5.
Hendricks RL, Fishman GA (1985). Lymphocyte subpopulations and S-antigen reactivity in Retinitis pigmentosa. Arch Ophthal 103:61-65.
Hise MK, Mantulin WW, Weinman EJ (1986). Fatty acyl chain composition in the determination of renal membrane order. J. Clin Invest 77: 768-773.
Lloyd-Still JD, Johnson SB, Holman RT (1981). Essential fatty acid status in cystic fibrosis and the effects of safflower oil supplementation. Amer J Clin Nutri 43:1-7.
Streilein JW, Wegmann TJ (1987). Immunologic privilege in the eye and the fetus. Immunol Today 8(12):362-366.
Tishon A, Southern PJ, Oldstone MBA (1988). Virus-lymphocyte interactions. J Immunol 140:1280-1284.
Williams LL, Doody DM, Horrocks LA (1988a). Serum fatty acid proportions are altered during the year following acute Epstein-Barr virus infection. Lipids 23:981-988.
Williams LL, Shannon BT, Leguire LE (1988b). Immune alterations associated with T lymphocyte activation and regulation in Retinitis pigmentosa patients. Clin Immunol Immunopath 49 (in press).
Williams LL, Leguire LE, Shannon BT (1988c). Unusual incidence of high anti-rubella antibodies in Retinitis pigmentosa (Manuscript submitted).

ACKNOWLEDGEMENTS

This study was supported by the Retinitis Pigmentosa Foundation, the Ohio Lion's Clubs, the Ohio Retintis pigmentosa Chapter, and the Children's Hospital Research Foundation. The expert laboratory assistance of Barbara Meyer, Richard Loomis, and Deborah Jacobs is much appreciated. We particularly thank Dr. Robert E. Anderson of the Cullen Eye Institute of Baylor College of Medicine, Houston TX for fatty acid measurments in the serum phospholipid compartment of these samples, for their statistical analyses, and for helpful discussions.

Inherited and Environmentally Induced
Retinal Degenerations, pages 57–68

STUDIES ON BLOOD FROM PATIENTS WITH DOMINANTLY-INHERITED RETINITIS PIGMENTOSA

Mary J. Voaden, Philip J. Polkinghorne,
Joanna Belin and Anthony D. Smith.
Departments of Visual Science (MV) and Clinical Ophthalmology (PP), Institute of Ophthalmology, London WC1H 9QS, England, U.K.
Department of Chemical Pathology (AS,JB), University College and Middlesex School of Medicine London W1P 6DB, England, U.K.

INTRODUCTION

Clinical, psychophysical and electrophysiological tests have shown that there are at least two major variants of dominantly-inherited retinitis pigmentosa (RP) - one in which the disease expresses diffusely through the retina, with losses of rod function extending into areas of good cone function, AD(D-type) (Lyness et al., 1985) or Type I of Massof and Finkelstein (1981), and the other in which the lesions are regionally distributed with both rods and cones apparently equally disturbed, AD(R-type) or Type II. Moreover, it is likely that AD(R-type) RP comprises several subtypes. In particular there are diseases that show complete (CP) and incomplete (IP) penetrance within families (Berson et al, 1969; Ernst and Moore, 1988). When penetrance is incomplete, approximately one third of persons with the affected gene appear not to express the disease, although there is some mild abnormality in retinal function: when it is fully expressed, however, the disease is severe and has early onset.

In addition to the major division into IP and CP diseases, sub-groups of families with slowed rates of dark-adaptation can be discerned in the CP group (Ernst and Moore, 1988). These have not been included in the studies reported here.

Little is known of the aetiology of RP, and whilst clinical and genetic assessments are progressively subclassifying the diseases and defining their prognoses, the location and selectivity of the photoreceptor lesions appears to preclude biochemical study. Nevertheless, there are many reports in the literature of changes in blood biochemistry in RP patients, and, as patient classification improves and 'purer' populations are investigated, it is predictable that some causes will be discerned from peripheral studies. Our own investigations have provided evidence for abnormalities in taurine and membrane homeostasis in patients with AD (R-type) RP that are not present in those with D-type disease. In particular, in some families in which the abnormal gene(s) show complete penetrance of expression, taurine uptake by blood platelets is reduced (Hussain and Voaden, 1987) and erythrocyte osmotic fragility increased (Hussain and Voaden, 1985).

Taurine is highly concentrated in photoreceptors, being essential for maintenance of structure and function: deficiency leads to cell death (Pasantes-Morales, 1986). Present evidence suggests that photoreceptor taurine is derived principally from extracellular sources and that specific carrier proteins are needed for its transport across photoreceptor and retinal pigment epithelium membranes (Voaden et al, 1981; Pasantes-Morales, 1986). Thus a defect in taurine uptake, if expressed in the eye, would be likely to lead to a selective loss of photoreceptor cells, such as occurs in RP.

There is evidence to suggest that a change in taurine homeostasis will, in turn, affect the phospholipid profile of membranes (Cantafora et al, 1986) and the osmotic fragility of erythrocytes (Masuda et al, 1984). Equally it is known that a change in the saturation of phospholipid fatty acids will affect the trans-membrane transport of taurine (Balcar et al., 1980; Yorek et al., 1984). Thus, if the alteration in taurine handling in patients with AD (R-type) RP is viewed alongside the fact that their erythrocytes also lyse more readily in hypo-osmotic media (Hussain and Voaden, 1985), we see the possibility of change(s) in membrane composition in these patients. In this paper, we present some preliminary observations on erythrocyte composition in AD (R-type) RP with complete penetrance: fatty acids, vitamin E and cholesterol have been investigated.

PATIENT SAMPLE

Patients attended Moorfields Eye Hospital, London, where the category and extent of retinal dystrophy were defined genetically, clinically and, where possible, by psychophysical and electrophysiological tests (Jay, 1982; Arden et al, 1983; Lyness et al, 1985; Ernst et al, 1986; Ernst and Moore, 1988). Three families with forms of complete penetrance AD (R-type) RP are being studied. Although they fit the criteria defining the category, they differ in loci of disease expression. The most typical, as regards RP in general, is family 1337, whose affected members have intraretinal pigmentation at about 20° of eccentricity. Family 53 has more peripherally located lesions, and family 479 tends to express sectorially, with the greatest intraretinal changes in the inferonasal quadrant of the retina, but with a more diffuse expression than that regarded as typical of sectorial RP (c.f. Heckenlively, 1988).

METHODS

Venous blood was collected from the donors into a heparinized tube. The cellular component was separated from the plasma 2-3 hours after withdrawal and the erythrocytes washed three times with phosphate buffered saline before analysis.

Fatty acids and cholesterol were extracted with isopropanol and chloroform, containing 0.005% butylated hydroxytoluene, as described by Saito et al (1983). Following passage into chloroform/methanol (1:1), some aliquots were assayed for cholesterol and its esters by the o-phthalaldehyde method (Zlatkis and Zak, 1969; Gottlieb, 1980) and others assayed for fatty acids using gas chromatography. After saponification with alcoholic KOH, the fatty acids were extracted from acidified samples with hexane, transferred into heptane and methylated with diazomethane. The methyl esters were then analysed, using programmed temperature and splitless injection, on a 25 m X 0.25 mm fused silica, open tubular, bonded Superox FA column, film thickness 0.2 µm (Altech Associates/Applied Science), with helium as the carrier gas. After FID, peak area was determined by electronic integration on a Hewlett-Packard 3390A integrator.

Vitamin E was extracted from erythrocytes with sodium dodecylsulphate, ethanol and heptane (Burton et al, 1985), purified by HPLC and quantified spectrophotometrically at 295 nm (Bieri et al, 1979). Internal and external standards of tocopherol acetate and tocopherol, respectively, were used.

The accumulation of tritiated taurine by platelets was monitored as described by Voaden et al (1982).

All results are expressed as the mean ± the standard error of the mean. Probability values were obtained by the t-test.

RESULTS

Fatty acids. Because of the potential importance of docosahexaenoate (22:6 ω 3; DHA) and arachidonate (20:4 ω 6; ADA) for the maintenance of photoreceptor structure and function (Neuringer and Connor, 1986; Bazan and Scott, 1987), we have specifically investigated to see if there is evidence for an abnormality in their concentrations in patients with AD (R-type) RP.

TABLE 1. Erythrocyte Fatty Acids in Patients with AD (R-type) RP with Complete Penetrance

Patients category	Patients number	20:4/16:0+18:0 X 10	22:6/16:0+18:0 X 10
normal'	22	3.9 ± .30	1.2 ± .05
AD (R-type) RP			
RP 53	5	3.8 ± .09	1.1 ± .07
RP 479	6	2.7 ± .16 $p < 0.02$	0.9 ± .04 $p < 0.005$
RP 1337	4	3.8 ± .06	1.3 ± .15

To eliminate the potential variable of total lipid, levels of ADA and DHA have been assessed relative to the contribution of the saturated fatty acids palmitate (16:0) and stearate (18:0). Our initial results (Table 1) show a significant decrease in the relative contributions of both ADA and DHA to the profile of erythrocyte fatty acids in family 479, whereas levels are normal in families 53 and 1337.

Vitamin E. Significant differences to normal have been observed in the levels of vitamin E in erythrocytes from patients belonging to families 53 and 1337 (Table 2), with the former decreased and the latter raised. However, both essentially lie within the normal range.

TABLE 2. The Concentration of Vitamin E in Erythrocytes from Patients with AD (R-type) RP with Complete Penetrance

Patient Category	n.mol. Vitamin E per ml. packed cells	range
'normal'	2.7 ± 0.2 (21)	1.5 - 4.7
AD (R-type) RP	2.7 ± 0.3 (11)	1.5 - 4.9
RP 53	1.9 ± 0.3 (5) $p < 0.025$	1.5 - 2.8
RP 479	2.5 (2)	
RP 1337	3.8 ± 0.4 (4) $p < 0.01$	3.1 - 4.9

Cholesterol. As shown in Table 3, the overall concentration of cholesterol was found to be increased in non-fasting plasma from patients with both incomplete and complete penetrance AD (R-type) RP - the raised values contrasting with the normal levels seen in AD (D-type), multiplex and X-linked RP. Moreover, increased values are apparent in young as well as older patients (Table 4).

TABLE 3. The Concentration of Plasma Cholesterol in Patients with Retinitis Pigmentosa

Patient Category	mg. cholesterol / 100 mls plasma
'normal'	191 ± 7 (44)
AD (D-type)	207 ± 14 (15)
AD (R-type) (complete penetrance)	240 ± 14 (22) $p < 0.001$
AD (R-type) (incomplete penetrance)	234 ± 27 (12) $p > 0.02$
X-hemizygote	184 ± 15 (24)
multiplex	203 ± 32 (5)

TABLE 4. Plasma Cholesterol in Patients with AD (R-type) RP

Patient Category	Age (yrs)	mg. cholesterol/100 ml. plasma	
'normal'	< 40	169 ± 7 (22)	
	> 40	217 ± 14 (12)	
AD (R-type) complete pen.	< 40	199 ± 11 (12)	$p < .01$
	> 40	295 ± 21 (8)	$p < .001$
AD (R-type) incomplete pen. (symptomatic)	< 40	213 ± 8 (5)	$p < .005$
	> 40	277 ± 25 (7)	$p < .001$

Potentially, an increase in circulating cholesterol will in turn raise eythrocyte median corpuscular fragility (MCF) via passive exchange (Gottlieb, 1980). Therefore, as we had previously observed increased MCF's in AD (R-type) RP

(Hussain and Voaden, 1985), the group consisting predominantly of people from families in which the disease showed complete penetrance, we measured the cholesterol content of patients' erythrocytes. The results (Table 5) showed an increase in the level of free plus esterified cholesterol in erythrocytes from family 479 but not in families 53 and 1337.

TABLE 5. Free and Esterified Cholesterol in Red Blood Cells from patients with AD (R-type) RP with Complete Penetrance

Patient Category	Cholesterol (mg per ml packed RBC)	
'normal'	1.44 ± .07 (16)	
AD(R-type) RP		
RP 53	1.07 ± .08 (4)	
RP 479	2.07 ± .10 (5)	$p < 0.001$
RP 1337	1.17 ± .14 (3)	

Taurine. We have previously observed a reduction in taurine uptake by platelets from patients with AD (R-type) RP (Hussain and Voaden, 1987) - the donors being predominantly from families in which the disease showed complete penetrance. However, our current results, outlined above, indicate biochemical differences between families, previously grouped into this category. It is, therefore, pertinent to ask whether the individual families also have reductions in platelet taurine uptake. Only families 57 and 479 have been investigated and both do (Table 6).

DISCUSSION

The biochemical differences found between families classified into the category of AD (R-type) RP, expressing with complete penetrance, emphasize not only that subtle

TABLE 6. The Accumulation of Tritiated Taurine by Platelets from Patients with AD (R-type) RP with Complete Penetrance

Patient Category	Taurine Uptake in Autologous Plasma (tissue/medium ratio)*	
'normal'	5.7 ± 0.2 (32)	
AD (R-type) RP		
RP 53	4.7 ± 0.4 (6)	$p < 0.05$
RP 479	4.0 ± 0.4 (4)	$p < 0.01$

* Platelet wet weights were estimated from the protein values by assuming a 10% protein content and, for calculations of T/M ratios, 1.0 mg wet weight was taken as equivalent to 1.0 µl incubation medium.

genetic heterogeneity may exist in RP subgroups, but also the need to concentrate investigations on large pedigree families.

We have observed reduced platelet taurine uptake in families 53 and 479 (family 1337 was not tested; Table 6), and increased erythrocyte MCF and hypercholesterolaemia (albeit in non-fasting plasma) in all three families (Tables 3 & 4; Hussain and Voaden, 1985 and unpublished). It is surprising, therefore, that only family 479 shows a reduction in the relative contribution of DHA and ADA to the total fatty acid profile, assessed in erythrocytes (Table 1). Thus our findings reveal not only intriguing similarities in the changes occurring in blood biochemistry in patients with AD (R-type) RP, including preliminary findings on patients with 'incomplete penetrance' disease (Tables 3 & 4; Hussain and Voaden, 1987) but also a striking difference as regards the fatty acids. The existence of interfamily variation in loci of disease expression (see 'Patient Sample' above) in itself suggests different primary pathogenetic mechanisms. It would appear likely, therefore, that diverse causes are leading to similar end results, not only as regards photoreceptor degeneration but also more generally. In this context, it may prove pertinent that a central, physiological action of taurine is conjugation with bile acids. Taurocholate formation is not only important in fat absorption, but also in the control of hepatic cholestolerol synthesis and, ultimately, of total body cholesterol. Thus abnormalities in taurine homeostasis might lead to changes in cholesterol metabolism. Equally, there are links between taurine homeostasis and the composition of membrane phospholipids (see Introduction). Clearly, it is now of considerable importance to investigate the causal relationships of the changes.

A strong association between hyperlipidaemia (including hypercholesterolaemia) and RP in middle aged patients has been noted previously by Converse et al (1983, 1985, 1987). However, a few with dominantly-inherited disease were hypocholesterolaemic . In addition, low levels of DHA and ADA have been observed previously in patients with AD, but otherwise untyped, RP (Converse et al, 1983; Anderson et al, 1987). Research into such abnormalities should be aided considerably by the availability of miniature poodles, suffering from a 'regional' retinal degeneration resembling human RP, and with lower levels of plasma DHA than normal (Anderson et al, 1988).

More people, both normal and affected, must be investigated before we can reach conclusions as regards the vitamin E status of the RP patients investigated here. Our present results show significant differences between affected families and our control population (Table 2) but the findings may simply reflect normal variation. Wide genetically-determined variation would, in itself, be of interest, however, as lower levels might predispose towards radiation and free radical induced damage, and this might have considerable significance for photoreceptor suscepti-bility to other damaging insults (see e.g. Tso, 1988).

ACKNOWLEDGEMENTS

The expert technical assistance of Mr. C. Jubb B.Sc. and Mr. D. Moore B.Sc. is gratefully acknowledged. In addition, we are indebted to all colleagues who have contributed to the classification of the patients investigated here. Essential financial support has been received from The American Retinitis Pigmentosa Foundation, The British Retinitis Pigmentosa Society, Fight For Sight and The Royal National Institute For The Blind. The high pressure liquid chromatograph was provided by The Wellcome Trust and The RNIB, and the gas chromatograph by The Headley Trust.

REFERENCES

Anderson RE, Maude MB, Lewis RA, Newsome DA, Fishman GA (1987). Abnormal plasma levels of polyunsaturated fatty acid in autosomal dominant retinitis pigmentosa. Exp Eye Res 44:155-159.

Anderson RE, Maude MB, Alvarez RA, Acland GM, Aguirre GD (1988). Plasma levels of docosahexaenoic acid in miniature poodles with an inherited retinal degeneration. Invest Ophthal Vis Sci 29 (Suppl):169.

Arden GB, Carter RM, Hogg CR, Powell DJ, Ernst WJK, Clover GM, Lyness AL, Quinlan MP (1983). Rod and cone activity in patients with dominantly inherited retinitis pigmentosa: comparisons between psychophysical and electroretinographic measurements. Brit J Ophthalmol 67:405-418.

Balcar VJ, Borg J, Robert J, Mandel P (1980). Uptake of L-glutamate and taurine in neuroblastoma cells with altered fatty acid composition of membrane phospholipids. J Neurochem 34:1678-1681.

Bazan NG, Scott BL (1987). Docosahexaenoic acid metabolism and inherited retinal degenerations. In Hollyfield JG, Anderson RE, LaVail MM (eds): "Degenerative retinal disorders: clinical and laboratory investigations," New York: Alan R. Liss, pp 103-118.

Berson EL, Gouras P, Gunkel RD, Myrianthopoulos NC (1969). Dominant retinitis pigmentosa with reduced penetrance. Arch Ophthalmol 81:226-234.

Bieri JG, Tolliver TJ, Catignani GL (1979). Simultaneous determination of α-tocopherol and retinol in plasma or red cells by high pressure liquid chromatography. Am J Clin Nutr 32:2143-2149.

Burton GW, Webb A, Ingold KU (1985). A mild, rapid and efficient method of lipid extraction for use in determining vitamin E/lipid ratios. Lipids 20:29-39.

Cantafora A, Mantovani A, Masella R, Mechelli L, Alvaro D (1986). Effects of taurine administration on liver lipids in guinea pigs. Experientia 42:407-408.

Converse CA, Hammer HM, Packard CJ, Shepherd J (1983). Plasma lipid abnormalities in retinitis pigmentosa and related conditions. Trans Ophthal Soc UK 103:508-512.

Converse CA, McLachlan T, Hammer HM, Packard CJ, Shepherd J (1985). Hyperlipidemia in retinitis pigmentosa. In LaVail MM, Hollyfield JG, Anderson RE (eds): "Retinal degeneration: experimental and clinical studies," New York: Alan R Liss, pp 93-101.
Converse CA, McLachlan T, Bow AC, Packard CJ, Shepherd J (1987). Lipid metabolism in retinitis pigmentosa. In Hollyfield JG, Anderson RE, LaVail MM (eds): "Degenerative retinal disorders: clinical and laboratory investigations," New York: Alan R. Liss, pp 93-101.
Ernst W, Kemp CM, Arden GB, Moore AT, Bryan S (1986). Functional heterogeneity in retinitis pigmentosa. In Agardhi E, Ehinger B (eds): "Retinal signals, systems, degenerations and transplants," Amsterdam: Elseveir, pp 207-222.
Ernst W, Moore AT (1988). Heterogeneity, anomalous adaptation and incomplete penetrance in autosomal dominant retinitis pigmentosa. In Zrenner E, Krastel H, Goebel H-H (eds): "Research in retinitis pigmentosa," Oxford: Pergamon Press, pp 115-120.
Gottlieb MH (1980). Rates of cholesterol exchange between human erythrocytes and plasma lipoproteins. Biochim Biophys Acta 600:530-541
Heckenlively JR (1988). "Retinitis Pigmentosa" Philadelphia: JB Lippincott.
Hussain AA, Voaden MJ (1985). Studies on retinitis pigmentosa in man II erythrocyte osmotic fragility. Brit J Ophthalmol 69:126-128.
Hussain AA, Voaden MJ (1987). Some observations on taurine homeostasis in patients with retinitis pigmentosa. In Hollyfield JG, Anderson RE, LaVail MM (eds): "Degenerative retinal disorders: clinical and laboratory investigations," New York: Alan R. Liss, pp 119-129.
Jay M (1982). On the heredity of retinitis pigmentosa. Brit J Ophthalmol 66:405-416.
Lyness AL, Ernst W, Quinlan MP, Clover GM, Arden GB, Carter RM, Bird AC, Parker JA (1985). A clinical, psychophysical and electroretinographic survey of patients with autosomal dominant retinitis pigmentosa. Br J Ophthalmol 69:326-339.
Massof RW, Finkelstein D (1981). Two forms of autosomal dominant retinitis pigmentosa. Doc Ophthalmol 51:289-346.
Masuda M, Horisaka K, Koeda T (1984). Influences of taurine on functions of rat neutrophils. Jap J Pharm 34:116-118.0
Neuringer M, Connor WE (1986). n-3 Fatty acids in brain and retina: evidence for their essentiality. Nutr Revs 44:285-294.
Pasantes-Morales H (1986). Current concepts on the role of

taurine in the retina. Prog Ret Res 5:207-229.
Saito M, Tanaka Y, Ando S (1983). Thin-layer chromatography-densitometry of minor acidic phospholipids: application to lipids from erythrocytes, liver and kidney. Analyt Biochem 132:376-383.
Tso MOM (1988). Photic injury to the retina and pathogenesis of age-related macular degeneration. In Tso MOM (ed): "Retinal diseases: biomedical foundations and clinical management," Philadelphia, JB Lippincott Co. pp 187-214.
Voaden MJ, Oraedu ACI, Marshall J, Lake N (1981). Taurine the retina. In Schaeffer S, Koscis JJ, Baskin S (eds): "The effects of taurine on excitable tissues," New York: Spectrum Publications, pp 145-160.
Voaden MJ, Hussain AA, Chan IPR (1982). Studies on retinitis pigmentosa in man I. taurine and blood platelets. Brit J Ophthalmol 66:771-775.
Yorek MA, Strom DK, Spector AA (1984). Effect of membrane polyunsaturation on carrier-mediated transport in cultured retinoblastoma cells: alterations in taurine uptake. J Neurochem 42:254-261.
Zlatkis A, Zak B (1969). Study of a new cholesterol reagent. Analyt Biochem 29: 143-148.

Inherited and Environmentally Induced
Retinal Degenerations, pages 69–75

GLYCOSYLTRANSFERASE ACTIVITIES IN LYMPHOCYTES OF PATIENTS WITH RETINITIS PIGMENTOSA

Kazuhiko Unoki, Fumiyuki Uehara, Norio Ohba and Takashi Muramatsu

Departments of Ophthalmology (K.U., F.U., N.O.) and Biochemistry (T.M.), Kagoshima University Faculty of Medicine, Kagoshima-shi 890, Japan

INTRODUCTION

Cell membrane-bound glycoconjugates are suggested to be involved in a variety of cellular processes including development and migration (Muramatsu 1988). The sugar residues of glycoconjugates have been localized and characterized in the vertebrate eye, including the cell surface of retinal photoreceptors and pigment epithelium (Nir et al, 1979; Uehara et al, 1983; Blanks et al, 1986; Sameshima et al, 1987). A number of experimental evidence suggest that carbohydrate residues of cell surface glycoconjugates are involved in the organization and maintenance of the retina, e.g. mediation of the phagocytosis of photoreceptor outer segments (O'Brien 1976; Colley et al, 1987). As regards the retinal pathology, sialic acid residues of glycoconjugates in the dystrophic RCS rat are reduced on the surface of rod outer segment debris accumulated in the subretinal space (Cohen et al, 1983). The terminal sialic acid of glycoconjugates in the C3H mouse retina might be replaced by galactose (Uehara et al, 1987). The oligosaccharide side-chain of cell surface glycoconjugates are synthesized by a stepwise process in such a manner that carbohydrates are successively added to a growing side-chain by a group of catalyzing enzymes called glycosyltransferases (Roth et al, 1972; Shur, 1982). Glycosyltransferase activities have been demonstrated in the peripheral blood and the eye (O'Brien et al, 1968; Unoki et al, 1988). Keeping these in mind, we thought it justified to measure the glycosyltransferases in the peripheral lymphocytes of patients with retinitis pigmentosa primarily affecting the retina.

MATERIALS AND METHODS

To assess the glycosyltransferase activities of peripheral lymphocytes, 27 systemically healthy patients with primary retinitis pigmentosa, 15 males and 12 females ranging in age 23-78 years, were used; their inheritance was autosomal recessive in 15 cases and sporadic in 12 cases. In addition, 3 cases of cone dystrophy, 1 male and 2 females ranging in age 24-64 years, were also used. A comparable number of healthy adults participated as a normal control group.

Peripheral blood (10 ml) was obtained from the cubital vein, with 1 ml heparin added. The lymphocyte rich fraction was obtaiend by Mono-poly resolving medium (Ficoll-Hypaque, Flow Lab.) gradient centrifugation, and washed twice with 0.1 M phosphate buffered saline. The lymphocytes were suspended in 25 mM Tris-HCl buffered solution (pH 7.2) containing 1 mM dithiothreitol and 0.1% Triton X-100, frozen at -80°C, melted, and homogenized manually by 20 strokes with loose pestle in a Wheaton glass homogenizer. Following sedimentation at 120 000 x *g* for 15 min, unbroken nuclei were discarded, and the supernatant fluid was used for the sample of glycosyltransferase enzyme assays by the following procedures. Protein concentration was determined by the method of Lowry et al (1951) with a bovine serum albumin as a standard.

Activities of three glycosyltransferases (sialyltransferase, fucosyltransferase and galactosyltransferase), were measured according to the following established procedures (Miyagi et al,1982). Sialyltransferase activity was assayed in the medium containing: 0.1 μCi CMP-sialic acid (sialic acid-4,5,6,7,8,9-^{14}C, 257.1 mCi/mmol, New England Nuclear) as a substrate, 0.3 mg asialofetuin (Sigma,type 1) as an acceptor, 50 mM Tris-HCl buffered solution (pH 7.2), 5 mM $MnCl_2$, and 50 μl lymphocyte extract (2 x 10^7 cells/ml). The total volume was adjusted to 90 μl with distilled water. Incubation was continued in a shaking water bath at 37°C for 30 min, and terminated by addition of 1 ml ice-cold 1% phosphotungric acid in 0.5 M HCl.Precipitates were obtained by centrifuging the sample at 150 x *g* for 10 min, spotted onto a 25 mm glass fiber disk (Whatman GF/A), washed three times with ice-cold mixture of 5% trichloroacetic acid and 1% phosphotungric acid, washed with ether-ethanol (1:1), and dried under vacuum. The dried disk was solubilized in 1 ml distilled water, and mixed with 0.4% 2,5-diphenyloxazole in toluen-Triton X-100 (2:1). The radioactivity was determined by a liquid scintillation counter (Aloca, Tokyo).

The activities of fucosyltransferase and galactosyltransferase were assayed by the same procedures as above except a separate set of substrate and acceptor. For the assay of fucosyltransferase, 0.1 μCi GDP-fucose (fucose-^{14}C, 238.7 mCi/mmol, New England Nuclear) was used as a substrate and 0.3 mg asialofetuin as an acceptor. For the assay of galactosyltransferase activity, 0.1 μCi of UDP-galactose (galactose-^{14}C, 333 mCi/mmol, New England Nuclear) was used as a substrate and 0.3 mg ovalbumin (Sigma, grade V) as an acceptor.

In all samples, endogenous glycosyltransferase activity was also assayed in the appropriate incubation medium devoid of exogenous substrate. The difference between the value with the exogenous glycoprotein acceptor and that without it was calculated to determine the exogenous enzyme activity.

RESULTS

Sialyltransferase Activity

Table 1 and Figure 1 illustrate the results of sialytransferase assays of the peripheral lymphocyte membranes. Both the endogenous and exogenous activity were found significantly reduced in patients with retinitis pigmentosa when compared with those in a control group, although there was a marked individual variability in activity in the two groups, with a considerable overlap of values. There was no

Table 1. Activity of lymphocyte sialyltransferase#

Subjects		acceptor(-)	acceptor(+)	difference
Retinitis pigmentosa	(n=27)	0.6±0.2*	11.8±4.6**	10.9±4.6**
Cone dystrophy	(n= 3)	1.2±0.2	13.5±5.2	12.3±5.0
Normal volunteer	(n=30)	0.9±0.6	16.3±6.3	15.5±5.9

#pmol radioactive-sialic acid incorporated/mg protein/30 min
The values in this and all other tables represent mean±s.d.
Significant difference from the normal, *$p<0.05$ **$p<0.005$

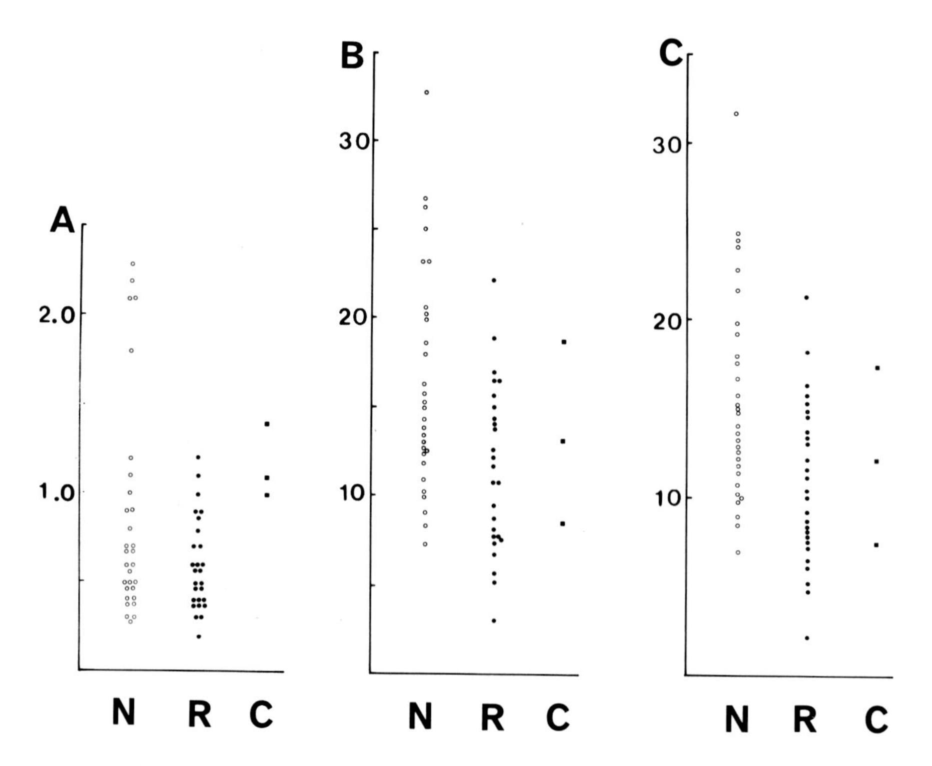

Figure 1 Sialyltransferase activities (pmol radioactive-sialic acid incorporated/mg protein/30 min) in lymphocytes. A: measurements of endogenous activity without glycoprotein acceptor asialofetuin in the incubation medium; B: in the presence of the acceptor; C: differential values B-A. N: normal adults; R: cases of retinitis pigmentosa; C: cases cone dystrophy (C).

difference in the sialyltransferase activitiy between the autosomal recessive form and sporadic form of retinitis pigmentosa. On the other hand, all of the three cases of cone dystrophy showed lymphocyte sialytransferase activity within a normal range.

Fucosyltransferase Activity

Table 2 summarizes the results of fucosyltransferase assays. There was no difference in the endogenous and exogenous activity among retinitis pigmentosa, cone dystrophy and the normal.

Galactosyltransferase Activity

Table 3 shows measurements of galactosyltransferase. There was no difference in the endogenous and exogenous galactosyltransferase activity between retinitis pigmentosa and the normal.

Table 2. Activity of fucosyltransferase in lymphocytes#

Subjects		acceptor(-)	acceptor(+)	difference
Retinitis pigmentosa	(n=10)	2.8±1.2	7.2±3.2	4.4±2.2
Cone dystrophy	(n= 3)	3.4±3.5	8.8±6.7	5.3±3.3
Normal volunteer	(n=10)	3.5±1.4	9.8±5.6	6.0±4.9

#pmol radioactive-fucose incorporated/mg protein/30 min

Table 3. Activity of lymphocyte galactosyltransferase#

Subjects		acceptor(-)	acceptor(+)	difference
Retinitis pigmentosa	(n=15)	1.2±1.2	16.7±8.3	15.4±7.4
Normal volunteer	(n=16)	1.7±0.8	19.9±7.5	18.2±7.4

#pmol radioactive-galactose incorporated/mg protein/30 min

DISCUSSION

The above results indicate a significant reduction in sialyltransferase activity of the membrane fraction of peripheral lymphocytes of patients with primary retinitis pigmentosa. The implication of this finding remains undefined, since the abnormality could be observed only on a statistical basis, with a considerable inter-individual variability. Although our patients with isolated ocular disease appeared homogeneous clinically and genetically. There might exist a spectrum of diseases associated with separate underlying etiology. It may also be remarkable that glycosyltransferase activities catalyzing fucose or galactose were within a normal range of values. Further studies will be need to make any conclusive statement whether glycosyltransferases are involved in the etiology of classical retinitis pigmentosa.

In our three patients with cone dystophy, there was no noticeable abnormality in the lymphocyte glycosyltransferases including fucosyltransferase. Patients with a certain form of cone dystrophy demonstrate abnormally low fucosidase activity in the serum and peripheral leukocytes (Hayasaka et al, 1985).

This study was supported by grant-in-aid for scientific research (No. 63771398) from the Ministry of Education, Science and Culture of Japan, and by special research grant for retinochoroidal dystrophy from the Ministry of Health and Welfare of Japan.

REFERENCES

Blanks JC, Johnson LV (1986). Selective lectin binding of the developing mouse retina. J Comp Neurol 221:31-41.

Cohen D, Nir I (1983). Cytochemical evaluation of anionic sites on the surface of cultured pigment epithelium cells from normal and dystrophic RCS rats. Exp Eye Res 37:575-582.

Colley NJ, Clark VH, Hall HO (1987). Surface modification of retinal pigment epithelial cells: effects on phagocytosis and glycoprotein composition. Exp Eye Res 44:377-392.

Hayasaka S, Nakazawa M, Okabe H, Masuda K, Mizuno K (1985). Progressive cone dystrophy associated with α-L-fucosidase activity in serum and leukocytes. Amer J Ophthalmol 99: 681-685.

Lowry OH, Rosenbrough NJ, Farr A, Randall RK (1951). Protein measurement with folin phenol reagent. J Biol Chem 193: 265-275.

Miyagi T, Tsuiki S (1982). Purification and characterization of ß-galactoside(α2-6)sialyltransferase from rat liver and hepatomas. Eur J Biochem 126:253-261.

Muramatsu T (1988). Developmentally regulated expression of cell surface carbohydrates during mouse embryogenesis. J Cell Biochem 36:13-26.

Nir I, Hall HO (1979). Ultrastructural localization of lectin binding sites on the surfaces of retinal photoreceptors and pigment epithelium. Exp Eye Res 29:181-194.

O'Brien PJ (1976). Rhodopsin as a glycoprotein: a possible role for the oligosaccharide in phagocytosis. Exp Eye Res 23:127-137.

O'Brien PJ, Muellenberg CG (1968). Properties of glycosyl transfer enzymes of bovine retina. Biochem Biophys Acta 167:268-273.

Roth S, White O (1972). Intercellular contact and cell surface galactosyltransferase activity. Proc Natl Acad Sci USA. 69:485-489.

Sameshima M, Uehara F, Ohba N (1987). Specialization of the interphotoreceptor matrices around cone and rod photoreceptor cells in the monkey retina, as revealed by lectin cytochemistry. Exp Eye Res 45:845-863.

Shur BD (1982). Evidence that galactosyltransferase is a surface receptor for poly(N)-acetyllactosamine glycoconjugates on embryonal carcinoma cells. J Biol Chem 257:6871-6878.

Uehara F, Sameshima M, Muramatsu T, Ohba N (1983). Localization of fluorescence-labeled lectin binding sites on photoreceptor cells of the monkey retina. Exp Eye Res 36:113-123.

Uehara F, Muramatsu T, Takumi K, Ohba N (1987). Two-dimensional gel electrophoretic analysis of lectin receptors in the degenerative retina of C3H mouse. In Hollyfield JG, Anderson RE, LaVail MM (eds): "Degenerative Retinal Disorders: Clinical and Laboratory Investigations", New York: Alan R. Liss, pp 219-227.

Unoki K, Uehara F, Muramatsu T (1988). Cell surface glycosyltransferase activities in bovine eyes. submitted for publication.

Inherited and Environmentally Induced
Retinal Degenerations, pages 77–82

MONAPTERIN IN BLOOD OF PATIENTS WITH RETINITIS PIGMENTOSA (RP)

G. Cremer-Bartels, H. Gerding, W. Gerlach,
W. Gremm, P. Guntermann, L. Hanneken, V. Kramer,
K. Krause, D. Seibt

Eye Hospital of the Westfaelian Wilhelms
University Muenster, F.R.G.

INTRODUCTION

In vertebrates pterins are found especially in tissues underlying light exposure such as skin and eyes. We suggested already 1962 that photoreactions of not identified pteridines may be involved in mammalian retinal metabolism. (Cremer-Bartels 1962, Cremer-Bartels and Krause, 1987). Wang et al. discovered in 1988 that the photolyase-repair of DNA in E. coli is photosensitized by a pterin derivate. This repair mechanism takes place in visible light.

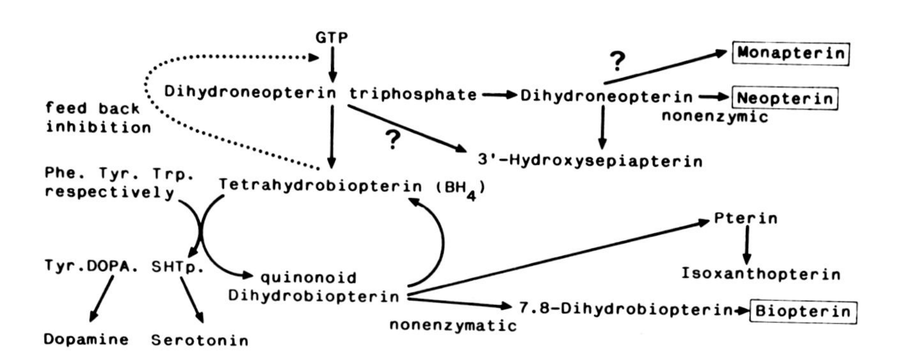

Figure 1. Biosynthetic pathways of some pterins according to Niederwieser et al. (1986).

It is well established that tetrahydrobiopterin is an essential cofactor in the synthesis of different neuro-transmitters and neuromodulators. In figure 1 the biosyn-thetic pathways of some pterins according to Niederwieser et al. (1986) are illustrated.

The biosynthetic pathway of monapterin is up to now not established. In figure 2 the chemical structure of monapterin, biopterin and neopterin is illustrated. The biological function of monapterin, which is a stereoisomer of neopterin is still unknown.

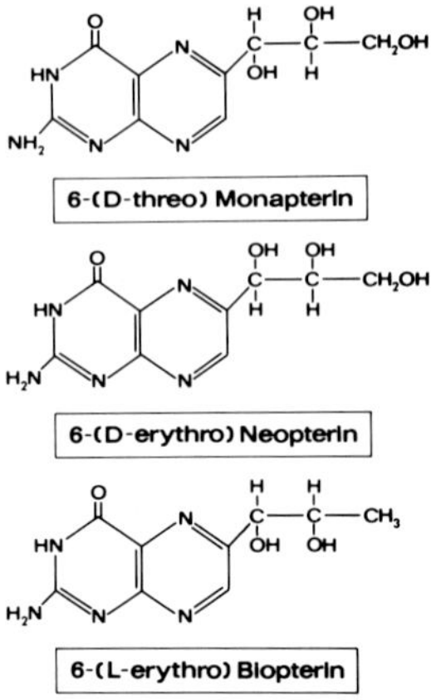

6-(D-threo) Monapterin

6-(D-erythro) Neopterin

6-(L-erythro) Biopterin

Figure 2. Chemical structure of Monapterin, Neopterin and Biopterin.

Elevated levels of neopterin can be found under the following conditions: 1. impaired renal function 2. certain variants of atypical phenylketonuria and 3. activation of the T-cell mediated immunereactions (Reibnegger et al., 1988).

The get insight into a possible role of monapterin in dystrophic retinal diseases we examined the content of monapterin, neopterin and biopterin in the human retina, serum and urine respectively. We determined the ratio of monapterin to neopterin and found the highest ratio in the retina (Cremer-Bartels et al., 1988). It may be suggested that this phenomenon is reflecting a metabolic role of monapterin or its precursors in the retina.

MATERIAL

We performed two experimental series. During a meeting of RP-patients in May 1986 we collected blood samples for analysis in lymphocytes, erythrocytes and serum respectively. The control sera were collected during the following two months (May - July).

During another meeting in November 1987 collection of blood samples of RP patients and healthy volunteers was performed simultanously. We determined the pterin pattern in serum.

METHODS

The oxidized forms of pterines show a strong fluorescence and can be detected in low concentrations by means of HPLC with a fluorescence detector.

The preparation of blood samples for determination of pterin content in erythrocytes and lymphocytes was performed as described (Cremer-Bartels et al., 1987). The sera in the first series were not oxidized and determination was performed after elimination of the high molecular weight fractions by filtration through Amicon membranes XM 300 and XM 100 under nitrogen.

Contrary to these preparations we determined in the second series the pterin pattern in the sera after Dowex column chromatography and iodine oxidization as described by Fukushima and Nixon (1980).

To get a more sufficient separation of monapterin we used a modified elution medium: 0,035 phosphoric acid (85%); 1,7 acetonitril; 98,265 aqua bidest in percent V/V.

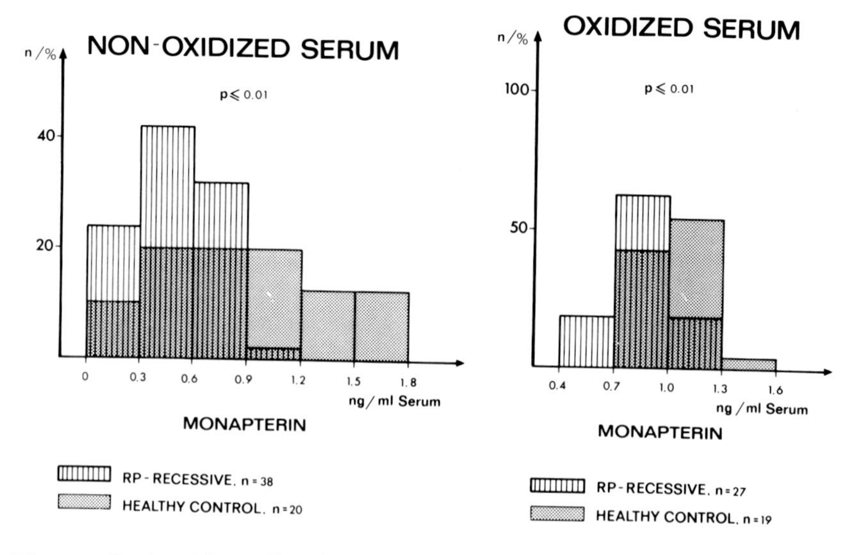

Figure 3a,b. Distribution of monapterin concentration in non oxidized (a) and oxidized serum (b) of RP-patients and healthy volunteers.

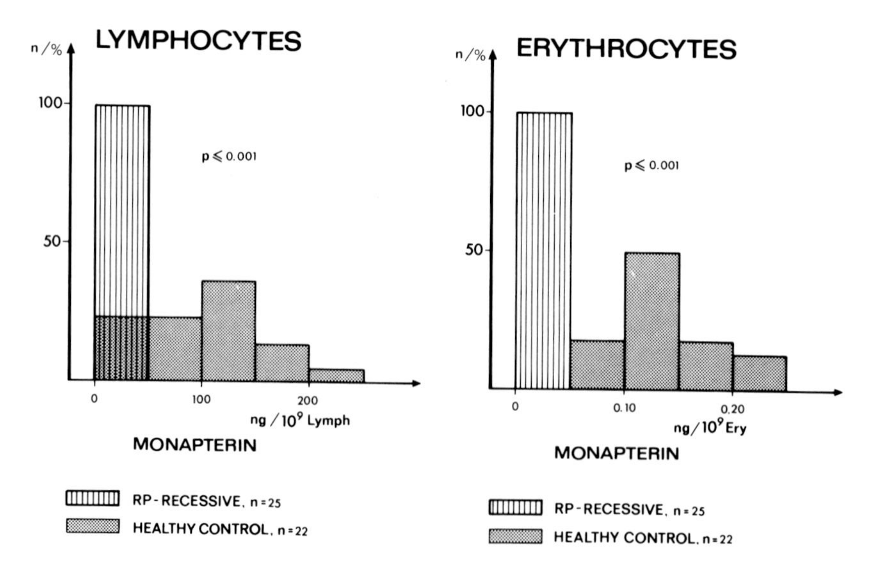

Figure 4a,b. Distribution of monapterin concentration in lymphocytes (a) and erythrocytes (b) of RP-patients and healthy volunteers.

RESULTS

In figure 3a the distribution of monapterin concentration in non oxidized ultrafiltrated sera of patients suffering of RP (autosomal recessive mode of inheritance) is demonstrated. This distribution for RP-patients is significantly shifted to lower levels compared to the controls. This decrease of monapterin content was significantly confirmed (figure 3b) in the second series of experiments where we applied the mentioned modifications of methods. This phenomenon is also demonstrated significantly measuring the monapterin concentrations in erythrocytes and lymphocytes as illustrated in figures 4a, b. The most impressive effect was the decrease of monapterin concentration in erythrocytes.

DISCUSSION

The four different and independently performed experiments clearly brought out that concentration of monapterin in blood of patients suffering of RP is decreased. The knowledge about monapterin is so far very poor. Neither the biosynthesis is established nor the biological function. Guroff and Rhoads (1969) proposed a cofactor function of reduced forms of monapterin for phenylalanin-hydroxylation in bacteria. According to Blau et al. (1987) high concentration of monapterin can inhibit the synthesis of tetrahydrobiopterin. As mentioned before monapterin is found in a high ratio compared to other pterins in the retina in comparison to blood or urine. Taking this observation into consideration a specific role of monapterin or its precursors in retinal function seems possible. Further experiments are necessary to elucidate a possible role of an imbalance of pterin metabolism as a causal factor in the pathogenesis of RP.

ACKNOWLEDGEMENTS

This work was financially supported by the Deutsche Forschungsgemeinschaft Cr 40-11-1, Cr 40-11-2.

REFERENCES

Blau N, Steinerstauch P, Redweij U, Pfister K, Schoeden G, Kierat L, Niederwieser A (1987). Dihydromonapterin Triphosphate: Occurance, Analysis and Effect on Tetrahydrobiopterin Biosynthesis in vivo and in vitro. In Curtius H-Ch, Blau N, Levine RA (eds): Unconjugated Pterins and Related Biogenic Amines.

Cremer-Bartels G (1962). Ueber das Vorkommen lichtempfindlicher fluoreszierender Substanzen in verschiedenen Geweben von Rinderaugen. Graefe's Arch. Klin Exp Ophthalmol 164:391-398.

Cremer-Bartels G, Krause K (1987). Retinal Pteridines. Advances in Biosciences 62:291-302.

Cremer-Bartels G, Gerding H, Gerlach W, Gremm W (1987). Neopterin and Monapterin in erythrocytes and lymphocytes of patients with retinitis pigmentosa. In Biochemical and Clinical Aspects of Pteridines 5:359-364.

Cremer-Bartels G, Gerding H, Krause K (1988). Monapterin and Neopterin in Human Retina, Sera and Urine. Proceedings of the international society for eye research V:120.

Fukushima T, Nixon JC (1980). Analysis of Reduced Forms of Biopterin in Biological Tissues and Fluids. Analytical Biochemistry 102:176-188.

Guroff G, Rhoads CA (1969). Phenylalanine Hydroxylation by Pseudomonas Species. J. Biol. Chem. 244:142-146.

Niederwieser A, Joller P, Seger R, Blau N, Prader A, Bettex JD, Luethy R, Hirschel B, Vetter U (1986). Neopterin in AIDS, other Immundeficiencies and Bacterial and Viral Infections. Klin. Wochenschrift 64:333-337.

Reibnegger G, Fuchs D, Hausen A, Werner ER, Wachter H (1988). Neopterin in clinical use. Biol. Chem. Hoppe-Seyler 269:534-535.

Wang B, Jordan SP, Jorns MS (1988). Identification of a Pterin Derivative in Escherichia coli DNA Photolyase. Biochemistry 27:4222-4226.

Inherited and Environmentally Induced
Retinal Degenerations, pages 83–97

X-LINKED RETINITIS PIGMENTOSA: A MOLECULAR GENETIC APPROACH TO ISOLATING THE DEFECTIVE GENES.

S Lindsay[1,2], M Jay[3], DJ Bower[2], G Adam[1], CFI Inglehearn[2], PG Sealey[2], SS Papiha[1] and SS Bhattacharya[1]

INTRODUCTION

Retinitis pigmentosa (RP) is observed as photoreceptor cell degeneration and invasion of the innervated layers of the retina by pigmented cells (Bird 1975). The exact cellular location of the primary defect is not yet known. There are several distinct, genetically-determined forms which can be inherited in an autosomal dominant, autosomal recessive or X-linked manner (Boughman et al 1980; Jay 1982). Initially, close genetic linkage between X-linked RP (XLRP) and a DNA probe L1.28 (DXS7, localised to band Xp11.3, Hofker et al 1986) was demonstrated in five British families (Bhattacharya et al 1984). Further linkage data from this group (Clayton et al 1986) and other researchers (Mukai et al 1985; Friedrich et al 1985) supported the early findings. However, Francke et al (1985) described a male patient (BB) with a deletion in XP21 and a pathology including Duchenne Muscular Dystrophy, chronic granulomatous disease and pigmentary retinopathy. This implied a locus for XLRP close to, or in, Xp21. Genetic linkage evidence for this second locus was presented by Denton et al (1988) who found no crossovers between the gene for ornithine transcarbamylase (OTC) and XLRP in his families. Wright et al (1987) presented more linkage data supporting localisation of XLRP between L1.28 and 58.1 (DXS14). There-

1 Molecular Genetics Unit, University of Newcastle-upon-Tyne
2 MRC Human Genetics Unit, Edinburgh, UK
3 Moorfields Eye Hospital, London, UK

fore it appears that, based on linkage data, there are at least two genes on the short arm of the X chromosome causing RP; one gene located proximal to L1.28 (RP2) and another close to Xp21 (RP3). Preliminary localisations of these two genes are indicated in Figure 1.

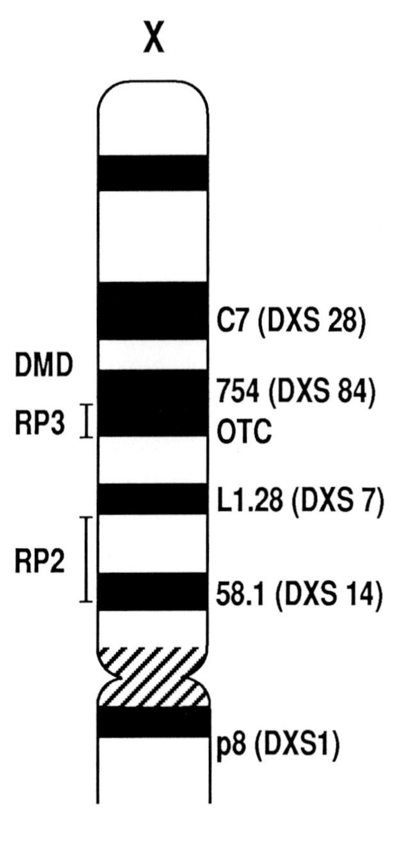

FIGURE 1
This is a diagrammatical representation of the short arm and part of the long arm of the human X chromosome. The position of loci C7, 754, OTC, L1.28, 58.1 and p8 are as indicated in Davies et al (1988). The preliminary positions of RP2 and RP3 are indicated by lines on the left hand side of the figure.

GENERAL APPROACH

X-linked retinitis pigmentosa has, as yet, no known biochemical or physiological basis and there are few ways of identifying the genes responsible for the disease amongst a number of candidates. Clearly, the more precisely that each disease gene is mapped to a specific region of the X chromosome, the fewer will be the number of potential candidate genes. Genetic linkage studies are, at present, our only way of assaying for and localising the XLRP gene and these have a limit of resolution of approximately one centimorgan (cM). A centimorgan is a unit of genetic distance and corresponds to a crossover frequency of one per cent between two adjacent loci (Ott 1985). In humans, 1cM approximates to one million base pairs, (1Mb), although this does vary considerably in different parts of the genome (Donnis-Keller et al

1987). The current estimate of the number of genes in the human genome is 50,000 to 100,000 (McKusick 1986). Since the X chromosome is approximately 6% of the human genome, and assuming an even distribution of genes among the chromosomes, then the number of genes located on the human X chromosome is of the order of 3,000 to 6,000. The genetic distance between Xpter and Xqter is estimated at about 200cM (Drayna and White 1985; Donnis-Keller et al 1987) and, therefore, a 1cM interval on the X chromosome will, on average, contain 15 to 30 genes. At present, with the problems of probable heterogeneity (see introduction) there are two candidate regions on the short arm of the X chromosome which could contain an XLRP gene. These are indicated as RP2 and RP3 in Figure 1. These regions comprise approximately 25cM. Our general approach is, therefore, to generate markers localised to these regions of the X chromosome and use them in linkage studies to more precisely map the XLRP genes. The results of these studies may indicate that the XLRP families in our collection fall into two classes and this will enable us to further resolve the question of heterogeneity.

In addition to linkage studies at the genetic level, we are constructing long-range physical maps of the same regions. These maps are constructed using pulsed-field gel (PFG) technology and this effectively alters the definition of saturation cloning; thus in a 10cM region 200 markers at 50kb intervals would be needed to provide a complete physical map using cosmids and conventional gel techniques whereas only 20 markers at 500kb intervals are needed to map the same region using pulse-field gel technology (Poutska and Lehrach 1986). Clearly, such orderly spacing of markers will not be achieved, however we aim to generate clusters of markers covering the relevant 25cM at approximately 1cM intervals. The physical and genetic maps will be closely compared, which may be interesting of itself, in order to precisely define the locations of the RP2 and RP3 genes.

Therefore, there are three strands to our approach which inter-relate and are carried on essentially simultaneously:

1) generating new markers
2) genetic and physical mapping
3) looking for genes.

These are dealt with separately in the following sections.

GENERATING NEW MARKERS

New markers could, in principal, be generated from any library containing human DNA inserts. Ideally, one would wish the library to contain as many relevant clones (i.e. those mapping proximal to L1.28 and around Xp21) and as few irrelevant ones as possible. We have available to us two X-chromosome libraries: one prepared from flow-sorted X chromosomes cloned into lambda gtWES (Cooke et al 1985) and the other is a cosmid library with insert DNA from a mouse-human hybrid cell line which has the X chromosome as the only human chromosome (Lindsay and Bird 1987). Both these libraries contain clones from the whole human X chromosome. In order to increase the proportion of relevant to irrelevant clones even further we are constructing a third library using insert DNA from a cell line which contains only the short arm of the X chromosome. This library is also a cosmid library as the large size of cosmid inserts (~40kb) has several advantages; it increases the likelihood of finding polymorphisms with the clones and also increases the probability of a gene sequence being present in a given number of clones. Furthermore, cosmids are more effective probes than smaller clones both for mapping by _in situ_ hybridisation to metaphase spreads and for hybridisation to pulse field gels.

Markers can be targeted to specific regions of chromosomes much more effectively if they are tested against appropriate DNAs. We are localising markers in two stages by hybridisation to different panels. The first panel checks for X chromosome specificity using DNAs from a lymphoblastoid cell line with a 49,XXXXY karyotype (4X), and from a male (1X). DNA markers which hybridise as a single band (or a small number of bands) and show the dosage relationship 4:1 related to the number of X chromosomes present will be taken to be X-chromosome specific. This panel also has DNAs from three cell lines derived from patients with deletions in particular regions of the X chromosome: markers which fail to give a hybridisation signal against these DNAs are very likely to be located in the deletion region and thus are of great interest. No deletions have so far been identified in XLRP patients, however we have one deletion cell line which is potentially relevant to RP2 from a patient affected by Norrie disease (Gal et al 1986). Norrie disease is an X-linked eye condition which is characterised by malformation of the retina and blindness from birth. This disease is also closely linked to

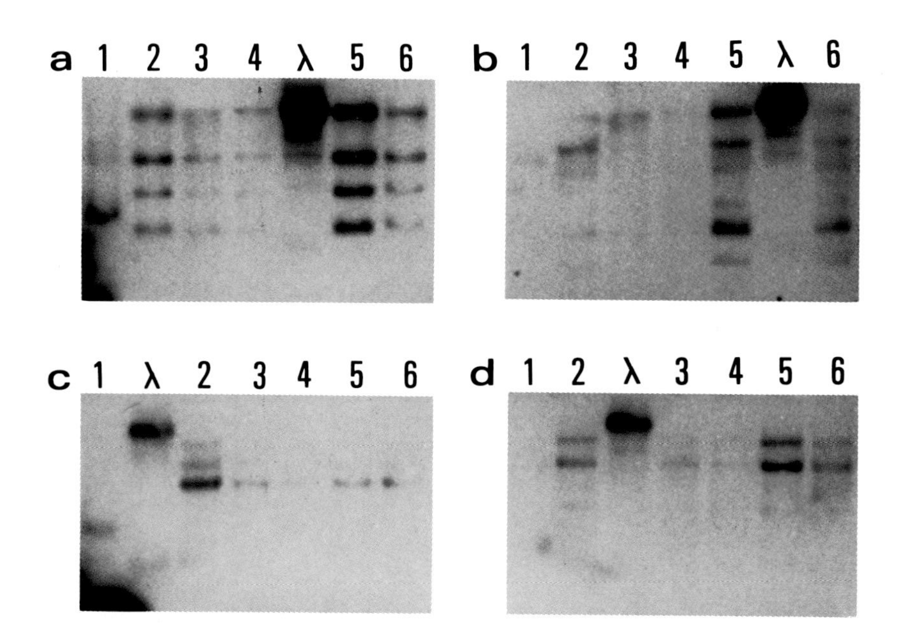

FIGURE 2
Here cosmids a,b,c and d have been hybridised to individual panels containing genomic DNAs which have been digested with the restriction enzyme Hind III. The panels were prepared using standard methods of DNA transfer (Southern 1975) to nylon membranes (Genescreen plus, Du Pont). The radioactively labelled (Feinberg and Vogelstein 1984) cosmid DNA was precompeted with sonicated human DNA to remove repeat sequences (Sealey et al 1985) before the hybridisation (Maniatis et al 1982). The genomic DNAs are 1) mouse; 2) cell line from a patient with choroideroemia (Hodgson et al 1987); 3) cell line from patient BB; 4) cell line from a patient with Norrie disease (Gal et al 1986); 5) from a cell line with the karyotype 49,XXXXY (TilI); 6) from a normal male karyotype 46,XY. λ- λ Hind III marker. The cos sites in the cosmid vector (pJB8) are hybridising to λ bands. It can be seen from the relative intensity of hybridisation to the 4XY and XY DNAs (approximately 4:1) that cosmids a,

L1.28 (Gal et al 1985) and therefore markers mapping to this deletion should also be tightly linked to RP2. A second deletion cell line was derived from a patient affected by Duchenne muscular dystrophy and RP (BB, Francke et al 1985 and see introduction). The deletion in the BB cell line extends from proximal to the chronic granulomatous disease (CGD) locus into the DMD gene and if the patient did have a form of XLRP then this region should contain the putative RP3 locus. The third cell line contains a deletion in Xq21.1 (Hodgson et al 1987). The final track in the panel has mouse DNA as a control. Figure 2 shows the hybridisation of four cosmids to such panels. Three of the cosmids are clearly X-linked (a,b and d) while the fourth (c), although human in origin, is autosomal. In Figure 3 cosmid e as well as being X-linked, has been localised to the RP3 region since it is absent in the BB cell line DNA.

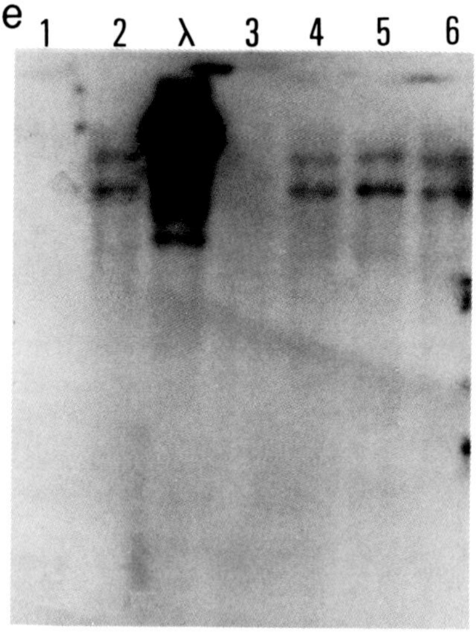

FIGURE 3
Cosmid e has been hybridised to a similar panel as described in the legend to Figure 2 and under the same conditions. It can be seen that this cosmid is located on the X chromosome (compare intensity of hybridisation in tracks 5 and 6) and that it has been more finely mapped to the BB deletion, since the signal is absent in track 3.

The second panel maps those markers which are X-linked and lie outwith the three deletion regions (the majority of clones). These are mapped using a series of

---

b and d are located on the X chromosome while cosmid c, although human is located on an autosome. Cosmids a and c also cross-hybridise to a single copy mouse sequence.

cell hybrids which contain different portions of the human X chromosome. Of particular relevance to the RP2 locus are two hybrids each containing an X-autosomal translocation chromosome; in one the breakpoint is in Xp11 and in the other it is an Xp11.3.

Markers mapping to appropriate regions will be used as described in the following sections. We are at present using these strategies with markers already generated by our own and other groups.

GENETIC AND PHYSICAL MAPPING

As stated above, no XLRP patients have been identified who have microscopically visible deletions. Therefore, markers mapping to the RP2 and RP3 regions are being used to screen DNA from affected individuals to look for submicroscopic deletions. This could be a rapid way of getting to the disease loci. Linkage studies will also be carried out with new markers which reveal RFLPs, to establish their genetic distances from the disease loci. For this we have more than thirty large XLRP families. Such a linkage study should identify which markers are closest to the disease loci. These can then be used in linking studies at the DNA level as described below.

An important recent advance, PFG electrophoresis, allows for the separation and accurate sizing of very large pieces of DNA - 50 to 5,000kb. Therefore markers which are genetically quite far apart (1-5cm) can now be physically linked by their hybridisation to the same PFG fragment. Restriction enzymes can be used to generate large fragments (50-1000kb) in high molecular weight genomic DNA and these fragments can then be separated using PFG electrophoresis (Schwartz and Cantor 1984; Carle and Olson 1985). This technique can be used in conjunction with cloned DNA markers to construct long-range physical maps of appropriate regions of the X chromosome. The restriction enzymes used are those which cut rarely in mammalian DNA, either because there are very few restriction sites for these enzymes (e.g. Sfi) or because the enzyme is methylation-sensitive and only a few of its sites are unmethylated. Particularly useful are enzymes whose sites are in HTF islands (see below) which are spaced, on average, 100kb apart (Brown and Bird 1986).

Examples of such enzymes are SacII (CCGCGG), EagI (CGGCCG), NaeI (GCCGGC), NarI (GGCGCC) and NotI (GCGGCCGC). Initially, because of differences in methylation pattern among cell lines, we are using the enzyme SfiI which is not sensitive to methylation at its restriction sites.

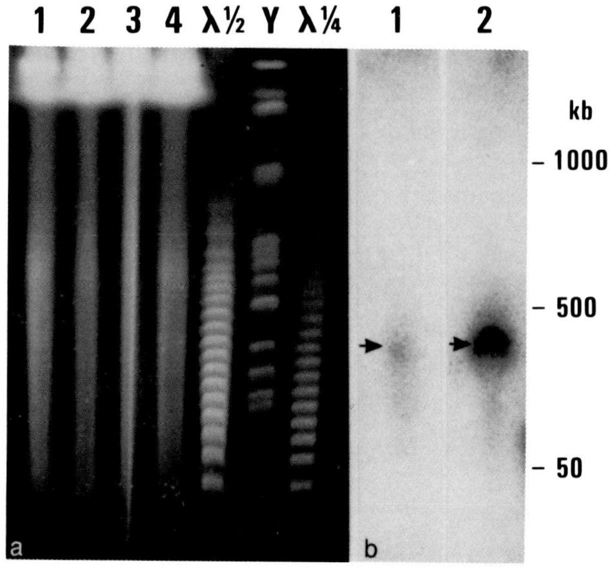

FIGURE 4
Panel a) shows an ethidium bromide stained 1% agarose CHEF (LKB) gel run for 25 hours at 170 volts in 0.5xTBE (0.045 M Tris-borate, 0.045M boric acid, 0.001M EDTA) running buffer. Lanes 1-4 are DNAs from several different cell lines which have been digested with the restriction enzyme Not I. λ½ and λ1/4 are λoligomer marker tracks at different concentrations. Y indicates a yeast (strain AB972) chromosomal marker lane.
Panel b) shows an autoradiogram of a pf gel probed with L1.28. Lane 1 contains DNA from a normal male (XY) while lane 2 contains DNA from the cell line TilI (4XY). Size markers are given in kb on the right hand side of the panel. L1.28 detects an approximately 400kb fragment in both lanes which clearly shows differential hybridisation intensity relating to the number of X chromosomes in the DNA in each lane.

Fragments between 50kb and 2000kb can be resolved as required, as is shown in Figure 4a. In Figure 4b, DNA from a similar gel has been transferred to a nylon membrane which has been hybridised to clone L1.28. It can be seen that a fragment of approximately 400kb is detected by L1.28. Lane 1 contains DNA from a male (46,XY), Lane 2 a 49,XXXXY karyotype (TilI). The relative hybridisation intensity between the two lanes is approximately 1:4 as expected for an X-linked clone. If a 100 random marker can be located in the regions of interest then several markers are likely to hybridise to the same SfiI PFG fragment. This, in conjunction with information from already characterised markers located in these regions, will allow us to build up a physical map.

LOOKING FOR GENES

There is a major practical difficulty here because only a small proportion of human DNA actually comprises coding sequences and there is no straightforward way of deciding where to look for them with an optimal chance of success. Ideally, we want ways of screening large numbers of cosmids simultaneously for sequences associated with genes. There are several approaches which we are currently using.

1) Screening for HTF islands. Recently, a small proportion of vertebrate DNA has been discovered to have an unusual composition (Cooper et al 1983). Bulk genomic DNA is approximately 40% guanine (G) and cytosine (C) nucleotides, has a five-fold deficiency of CpG and is heavily methylated. In contrast HTF-island DNA is more than 60% G and C, has no deficiency of the dinucleotide CpG and is unmethylated in all tissues. These differing properties, in particular the contrasting methylation statuses, make HTF-island DNA highly distinguishable from bulk genomic DNA (Bird et al 1985) when using certain restriction enzymes. These enzymes are methylation-sensitive and their recognition sequences consist of only G and C and contain one or more CpG's (C-G restriction enzymes). HTF islands are distributed approximately 100kb apart in bulk genomic DNA (Brown and Bird 1986).

HTF islands have been found at the 5' ends of all polymerase II housekeeping genes characterised to date and also several tissue-specific genes (Bird 1986). It has now been shown that finding certain C-G restriction enzyme sites (e.g. SacII) in libraries of cloned DNA is an efficient way of finding HTF islands and that there is a high probability that these islands are associated with genes (Lindsay and Bird 1987). We are, therefore, screening all X-chromosome clones with SacII and those with sites we are further analysing with a battery of restriction enzymes diagnostic for the presence of HTF islands. Figure 5 shows a number of cosmid clones digested with HindIII and double digested with HindIII and SacII. It can be seen that clones 1,2 and 4 to 9 do not contain SacII sites while clone 3 has at least two since a minimum of three extra bands are present in track l, and one band has disappeared, compared to track c; these are indicated with arrows in Figure 5. We have found empirically that this is the pro-

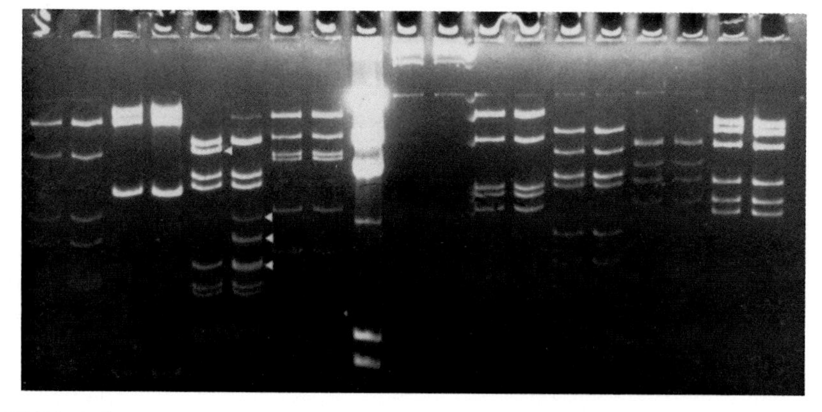

FIGURE 5
Cosmids 1-9 have been digested with HindIII (lanes A-I) or HindIII and SacII (lanes J-R) and electrophoresed on an agarose gel. One cosmid (3) has several SacII sites, with the result that a HindIII fragment is cleaved giving rise to three smaller fragments (arrowed in lanes C and L respectively). Laneλ is a λHindIII size marker.

portion of X chromosome cosmids which contain a SacII site (ie one in eight or nine).

2) Screening for conserved regions. A proportion of genes will not have islands and these must be searched for in much more laborious fashions; including looking for those cosmids which contain sequences which cross-hybridise to DNA from different animal species. These are likely to be conserved sequences and hence very likely to be genes. We are also identifying DNA motifs which are associated with genes and will search the cosmids for the presence of these sequences. These, and the HTF-island containing sequences are being tested against RNA from different tissues (liver, muscle, brain and retina) to see if they are expressed and to check their tissue-specifities. This is somewhat easier now that whole clones can be used to hybridise to RNA, repeat sequences are much less of a problem because they can be competed out (Sealey et al 1985).

There are two ways in which we can assess whether genes which we find are candidates for the disease loci:
i) if the gene sequence is present in DNA from unaffected people and absent or altered in DNA from affected individuals.
ii) from genetic studies. Firstly, in the case of XRLP, we have DNA from several patients who have XLRP and mental retardation and who have no visible deletion of their X chromosomes. If a candidate gene is absent in these patients it is likely to be very close to RP. Secondly, RFLPs will be identified and linkage disequilibrium with the disease locus looked for in family studies for all the disease loci.

CONCLUSIONS

The new markers generated will have an immediate benefit at the clinical level as more accurate tools for carrier detection and for prenatal diagnosis. The long-range physical maps can be compared to the existing genetic maps of these regions which should lead to a better understanding of recombinational events affecting these specific regions and, by analogy, the genome as a whole. Further, such maps should be an invaluable resource for identifying and isolating other genes, including those for other inherited diseases, which may be genetically mapped to these regions.

The search for the RP genes amongst suitable candidates is a not inconsiderable task (as has proved to be the case for the cystic fibrosis gene). However, the techniques of "reverse genetics" have enabled us and others to progress much more rapidly than was previously possible towards the isolation of the defective gene. It is to be hoped that future advances in the identification and categorisation of genes will allow a similar increase in the speed with which the defective gene can be distinguished from amongst a number of candidates.

ACKNOWLEDGEMENTS

We are grateful to Professor Ropers (Nijmegen University, Holland) for generously providing us with the cell line from the Norrie disease patient and to Dr Hans Ochs for similarly providing us with the BB cell line. We thank the Wellcome Trust, the Medical Research Council, University of Newcastle upon Tyne and The Scottish National Institute for the War-blinded for supporting our work. We would also like to thank Helen Ackford and Sylvia Pierce for excellent technical support and Evalyn Kay and Ann Kenmure for showing considerable patience in typing the manuscript.

REFERENCES

Bhattacharya SS, Wright AF, Clayton JF, Price WH, Phillips CI, McKeown CME, Jay M, Bird AC. Pearson PL, Southern EM, Evans HJ (1984). Close genetic linkage between X-linked retinitis pigmentosa and a restriction fragment length polymorphism identified by recombinant DNA probe L1.28. Nature 309: 253-255.

Bird AC (1975). X-linked retinitis pigmentosa. Brit. J. Ophthalmol. 59: 177-199.

Bird AP (1986). CpG-rich islands and the function of DNA methylation. Nature 321: 209-213.

Bird AP, Taggart MH, Frommer M, Miller OJ, Macleod D (1985). A fraction of the mouse genome that is derived from islands of nonmethylated, CpG-rich DNA. Cell 40: 91-99.

Boughman JA, Conneally PM, Nance WE (1980). Population genetic studies of retinitis pigmentosa. Am. J. Hum. Genet. 32: 223-235.

Brown WRA, Bird AP (1986). Long-range restriction site mapping of mammalian genomic DNA. Nature 322: 477-481.

Carle GF, Olsen MV (1984). Separation of chromosomal DNA molecules from yeast by orthogonal-field-alteration gel electrophoresis. Nucleic Acids Res. 12: 5647-5664.

Clayton JF, Wright AF, Jay M, McKeown CME, Dempster M, Jay BS, Bird AC, Bhattacharya SS (1986). Genetic linkage between X-linked retinitis pigmentosa and DNA probe DXS7 (L1.28): further linkage data, heterogeneity testing, and risk estimation. Hum. Genet. 74: 168-171.

Cooke H, Bhattacharya SS, Fantes JA, Green DK, Evans HJ (1985). Preparation of X-chromosome specific probes from a flow-sorted library. Cytogenet. Cell Genet. 40: 607.

Cooper DN, Taggart MH, Bird AP (1983). Unmethylated domains in vertebrate DNA. Nucleic Acids Res. 11: 647-658.

Davies KE, Mandel JL, Weissenbach J, Fellows M (1988). In "Human Gene Mapping 9". Cytogenet. Cell Genet. 46: 277-315.

Denton MJ, Chen JD, Serraville S, Colley P, Halliday FB, Donald J (1988). Analysis of linkage relationships of X-linked retinitis pigmentosa with the following Xp loci: L1.28, OTC, 754, XJ-1.1, pERT87, and C7. Hum. Genet. 78: 60-64.

Donis-Keller H, Green P, Helms C, Cartinhour S, Weiffenbach B, Stephens K, Keith TP, Bowden DW, Smith DR, Lander ES, Botstein D, Akots G, Rediker KS, Gravius T, Brown VA, Rising MB, Parker C, Powers JA, Watt DE, Kauffman ER, Bricker A, Phipps P, Kahle HM, Fulton TR, Ng S, Schumm JW, Braman JC, Knowlton RG, Barker DF, Crooks SM, Lincoln SE, Daly MJ, Abrahamson J (1987). A genetic linkage map of the human genome. Cell 51: 319-337.

Drayna D, White R (1985). The genetic linkage map of the human X chromosome. Science 230: 753-758.

Francke U, Ochs HO, de Martinville B, Giacalone J, Lindgren V, Disteche C, Pagon RA, Hofker MH, van Ommen GJB, Pearson PL, Wedgewood RJ (1985). Minor Xp21 chromosome deletion in a male associated with expression of Duchenne muscular dystrophy, chronic granulomatous disease, retinitis pigmentosa and McLeod syndrome. Am. J. Hum. Genet. 37: 250-267.

Friedrich U, Warburg M, Wieacker P, Wienker TF, Gal A, Ropers HH (1985). X-linked retinitis pigmentosa: linkage with the centromere and a cloned DNA sequence from the proximal short arm of the X chromosome. Hum. Genet. 71: 93-99.

Feinberg AP, Vogelstein B (1984). A technique for radiolabelling DNA restriction endonuclease fragments to high specific activity. Analyt. Biochem. 137: 226-267.

Gal A, Stolzenberger C, Wienker TF, Wieacker PF, Ropers HH, Friedrich U, Bleeker-Wagemakers EM, Pearson P, Warburg M (1985). Norrie's disease: close genetic linkage with genetic markers from the proximal short arm of the X chromosome. Clin. Genet. 27: 282-283.
Gal A, Wieringa B, Smeets DFCM, Blecker-Wagemakers L, Ropers HH (1986). Submicroscopic interstitial deletion of the X chromosome explains a complex genetic syndrome dominated by Norrie disease. Cytogenet. Cell Genet. 42: 219-224.
Hodgson SV, Robertson ME, Fear CN, Goodship J, Malcolm S, Jay B, Bobrow M, Pembrey ME (1987). Prenatal diagnosis of X-linked choroideremia with mental retardation, associated with cytologically detectable X-chromosome deletion. Hum. Genet. 75: 286-290.
Hofker MH, Wapenaar MC, Goor N, Bakker E, van Ommen GJB, Pearson PL (1985). Isolation of probes detecting restriction fragment length polymorphisms from X chromosome-specific libraries: potential use for diagnosis of Duchenne muscular dystrophy. Hum. Genet 70: 148-156.
Jay M (1982). On the heredity of retinitis pigmentosa. Brit. J. Ophthalmol. 66: 405-416.
Lindsay S, Bird AP (1987). Use of restriction enzymes to detect potential gene sequences in mammalian DNA. Nature 327: 336-338.
Maniatis T, Fritsch EF, Sambrook J (1982). In "Molecular cloning: a laboratory manual". Cold Spring Harbor pp 387-389.
McKusick VA (1986). "Mendelian Inheritance in Man". Baltimore: Johns Hopkins University Press, ed 7. ppXVII-XVIII.
Mukai S, Dryja TP, Bruns GAP, Aldridge JF, Berson EL (1985). Linkage between the X-linked retinitis pigmentosa locus and the L1.28 locus. Am. J. Ophthamol 100: 225-229.
Ott J (1985). "Analysis of Human Genetic Linkage". Baltimore: Johns Hopkins University Press.
Poustka A, Lehrach H (1986). Jumping libraries and linking libraries: the next generation of molecular tools in mammalian genetics. TIG 2: 174-179.
Schwartz DC, Cantor CR (1984). Separation of yeast chromosome sized DNAs by pulsed field gradient gel electrophoresis. Cell 37: 67-75.
Sealey PG, Whittaker PA, Southern EM (1985). Removal of repeated sequences from hybridisation probes. Nucleic Acids Res. 13: 1905-1922.

Southern EM (1975). Detection of specific sequences among DNA fragments separated by gel electrophoresis. J. Mol. Biol. 98: 503-517.

Wright AF, Bhattacharya SS, Clayton JF, Dempster M, Tippet P, McKeown CME, Jay M, Jay B, Bird AC (1987). Linkage relationships between X-linked retinitis pigmentosa and nine short-arm markers and localisation to between DXS7 and DXS14. Am. J. Hum. Genet. 41: 635-644.

Inherited and Environmentally Induced
Retinal Degenerations, pages 99–111

MOLECULAR GENETICS OF ORNITHINE AMINOTRANSFERASE DEFECT IN GYRATE ATROPHY

George Inana, Yoshihiro Hotta, Carmelann Zintz, Akira Nakajima, Takashi Shiono, Nancy G. Kennaway and Richard G. Weleber

National Eye Institute, NIH, Bethesda, Maryland 20892 (G.I., Y.H., C.Z.), Departments of Ophthalmology (R.G.W.) and Medical Genetics (N.G. K.), The Oregon Health Sciences University, Portland, Oregon 97201, Department of Ophthalmology, Juntendo University School of Medicine, Tokyo, Japan (A.N.) and Department of Ophthalmology, Tohoku University of School of Medicine, Sendai, Japan (T.S.)

Gyrate atrophy (GA) is a rare autosomal recessive degenerative disease of the retina and choroid of the eye that leads to blindness (Valle and Simell, 1983). Simell and Takki demonstrated the hyperornithinemia present in patients with GA in 1973 (Simell and Takki, 1973), and a generalized deficiency in the mitochondrial enzyme, ornithine aminotransferase (OAT), was reported in 1977 (Valle et al., 1977; Trijbels et al., 1977; O'Donnell et al., 1977). The obligate heterozygous carriers have 50% of the normal level of OAT activity in their cells, in agreement with the autosomal recessive pattern of inheritance observed. The deficiency of OAT activity in the fibroblasts of two GA patients was shown by an immunoassay technique to be due to a decreased concentration of the enzyme (Ohura et al., 1984). Despite the generalized nature of the OAT deficiency in GA, the eye appears to be the only site of significant pathology in this disease (Valle and Simell, 1983).

Ornithine aminotransferase is a nuclear-encoded, pyridoxal phosphate-requiring, mitochondrial matrix enzyme that catalyzes the interconversion of ornithine, glutamate, and proline (Peraino and Pitot, 1963). OAT is the main catabolic enzyme of ornithine, and a defect in OAT would result in hyperornithinemia as seen in GA. The enzyme is present in many mammalian tissues and has been purified from various tissue sources, including rat liver (Shiotani et al., 1977), kidney (Sanada et al., 1970), brain (Deshmukh and Srivastava, 1984), and human liver in a crystalline-pure form (Ohura et al., 1982). The rat liver and kidney OAT has been shown to be synthesized as a 49-kilodalton precursor molecule in the cytoplasm, which is then transported to the mitochonrion where it is processed (Mueckler et al., 1982).

Thus, GA is unique among the hereditary retinal degenerative diseases in that the underlying biochemical defect is known. Although the exact mechanism by which the OAT deficiency and hyperornithinemia lead to the chorioretinal degeneration is not known, this disease still offers a great opportunity to study the molecular basis of the genetic defect in OAT using a molecular biological approach. Elucidation of the molecular genetics of GA may yield information that may be useful for genetic counseling and future gene therapy for this disease.

In order to study GA at the gene level, we had constructed a molecular probe for the human OAT in the form of a cDNA clone (Inana et al., 1986). We utilized the λgt11 expression cloning system developed by Young and Davis for cloning the OAT cDNA (Young and Davis, 1983). We were able to use the λgt11 approach through a collaboration with a Japanese group who had purified the human OAT and provided it to us (Ohura et al., 1982). An anti-human OAT antibody was prepared with the pure enzyme and used to isolate the OAT cDNA clone.

The OAT cDNA consists of 2073 nucleotides and codes for a 439 amino acid OAT precursor protein with a molecular mass of 48,534 dalton. Analysis of the cDNA-derived OAT sequence revealed a leader sequence, similar to those found in other mitochondrial proteins of cytoplasmic origin, in the precursor protein. The OAT protein sequence was also shown to be 91% homologous to the rat OAT and 27% similar to the chicken aspartate aminotransferase with a conservation of the pyridoxine-binding lysine residue.

A genomic Southern analysis using the human OAT cDNA as a probe indicated that the OAT gene is a very complicated gene family consisting of multiple copies of OAT-related

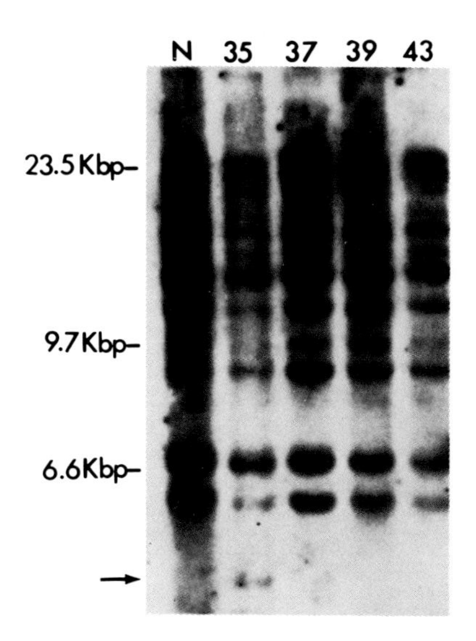

Figure 1. Southern blot analysis of genomic DNA from GA patients. The hybridization patterns of Eco RI OAT genomic fragments are shown for four representative GA patients (35, 37, 39 and 43) and normal control (N). Arrow points to the novel 5 Kbp fragment in patient 35.

gene sequences (Fig.1, lane N). The results of a hybridization experiment using specific probes and varying temperatures indicated the presence of at least four copies of OAT or OAT-related gene sequences, one of which is the functional OAT gene corresponding to the cDNA (data not shown). We have mapped the OAT gene sequences to chromosomes 10 and X, and confirmed that the functional OAT gene is on the former (Barrett et al., 1987). The chromosomal mapping result appeared to confirm the presence of the OAT gene family and the autosomal recessive inheritance of GA. The analysis of OAT gene clones demonstrated a multi-exon functional gene on chromosome 10 and at least one processed pseudogene on the X chromosome (data not shown).

We used the OAT probe to investigate the status of the OAT gene and its expression in tissues from GA patients (Inana et al., 1988). High-molecular-weight genomic DNAs of leukocytes or skin fibroblasts from 14 GA patients and normal subjects were subjected to a Southern analysis using the OAT cDNA probe. The results for four representative cases and a control are shown in Figure 1. The hybridization analysis of the OAT gene sequences in the patients identified a case containing a novel 5 Kbp Eco RI genomic fragment, not found in normal subjects or other GA patients (Fig.1, Patient 35). The 5 Kbp fragment appears to be a truncated form of one of the two 5.7 Kbp allelic fragments containing the functional OAT gene sequences (Barrett et al., 1987) since the latter shows a 50% decrease in relative hybridization intensity. That this case represents a partial heterozygous deletion of the OAT gene was supported by the additional demonstration of a truncated form of a Hind III, Bgl II, Sph I and Msp I functional OAT gene fragment in this patient (data not shown) and the failure to detect any of the altered forms in up to 30 subjects, making it unlikely for them to be restriction fragment length polymorphisms (RFLP).

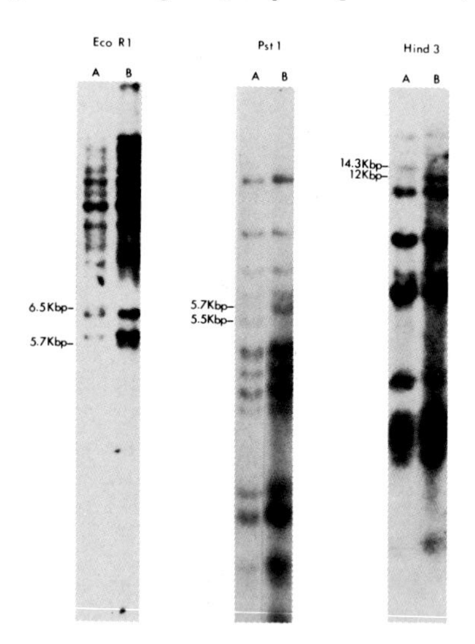

Figure 2. Restriction fragment length polymorphisms of OAT gene sequences. For each enzyme digestion, (A) represents the restriction fragment pattern usually observed, and (B) represents a less common variant containing a restriction fragment length polymorphism (RFLP). RFLPs with Eco RI (6.5 Kbp and 5.7 Kbp fragments), Pst I (5.5 Kbp and 5.7 Kbp fragments), and Hind III (14.3 Kbp and 12 Kbp fragments) are shown.

RFLPs were observed with Eco RI (6.5 Kbp and 5.7 Kbp fragments), Hind III (14.3 Kbp and 12 Kbp fragments) and Pst I (5.5 Kbp and 5.7 Kbp fragments) in the OAT gene sequences (Fig.2). The approximate frequency of occurrence of the less common variant in each of the RFLP sets was 13% for the 5.7 Kbp Eco RI fragment, 5% for the 12 Kbp Hind III fragment, and 13% for the 5.7 Kbp Pst I fragment among the subjects. The Eco RI polymorphism is present in the functional OAT gene sequence located on chromosome 10 whereas the Hind III and Pst I polymorphisms are present in the OAT-related gene sequences located on the X chromosome, as judged by the previous chromosomal mapping of these gene sequences (Barrett et al., 1987).

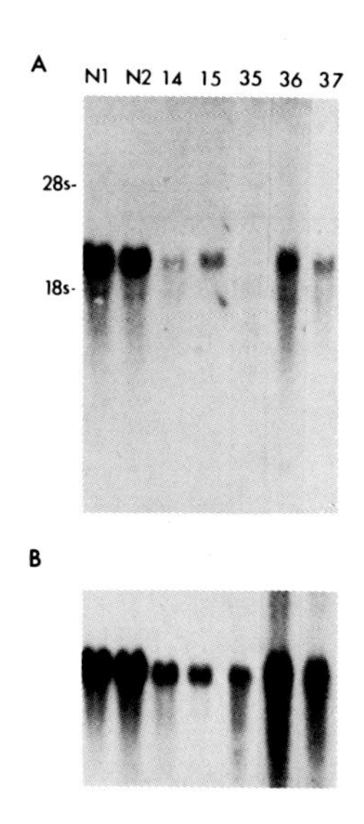

Figure 3. Northern blot analysis of fibroblast RNA from GA patients. (A) OAT mRNA. (B) Actin mRNA. The levels of specific mRNA are shown for five representative GA patients (14, 15, 35, 36 and 37) and normal controls (N1, N2). Position of the 18s and 28s ribosomal RNA are indicated.

Skin fibroblasts were obtained from the GA patients, and mRNAs were isolated and subjected to a northern analysis. The results for five representative cases and two controls are shown in Figure 3. Patient 35 is the case with the partial heterozygous deletion of the OAT gene described above. Patient 36 is a B_6-responder, and the rest of the cases are B_6-nonresponders (Weleber et al., 1982). The northern blot analysis of the fibroblast RNAs with the OAT probe indicated the presence of OAT mRNA, similar in size and amount to that found in normal control, in all cases except one (Patient 35), in whom no OAT mRNA was detectable (Fig.3A).

The level of OAT protein had been shown to be markedly decreased in patient 14 by immunoassay previously, as mentioned (Ohura et al., 1984). The level of OAT protein present in fibroblasts from patients 35, 36 and 37 was also determined by western blot analysis using the anti-human OAT antibody. The result indicated the presence of variably reduced levels of immunoreactive OAT protein in these cases with patient 35 showing the lowest level, if any, of immunoreactive protein (data not shown).

The finding of an OAT gene, mRNA, and protein defect in one GA case constitutes the first real demonstration of the OAT defect in GA at the gene level and provides confirmation for the primary role the OAT gene is thought to play in this disease. The absence of OAT mRNA in patient 35 (patient 4 in Weleber et al., 1982) is most likely due to the partial deletion of one of the OAT gene alleles and a mutation in the regulatory region of the other OAT gene allele which results in nontranscription of the gene and/or a mutation in the gene itself, which in turn results in lability or defective processing and rapid degradation of the transcribed mRNA. Examples of both types of gene mutation have been demonstrated in thalassemia (Orkin and Nathan, 1981). The presence of apparently normal OAT mRNA and a variable amount of immunoreactive OAT protein in three GA patient's fibroblasts, despite the lack of OAT activity in these cells, indicate that a subtle defect, such as a point mutation, is most likely present in the mRNA. Such a defect may result in poor translation of the message, lability or defective transport of the translated protein to the mitochondria, or inactive OAT protein. In fact, a point mutation which results in an amino acid change in the OAT protein and defective processing of the precursor protein has been demonstrated in another case of GA (Inana et al., submitted for publication). The finding of normal-appearing mRNA as seen here is apparently more often the rule in inherited diseases, as demonstrated in patients with ornithine transcarbamylase deficiency (Saheki et al., 1984), argininosuccinate synthetase deficiency (Kobayashi et al., 1986), juvenile form of Sandhoff disease (O'Dowd et al., 1986), partial adenosine deaminase deficiency (Daddona et al., 1985), and thalassemia (Orkin and Nathan, 1981). The finding of a normal pattern of OAT gene sequences in most of the GA patients tested by genomic Southern analysis is also consistent with the likely presence of a subtle mutation, such as a point mutation, instead of gross deletions or rearrangements, in most cases of GA. This has also been found to be the case in patients with ornithine transcarbamylase deficiency (Rozen et al., 1985) and β-thalassemias (Orkin and Nathan, 1981). A subtle mutation in the OAT gene would also be more likely to allow for synthesis of OAT mRNAs from it, as actually demonstrated. The results on the status of OAT gene expression in GA patients appear to correlate well with the clinical heterogeneity that is observed for GA. It is likely that a complete lack of OAT gene expression as essentially found in patient 35 represents those cases that have no detectable OAT activity, nonresponsiveness to B_6 therapy, and a more rapid progression to blindness. Expression of the OAT gene and synthesis of mRNA and a decreased amount of OAT protein as found patients 14, 15, 36 and 37 may represent those cases that have residual OAT activity (up to 19% of normal), responsiveness to B_6 therapy, and a milder course of the disease.

The case 35 represents a complete loss of expression of both alleles of the OAT gene by different mechanisms resulting in no mRNA. If each allele of the OAT gene contributes half of the total OAT expression, as indicated by the OAT activity in heterozygous carriers, non-expression of one allele should be reflected at the mRNA level in the obligate heterozygous carriers in this patient's family, and this defect should be inherited in a stable manner. To investigate the inheritance of the OAT gene defects in this family, cellular DNA, RNA, and proteins were isolated from skin fibroblasts of the family members of patient 35 and analyzed (Hotta et al., in press).

Southern blot analysis of the OAT gene in the family members indicated that the patient had inherited the partially deleted allele of the OAT gene, represented by the 5 Kbp Eco RI fragment, from his father and had passed it on to his daughter and a son (Fig.4). Northern blot

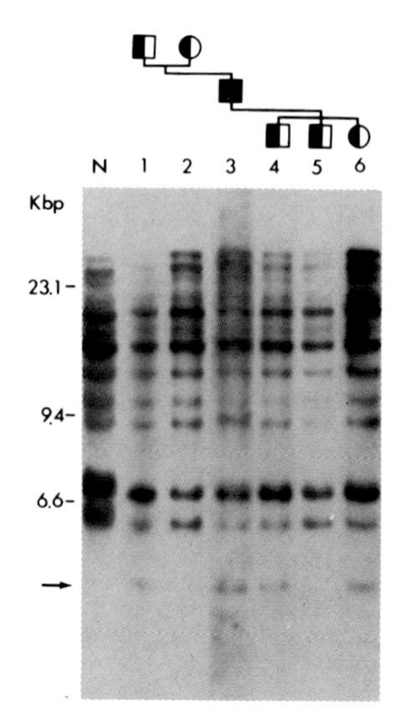

Figure 4. Southern blot analysis of the OAT gene in the GA family members. (N), DNA from normal human liver. (3), fibroblast DNA from the GA patient; (1), his father; (2), mother; (4,5), two sons; and (6), daughter. Eco RI OAT gene fragments are shown. The 6.7, 6.6 and 5.7 Kbp bands represent the functional OAT gene (Barrett et al., 1987). The arrow at 5 Kbp points to the band corresponding to the partially deleted allele of the 5.7 Kbp functional OAT gene band which shows a 50% decrease in hybridization intensity. Size markers are Hind III fragments of bacteriophage λ DNA.

analysis of the fibroblast RNAs demonstrated the complete absence of OAT mRNA in the patient, as described above, and decreased levels of OAT mRNA in the father, mother, two sons and daughter (Fig.5). Densitometric standardization of the mRNA levels using the amount of actin mRNA present in each lane indicated that the level of OAT mRNA present in the patient's parents and children, who are all obligate heterozygous carriers, is between 41% and 63% of normal (Table 1). The fibroblast cytosol preparations from the family members were used for assay of OAT activity. The OAT activity in the patient's cells was 3 nmol/hr/mg protein and in the other family members was 170 to 300 nmol/hr/mg protein, which is 35 to 61% of normal (490 ± 98 nmol/hr/mg protein).

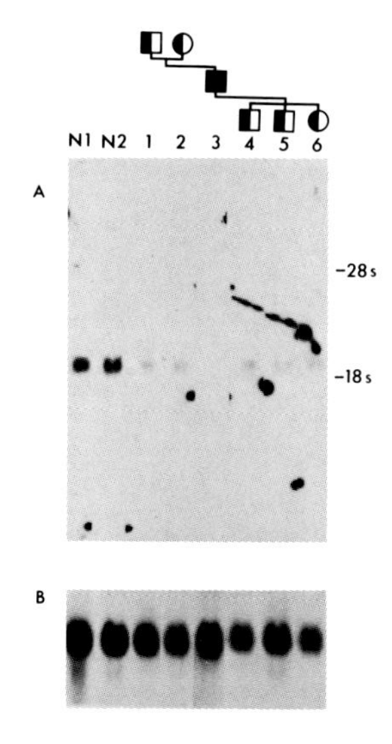

Figure 5. Northern blot analysis of the OAT mRNA in the GA family members. A. Probing for the OAT mRNA with OAT cDNA: (N1,N2), fibroblast RNAs from two normal individuals. (1-6), fibroblast RNAs from the GA family members. Numbers 1-6 are the same as described for Figure 4. B. Probing for the actin mRNA with actin cDNA for standardization of the quality and quantity of RNAs present. Position of the 28s and 18s ribosomal RNA are indicated.

Most of the cases with GA examined so far at the molecular level have shown grossly intact OAT genes and mRNAs, suggesting that subtle point mutations are probably the most common defects in GA, as is usually the case in genetic diseases (Ramesh et al., 1986; Inana et al., 1988). The patient 35 is unique in that gross abnormalities, including a partial heterozygous deletion of the OAT gene and a complete absence of the OAT mRNA, are present. Because the molecular genetic defects of OAT are relatively easily identifiable in this patient, the case offered an opportunity to confirm the autosomal recessive inheritance of OAT gene defects in GA in addition to examining the gene dose-product relationship of the OAT gene by examining the status of OAT gene expression in the patient's family members.

Table 1.

Densitometric Analysis of Northern Blot Hybridization

probe	control	1	2	3	4	5	6
OAT	1*	0.447	0.353	0	0.393	0.440	0.353
Actin	1	0.994	0.869	0.493	0.670	0.768	0.560
OAT/Actin x100 (%)	100	45.0	40.6	0	58.7	57.3	63.0

* Band intensities on the northern blot were measured by a densitometer, and the areas were calculated. Each value is presented as a ratio to control (normalized to 1) which is an average of two normal human mRNA levels shown in Fig. 2.

The result of the gene analysis identified the patient's father as the source of the partially deleted allele of the OAT gene. The defective allele showed stable inheritance in the family, and the partial deletion could very likely result in inactivation of this allele of the OAT gene as mentioned. The other defective allele coming from the patient's mother was not readily identifiable, but its presence was clearly indicated by the half-normal expression of OAT at both the mRNA and enzyme level. The defect in this allele may involve a point mutation, deletion, or insertion in the gene, especially in the regulatory region, that results in non-expression.

The results of the mRNA and enzyme analysis clearly demonstrated the presence of two distinct heterozygous defects in OAT gene expression in the parents of the patient, their coming together to result in the loss of both functional alleles in the patient, and their transmission to the children to establish the separate heterozygous states. The data demonstrated the stable inheritance of the expression defects of the OAT gene in this family and confirmed the autosomal recessive nature of this disease. The occurrence of two distinct molecular defects, both resulting in lack of expression of the OAT gene in this family, is so far unique among patients with gyrate atrophy.

The data also pointed out that a normal active allele of the OAT gene is responsible for 50% of the total OAT mRNA and enzyme activity in a cell in agreement with the previous demonstration of half-normal level of OAT activity in heterozygous carriers (Trijbels et al., 1977; Valle et al., 1977; O'Donnell et al., 1977). In this regard, the OAT gene allele is similar to many other mammalian gene alleles that show a co-dominant mode of action such as in the hemoglobin gene (Honig and Adams, 1986).

REFERENCES

Barrett DJ, Bateman JB, Sparkes RS, Mohandas T, Klisak I, Inana G (1987). Chromosomal localization of human ornithine aminotransferase gene sequences. Invest Ophthalmol Vis Sci 28:1037-42.

Daddona PE, Davidson BL, Perignon JL, Kelley WN (1985). Genetic expression in partial adenosine deaminase deficiency. J Biol Chem 260:3875.

Deshmukh DR, Srivastava SK (1984). Purification and properties of ornithine aminotransferase from rat brain. Experientia 40:357-359.

Honig RH, Adams JG (1986). "Human Hemoglobin Genetics." New York: Springer-Verlag/Wien.

Hotta Y, Kennaway NG, Weleber RG, Inana G (1989). Inheritance of ornithine aminotransferase gene, mRNA and enzyme defect in a family with gyrate atrophy of the choroid and retina. Am J Hum Genet (in press).

Inana G, Totsuka S, Redmond M, Dougherty T, Nagle J, Shiono T, Ohura T, Kominami E, Katunuma N (1986). Molecular cloning of human ornithine aminotransferase mRNA. Proc Natl Acad Sci (USA) 83:1203-1207.

Inana G, Hotta Y, Zintz C, Takki K, Weleber RG, Kennaway NG, Nakayasu K, Nakajima A, Shiono T (1988). Expression defect of ornithine aminotransferase gene in gyrate atrophy. Invest Ophthalmol Vis Sci 29:1001-5.

Kobayashi K, Saheki T, Imamura Y, Noda T, Inoue I, Matuo S, Hagihara S, Nomiyama H, Jinno Y, Shimada K (1986). Messenger RNA coding for argininosuccinate synthetase in citrullinemia. Am J Hum Genet 38:667.

Mueckler MM, Himeno M, Pitot HC (1982). In vitro synthesis and processing of a precursor to ornithine aminotransfease. J Biol Chem 257:7178-80.

O'Donnell JJ, Sandman RP, Martin SR (1977). Deficient L-ornithine: 2-oxoacid aminotransferase activity in cultured fibroblasts from a patient with gyrate atrophy of the retina. Biochem Biophys Res Commun 79:396-399.
O'Dowd BF, Klavins MH, Willard HF, Gravel R, Lowden JA, Mahuran DJ (1986). Molecular heterogeneity in the infantile and juvenile forms of Sandhoff Disease (o-variant G μ2 gangliosidosis). J Biol Chem 261:12680.
Ohura T, Kominami E, Tada K, Katunuma N (1982). Crystallization and properties of human liver ornithine aminotransferase. J Biochem 92:1785-1792.
Ohura T, Kominami E, Tada K, Katunuma N (1984). Gyrate atrophy of the choroid and retina: decreased ornithine aminotransferase concentration in cultured skin fibroblasts from patients. Clin Chim Acta 136:29-37.
Orkin SH, Nathan DG (1981). The molecular genetics of thalassemia. In Harris H, Hirschhorn K (eds): "Advances in Human Genetics," New York: Plenum, pp.233-80.
Peraino C, Pitot HC (1963). Ornithine-δ-transaminase in the rat: I. Assay and some general properties. Biochim Biophys Acta 73:222-231.
Ramesh V, Shaffer MM, Allaire JM, Shih VE, Gusella JF (1986). Investigation of gyrate atrophy using a cDNA clone for human ornithine aminotransferase. DNA 5:493-501.
Rozen R, Fox J, Fenton WA, Horwich AL, Rosenberg LE (1985). Gene deletion and restriction fragment length polymorphisms at the human ornithine transcarbamylase locus. Nature 313:815.
Saheki T, Imamura Y, Inoue I, Miura S, Mori M, Ohtake A, Tatibana M, Katsumata N, Ohno T (1984). Molecular basis of ornithine transcarbamylase deficiency lacking enzyme protein. J Inher Metab Dis 7:2.
Sanada Y, Sunemori I, Katunuma N (1970). Properties of ornithine aminotransferase from rat liver, kidney and small intestine. Biochim Biophys Acta 220:42-50.
Shiotani T, Sanada Y, Katunuma N (1977). Studies on the structure of rat liver ornithine aminotransferase. J Biochem 81:1833-1838.
Simell O, Takki K (1973). Raised plasma ornithine and gyrate atrophy of the choroid and retina. Lancet 1:1030-1033.

Trijbels JMF, Sengers RCA, Bakkeren JAJM, DeKort AFM, Deutman AF (1977). L-ornithine-ketoacid-transaminase deficiency in cultured fibroblasts of a patient with hyperornithinemia and gyrate atrophy of the choroid and retina. Clinica Chimica Acta 79:371-377.

Valle D, Kaiser-Kupfer MI, DelValle LA (1977). Gyrate atrophy of the choroid and retina: Deficiency of ornithine aminotransferase in transformed lymphocytes. Proc Natl Acad Sci (USA) 74:5159-5161.

Valle D, Simell O (1983). The hyperornithinemias: gyrate atrophy of the choroid and retina. In Stanbury JB, Wyngaarden JB, Fredrickson DS, Goldstein JL, Brown MS (eds): "The Metabolic Basis of Inherited Disease", New York: McGraw Hill, pp. 389-396.

Weleber RG, Wirtz MK, Kennaway NG (1982). Gyrate atrophy of the choroid and retina: Clinical and biochemical heterogeneity and response to vitamin B_6. Birth Defects Original Article Series 18:219-30.

Young RA, Davis RW (1983). Efficient isolation of genes by using antibody probes. Proc Natl Acad Sci (USA) 40:1194-8.

Inherited and Environmentally Induced
Retinal Degenerations, pages 113–129

VITAMINS A AND E IN RPE LIPOFUSCIN FORMATION AND IMPLICATIONS FOR AGE-RELATED MACULAR DEGENERATION

Graig E. Eldred

Department of Ophthalmology, University of Missouri, School of Medicine, Columbia, Missouri 65212

INTRODUCTION

Current understanding of the pathophysiology of age-related macular degeneration (AMD) has been reviewed recently in a pair of excellent review articles (Young 1987, 1988). It is the purpose of this article to incorporate our recent research findings into this body of evidence.

THE CURRENT THEORY OF AMD PATHOPHYSIOLOGY

Many empirical gaps exist in our knowledge of the events leading to the clinical entity known as age-related macular degeneration, yet a theory has emerged from clinical and morphological observations of the progression of the disease. Funduscopically, discrete areas of RPE hyperpigmentation and numerous macular drusen are hallmarks of the disease. There are two primary forms of lesions: disciform and geographic. The disciform lesions are marked by fluid leakage from choroidal vessels that invade Bruch's membrane especially in the vicinity of large, confluent drusen. The geographic lesions, as the name implies, occur in traceable patterns that are defined by regions of RPE cell death. In both cases, the death of the RPE precedes the loss of the visual cells ultimately resulting in irreversible visual field deficits.

The primary site of physiological disfunction is believed to be the retinal pigment epithelium (RPE). The earliest age-related morphological change in the retina occurs in this cell layer. Over time the cells becomes filled with lipofuscin pigment granules, also known as age pigments.

Lipofuscin granules are known to be residual accretions of the cellular phagolysosomal system. Enzyme histochemistry and biochemistry experiments have demonstrated the presence of lysosomal enzymes within these granules. The materials come to fill the granules either because there is a reduction in the production of lysosomal enzymes with age (Wilcox, 1988), and/or because the substrates are incapable of being degraded by the normal array of lysosomal enzymes because they have been altered or damaged to such an extent that they are no longer recognizable as substrates for the enzymes (Feeney-Burns et al., 1980). Most of the granule contents are classified as lipids by virtue of the fact that they are histochemically sudanophilic (Streeten, 1961) and largely chloroform: methanol soluble (Feeney, 1978). Many of the molecular residues are autofluorescent, and their combined golden yellow emission is pathognomonic for these granules (Feeney, 1978).

The main source of materials accumulating in the lipofuscin granules has long been suspected to be the lipids of the photoreceptor outer segments (Hogan, 1972) that enter the RPE in the daily phagocytosis of shed packets of photoreceptor outer segment disc membranes (Young and Bok, 1969). The primary cause of molecular damage is thought to be lipid peroxidation reactions within these disc membranes (Feeney and Berman, 1976). The oxidative degradation may be initiated by photodynamic activation with the visual pigment chromophores acting as the sensitizing agents (Delmelle, 1977). In fact, although many factors might contribute to the onset of the disease, a leading factor may be a lifetime exposure of the central retina to solar radiation (Young, 1988). Aldehydic fragments of the damaged lipids can theoretically act as crosslinkers. Thus, it is thought that molecules, especially amine-containing compounds, are bound together into complexes that cannot be enzymatically degraded, and that are fluorescent by virtue of their crosslinking bond (Tappel, 1975).

As the RPE cell becomes filled with these granules with advanced age, function is thought to be compromised. One indication of this is that the cell is stimulated to dispose of its excess burden via a cytoplasmic shedding process (apoptosis) through its basal surface to form drusen and basal linear deposits in Bruch's membrane (Burns and Feeney-Burns, 1980). Apparently, the lipofuscin granules themselves cannot be ejected in this manner, however. The thickening of Bruch's membrane in turn further impedes the passage of nutrients and metabolites between the photoreceptor cells and the choriocapillaris, a process normally regulated by the retinal pigment epithelium. In advanced stages, the RPE dies and the overlying retina starts to atrophy. At some point blood vessels may invade Bruch's membrane and the subretinal space. Leakage or hemorrhage then causes the disciform form of the disease (Sarks et al., 1980).

At present, there is no way to clinically or experimentally reduce the lipofuscin granule burden in the RPE to test whether the onset of AMD can be delayed or eliminated. To do this, a much better knowledge of the molecular identity of the contents of the lipofuscin granule will be required. Our research has been devoted to identifying the autofluorescent components of RPE lipofuscin in order to verify the role of lipid peroxidation in RPE lipofuscinogenesis. The results of these studies suggest that lipid peroxidation may not be as important as suspected, and that the alterations of vitamin A itself may be more important in the formation of RPE lipofuscin. The evidence for these conclusions and some speculations as to the significance of these findings for our understanding of age-related retinopathies constitutes the remainder of this paper.

CRITERIA FOR FLUOROPHORE COMPARABILITY

Lipid peroxidation results in the formation of autofluorescent byproducts (Tappel, 1975). This is one of the key pieces of evidence used in invoking a role for lipid peroxidation in lipofuscinogenesis (Chio et al., 1969). For this to be true, the fluorophores of age pigments should be comparable to these byproducts. We have developed three criteria for comparability that must be met to conclude that there may be a link between the two chemistries.

First is the criterion of extractability and solubility. The fluorophores of RPE lipofuscin are completely solubilized in 2:1, chloroform:methanol. If this solution is washed with water, some of the fluorophores reprecipitate at the interface, and the amount of material precipitating increases with age (Eldred and Katz, 1988a). The water wash is used initially to remove extracted proteins, flavins, etc. from the extracts. It has been found that these interfacial fluorophores may be resolubilized in the chloroform:methanol solution. Comparable byproducts of lipid peroxidation should behave similarly.

Second is the criterion of spectral similarity. Both absorbance and fluorescence spectra are useful in this regard. As will be discussed below, it has been found that spectral correction for bias in the fluorescence spectrometric instrumentation is essential in these comparisons (Eldred et al., 1982; Stark et al., 1984; Katz et al., 1984), as is adequate sample dilution to eliminate the spectral shifts that could occur as a result of internal filter effects (Eldred, 1987). We have since refined the recommended procedures for spectral correction when analyzing lipofuscin fluorescence by applying Basic Blue 3, a laser dye, as a quantum counter (Eldred and Katz, 1988a).

While spectral comparisons of crude extracts are useful, fractionation of the extracts into semipurified preparations has been more insightful, which leads to the third criterion for comparability: that of chromatographic mobility. A thin-layer chromatography system has been developed that is capable of resolving both the nonpolar and more polar fluorophores of the chloroform:methanol extracts (Eldred and Katz, 1988a). When testing unknown samples this system has the advantage over HPLC techniques of being able to observe the entire applied sample during chromatography. HPLC in this case can prove to be a "black box" in which components applied may never appear again. With TLC, chromatograms are photographable, and individual fluorescent fractions are elutable for further analysis.

Using these criteria it has been found that human RPE lipofuscin is essentially identical to rat RPE lipofuscin, thereby verifying the rat as a good model for the study of age pigment formation. It has also been found that the fluorophores of human RPE lipofuscin are not comparable to those of human heart lipofuscin (Eldred, 1987), or human or canine ceroid lipofuscinosis pigments (Katz et al., 1988). Other pertinent comparisons with dietary deficiency pigments and in vitro lipid oxidation products have been performed and will be discussed below.

THE RETINAL PIGMENT EPITHELIUM LIPOFUSCIN FLUOROPHORES

All of the chloroform:methanol soluble fluorophores isolated from whole RPE extracts are present in extracts of isolated RPE lipofuscin granules (Feeney-Burns and Eldred, 1984). Ten fluorescent fractions have now been separated from human RPE lipofuscin (Eldred and Katz, 1988a). They fall into four categories based upon their fluorescent excitation/emission characteristics: a pair of green-emitting fractions (330 nm Ex; 520 nm Em) that cochromatograph with retinol and retinyl esters (Fig. 1); three yellow/green-emitting fractions (280, 330 nm Ex; 568 nm Em); one golden yellow-emitting fraction (280, 330 nm Ex; 585 nm Em); and four orange/red emitting fractions (285, 335, 420 nm Ex; 605, 633, 670 nm Em). It is these fluorescent fractions that are used as the basis for comparisons with the various dietary and experimental treatments discussed below. It should be noted that erratic blue emitting fluorophores occur in human whole cell extracts only occasionally, and have been traced to pharmaceuticals administered to the patients prior to their deaths (Eldred et al., 1982; Eldred and Katz, 1988a). These pharmaceuticals are most likely bound to the RPE melanin prior to their extraction (Potts, 1964).

TESTS FOR A ROLE OF LIPID OXIDATION IN THE GENERATION OF LIPOFUSCIN AUTOFLUORESCENCE

The key papers that served as the basis for the current theory of lipofuscinogenesis (Chio et al., 1969; Chio and Tappel, 1969) relied on the apparent similarities between the uncorrected fluorescence spectra of several samples. First, lipid peroxidation reactions were induced in suspensions of subcellular organelles by incubation with oxygen. Fluorescence emission spectra of the byproducts were stated to be similar to previously published uncorrected fluorescence spectra of age pigment extracts. Then, based on the supposition that the fluorophores originated from a crosslinking reaction between one possible product of lipid peroxidation (malonaldehyde) and amine-containing compounds, these investigators synthesized several imine-conjugated Schiff bases from malonaldehyde and amino acids. Again, the uncorrected emission spectra

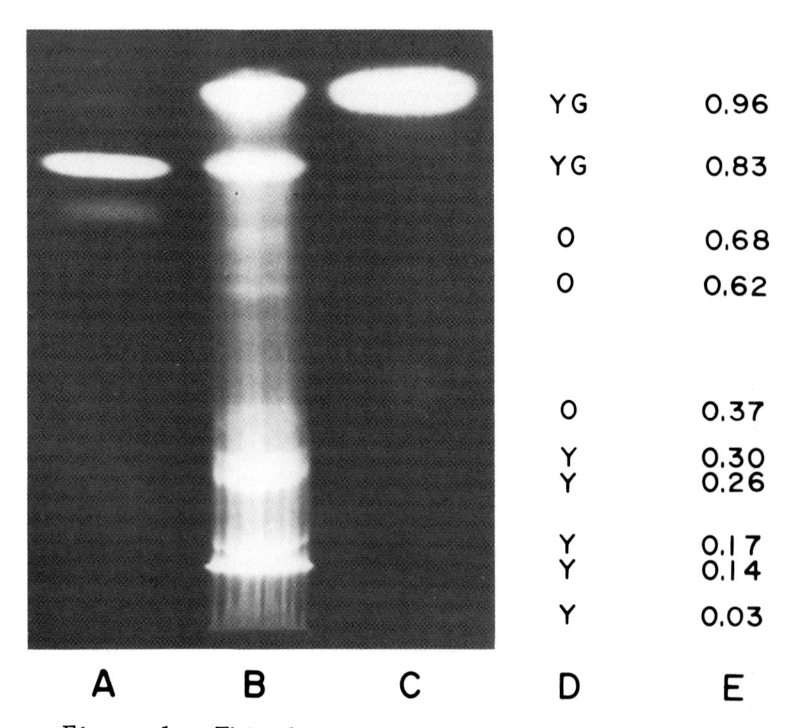

Figure 1: Thin-layer chromatogram of isolated lipofuscin granules from the human RPE (B) cochromatographed with trans-retinol (A) and trans-retinyl palmitate (C) standards. Colors seen under 366 nm UV illumination (D) and measured Rf values (E) are also shown (YG = bright green; O = orange/red; Y = yellow/green and G = gold).

were deemed comparable to those of age pigment extracts. But these spectra all showed blue peak emissions whereas the emission from lipofuscin has long been described to be golden yellow.

This discrepancy was resolved when it was demonstrated that instrumental bias caused a shift in lipofuscin extract spectra resulting in the recording of blue emission peaks from yellow-emitting mixtures of fluorophores (Eldred et al., 1982). We went on to synthesize the model

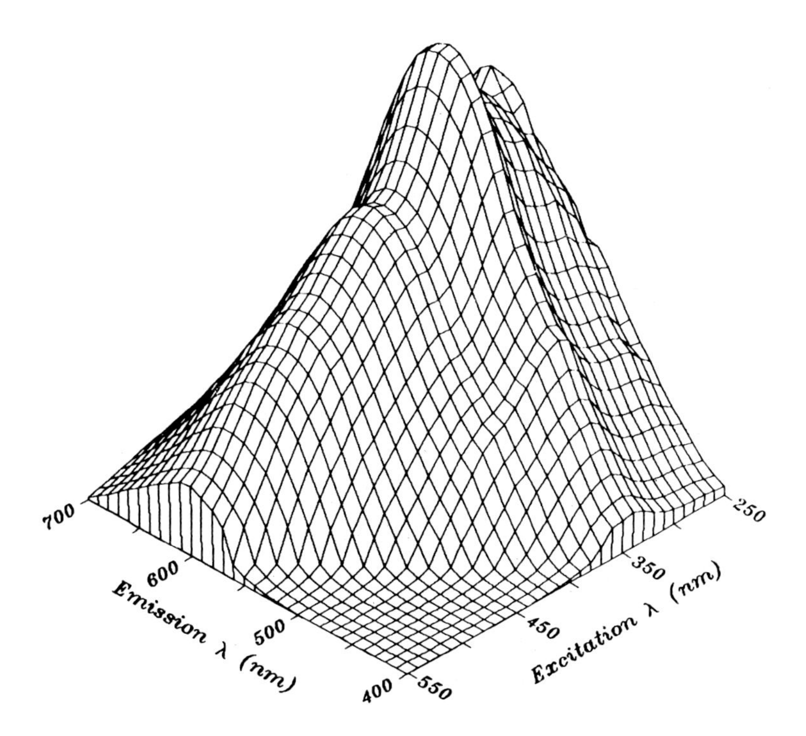

Figure 2: Corrected total luminescence spectrum of the chloroform extract of a lipofuscin-laden human RPE sample from a 73 year old donor. Vitamin A-related (green emitting) fluorophores have been removed via a solid phase extraction technique. Only the lipofuscin-related (yellow-green through orange-red emitting) fluorophores remain.

malonaldehyde-amino acid reaction products and demonstrated that they are truly blue-emitting (Eldred, 1987) (Figures 2 and 3). They also proved to be much more polar than the age pigment fluorophores.

Other authors had conjectured that polymerization products of malonaldehyde, rather than the Schiff base products were responsible for age pigment fluorescence (Siakotos and Munkres, 1981). In synthesizing these products and comparing them both chromatographically and spectrally with RPE age pigment fluorophores, they were found not to be similar (Eldred, 1987).

Retinal and RPE homogenates have also been subjected to conditions commonly used to initiate lipid oxidation

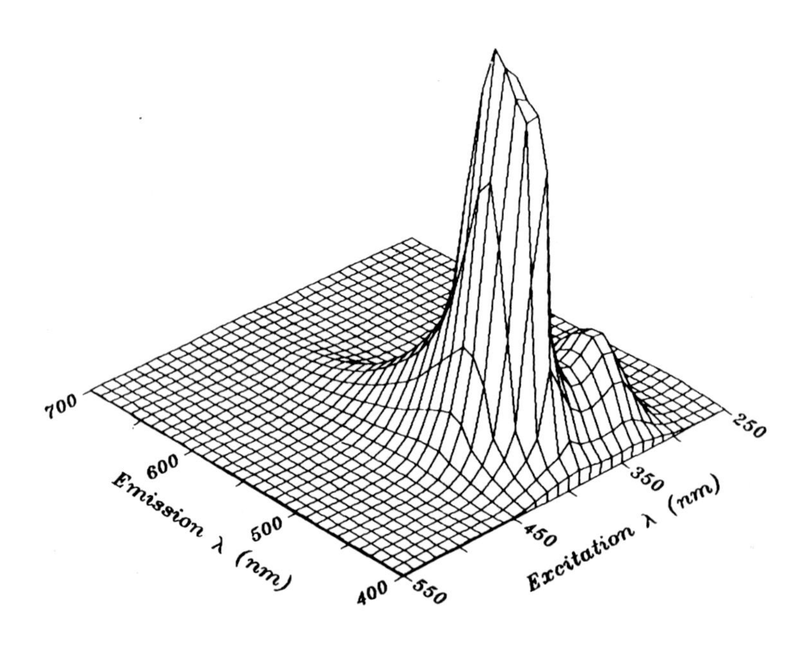

Figure 3: Corrected total luminescence spectrum of the chloroform extract of diglycinyl imine-conjugated Schiff base reaction product(s).

reactions in vitro. Typical blue-emitting lipid oxidation byproducts were generated, but nothing resembling the fluorophores extracted from RPE lipofuscin were seen (Eldred and Katz, 1988b). Thus, the concept of lipofuscin fluorophores being the direct products of lipid peroxidation reactions is cast into doubt.

ROLE OF VITAMIN E

Another line of evidence that suggests that lipofuscin fluorophores may arise as products of lipid peroxidation reactions is the fact that dietary vitamin E deficiency and various other antioxidant deficiencies lead to the accumulation of golden yellow-emitting autofluorescent granules in a variety of tissues including the RPE (Katz et al., 1978). Yet, of the tissues examined, those accumulating vitamin E deficiency pigments are not the same tissues that accumulate age pigment with the sole exception of the retinal pigment epithelium (Katz et al., 1984). Furthermore, we have demonstrated that vitamin E deficiency pigment of the RPE differs from the RPE lipofuscin in both its extractability and chromatographic characteristics (Eldred and Katz, 1988c). Until the vitamin E-deficiency pigment of RPE can be solubilized and chemically identified, it is impossible to deduce the chemical structures of the fluorophores responsible for their fluorescence. However, at the present time we cannot conclude that these pigments are related to the readily extractable materials in RPE age pigment.

ROLE OF PHOTORECEPTOR OUTER SEGMENTS

The question of the identity and mechanism of formation of RPE lipofuscin fluorophore therefore remains open. Our investigations have produced evidence that the photoreceptor outer segments serve as the primary source of materials accumulating in RPE lipofuscin. When photoreceptors are eliminated by photic damage, the RPE does not accumulate the same amount of lipofuscin as seen in dim light reared animals (Katz and Eldred, 1988). Also the RPE of RCS rats that are unable to phagocytose POS except at very low levels do not accumulate the same amount of lipofuscin as seen in congenic age matched control animals (Katz et al., 1986). Thus, the photoreceptor outer segments appear to play a primary role in determining RPE lipofuscin content.

ROLE OF VITAMIN A

Robison and coworkers have clearly established a role for vitamin A in determining the lipofuscin content of the RPE. Dietary retinol deficiency dramatically reduced the number of lipofuscin granules in rats raised on both vitamin E sufficient and deficient diets (Robison et al., 1979; Robison et al., 1980). Even in retinas lacking photoreceptor cells, vitamin A deficiency resulted in greatly reduced amounts of lipofuscin (Robison et al., 1982), indicating a probable incorporation of vitamin A metabolites into the lipofuscin of the RPE through pathways other than outer segment phagocytosis (Robison and Katz, 1987).

RELATIONSHIP OF VITAMIN A TO THE INDIVIDUAL FLUOROPHORES OF RPE LIPOFUSCIN

Among the fluorophores seen in these pigments are some that have the same corrected fluorescence spectra and chromatographic mobilities as retinol and retinyl esters (Fig. 1). It may be that rather than actively accumulating these vitamin A derivatives in lipofuscin, the lipophilic environment of the granules may provide a suitable phase for the vitamin A to separate into. If so, it would be important to know whether this pool is an exchangeable pool for vitamin A as it cycles between the retina and RPE.

Another predominant component separated from the lipofuscin granules is an orange-emitting fluorophore (Eldred and Katz, 1988a). This fluorophore is virtually eliminated in rats raised on a vitamin A deficient diet.

Evidence that this orange-emitting fluorophore is derived from materials originating in the photoreceptor outer segments comes from experiments performed on RCS rats. The photoreceptor outer segment debris accumulating above the RPE in RCS rats evolves a fluorescence similar to lipofuscin (Katz et al., 1987). When the debris is extracted and chromatographed an orange-emitting fluorophore similar in fluorescent emission characteristics, but differing in chromatographic mobility was observed (Katz et al., 1986). When RCS rats are raised on vitamin A deficient diets, this orange-emitting fluorophore in the photoreceptor outer segment debris is greatly reduced (Katz et al., 1987).

There is also some evidence that the other yellow-emitting fluorophores of RPE lipofuscin are chemically related to the orange-emitting fluorophore. Spectrally, all of the fluorophores appear to have absorbance peaks at approximately 280 nm and 330 nm. The orange-emitting fluorophores have an additional peak at 420 nm (Eldred and Katz, 1988a). Additionally, when the orange-emitting fluorophore degrades on storage for extended periods of time, yellow-emitting fluorophores similar to some of the other age pigment fluorophores are seen.

Additional evidence that vitamin A may serve as a precursor to the lipofuscin fluorophores arises from the results of preliminary analytical spectroscopy on purified samples of the human RPE lipofuscin orange-emitting flurophore. The results are consistent with a linear polyenic structure for the molecule such as might be expected of a retinoid compound (Eldred and Katz, 1988d). More recent mass spectral evidence suggests that the orange-emitting fluorophore may be a photodimerization product of vitamin A.

These experiments lead us to believe that many of the lipofuscin fluorophores are either primary or secondary metabolites of retinoids originating in the photoreceptor outer segments. The possibility that they may be cooxidation products of both vitamin A and lipids has not yet been ruled out, although we have never observed any blue-emitting products typical of lipid peroxidation products.

SUMMARY OF THE IMPORTANCE OF THE FINDINGS

The conclusions that emerge from these findings are that:

(1) Contrary to current theory, RPE lipofuscin fluorophores are not characteristic of the direct byproducts of lipid oxidation.

(2) Vitmain E-deficiency is not a good model for the study of RPE lipofuscinogenesis.

(3) Photoreceptor outer segments do provide the major source of precursors for autofluorescent compounds that accumulate in RPE lipofuscin.

(4) Lipofuscin-like fluorophores can arise in POS prior to phagocytosis by the RPE.

(5) Some degree of modification of the POS lipofuscin-like fluorophores occurs after phagocytosis by the RPE.

(6) Both the most prominent POS lipofuscin-like fluorophores and the most prominent RPE lipofuscin fluorophores are virtually eliminated when retinyl palmitate is eliminated from the diet.

(7) All of the fluorophores of human RPE age pigments appear structurally related, and the analytical data are consistent with a polyenic structure as might be expected of modified retinoids.

ROLE FOR SOLAR RADIATION IN ETIOLOGY OF AMD

The evidence for a possible involvement of retinal exposure to solar radiation as a contributing factor in AMD is compelling. That lipids can be autoxidized under experimental light damage conditions is also compelling. The fluorophores that accumulate in RPE lipofuscin, however, do not appear to be the byproducts of lipid peroxidation reactions, but rather appear to be vitamin A metabolites. Photochemical transformations of vitamin A yield isomerized, dimerized and oxidized products (Mousseron-Canet, 1971). It is interesting to note that vitamin A oxidation products are very similar to lipid oxidation products in that a variety of aldehydes, alcohols, carboxylic acids, etc. arise (Suyama et al., 1983). Others have suggested the possibility that retinal vitamin A should be susceptible to oxidation (Daemen, 1973; Hayes, 1974; Robison and Katz, 1987). Yet to date, no investigations have inquired into the possibility that vitamin A oxidation products might be generated in vivo upon exposure to light. Our findings in the RPE lead us to believe that the fluorophores of lipofuscin may be the first indication that such reactions do occur.

QUESTIONS NEEDING FURTHER STUDY

If vitamin A oxidation products accumulate in RPE lipofuscin, it becomes important to question whether these modified vitamin A molecules might be recognizable by the various retinol binding proteins and whether they enter the vitamin A cycle between the retina and RPE and disrupt normal function. Some evidence exists that the materials in lipofuscin may not be stored there permanently. The rate of accumulation is great up to age 20 and thereafter the rate declines significantly (Feeney-Burns et al., 1984). The RPE does not appear to eject lipofuscin granules during apoptosis (Burns and Feeney-Burns, 1980). Presumably the rate of photoreceptor turnover and phagocytosis does not decline with age. Why then does the rate of accumulation of lipofuscin not continue to increase? Could it be that there is some removal of materials at a molecular level? Could this removal be mediated by the retinoid binding proteins? Could these molecules interfere with the vitamin A cycle (Bridges et al., 1983), or the glycosyl transfer reactions which are essential for the initial glycosylation of asparagine linked glycoproteins (DeLuca, 1977)?

One clinical test for AMD is the macular photostress recovery test (Lovasik, 1983). In this test the macula is illuminated by a relatively bright light and the time required for subjectively measured visual acuity to return to pretest levels is compared to the normal range of recovery times. Recovery is dependent on the rate of photopigment resynthesis and functional relationship between photoreceptors and the RPE. The test is useful for differentiating age-related maculopathy from optic neuropathy. While many factors could come into play in reducing the recovery time with age, might one of the factors be that the binding proteins are picking up visually useless oxidation metabolites of vitamin A that are competing for normal vitamin A in the vitamin A cycle?

Similarly, rhodopsin recovery times have been used in assays for photic damage in the rat retina (Organisciak et al., 1988). Could part of the damage in recovery times be due to a competition between normal vitamin A and oxidized metabolites of vitamin A for sites on the rhodopsin molecule or sites on the carrier and enzymatic pathways leading to the regeneration of retinal?

Evidence indicates that retinol and retinyl esters are present within the lipofuscin granule. It will be important to know how these pools of normal vitamin A affect this cycle, and how these potential problems might be related to age-related changes in the retina and ultimately to the etiology of age-related macular degeneration.

In conclusion, we believe that photochemically transformed vitamin A metabolites are responsible for lipofuscin fluorescence and that it is now important to positively identify them, and to test what effect their accumulation might have on the vitamin A cycle between the RPE and photoreceptors. Such tests may yield further insights into the mechanisms of age-related maculopathies.

REFERENCES

Bridges CDB, Fong S-L, Liou GI, Alvarez RA, Landers RA (1983). Transport, utilization and metabolism of visual cycle retinoids in the retina and pigment epithelium. Prog Retinal Res 2:137-162.

Burns RP, Feeney-Burns (1980). Clinico-morphologic correlations of drusen of Bruch's membrane. Tr Am Ophth Soc 78:206-255.

Chio KS, Reiss U, Fletcher B, Tappel AL (1969). Peroxidation of subcellular organelles: Formation of lipofuscinlike fluorescent pigments. Science 166:1535-1536.

Chio KS, Tappel AL (1969). Synthesis and characterization of the fluorescent products derived from malonaldehyde and amino acids. Biochemistry 8:2821-2827.

Daemen FJM (1973). Vertebrate rod outer segment membranes. Biochim Biophys Acta 300:255-288.

Delmelle M (1977). Retinal damage by light: Possible implication of singlet oxygen. Biophys Struct Mechanism 3:195-198.

DeLuca LM (1977). The direct involvement of vitamin A in glycosyl transfer reactions of mammalian membranes. Vitam Horm 35:1-57.

Eldred GE (1987). Questioning the nature of the fluorophores in age pigments. In Totaro EA, Glees P, Pisanti FA (eds): "Advances in Age Pigments Research," Oxford: Pergamon Press, pp 23-36.

Eldred GE, Katz ML (1988a). Fluorophores of the human retinal pigment epithelium: Separation and spectral characterization. Exp Eye Res 47:71-86.

Eldred GE, Katz ML (1988b). Lipofuscin fluorophores of the retinal pigment epithelium differ from in vitro autoxidation products of retinal constituents. (submitted).

Eldred GE, Katz ML (1988c). Vitamin E-deficiency pigment and lipofuscin (age) pigment from the retinal pigment epithelium differ in fluorophoric composition. Invest Ophthalmol Vis Sci 29(suppl):92.

Eldred GE, Katz ML (1988d). Possible mechanism for lipofuscinogenesis in the retinal pigment epithelium and other tissues. In Zs.-Nagy I (ed): "Lipofuscin --- 1987: State of the Art," Budapest and Amsterdam: Akademiai Kiado and Elsevier Science Publishers, pp 185-211.

Eldred GE, Miller GV, Stark WS, Feeney-Burns L (1982). Lipofuscin: Resolution of discrepant fluorescence data. Science 216:757-759.

Feeney L (1978). Lipofuscin and melanin of human retinal pigment epithelium. Invest Ophthalmol Vis Sci 17:583-600.

Feeney L, Berman ER (1976). Oxygen toxicity: Membrane damage by free radicals. Invest Ophthalmol 15:789-792.

Feeney-Burns L, Berman ER, Rothman H (1980). Lipofuscin of human retinal pigment epithelium. Am J Ophthalmol 90:783-791.

Feeney-Burns L, Eldred GE (1984). The fate of the phagosome: Conversion to 'age pigment' and impact in human retinal pigment epithelium. Trans Ophthalmol Soc UK 103:416-421.

Feeney-Burns L, Hildebrand ES, Eldridge S (1984). Aging human RPE: Morphometric analysis of macular, equatorial, and peripheral cells. Invest Ophthalmol Vis Sci 25:195-200.

Hayes KC (1974). Retinal degeneration in monkeys induced by deficiencies of vitamin E or A. Invest Ophthalmol Vis Sci 13:499-510.

Hogan MJ (1972). Role of the retinal pigment epithelium in macular disease. Trans Am Acad Ophthalmol Otolaryngol 76:64-80.

Katz ML, Drea CM, Eldred GE, Hess HH, Robison WG Jr (1986). Influence of early photoreceptor degeneration on lipofuscin in the retinal pigment epithelium. Exp Eye Res 43:561-573.

Katz ML, Eldred GE (1988). Retinal light damage reduces autofluorescent pigment deposition in the retinal pigment epithelium. Invest Ophthalmol Vis Sci (in press).
Katz ML, Eldred GE, Robison WG Jr (1987). Lipofuscin autofluorescence: Evidence for vitamin A involvement in the retina. Mech Ageing Dev 39:81-90.
Katz ML, Eldred GE, Siakotos AN, Koppang N (1988). Characterization of disease-specific brain fluorophores in ceroid-lipofuscinosis. Am J Med Genetics, Supplement 5:253-263.
Katz ML, Stone WL, Dratz EA (1978). Fluorescent pigment accumulation in retinal pigment epithelium of antioxidant-deficient rats. Invest Ophthalmol Vis Sci 17:1049-1058.
Katz ML, Robison WG Jr, Herrmann RK, Groome AB, Bieri JG (1984). Lipofuscin accumulation resulting from senescence and vitamin E deficiency: Spectral properties and tissue distribution. Mech Ageing Dev 25:149-159.
Lovasik JV (1983). An electrophysiological investigation of the macular photostress test. Invest Ophthalmol Vis Sci 24:437-441.
Mousseron-Canet M (1971). Photochemical transformation of vitamin A. Meth Enzymol 18C:591-615.
Organisciak DT, Jiang Y-L, Wang H-M (1988). Ascorbic acid protective effect in type I retinal light damage. Invest Ophthalmol Vis Sci 29(Suppl):124.
Potts AM (1964). The reaction of uveal pigment in vitro with polycyclic compounds. Invest Ophthalmol 3:405-416.
Robison WG Jr, Katz ML (1987). Vitamin A and lipofuscin. In Sheffield JB, Hilfer SR (eds): "The Microenvironment and Vision (Cell and Developmental Biology of the Eye)," New York: Springer-Verlag, pp 95-122.
Robison WG Jr, Kuwabara T, Bieri JG (1979). Vitamin E deficiency and pigment epithelial changes. Invest Ophthalmol Vis Sci 18:683-690.
Robison WG Jr, Kuwabara T, Bieri JG (1980). Deficiencies of vitamins E and A in the rat, retinal damage and lipofuscin accumulation. Invest Ophthalmol Vis Sci 19:1030-1037.
Robison WG Jr, Kuwabara T, Bieri JG (1982). The roles of vitamin E and unsaturated fatty acids in the visual process. Retina 2:263-281.
Sarks SH, van Driel D, Maxwell L, Killingsworth M (1980). Softening of drusen and subretinal neovascularization. Trans Ophthalmol Soc UK 100:414-422.

Siakotos AN, Munkres KD (1981). Purification and properties of age pigments. In Sohal RS (ed): "Age Pigments," New York: Elsevier/North-Holland Biomedical Press, pp 181-202.
Stark WS, Miller GV, Itoku KA (1984). Calibration of microspectrophotometers as it applies to the detection of lipofuscin and the blue- and yellow-emitting fluorophores in situ. Meth Enzymol 105:341-347.
Streeten BW (1961). The sudanophilic granules of the human retinal pigment epithelium. Arch Ophthalmol 66:391-398.
Suyama K, Yeow T, Nakai S (1983). Vitamin A oxidation products responsible for haylike flavor production in nonfat dry milk. J Agric Food Chem 31:22-26.
Tappel Al (1975). Lipid peroxidation and fluorescent molecular damage to membranes. In Trump BF, Arstila AV (eds): "Pathobiology of Cell Membranes, Vol 1," New York: Academic Press, pp 145-173.
Wilcox DK (1988). Vectorial accumulation of cathepsin D in retinal pigmented epithelium: Effects of age. Invest Ophthalmol Vis Sci 29:1205-1212.
Young RW (1987). Pathophysiology of age-related macular degeneration. Surv Ophthalmol 31:291-306.
Young RW (1988). Solar radiation and age-related macular degeneration. Surv Ophthalmol 32:252-269.
Young RW, Bok D (1969). Participation of the retinal pigment epithelium in the rod outer segment renewal process. J Cell Biol 42:392-403.

Inherited and Environmentally Induced
Retinal Degenerations, pages 131–139

ZINC CONTENT IN MELANOSOMES OF DEGENERATING RPE AS MEASURED BY X-RAY MAPPING

Robert J. Ulshafer, Ph.D.

Department of Ophthalmology, University of Florida, Gainesville, Florida 32610 USA.

INTRODUCTION

Several human retinal diseases, such as age-related macular degeneration (ARMD), are associated with degenerative changes in the retinal pigment epithelium (RPE). The RPE normally forms a blood-retinal barrier between the choroidal circulation and the photoreceptor cells of the outer retina and when it malfunctions, physiological or pathological changes frequently occur in the photoreceptors as well. ARMD is characterized by loss of visual acuity with advancing age (Liebowitz et al., 1980). The fundus is seen to have small, yellowish hyperreflective spots in the macular region; these are called drusen. Under the electron microscope, drusen appear as deposits of debris occurring within Bruch's membrane, the pentilaminar structure separating RPE from choroidal capillaries. The deposits are composed of membranous profiles, granules, vesicles, abnormal arrangements of collagen fibrils and occasionally macrophage-like cells (Burns and Feeney-Burns, 1980; Feeney-Burns and Ellersieck, 1985; Ulshafer et al., 1987). It is believed by some that the drusen form a physical barrier between RPE and the vasculature, thereby preventing the necessary 2-way exchange of metabolites, waste products, and other substances between the two.

The RPE is a rich source of trace metals in the human body. Concentrations of zinc and copper in the RPE are among the highest per dry tissue weight all body tissues (Bowness et al., 1982; Eckhert, 1979; Karcioglu, 1982).

Recently, investigators have identified abnormal concentrations of several trace metals, including Zn and Cu, in the blood of patients suffering from ARMD. Miller et al. (1986) and Newsome et al. (1988) reported reduced zinc content in plasma of ARMD victims.

Rats (Leurre-du-Pree, 1981; Leurre-du-Pree and McClain, 1982) and pigs (Whitley et al., 1988) fed zinc deficient diets exhibit pathological changes at the ultrastructural level in both photoreceptors and RPE.

It is believed that in both humans with ARMD and the animal models, that reduction in Zn content in RPE and retina may cause the reduction in visual acuity and pathological changes in retina and RPE.

The purpose of the current study is to evaluate Zn content of RPE in aging human eyes having pathological features of ARMD. The technique that we use is energy dispersive x-ray microanalysis (EDX).

EDX

EDX works by detecting characteristic x-rays of atoms that are bombarded by an electron beam (Reviewed by Chandler, 1973). Each atomic species has a characteristic set of energies that are released when an electron is ejected from its orbital and replaced by an electron from an outer shell. The EDX unit usually displays these energies as an energy spectrum (KeV vs counts) and the peaks are matched with known x-ray energies of all the elements above the atomic number of sodium. Elements under Na in the periodic chart cannot be measured using conventional detectors because their electrons emit x-rays whose energies are too low. The primary advantage of EDX is that the structure being analyzed can be viewed in the electron microscope at the same time.

The intensity of X-rays produced is proportional to the number of atoms of the element present (Russ, 1973). The minimum concentration of ion that can be detected is in the range of 100-500 ppm (0.01-.05%) for most elements. Since the total mass that is usually analyzed in a thin section is relatively small (ca. 10^{-15} to 10^{-16} grams), this limit of concentration would imply that an elemental mass of about 10^{-19} grams could be detected (Russ, 1973). The limit of resolution is also determined by the statistical difference between an energy peak and the background radiation.

The EDX unit can also map a section for particular ionic species by performing a point-by-point raster of the section and recording only those points which emit the energy level(s) of interest. The Kevex model 8000 used in the current study can map up to 16 atomic species simultaneously. In the current report, we our data of is in the form of x-ray maps.

METHODS

The eyes analyzed in the current study were obtained from the North Florida Lions Eye Bank. They were from a 66 year old female. To our knowledge no previous diagnosis of ARMD had been made. However, upon dissecting away the anterior third of the globe and removing the vitreous and neural retina, we noted several large drusen in the macular region of the posterior globe.

Pieces of RPE-choroid were cut from the overlying sclera. These were placed on the surface of highly polished copper plates of cryopliers that had been cooled to liquid nitrogen temperature. The pliers were immediately closed on the specimen and plunged into liquid nitrogen. Frozen tissues were then lyophilized, exposed to acrolein vapors, and infiltrated in epoxy resin. Following identification of regions containing drusen in LM sections, the blocks were trimmed and 150 nm thick sections were cut on an ultramicrotome and mounted on copper slot grids covered with formvar. Specimens were examined in a Hitachi H7000 TEM and then switched to STEM for photography and analysis in the Kevex 8000 EDX unit.

RESULTS

A typical low magnification STEM micrograph of the RPE and choriocapillaris in a region containing a large druse appears in Fig. 1. A degenerating RPE cell is seen in the left of the micrograph; note the lack of apical processes and the washed-out appearance of the cytoplasm. The cell in the right quarter of the micrograph appears normal. The cell in the middle of the micrograph overlies a large druse. The cell appears stretched over the druse causing a significant thinning of cytoplasm. The choriocapillaris appears normal.

An x-ray map from a normal appearing RPE cell that is not associated with a druse appears in Fig. 2. Fig. 2a shows the digitized image of the cell. Figure 2b shows x-ray the map for sulphur and in 2c is the x-ray map for zinc. Note that both S and Zn are in highest concentrations over the melanin granules and in lesser amounts over the cytoplasm. All normal appearing RPE cells that were not associated with drusen mapped in this way for S and Zn.

In Fig 3, similar maps are shown for a degenerating RPE cell, similar to the left cell in Fig. 1. In the center of the digitized image (Fig 3a) is a large vacuole surrounded by normal appearing melanosomes and lipofuscin granules. The x-ray map for S shows high levels of this element primarily in the melanosomes with less in lipofuscin, cytoplasm, and the degeneration vacuole. The pattern of S localization is similar to that in normal RPE. Zinc, however, does not appear to localize in melanosomes at a level much greater than that in the cytoplasm or in the vacuole (Fig 3c). This pattern of S and Zn localization was noted in all degenerating RPE cells which were mapped.

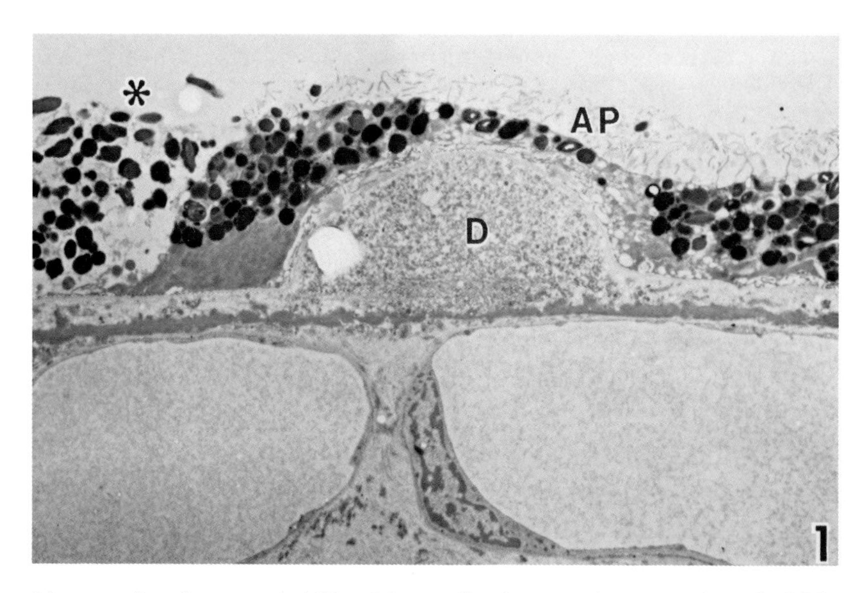

Figure 1. Low magnification electron micrograph of RPE-Choroid. A large druse (D) lies in Bruch's membrane causing the overlying RPE cell to be thin. Both that cell and the cell to its right have apical processes (AP) extending from their surfaces. The RPE cell on the left is undergoing degenerative changes. Note the lack of apical processes (*) and the vacuolated cytoplasm in the degenerating cell to the left.

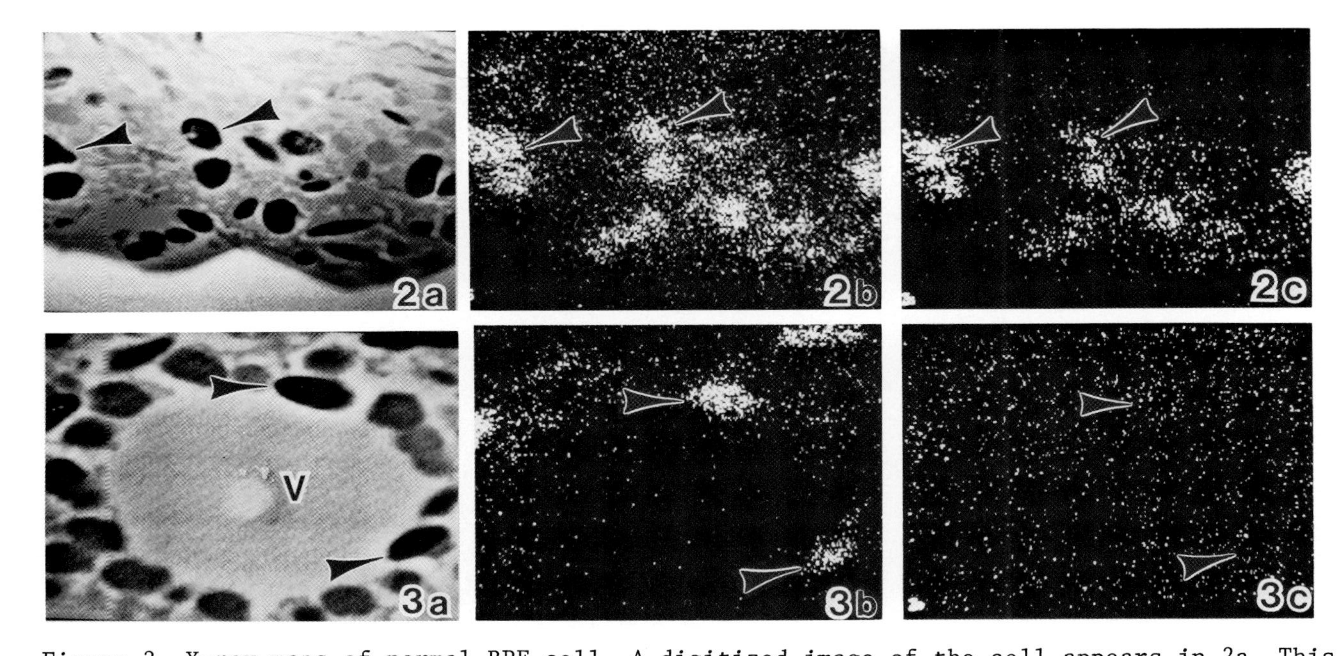

Figure 2. X-ray maps of normal RPE cell. A digitized image of the cell appears in 2a. This corresponds to the maps for sulphur and zinc in 2b and 2c, respectively. The apical surface is facing down. Note the same electron dense melanosomes (arrowheads) in all three figures. Sulphur (2b) and zinc (2c) occur predominantly in melanosomes with some in cytoplasm.

Figure 3. Degenerating RPE. A large vacuole (V) appears in the digitized image of the cell in 3a. Melanosomes (arrowheads) containing high amounts of S (3b) lack Zn (3c). The degeneration vacuole contains Zn but lacks appreciable amounts of S.

DISCUSSION

EDX has been used in other studies of visual system function and disease. Panessa and Zadunaisky (1981) analyzed Ca reservoirs in ocular tissues and reported that cells containing pigment granules (such as the iris, ciliary body, choroid and RPE) sequester Ca under physiological conditions and that melanosomes in these tissues contain up to 10 times more Ca than adjacent cytoplasm. Staple et al. (1987) and Fain and Schroder (1985) measured Ca levels in photoreceptor and both found that Ca levels do not change in response to illumination. However, in the octopus photoreceptor cell, Takagi et al. (1987) did find slight increases of Ca concentration with light stimulation but larger changes in levels of Na, K, and Cl. Ulshafer et al. (1987) found that the small spherical bodies in human drusen are composed of Ca and P, while pleomorphic bodies have a composition similar to cell cytoplasm. Allen et al. (1987) used EDX to document progression of intraocular ossification in pthysis bulbi in end-stage retinal degeneration in an animal model of hereditary blindness. Most of these studies employed EDX with a scanning electron microscope (SEM).

Recent findings that Zn levels are reduced in patients with ARMD prompted Newsome and collaborators (1988) to perform a trial using zinc supplement in patients with ARMD. They reported that Zn does indeed improve visual acuity in many cases of ARMD. Other studies have linked retinitis pigmentosa to altered trace elements in the blood (Bastek et al., 1977; Gahlot and Khosla, 1976; Rao et al., 1981).

The current study was performed on semi-thin sections in a STEM. Both S and Zn localized mainly to melanosomes in normal RPE while Zn levels were notably less in melanosomes of degenerating cells, despite normal levels of S. These data therefore support the hypothesis that zinc levels are lower in degenerating RPE cells. However, whether Zn can be restored in affected cells, has yet to be shown.

ACKNOWLEDGEMENTS

This study was supported in part by a Non-restrictive Departmental Grant from Research to Prevent Blindness, Inc. Dr Ulshafer is a Research to Prevent Blindness Scientific Investigator.

REFERENCES

Allen CB, Ulshafer RJ, Ellis EA and Woodard JC. 1987. SEM analysis of intraocular ossification in advanced retinal disease. Scanning Microsc 1:233-239.

Bastek J, Bogden J, Cinotti A, TenHove W, Stephens G, Markopoulos M and Charles J. 1977. Trace metals in a family with sex-linked retinitis pigmentosa. In: Retinitis Pigmentosa: Clinical Implications of Current Research: Advances in Experimental Medicine and Biology, Landers MB III, Wolbaraht ML, Dowling JE, and Laties AM, Eds. Plenum Press, New York, pp 43-50.

Bowness JM, Morton RA and Shakir MH. 1982. Distribution of Copper and Zinc in mammalian eyes. Occurrence of metals in melanin fracture from eye tissue. Biochem J 51:521.

Burns RP and Feeney-Burns L. 1980. Clinico-morphologic correlations of drusen of Bruch's membrane. Tr Am Ophthalmol Soc LXXCIII:206-225.

Chandler JA (1973). The use of wavelength dispersive X-ray-microanalysis in cytochemistry. In Wisse E, Daems W Th, Molenaar I, van Duijn P (eds): "Electron Microscopy and Cytochemistry". North Holland Publishing Co., Amsterdam. pp. 203-222.

Eckhert CD. 1979. A comparative study of the concentrations of Ca, Fe, Zn, and Mn in ocular tissues. Fed Proc 38:872.

Fain GL and Shroder WH. 1985. Calcium content and calcium exchange in dark-adapted toad rods. J Physiol 368:641-665.

Feeney-Burns L and Ellersieck MR. 1986. Age related changes in the ultrastructure of Bruch's membrane. Am J Ophthalmol 101:686-697.

Gahlot DK and Khosla PK. 1976. Copper metabolism in retinitis pigmentosa. Br J Ophthalmol 60:770-774.

Galin MA, Nano HD and Hall T. 1962. Ocular zinc concentrations. Invest Ophthalmol 1:142.

Gilkey JC and Staehelin LA. 1986. Advances in ultrarapid freezing for the preservation of cellular ultrastructure. J Elec Microsc Tech 3:177-210.

Karcioglu ZA. 1982. Zinc in the eye. Surv Ophthalmol 27:114.

Leurre-du-Pree AE. 1981. Electron-opaque inclusions in the rat retinal pigment epithelium after treatment with chelators of zinc. Invest Ophthalmol Vis Sci 21:1-9.

Leurre-du-Pree AE and McClain CJ. 1982. The effect of severe zinc deficiency on the morphology of the rat retinal pigment epithelium. Invest Ophthalmol Vis Sci 23:425-434.

Liebowitz HM, Krueger DE and Maunder LR. 1980. Framington Eye Survey. VI. Macular degeneration. Surv Ophthalmol 24:428-435.

Miller ED, Swartz M, Leone NC, Bennecoff TA and Newsome DA. 1986. Serum zinc concentrations in macular degeneration. Invest Ophthalmol Vis Sci 27(Suppl):20.

Newsome DA, Swartz M, Leone NC, Elston RC, and Miller E. 1988. Oral zinc in macular degeneration. Arch Ophthalmol 106:192-198.

Panessa BJ and Zadunaisky JA. 1981. Pigment granules: A calcium reservoir in the vertebrate eye. Exp Eye Res 32:593-604.

Rao SS, Satapathy M and Sitaramayya A. 1981. Copper metabolism in retinitis pigmentosa. Br J Ophthalmol 65:127-130.

Russ JC (1973). Microanalysis of thin sections in the TEM and STEM using energy-dispersive X-ray analysis. In Wisse E, Daems W Th, Molenaar I, van Duijn P (eds): "Electron Microscopy and Cytochemistry". North-Holland Publishing Co., Amsterdam. pp. 223-228.

Staple J, Kisly A and Mata M. 1987. Ultrastructural distribution of calcium within the rat retinal rod outer segment. Invest Ophthalmol Vis Sci 28(Suppl):342.

Takagi M, Salehi SA, Nakagaki I and Sasaki J. 1987. X-ray microanalysis of the effect of illumination on the ionic composition of the octopus photoreceptors. Photochem and Photobiol 45:651-656.

Ulshafer RJ and Allen CB. 1988. Elemental mapping of human RPE and Bruch's membrane via X-ray microanalysis and STEM. Invest Ophthalmol Vis Sci 29(Suppl):In Press.

Ulshafer, RJ, Allen CB, Nicolaissen J and Rubin ML. 1987a. Scanning electron microscopy of human drusen. Invest Ophthalmol Vis Sci 28:683-689.

Whitley RD, Samuelson DA, Hendricks DG, Olsen AE, Shupe JL, and Leone NC. 1988. Ocular effects of chronic low zinc diet in the pig. Invest Ophthalmol Vis Sci 29 (Suppl):289.

II. INHERITED RETINAL DEGENERATIONS IN LABORATORY ANIMALS

In the introduction to the first section, it was mentioned that the strengths of mutant laboratory animals as models of human retinal degenerations have made them important choices for experimental investigations. The twenty-three papers that constitute this largest section illustrate the diversity, not only of the experimental approaches currently being applied, but also of the models themselves. Investigators use the tools of cytopathology, cytochemistry, immunocytochemistry, biochemistry, electrophysiology, environmental modification, administration of drugs, and molecular genetics to study eight different forms of inherited retinal degeneration in mice, rats, dogs, cats, chicks, and *Drosophila*.

Many of the papers describe comparative studies of two or three of the mutants. Such studies are important because each phenotypic characteristic and degeneration mechanism ultimately needs to be compared to those of the many forms of retinal degeneration in human patients. However, the comparative studies and the overall diversity of these papers make the section difficult to arrange in a simple, coherent fashion. As far as possible, studies on the same mutants are grouped together. Those papers describing investigations on rodent models are presented first, followed by those on dogs and cats, chicks, and *Drosophila*. It is hoped that at some point the section on human retinal degenerations will be equally difficult to arrange.

Inherited and Environmentally Induced
Retinal Degenerations, pages 143–153

ASSIGNMENT TO MOUSE CHROMOSOMES OF CANDIDATE GENES FOR THE rd MUTATION

Debora B. Farber[*], Christine A. Kozak[#] and Michael Danciger[*+]

[*]Jules Stein Eye Institute, UCLA School of Medicine, Los Angeles, CA 90024, [+]Loyola Marymount University, Los Angeles, CA 90045, and [#]National Institute of Allergy and Infectious Diseases, Bethesda, MD 20891

INTRODUCTION

Mice carrying the rd gene are affected with an autosomal recessive disease which causes the degeneration of the retinal photoreceptor cell layer. The rd gene is a single gene located on chromosome 5 (Sidman and Green, 1965). The first morphological sign of the disease, swelling of the mitochondria of the photoreceptor inner segments, appears at postnatal day 8 (Lasansky and DeRobertis, 1960; Shiosi and Sonohara, 1969), a time when the visual cells have already begun to form outer segments. From this point on the degeneration process continues rapidly, and by 3-4 weeks of life most of the photoreceptor cells have died (LaVail and Sidman, 1974).

As early as postnatal day 6, the level of cGMP in the rd retina begins to rise above normal (Farber and Lolley, 1974). This is due to a deficiency in the activity of the rod-specific enzyme cGMP-phosphodiesterase (cGMP-PDE) (Farber and Lolley, 1976), a heterotrimer whose activation and deactivation involves several different proteins of the visual cell (for a review, see Farber and Shuster, 1986). The recognition of the abnormal cGMP-PDE activity has led to the proposal that a lesion in the gene for one of the three subunits of the enzyme itself, or a lesion in the gene for one of the proteins involved directly or indirectly in the activation of cGMP-PDE, may be the cause of the rd photoreceptor degeneration. Consequently, we have been screening for the rd gene by determining the chromosome location in the mouse of the genes for several proteins associated with the photoreceptor cGMP cascade. Any gene assigned to mouse chromosome 5 would be a candidate for the site of the rd mutation and would require further study.

In this report we describe the mouse chromosome mapping of cDNAs for the γ-subunit of cGMP-phosphodiesterase (cGMP-PDEγ), the 48KDa protein or S-antigen (48K) and the α-subunit of rod transducin (Tα1).

MATERIALS AND METHODS

Hamster-mouse somatic cell hybrids were derived from the fusion of E36 Chinese hamster cells with peritoneal or spleen cells from BALB/c or NFS.Akv-2 mice. The chromosome content of most hybrids was determined by trypsin-Giemsa banding followed by staining with Hoechst 33258; all hybrids were typed for specific marker loci on up to all 20 of the mouse chromosomes (Kozak et al., 1975; Kozak et al., 1977; Kozak and Rowe, 1979; Hoggan et al., 1988).

The rat-mouse hybrid F(11)J is a hybrid between a rat hepatoma line and mouse microcells. This line retains only mouse chromosome 11 as determined by staining with Hoechst 33258 and was kindly provided by R. E. K. Fournier (Killary and Fournier, 1984).

The cGMP-PDEγ cDNA was cloned from a mouse retinal library, has 482 base pairs (bp) and is comprised of a 261 bp coding region flanked by a 121 bp untranslated region at the 5′ end and a 100 bp untranslated region at the 3′ end (Tuteja and Farber, 1988). The Tα1 cDNA is of bovine origin and has 2,190 bp with a 1,050 bp coding region followed by a 1,140 bp untranslated region at the 5′ end, and was kindly provided by Dr. Melvin I. Simon, California Institute of Technology. This clone has been described previously (Tanabe et al., 1985). The 48K cDNA has 1,532 bp with a 1,209 bp coding region flanked by a 159 bp untranslated region at the 3′ end, and a 164 bp untranslated region at the 5′ end, and was cloned from a mouse retinal cDNA library in the laboratory of Dr. Toshimichi Shinohara (Tsuda et al., 1988).

Labeling of the cDNA probes was with [α^{32}P]-dCTP (3000 Ci/mole) by the random priming method (Feinberg and Vogelstein, 1983).

DNAs from the Chinese hamster-mouse hybrids and the rat-mouse hybrids were prepared by published procedures (Linnenbach et al., 1980) and digested with the appropriate restriction endonuclease. 6 µg of DNA per lane were electrophoresed in 1.2% agarose gels and then transblotted to a nylon membrane by the technique of Southern (1975).

For Tα1 and 48K the blots were prehybridized for 3 to 6 hours at 65°C in a solution containing 7.5% SDS (sodium dodecyl sulfate), 0.5 M phosphate buffer, pH 7.0, 1 mM EDTA and 1% bovine serum albumin. Hybridization was carried out at 60°C overnight in the same solution as that for prehybridization with the addition of radioactive probe. The times and temperatures used in the prehybridization and hybridization with cGMP-PDEγ were the same as above, but the prehybridization solution contained 3x SSC (1x SSC is 150 mM NaCl and 15 mM sodium citrate, pH 7.0), 2x Denhardt's solution, 0.2% SDS and 500 µg/ml boiled and sheared salmon sperm DNA, and the hybridization solution contained 6x SSC, 40 mM Tris pH 7.4, 4x Denhardt's, 0.2% SDS and 200µg/ml boiled and sheared salmon sperm DNA.

After hybridization, the blots were washed for 30 min at 37° C in 2x SSC + 0.1% SDS, two times for 20 min at 57°C in 2x SSC + 0.1% SDS, and two times for 20 min at 57° C in 0.2x SSC + 0.1% SDS (for cGMP-PDEγ and 48K) or 0.4x SSC + 0.1% SDS (for Tα1). The blots were then exposed to X-ray film with intensifier screens at -80° C for 3 to 14 days.

RESULTS AND DISCUSSION

A panel of 19 independent hamster-mouse somatic cell hybrid DNAs was first typed by Southern blot hybridization using the cGMP-PDEγ cDNA probe. Digestion of the control mouse and hamster DNAs with HindIII and hybridization with ^{32}P-labeled probe gave two bands of 2.1 and 6.6 kilobases (kb) in the mouse and bands of 4.6 and 8.0 kb plus a weak band of 6.6 kb in the hamster (Figure 1, lanes 1-3). The 2.1 kb band was used to score for the presence of cGMP-PDEγ sequences in mouse chromosomes. The results in lanes 4-10 of Figure 1 are representative of all 19 of the somatic

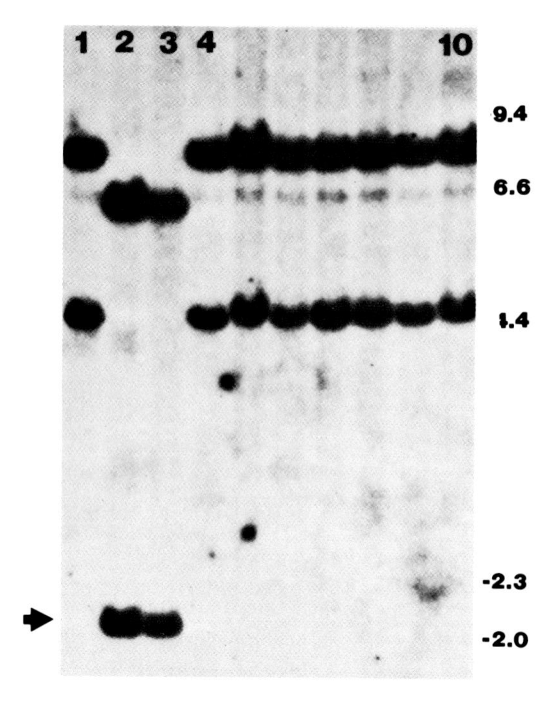

Figure 1. Autoradiogram of a Southern blot of Chinese hamster-mouse somatic cell hybrid DNAs digested with HindIII and hybridized with the cGMP-PDEγ cDNA probe; each lane has 6 μg of DNA. The hamster control is in lane 1, the NFS.Akv-2 mouse control is in lane 2 and the BALB/c mouse control is in lane 3. Seven representative somatic cell hybrid DNAs are shown in lanes 4-10. The numbers on the right (in kb) mark the positions of λ DNA fragments produced by digestion with HindIII. The arrow points to the 2.1 kb mouse band which was used to score for the presence of cGMP-PDEγ sequences in the hybrids. Reprinted with permission from Danciger et al., 1989; Copyright Academic Press Limited.

cell hybrid DNAs analyzed; none shows the unique 2.1 kb mouse band. Analysis of these data shows perfect concordance for the presence of cGMP-PDEγ sequences on chromosome 11, while all other mouse chromosomes show at least three discordancies. However, the only mouse chromosome not present in any of the hybrids is chromosome 11, which is preferentially lost in hamster-mouse cell hybrids (Kozak and Ruddle, 1977).

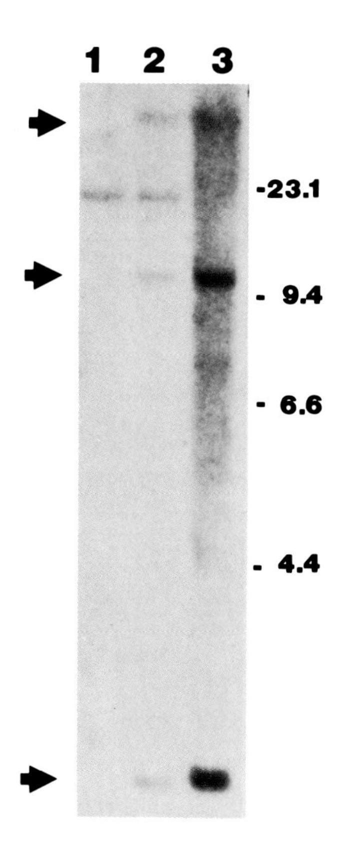

Figure 2. Autoradiogram of a Southern blot of a rat-mouse hybrid DNA digested with BamHI. The rat control DNA is in lane 1 and the C57Bl mouse control DNA is in lane 3. The hybrid DNA containing only mouse chromosome 11 is shown in lane 2. The numbers on the right (in kb) mark the positions of λ DNA fragments produced by digestion with HindIII. The arrows point to the bands present in the mouse and hybrid DNAs, but absent from the rat DNA. Reprinted with permission from Danciger et al., 1989; Copyright Academic Press Limited.

In order to confirm the chromosome assignment of cGMP-PDEγ, we used a rat-mouse somatic cell hybrid whose only mouse chromosome is 11. Figure 2 shows that when digested with BamHI and hybridized with ^{32}P-labeled probe, the rat-mouse hybrid shows the presence of cGMP-PDEγ sequences. Specifically, the rat control DNA (lane 1) showed only one band of 22 kb; the mouse control DNA (lane 3) showed three bands of >23 kb, 10.5 kb and 3.0 kb; and the hybrid DNA showed the rat band and the three mouse bands. This confirms the assignment of the cGMP-PDEγ to mouse chromosome 11.

Twenty three hamster-mouse somatic cell hybrid DNAs were typed by blot hybridization using the Tα1 cDNA probe. Digestion of the control mouse and hamster DNAs with EcoRI and hybridization with ^{32}P-labeled probe gave one band in hamster DNA of 23 kb and one band in mouse DNA of 5.8 kb (Figure 3, lanes 1 and 2). Typical results of hybrid DNAs, either positive or negative for the presence of Tα1 sequences, are shown in lanes 3-5 of Figure 3. Eight of 23 hybrids tested contained the 5.8 kb mouse band. With these data we found no discordancies for the presence of Tα1 on mouse chromosome 9, and at least 3 discordancies for its presence on any other mouse chromosome.

Twenty five hamster-mouse somatic cell hybrid DNAs were digested with HindIII and analyzed for sequence homology to the 48K cDNA by blot hybridization. The Chinese hamster control DNA showed 6 bands of 10.0, 6.4, 5.5, 3.4, 2.9 and 1.6 kb, while both of the mouse control DNAs showed 6 bands of 19.0, 7.0, 6.0, 4.8, 3.1 and 1.6 kb and a seventh faint band of 1.4 kb (Figure 4, lanes 1-3). The results in lanes 4-7 of Figure 4 are representative of the somatic cell hybrids tested. All mouse bands except the 1.6 and 1.4 kb mouse bands could be used to score for the presence of mouse 48K DNA sequences in the hybrids; the 1.6 kb band could not be distinguished from the hamster 1.6 kb band and the 1.4 kb band was often too faint to score. Ten of the 25 hybrids were positive for the presence of mouse 48K DNA sequences and showed all 5 scoreable mouse bands (top five arrows in Figure 4). With these results we found at least 6 discordancies for the presence of 48K DNA sequences in every mouse chromosome except chromosome 1 for which there are two discordancies. Both discordancies occurred in hybrids that had chromosome 1, but were negative for 48K DNA sequences. However, both hybrid clones

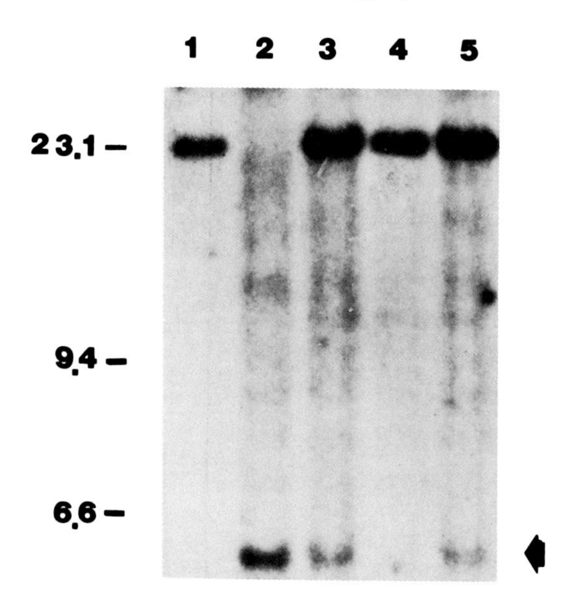

Figure 3. Autoradiogram of a Southern blot of hamster-mouse somatic cell hybrid DNAs digested with EcoRI and hybridized with the Tα1 cDNA probe; each lane has 6 μg of DNA. The Chinese hamster control is in lane 1 and one of the mouse controls, NFS.Akv-2, is in lane 2. Three representative somatic cell hybrid DNAs are shown in lanes 3-5. The arrow points to the 5.8 kb mouse band which was used to score for the presence of Tα1 sequences in the hybrids. The numbers on the left (in kb) mark the positions of λ DNA fragments digested with HindIII.
Reprinted with permission from Danciger et al., 1989a; Copyright Academic Press Limited.

had only fragments of mouse chromosome 1 with a common region missing, the first one third near the centromere (Kozak, 1983). Consequently, the locus of the gene for 48K is on mouse chromosome 1.

In summary, our studies have clearly assigned the gene for the γ-subunit of photoreceptor-specific cGMP-PDE to mouse chromosome 11, the gene for the α-subunit of retinal

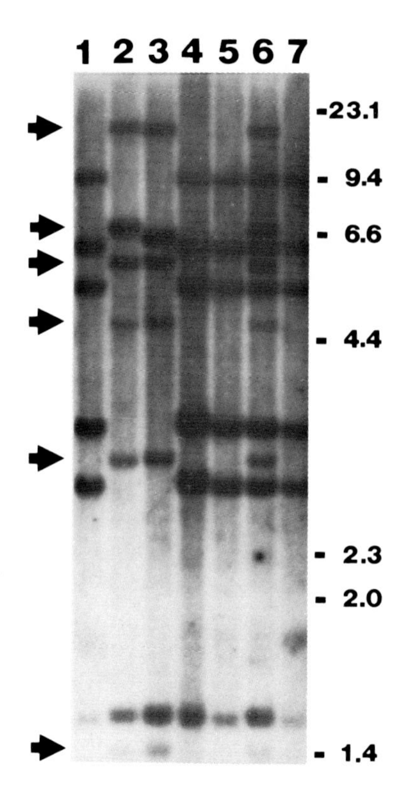

Figure 4. Autoradiogram of a Southern blot of hamster-mouse somatic cell hybrid DNAs digested with HindIII and hybridized with the mouse 48K cDNA probe; each lane has 6 μg of DNA. The Chinese hamster control is in lane 1 and the BALB/c and NFS.Akv-2 mouse controls are in lanes 2 and 3, respectively. Four representative somatic cell hybrid DNAs are shown in lanes 4-7. The top five arrows point to the 19.0, 7.0, 6.0, 4.8, and 3.1 kb mouse bands used to score for the presence of S-antigen sequences in the hybrids. The sixth arrow points to the 1.4 kb mouse band which is faint and, therefore, has not been used to score. The numbers on the right (in kb) mark the positions of λ DNA fragments produced by digestion with HindIII, and the one ΦX-174 DNA fragment (1.4 kb) produced by digestion with HaeIII. There is a slight polymorphism between the two mouse controls in the second mouse band from the top: the BALB/c band is 7.0 kb while the NFS.Akv-2 band is 6.6 kb.

rod transducin to mouse chromosome 9, and the gene for the retinal 48KDa protein (S-antigen) to mouse chromosome 1. Since the gene containing the rd mutation is on mouse chromosome 5, none of the three genes we mapped is the site of this mutation. We are currently studying the cDNAs of several other proteins involved in cGMP metabolism to determine if any of these maps to mouse chromosome 5.

ACKNOWLEDGEMENTS

This work was supported by National Institutes of Health grants EY02651 (D.B.F.) and Core Grant EY00331; a Center grant from the Retinitis Pigmentosa Foundation Fighting Blindness; and a grant from the George Gund Foundation.

REFERENCES

Danciger M, Tuteja N, Kozak CA, Farber DB (1989). The gene for retinal cGMP-phosphodiesterase is on mouse chromosome 11. Exp Eye Res 48, in press.

Danciger M, Kozak CA, Farber DB (1989a). The gene for the α-subunit of retinal rod transducin is on mouse chromosome 9. Genomics 4, in press.

Farber, DB, Lolley RN (1974). Cyclic guanosine monophosphate: Elevation in degenerating photoreceptor cells of C3H mouse retina. Science 186:449-451.

Farber DB, Lolley RN (1976). Enzymatic basis for cyclic GMP accumulation in degenerative photoreceptor cells of mouse retina. J Cyclic Nucleotide Res 2:139-148.

Farber DB, Shuster TA (1986). Cyclic Nucleotides in Retinal Function and Degeneration. In Adler R, Farber D (eds): "The Retina: A Model for Cell Biology Studies, Part 1", Orlando, FL: Academic Press, pp 239-96.

Feinberg AP, Vogelstein B (1983). A technique for radiolabeling restriction endonuclease fragments to high specific activity. Anal Biochem 132:6-13.

Hoggan MD, Halden NF, Buckler CE, Kozak CA (1988). Genetic mapping of the mouse c-fms proto-oncogene to chromosome 18. J Virol 62:1055-1056.

Killary AM, Fournier REK (1984). A genetic analysis of extinction: trans-Dominant loci regulate expression of liver-specific traits in hepatoma hybrid cells. Cell 38:523-534.

Kozak CA (1983). Genetic mapping of a mouse chromosomal locus required for mink cell focus-forming virus replication. J Virol 48:300-303.
Kozak CA, Lawrence JB, Ruddle FH (1977). A sequential staining technique for the chromosomal analysis of the interspecific mouse/hamster and mouse/human somatic cell hybrids. Exp Cell Res 105:109-117.
Kozak C, Nichols E, Ruddle FH (1975). Gene linkage in the mouse by somatic cell hybridization: assignment of adenosine phosphoribosyl-transferase to chromosome 8 and alpha-galactosidase to the X chromosome. Somatic Cell Genet 1:371-382.
Kozak CA, Rowe WP (1979). Genetic mapping of the ecotropic murine leukemia virus-inducing locus of BALB/c mouse to chromosome 5. Science 204:69-71.
Kozak CA, Ruddle FH (1977). Assignment of the genes for thymidine kinase and galactokinase to *Mus musculus* chromosome 11 and the preferential segregation of this chromosome in Chinese hamster/mouse somatic cell hybrids. Somatic Cell Genet 3:121-133.
Lasansky A, DeRobertis E (1960). Submicroscopic analysis of the genetic dystrophy of visual cells in C3H mice. J. Biophys Biochem Cytol 7:679-684.
LaVail MM, Sidman RL (1974). C57Bl/6J mice with inherited retinal degeneration. Arch Ophthalmol 91:394-400.
Linnenbach A, Huebner K, Croce CM (1980). DNA-transformed murine teratocarcinoma cells: regulation of expression of simian virus 40 tumor antigen in stem versus differentiated cells. Proc Natl Acad Sci 77:4875-4879.
Shiosi Y, Sonohara O (1969). Studies on retinitis pigmentosa. XXVI. Electron microscopic aspects of the early retinal changes in inherited dystrophic mice. Jap J Ophthalmol 72:299-313.
Sidman RL, Green MC (1965). Retinal degeneration in the mouse. J Hered 56:23-29.
Southern EM (1975). Detection of specific sequences among DNA fragments separated by gel electrophoresis. J Mol Biol 98:503-517.
Tanabe T, Nukada T, Nishikawa Y, Sugimoto K, Suzuki H, Takahashi H, Noda M, Haga T, Ichiyama A, Kangawa K, Minamino N, Matsuo H, Numa S (1985). Primary structure of the α-subunit of transducin and its relationship to *ras* proteins. Nature 315:242-245.
Tsuda M, Syed M, Burga K, Whelan JP, McGinnis JF, Shinohara T (1988) Structural analysis of mouse S-antigen cDNA. Gene, in press.

Tuteja N, Farber DB (1988) γ-Subunit of mouse retinal cyclic-GMP phosphodiesterase: cDNA and corresponding amino acid sequence. FEBS Lett 232:182-186.

Inherited and Environmentally Induced
Retinal Degenerations, pages 155–168

FAILED ASSEMBLY OF PHOSPHODIESTERASE COMPLEX IN DEVELOPING PHOTORECEPTORS OF rd MICE

Richard N. Lolley and Rehwa H. Lee

UCLA School of Medicine, Jules Stein Eye
Institute, 90024, and VA Medical Center,
Sepulveda, CA 91343

INTRODUCTION

The rd (retinal degeneration) mouse carries a recessive gene on chromosome 5 (Sidman and Green, 1965) which causes the rod photoreceptor population of the retina to degenerate during the postnatal period when visual cells differentiate (Noell, 1965). Expression of the rd gene produces cellular degeneration only within the retina and preferentially to rod photoreceptors. Coincident with rod degeneration, the cones (approximately 3% of the photoreceptor population) become changed morphologically, but they survive beyond the postnatal period (Carter-Dawson et al., 1978). The neurons of the inner retina too become altered upon the death of rod photoreceptors and some degenerate with advancing age.

Activation and transcription of the rd gene probably occurs in rod photoreceptors in the days which immediately precede the onset of morphological deterioration, but this has yet to be demonstrated. Even though there have been several attempts to clarify the role of the rd gene and to identify its expression product, these have mostly been able to demonstrate what the rd does not do. However, there are indications that the rd gene has an adverse affect on the phototransduction cascade.

The earlier discovery that cyclic GMP accumulates in rd photoreceptors from the time they begin to differentiate (Farber and Lolley, 1974) is now refocused onto what causes this accumulation. One explanation would be that the phosphodiesterase (PDE) enzyme is normal, but it is not activated by the phototransduction cascade and inactivity of the phosphodiesterase enzyme allows cyclic GMP to accumulate. An alternative explanation would be that the phosphodiesterase enzyme is defective (or not produced) and its inactivity allows cyclic GMP to accumulate. We and others have investigated the presence and apparent functional capabilities of the individual components which initiate the phototransduction cascade, but, with the exception that bleached rhodopsin is not phosphorylated in a normal manner, in vitro (Shuster and Farber, 1986), the transcription and protein content of transducin, the 48kDa protein and rhodopsin are not noticeably abnormal when the smaller cellular volume of rd photoreceptors is considered (Navon et al., 1987; Bowes et al., 1988).

Should the bleaching of rhodopsin and its activation of transducin prove as operative as we suggest, then the molecular composition and functional capabilities of the phosphodiesterase complex need examination. Such is the theme of this review which chronicles our investigations of the phosphodiesterase complex of normal photoreceptors and how PDE-complex assembly fails in rd photoreceptors.

PDE COMPLEX IN DEVELOPING AND MATURE PHOTORECEPTORS OF THE NORMAL MOUSE RETINA

Attention has focused on the phosphodiesterase (PDE) of normal photoreceptors since the PDE complex was identified as a component of the proposed phototransduction cascade which links photon-capture by rhodopsin with ion-channel closure in the rod plasmalemma (Stryer, 1986). The bleaching of rhodopsin activates the GTP-binding protein, transducin, which interacts with the phosphodiesterase complex. GTP-alpha transducin binds apparently to the inhibitory subunit of the PDE complex and the interaction relieves inhibitory constraints on the enzyme. Released from inhibition, the PDE complex rapidly hydrolyzes cyclic GMP. The fall in cyclic GMP concentration results in the removal of cyclic GMP from binding sites on ion-channels of the plasma membrane, closure of the channels and blockage of ion movement into illuminated photoreceptors.

The blockage of ion movement into illuminated

photoreceptors is brief when the stimulus is a low intensity flash of light, lasting only a few seconds. The duration of the response depends on several factors including the down-regulation of the activated PDE complex, but the turn-off of the light response is less well understood than the activation phase. A two-stage control mechanism is proposed in which phosphorylation of bleached rhodopsin (Kuhn and Dreyer, 1972) diminishes the production of GTP-alpha transducin and thus reduces the amount of activated transducin for interaction with the PDE complex. The phosphorylation of rhodopsin may also favor the binding of a 48kDa protein (Kuhn et al., 1984) which could interfere with the binding and activation of transducin. Secondly, the time during which activated transducin remains associated with the PDE complex is limited by the intrinsic GTPase activity of alpha transducin (Fung, 1983). The conversion of GTP to GDP leads to the dissociation of alpha transducin from the PDE complex and the reconstitution of inhibitory constraints within the PDE complex. Together, rhodopsin phosphorylation and transducin GTPase activity may be sufficiently rapid to explain the down regulation of PDE activation.

The PDE complex is a highly asymmetrical aggregate with an axial ratio of about 9:1 (Gillespie and Beavo, 1988). It is composed of two large and two small polypeptides with the largest subunit (88kDa) designated alpha, the other large subunit (84kDa) beta and the smallest subunits (11kDa) gamma. The PDE complex is apparently associated with ROS membranes, in situ, and ROS membrane is essential for activation by transducin. The PDE complex can be solubilized from ROS membranes in hypotonic (usually magnesium free) buffers (Baehr et al., 1979) in a mostly inactive form. Several methods can be utilized to activate the soluble PDE complex, including incubation with purified GTP-alpha transducin, dilution, partial proteolysis and incubation with polyanions or polycations (Hurwitz et al., 1985). Each of these procedures appear to relieve inhibitory constraints that are imposed by the gamma subunit by altering the conformation of the complex or by degrading the 11kDa inhibitory subunit. Much of our understanding of the PDE complex comes from work with bovine ROS from which milligram quantities of complex can be obtained by standard biochemical techniques.

The PDE complex of cone photoreceptors is similar in size and asymmetry to that of rod photoreceptors (Booth et al, 1986; Gillespie and Beavo, 1988). However, the cone PDE complex is a homo-dimer of only one large subunit (94kDa). Three other small polypeptides (15kDa, 13kDa and 11kDda) are associated with the complex, one of which (11kDa) may be identical to the gamma subunit of the rod PDE complex (Gillespie and Beavo, 1988). The cone PDE complex appears also to share functional characteristics with the rod PDE complex because it hydrolyzes preferentially cyclic GMP and it can be activated, in vitro, with GTP-alpha transducin from bovine rod photoreceptors as well as by histone and partial proteolysis (Hurwitz et al., 1985). Since cones hyperpolarize upon expose to light, it is proposed that cones utilize a cyclic GMP cascade that is analogous to that described for rods.

The genes for the individual PDE subunits of the rod PDE complex are activated in mice during the postnatal period when rod photoreceptors differentiate morphologically (6 to 8 postnatal days) and begin to form rod outer segments. Expression of the genes is implicated both by the onset and subsequent increase in the activity of PDE (Farber and Lolley, 1976) and by the detection and subsequent increase in immunoreactivity with antibodies against the PDE complex of bovine ROS (Fig. 1). The immunoreactive band at 84-88kDa represents the unresolved alpha and beta subunits of the PDE complex which can be resolved under different conditions of electrophoresis. The increase in 84-88kDa immunoreactivity parallels closely the developmental increase in basal and histone-activated PDE activity as measured in retinal homogenates (Lee et al., 1985). Whereas, the accumulation of gamma-subunit was not readily demonstrated in these studies, its association with the PDE complex is implicated at all ages by the ability of histone to stimulate PDE activity; by immunoreactivity with the monoclonal antibody, ROS-1, which recognizes the PDE complex only when the gamma subunit is present (Lee et al., 1985; Hurwitz et al., 1984); and by the normal size (170kDa) and subunit composition of the PDE complex from normal mouse retinas (See Fig. 2 and Fig. 3).

The apparent molecular mass of the native PDE complex of developing and mature photoreceptors of normal mice was determined by sucrose density-gradient centrifugation (Lee

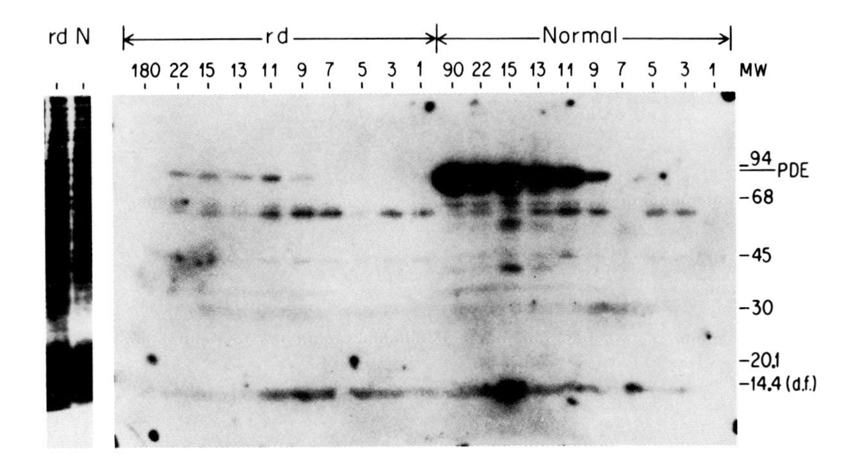

Figure 1. Autoradiogram from the immunoblot of developing retinas of normal and rd mice. Samples containing 40 microgram of protein from retinal homogenates of developing normal or rd mice were subjected to SDS polyacrylamide gel electrophoresis and Western blot analysis using our anti-PDE antibodies. The PDE-immunoreactive peptides were visualized by incubation of the immunoblot with radioactive protein A followed by autoradiography. The postnatal age (days) of the retinas used in each sample is indicated above each lane. The molecular weights (MW) of the proteins were calibrated with standards purchased from Pharmacia. d.f. = dye front. The two left lanes contain 40 microgram of coomassie blue stained retinal proteins from 11-day normal (N) or rd (rd) mice.

et al. 1988). In each gradient, the sedimentation profile for PDE was established by measuring in each fraction of the gradient both histone-activated PDE activity and PDE immunoreactivity with our polyclonal antibodies against bovine-ROS PDE. Histone-activated PDE activity was found exclusively with PDE immunoreactive peptides that sedimented as a complex with an apparent molecular mass of 170kDa (Figs. 2 and 3). Furthermore, the amount of 170kDa complex and the level of PDE activity increased in parallel during the maturation of photoreceptor cells in the normal mouse retina. Therefore, in normally developing retinas,

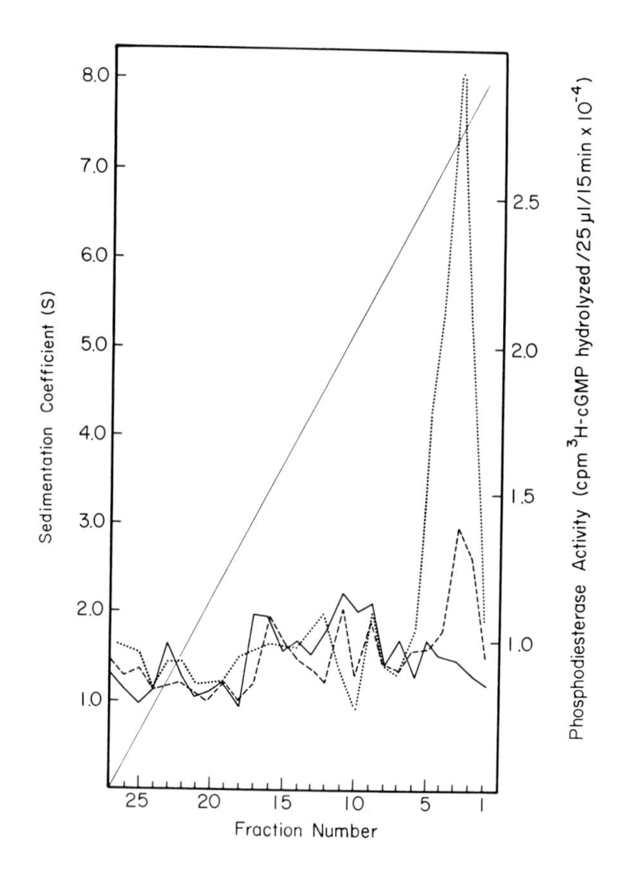

Figure 2. Sedimentation profile of histone-activated PDE activities in the retinas of normal and *rd* mice. Hypotonic extracts from the retinas of adult normal (·····), 9-days old normal (– –), and 9-days old *rd* (——) mice were centrifuged on 5-20% linear sucrose gradients. Each gradient fraction was assayed for histone-activated PDE activity by measuring the hydrolysis of ^{3}H-cyclic GMP in the presence of 1 mg/ml histone. Fraction #1 is the bottom fraction of the gradient.

the postnatal increase in photoreceptor PDE activity results from the accumulation of the 170kDa PDE complex.

During the above sedimentation studies, we observed that a small fraction of PDE-immunoreactive polypeptides migrated with an apparent molecular mass that was less than that of the 170kDa PDE complex (Lee et al., 1988). The

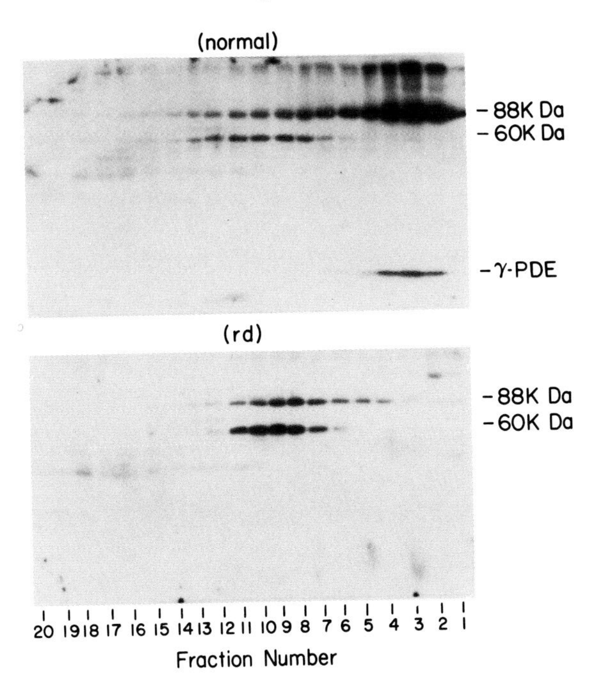

Figure 3. Sedimentation profile of the PDE-immunoreactive peptides in normal and rd retinas. Hypotonic extracts were prepared from six 9-days old normal and six 9-days old rd retinas. Then, each individual extract was centrifuged on identical 5-20% sucrose density gradients. Corresponding fractions from each of the normal or rd gradients were pooled and the proteins in each pooled fraction were precipitated by trichloroacetic acid before being subjected to SDS electrophoresis and Western blot analysis using our anti-PDE antibodies. The PDE-immunoreactive peptides were visualized by autoradiography after incubation of the immunoblot with radioactive protein A. Fraction #1 is the bottom fraction of the gradient.

average molecular mass of these PDE-immunoreactive polypeptides was 105kDa and high resolution SDS gel electrophoresis showed the presence in this fraction of both the alpha and beta subunits of PDE. Usually, the slower-sedimenting polypeptides do not resolve as an individual peak, but, on occasion, they separate from the 170kDa complex to form a distinct band with apparent mass of about 105kDa. The possibility that the material is only PDE complex which lagged during centrifugation was excluded since no histone-activated PDE activity was associated with the slower-sedimenting PDE-immunoreactive peptides (Fig.2 and 3) Interestingly, the amount of 105kDa polypeptide remains essentially constant in the immature and adult retinas even though the amount of PDE complex (170kDa) and PDE activity has increased substantially in the adult retina. These observations suggest that the 105kDa peptides are not degradation products of the larger PDE complex and that they probably represent a small pool of unassembled PDE polypeptides. Moreover, it is possible that the alpha and beta polypeptides in the 105kDa pool act as precursors in the normal process of PDE complex assembly.

FAILED ASSEMBLY OF PDE COMPLEX IN DEVELOPING PHOTORECEPTORS OF rd RETINA

Developing rd photoreceptors accumulate cyclic GMP, apparently as the result of a low rate of cyclic GMP degradation rather than from a high rate of cyclic GMP synthesis by guanylate cyclase (Farber and Lolley, 1976). The level of PDE activity in developing rd photoreceptors has not been established although it appears to be minimal. Biochemical analysis of hypotonic extracts and homogenates from immature rd retinas was unable to detect any measurable histone-activated PDE activity (Lee et al., 1985). The absence of histone-activated PDE activity implied either that the rd PDE was enzymatically inactive or the rd PDE was already in an activated state and, therefore, unresponsive to histone activation. Such a low level of histone-activated PDE activity would not be expected if the rd defect was confined to either rhodopsin or transducin and the PDE complex was normal. To analyze the rd defect further, it seemed appropriate to focus on the phosphodiesterase of immature rd photoreceptors and to determine particularly whether it has normal activity and subunit composition.

Developing rd photoreceptors synthesize polypeptides which contain antigenic determinants which are recognized by our monospecific polyclonal antibodies generated against the PDE complex of bovine ROS (Lee et al., 1988). These immunoreactive polypeptides which were identified by Western blot analysis, co-migrate on SDS gels with the alpha and beta subunits of PDE complex from normal mouse retinas (Fig. 1). The appearance and disappearance of these immunoreactive polypeptides in the developing rd retinas coincides with the morphological differentiation and subsequent degeneration of the rod photoreceptors. PDE immunoreactivity reaches the highest levels in the rd retina between 11-13 postnatal days when the rod photoreceptors still possess rudimentary outer segments. Nevertheless, PDE immunoreactivity in rd retinas is only about 10% of the levels in normal retinas of comparable age. The gamma subunit is not detectable under these conditions, but other studies suggest that the gamma subunit is probably synthesized because the gene which encodes this protein is transcribed in developing rd photoreceptors and the level of gamma-subunit mRNA is nearly normal (Tuteja and Farber, 1988). The low level of PDE immunoreactivity in rd retinas could indicate either that the rd photoreceptors actually contain low levels of the alpha and beta subunits of PDE or that the PDE polypeptides in rd photoreceptors show low crossreactivity with the PDE antibodies tested. We believe the low level of PDE immunoreactivity is, at least in part, due to reduced levels of PDE peptides since we have observed essentially the same level of immunoreactivity using two other preparations of polyclonal antibodies against PDE.

Do the PDE-immunoreactive polypeptides which do accumulate in rd photoreceptors assemble into a PDE complex with normal composition and hydrolytic activity? Our sucrose density-gradient experiments indicate that they do not (Lee et al., 1988). As expected from earlier studies which measured PDE activity in developing rd retinas (Lee et al., 1985), no measurable histone-activated PDE activity was found within any fraction of the sucrose gradient (Fig. 2). Furthermore, no PDE-immunoreactive polypeptides were found within the gradient at the position where the PDE complex (170kDa) from normal mouse retinas or that from bovine ROS sediments, indicating that a normal PDE complex does not accumulate in developing rd photoreceptors (Fig. 3). The absence in rd photoreceptors of a normal 170kDa PDE complex is not unexpected, since our earlier studies (Lee et al., 1985) showed that the ROS-1 monoclonal antibody, which recognizes only the PDE complex containing the gamma subunit, was unable to recognize and immunoprecipitate the PDE polypeptides from immature rd retinas. Furthermore, the absence of a 170kDa complex indicates that the PDE-immunoreactive peptides cannot be ascribed to the PDE complex of rd cones.

The PDE immunoreactive peptides of immature rd retinas were found to sediment more slowly than the 170kDa PDE complex of normal photoreceptors and to migrate within the sucrose gradient with an apparent molecular mass of 105kDa. It seems more than a coincidence that the PDE-immunoreactive polypeptides of rd photoreceptors are found in a fraction with apparent mass of 105kDa. A PDE-immunoreactive fraction of about this size was observed in developing and mature retinas of normal mice and we suggested that it might represent a precursor pool from which peptides are drawn for the assembly of the PDE complex. The presence of the 105kDa proteins in developing rd photoreceptors and the absence of a 170kDa PDE complex seems to support the hypothesis that the 105kDa pool of proteins could serve as precursors for PDE complex assembly. Whereas, this hypothesis requires verification, it remains clear that a normal PDE complex fails to accumulate in developing rd photoreceptors.

The accumulation of only low levels of PDE subunits may be related to the failure of rd photoreceptors to produce a normal PDE complex. Since immature rd retinas contain rhodopsin and transducin levels which are about 80% of normal, the low level of PDE polypeptides does not seem to reflect a generalized depletion of photoreceptor-specific proteins. Instead, a specific defect in the assembly and accumulation of the PDE complex may set the stage for the PDE polypeptides to be degraded when PDE complex assembly fails. Conceptually, the processing and packaging of the PDE enzyme can be broken down into three steps: synthesis of PDE subunits, assembly of subunits into PDE complex and accumulation of PDE complex. Synthesis of the PDE subunits normally leads to the assembly of a PDE complex and its sequestration by ROS membranes (Baehr et al, 1979). This chain of events is broken in the rd disorder. Following PDE polypeptide synthesis, PDE complex assembly fails and without a PDE complex there is no accumulation of PDE. Since the 105kDa pool of PDE-immunoreactive peptides do not accumulate in rd retinas above that of normal retinas, the rate of synthesis of these peptides must be coupled to their utilization or degradation. With PDE complex assembly blocked in immature rd photoreceptors, the PDE-immunoreactive polypeptides most likely are subjected to proteolysis when complex assembly fails. This idea is in agreement with the reported observation that the rate of synthesis of the alpha and beta PDE-subunits is nearly normal in immature rd retinas and that the PDE subunits appear to undergo rapid degradation (Farber et al. 1988).

It is informative to compare the rd defect with another mouse disorder (rds) which causes photoreceptor degeneration. The rds disorder is a different genetic abnormality from that of the rd mouse. In the rds disorder, photoreceptors fail to form ROS membranes and they exhibit only low levels of PDE activity (Fletcher et al., 1986). The PDE complex that is available appears to be normal since the activity of the solubilized enzyme is enhanced by histone. Moreover, the light-activated PDE cascade is operative since the rds photoreceptors exhibit light-evoked reductions in cyclic GMP (Cohen, 1983). An absence of ROS membranes seems to limit the amount of PDE complex which can accumulate in rds photoreceptors since heterozygous rds photoreceptors develop shorter than normal ROS (Sanyal, 1987) and they accumulate intermediate levels of PDE activity (Fletcher, 1986). The rds experiments as well as developmental studies of normal retina suggest that the amount of PDE complex which accumulates in a photoreceptor cell is related directly to the size of its ROS. Whereas, rd photoreceptors develop rudimentary ROS membranes, they are an exception to this rule; the rd photoreceptors deviate from the normal pattern because they fail to assemble the PDE complex.

Failure of rd photoreceptors to form a normal PDE complex could be related directly to structural abnormalities in the PDE subunits or indirectly through a defect in the mechanisms that control PDE complex assembly. If the rd gene encodes one of the PDE subunits and its expression results in a protein of abnormal structure, PDE complex assembly could be understandably aborted. Whereas, we and other laboratories have shown the accumulation of immunoreactive subunits of the PDE complex, it cannot be ruled out that a point mutation in the rd gene produces structural abnormalities in a PDE subunit which are critical for the formation of a functionally and structurally normal enzyme complex. On the other hand, the PDE subunits of rd photoreceptors might be normal and the mechanisms which regulate PDE complex assembly might be disrupted by the rd gene product. Almost nothing is known about PDE complex assembly. It presents a major challenge which may have to be addressed before we understand completely the etiology of the rd disorder. We are now focused on the PDE subunits and the mechanisms which control PDE complex assembly because the direct action of the rd gene product is expected to be found within this arena.

REFERENCES

Baehr W, Devlin MJ, Applebury ML (1979). Isolation and characterization of cGMP phosphodiesterase from bovine rod outer segments. J Biol Chem 254:11669.

Booth DP, Hurwitz RL, Lolley RN (1986). Characterization of the cyclic nucleotide phosphodiesterase of lizard cone photoreceptors. Invest Ophthalmol Vis Sci Suppl 27:217.

Bowes C, Van Veen T, Farber DB (1988). Opsin, G-protein and 48-kDa protein in normal and rd mouse retinas: Developmental expression of mRNSs and proteins and light/dark cycling of mRNAs. Exp Eye Res 47:369.

Carter-Dawson LD, LaVail MM, Sidman RL (1978). Differential effect of the rd mutation on rods and cones in the mouse retina. Invest Ophthalmol Vis Sci 17:489.

Cohen AI (1983). Some cytological and initial biochemical observations on photoreceptors in retinas of rds mice. Invest Ophthalmol Vis Sci 24:832

Farber DB, Lolley RN (1974). Cyclic guanosine monophosphate: Elevation in degenerating photoreceptor cells of the C3H mouse retina. Science 198:449.

Farber DB, Lolley RN (1976). Enzymatic basis for cyclic GMP accumulation in degenerative photoreceptor cells of mouse retina. J Cyclic Nucleotide Res 2:139.

Farber DB, Park S, Yamashita C (1988). Cyclic GMP-phosphodiesterase of rd retina: Biosynthesis and content. Exp Eye Res 46:363.

Fletcher RT, Sanyal S, Krishna G, Aguirre G, Chader GJ

(1986). Genetic expression of cyclic GMP phosphodiesterase activity defines abnormal photoreceptor differentiation in neurological mutants of inherited retinal degeneration. J Neurochem 46:1240.

Fung BK-K (1983). Characterization of transducin from bovine rod outer segments. Separation and reconstitution of the subunits. J Biol Chem 258:10495.

Gillespie PG, Beavo JA (1988). Characterization of bovine cone photoreceptor phosphodiesterase purified by cyclic GMP-sepharose chromatography. J Biol Chem 263:8133.

Hurwitz RL, Bunt-Milan AH, Beavo NA (1984). Immunologic characterization of the photoreceptor outer segment cyclic GMP phosphodiesterase. J Biol Chem 259:8612.

Hurwitz RL, Bunt-Milan AH, Chang ML, Beavo NA (1985). Cyclic GMP phosphodiesterase in rod and cone outer segments of the retina. J Biol Chem 260:568.

Kuhn H, Dreyer WJ (1972). light dependent phosphorylation of rhodopsin by ATP. FEBS Lett 20:1.

Kuhn H, Hall SW, Wilden U (1984). Light-induced binding of 48-kDa protein to photoreceptor membranes is highly enhanced by phosphorylation of rhodopsin. FEBS Lett 176:473.

Lee RH, Lieberman BS, Hurwitz RL, Lolley RN (1985). Phosphodiesterase-probes show distinct defects in rd mice and Irish setter dog disorders. Invest Ophthalmol Vis Sci 26:1569.

Lee RH, Navon SE, Brown BM, Fung B K-K, Lolley RN (1988). Characterization of a phosphodiesterase-immunoreactive polypeptide from rod photoreceptors of developing rd mouse retinas. Invest Ophthalmol Vis Sci 29:1021.

Navon SE, Lee RH, Lolley RN, Fung B K-K (1987). Immological determination of transducin content in retinas exhibiting inherited degeneration. Exp Eye Res 44:115.

Noell WK (1965). Aspects of experimental and hereditary retinal degeneration. In Biochemistry of the Retina (Ed. Graymore, CN). pp 51-72. Academic Press: New York.

Sanyal S (1987). Cellular site of expression and genetic interaction of the rd and the rds loci in the retina of the mouse. In Degenerative Retinal Disorders: Clinical and Laboratory Investigations (Eds Hollyfield JG, Anderson RE, LaVail MM) pp 175-194, Alan R. Liss, New York.

Shuster TA, Farber DB (1986). Rhodopsin phosphorylation in developing normal and degenerative mouse retinas. Invest Ophthalmol Vis Sci 27:264.

Sidman RL, Green MC (1965). Retinal degeneration in the mouse: location of the rd locus in linkage group XVII. J Hered 56:23
Stryer L (1986). Cyclic GMP cascade of vision. Ann Rev Neurosci 9:87.
Tuteja N, Farber DB (1988). Cloning and northern blot analysis of γ-subunit of cGMP-PDE in normal and rd mouse retina. Invest Ophthalmol Vis Sci Suppl 29:384.

ACKNOWLEDGMENT

We wish to express our appreciation for the continued support by USPHS Grant EY-00395, by the National Retinitis Pigmentosa Foundation and by the Medical Research Service of the Veterans Administration.

Inherited and Environmentally Induced
Retinal Degenerations, pages 169–181

EXPRESSION OF A "SURVIVAL CRISIS" BY NORMAL AND RD/RD MOUSE PHOTORECEPTOR CELLS IN VITRO

Ruben Adler and Luis Politi

Retinal Degenerations Research Center
Department of Ophthalmology, Johns Hopkins
University School of Medicine
Baltimore, Maryland 21205

I. SELECTIVE NEURONAL DEATH AND TROPHIC FACTORS

A. Pathological cell death in the adult may be related to normal developmental neuronal death

Human retinitis pigmentosa and related retinal degenerations belong to a family of pathological conditions having the occurrence of selective neuronal death as common denominator. This phenomenon is characterized by the degeneration of defined populations of neurons at particular times during the life of an organism, in the absence of degeneration in neighboring neuronal populations. Extensive neuronal death is usually associated with pathological conditions in the adult, but it occurs normally during embryonic development (rev: Hamburger and Oppenheim, 1982). The concept is gaining recognition that both types of neuronal death may share some common mechanisms, and some of the information derived from studies of the developmental death phenomenon may be relevant for understanding, preventing or treating neuronal degenerations in the adult (Appel, 1981; Varon and Adler, 1981; Adler, 1986, 1988). A key concept is that neuronal survival is a regulated behavior which requires not only a normal supply of nutritional elements, but also depends upon the availability of specific survival-promoting agents called neuronotrophic factors. These factors appear to derive from cells with which the neurons are in close contact, including pre- and postsynaptic partners as well as "satellite" cells such as glia.

B. The role of trophic factors

At specific stages of embryogenesis, as many as 50-80% of the neurons undergo normal developmental death in various regions of the nervous system. The neurons which are spared from this degeneration seem to be those which succeed in establishing adequate contacts with other cells acting as sources of trophic support. Neurons which are less successful in competing for these trophic agents undergo cell death.

The best characterized trophic factor is nerve growth factor (NGF), initially identified through its effects upon sympathetic and dorsal root ganglionic neurons (rev: Levi Montalcini, 1987). Its role in preventing the death of these neurons has been demonstrated by many studies, both in vitro and in vivo. What is particularly relevant for the topic of this article is that some neurons appear to require NGF not only during embryonic development, but also in the adult. The recent discovery that NGF is also active upon the cholinergic neurons that degenerate in Alzheimer's disease has raised the possibility that defects in an NGF-mediated trophic mechanism could be involved in the pathogenesis of this disease (rev: Korsching, 1986).

Trophic factors show both similarities and differences with the better known family of molecules known as hormones. Studies with NGF suggest that, as is the case with most regulatory substances, trophic factors act through receptor- and second messenger-mediated mechanisms. Thus, trophic factor-dependent cell behaviors can theoretically be perturbed by genetic mutations or pathological substances at many levels, affecting not only the synthesis and secretion of the factor by one cell type, but also the receptors and second messenger systems of responding cells. Therefore, the presence of alterations in second messenger mechanisms in some retinal degeneration animal models such as the "rd" mouse (rev: Lolley, 1982; Farber and Shuster, 1986) is not incompatible with a hypothetical role for trophic factors in the same pathological processes.

On the other hand, a most important difference between hormones and trophic factors is that the latter are usually involved in short range interactions, involving cells which are in contact with each other. An important practical corollary of this situation is that the investigation of the trophic requirements of neurons usually requires their isolation from neighboring cells (the likely sources of trophic factor support). This is usually accomplished by growing the neurons in dissociated cell cultures. The feasibility of this approach is based on the observation that cultured neurons frequently express trophic requirements similar to those observed in the intact organism (rev: Adler, 1987a).

C. The concept of "survival crisis"

One of the most remarkable features of the developmental neuronal death phenomenon is its rather abrupt onset, which seems to indicate that neurons undergo a developmentally-regulated "survival crisis." Thus, up to the stage when the neurons and target cells become synaptically connected, the survival and development of nerve cells is target cell-independent. This situation appears to change rather abruptly around the time of synaptogenesis when neurons undergo massive degeneration if they fail to interact with target cells. Similarly, isolated neurons grown in culture only demonstrate a dependency for special trophic support after the stage when they first express target dependency in vivo (Manthorpe et al., 1981; Adler and Varon, 1982; Berg, 1982). It is unfortunate that the nature of the "biological clocks" responsible for these developmental stage-dependent changes is not understood, because similar mechanisms could potentially trigger the onset of new trophic requirements in adult neurons, creating the possibility that neurons could degenerate if these requirements are not satisfied.

II. DO MOUSE PHOTORECEPTOR CELLS UNDERGO A "SURVIVAL CRISIS"?

Until recently, literature dealing with trophic interactions in the retina was largely limited to retinal ganglion cells, which have been shown to undergo a developmentally stage-dependent survival crisis in several species. In the intact animal this crisis seems to reflect the need for interactions with postsynaptic target elements in the midbrain (i.e., Hughes and McLoon, 1979; Sengelaub and Finlay, 1981). The brain derived neuronotrophic factor (BDNF) can substitute for target cells as a survival-promoting factor for these neurons in vitro (Johnson et al., 1986). It has also been shown that glial cells have a profound effect upon the capacity of adult retinal ganglion cells to regenerate their axons (Vidal-Sanz et al., 1987).

The trophic requirements of other retinal cell types, including photoreceptors, have been much less well studied (cf Adler, 1986). Some technical reasons may explain this deficit. Retinal ganglion cells are large, can be easily identified, and have a long and experimentally accessible axonal projections to other regions of the brain. On the other hand, cells located in the retinal inner and outer nuclear layers have smaller cell bodies, are intermingled with other cell types, and their pre- and postsynaptic connections involve other cells located within the retina itself. In spite of these complicating factors, there is some evidence suggesting that mouse photoreceptors do undergo a survival crisis similar to that described for other neurons. Quantitative analysis of the developing mouse retina demonstrated the abundance of pyknotic nuclei in the outer nuclear layer of the developing mouse retina during the first postnatal week (Young, 1984). This observation led Young to suggest that photoreceptor cell survival could depend upon the development of appropriate synaptic connections with other retinal neurons, a process that is very active at this stage of mouse retinal development. A second line of evidence suggesting the occurrence of a survival crisis derives from the extensive body of literature accumulated over the last few years regarding photoreceptor degeneration in the rd mouse. Although some abnormalities may be detectable in the rd mouse retina very early in postnatal life (Sanyal, 1982), microscopical, ultrastructural and biochemcal studies indicate that most aspects of rod photoreceptor cell development are normal during the first postnatal week in this mutant. Moreover, the onset of overt photoreceptor degeneration can be traced to the end of the first and the beginning of the second postnatal week in the rd mouse (Lasanski and DeRobertis, 1960; Blanks et al., 1974; Farber and Lolley, 1974; Fletcher et al., 1986; Lolley, 1982; Farber and Shuster, 1986). Interestingly, this "survival crisis" has been found to correlate with a failure in the final phases of synaptogenesis between rods and second order neurons, recognizable in the incomplete formation of the postsynaptic "triad" (Blanks et al., 1974). Although largely circumstancial, this evidence is consistent with the possibility that mouse photoreceptors may undergo a survival crisis at the time of synaptogenesis (Young, 1984), a possibility that, based on previous experience with many other neuronal types, can best be tested using dissociated cell culture systems.

III. AN IN VITRO SYSTEM FOR MOUSE RETINAL NEURONS AND PHOTORECEPTORS

A. Culture method

The possibility of growing isolated photoreceptor cells in low density cultures was first demonstrated using chick embryo retinal cells (Adler et al., 1984; Araki, 1984; rev. Adler, 1987b). More recently, a method was developed which is applicable to rod photoreceptors from both normal and rd/rd mouse retinas (Politi et al., 1988). The technique is applicable to newborn mouse retinas between postnatal days (PD) 1 and 5. On PD2 (the stage most frequently used for our experiments) the mouse retina still appears largely undifferentiated, showing a thick neuroepithelium with abundant mitotic figures (Fig. 1). There is some neuronal differentiation at the level of the retina ganglion cell layer and the inner portion of the inner nuclear layer, but photoreceptor differentiation is not detectable, at least by light microscopy (Fig. 1). The retina can be cleanly dissected from other intraocular tissues including the pigment epithelium, and dissociated into a suspension of single cells following mild trypsination and rinsing with soy bean trypsin inhibitor.

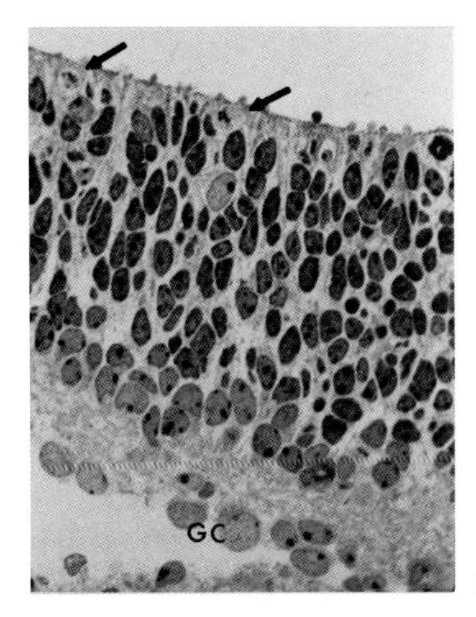

Figure 1. Toluidine blue-stained, 1 um plastic section of a 2-day old mouse retina, dissected from the pigment epithelium and other eye tissues. Although retinal ganglion cells (GC) are already evident, the inner and outer nuclear layers have not yet become separated, and photoreceptor differentiation is not yet obvious. Note presence of mitotic figures (arrow). Reprinted with permission from Politi et al., 1988; copyright Association for Research in Vision and Ophthalmology.

The dissociated cells are suspended in a serum free-chemically defined medium which is based on the high-pyruvate formulation of Dulbecco's modified Eagle's Medium (DME), supplemented with 2-fold concentrations of the "N1" formulation of Bottenstein and Sato (1979), as well as with CDP-choline, CDP-ethanolamine and hydrocortisone (Politi et al., 1988). In order to promote the development of nerve processes, retinal cells are grown on tissue culture plastic dishes sequentially coated with polyornithine and with Schwannoma conditioned medium which contains the neurite-promoting factor PNPF (Adler, 1982).

B. In vitro development of neurons and photoreceptors

The development of wild type (+/+), heterozygous rd/+ and homozygous rd/rd retinal cells is very similar under these culture conditions (Politi et al., 1988; Politi and Adler, 1988). Dissociated mouse retinal cells attach as individual units to this highly adhesive substratum, and begin to extend nerve fibers within a few hours after culture onset. As the cells express differentiated properties, the complexity of the cultures increases markedly during the next 3 or 4 days, and three different cell types can then be recognized (Fig. 2A). Some of the cells retain a process-free, morphologically undifferentiated appearance. Other cells differentiate as neurons, showing a large cell body and one or more long neurites. The majority (90%) of these neurons can be immunostained with the amacrine cell-specific monoclonal antibody HPC-1 (Politi et al., 1988) and show a very active high affinity uptake for the inhibitory neurotransmitter GABA (Politi and Adler, 1988; Abrams et al., 1988). The third cell type present in the cultures are the rod photoreceptor cells (Fig. 2A). Rod photoreceptors can be recognized morphologically on the basis of a small cell body, with a short neurite which terminates in a spherule-like body. There is also a short cilium at the opposite end of these cells. Rod cell identity was corroborated in these cultures by immunocytochemistry (Fig. 2B), using a polyclonal antiserum against the visual pigment opsin (from Papermaster and Schneider, 1982), the rod-specific monoclonal antibody RET-P1 (Barnstable, 1980; Akagawa and Barnstable, 1986), and an affinity purified antibody against bovine IRBP (from Wiggert et al., 1986). Biochemical and immunochemical studies have shown that IRBP is both synthesized and secreted by these cultured cells (Politi et al., 1988). Cultured rod photoreceptor cells do not take up GABA, but they appear very heavily labeled in cultures incubated with either tritiated glutamate or tritiated aspartate (Politi and Adler, 1988; Abrams *et al.*, 1988), excitatory amino acids which have been postulated as photoreceptor neurotransmitters. The diverging differentiation of photoreceptor cells and neurons can also be demonstrated using the neurotoxin kainic acid (KA; Abrams *et al.*, 1988), which, as has also been shown by *in vivo* studies, is toxic for amacrine cells and other inner retinal neurons, but not for photoreceptors (Morgan, 1983). Multipolar neurons begin to show susceptibility to the toxin towards *in vitro* day 3, and this sensitivity

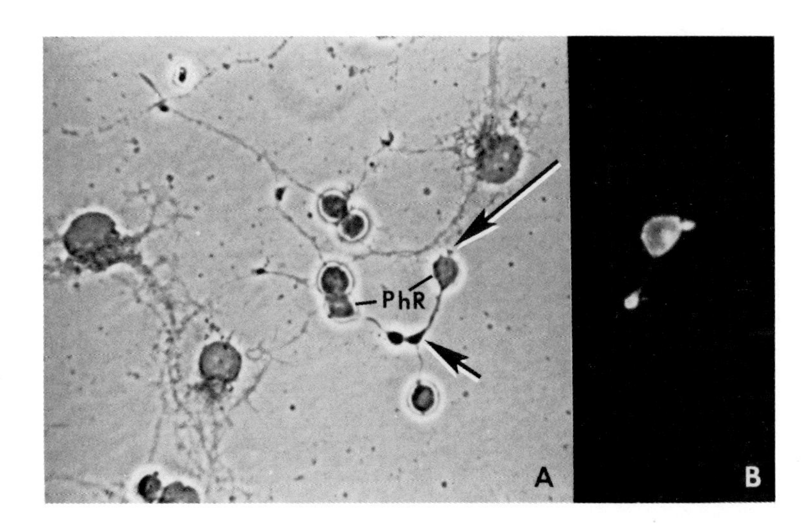

Figure 2. a) Mouse retinal cells after 7 days *in vitro*. Photoreceptor-like cells (PhR) show a short neurite ending in a spherule-like body (short arrow) and an apical structure (long arrow) that EM studies show to be a cilium (long arrow). Multipolar neurons show different morphologies, and in some cases show several long processes. b) Photoreceptor-like cell showing immunoreactivity with an antibody (Barnstable, 1980; Akagawa and Barnstable, 1986) against the rod-specific antigen RET-P1. Multipolar neurons were consistently negative with this antibody. Reprinted with permission from Politi *et al.*, 1988; copyright Association for Research in Vision and Ophthalmology.

increases upon further development of the cultures. The toxic effects of KA are time- and concentration-dependent, and are specific for multipolar neurons, since photoreceptor losses are minimal in cultures treated with KA concentrations that caused the loss of over 90% of the neurons (Abrams et al., 1988).

C. A photoreceptor "survival crisis" in vitro

Quantitative analysis of the cultures shows that the number of cells which can be recognized as either multipolar neurons or photoreceptors increases to a plateau as cell differentiation proceeds during the first week in culture (Fig. 3). However, the behavior of these two cell types is completely

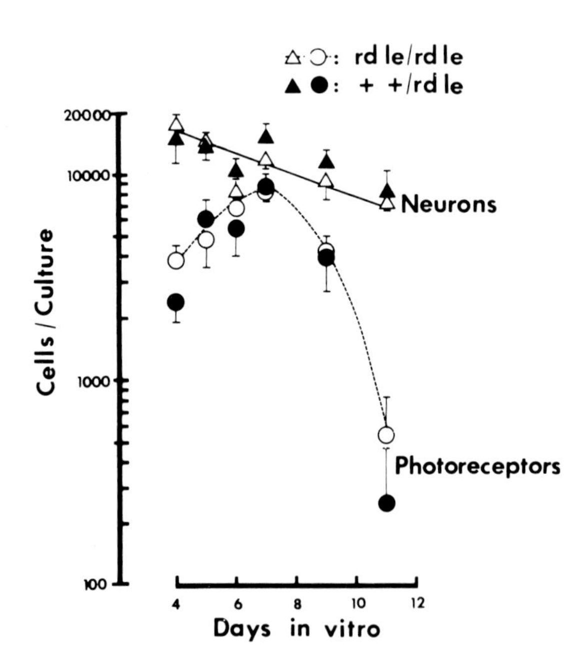

Figure 3. Quantitative analysis of the frequency of multipolar neurons and photoreceptor cells in rdle/rdle and rdle/++ retina cell cultures. During the first few days in vitro there is a small increase in the number of photoreceptor cells from cultures of both genotypes during the second week in vitro. Reprinted with permission from Politi and Adler (1988); Copyright Academic Press Ltd.

different during the second week in culture. Multipolar neurons seem to survive fairly normally, with only minor losses being detected by the end of this second week in vitro. As shown on Fig. 3, however, there is an abrupt decline in the number of photoreceptor cells during this period, so that by days 12-14 only approximately 10% of the original number of photoreceptor cells still remain in the cultures (Politi et al., 1988; Politi and Adler, 1988). This decline can also be documented by immunocytochemical analysis of photoreceptor-specific antigens. Interestingly, similar rates of photoreceptor degeneration in vitro were observed with wild type retina cells and with cells isolated from either heterozygous or homozygous rd mice.

IV. INTERPRETATION OF THE IN VITRO SURVIVAL CRISIS FROM THE PERSPECTIVE OF THE TROPHIC HYPOTHESIS

There is a striking correlation between the timing of photoreceptor cell death in culture, the occurrence of developmental death among photoreceptors in the normal retina (Young, 1984), and the onset of photoreceptor degeneration in the rd/rd retina in vivo (See Section II). Of course, this correlation could be just coincidental. However, photoreceptor cell behavior is consistent with a hypothetical neuronotrophic mechanism similar to the one which has been described for other neuronal systems. It could be hypothesized that the end of the first week of in vivo development represents a critical stage at which photoreceptors become dependent upon trophic support from other retinal cells. In normal animals, cell death would be limited to those photoreceptors which fail to establish adequate contacts with the sources of this hypothetical trophic factor. Massive degeneration would occur in rd retinas because this hypothetical mechanism would be qualitatively or quantitatively defective, and/or because the photoreceptors would be insensitive to it. When isolated cells are grown in vitro in chemically defined medium in the absence of pigment epithelial and glial cells, this hypothetical trophic factor would either be absent, or it would be present at suboptimal concentrations. This would explain why genetically normal (+/+ and rd/+) as well as mutant (rd/rd) photoreceptors degenerate in vitro after reaching the critical postnatal day 8-10 stage.

This model is obviously hypothetical, and is based largely on circumstancial evidence. However, it raises questions that are amenable to experimental verification. Putative trophic factors derived from retinal and pigment epithelial cells are likely to play a role together with other genetic and environmental influences affecting photoreceptor survival. The likelihood that trophic factors active on photoreceptor cells could actually exist is suggested by recent studies showing that interphotoreceptor matrix preparations contain a survival-promoting factor for chick cone photoreceptors (Lindsey et al., 1988). Further studies along these lines can be expected to increase our understanding of the mechanisms involved in selective cell death in various retinal pathologies, thus opening new avenues of investigation regarding the prevention and treatment of retinal degenerations.

ACKNOWLEDGMENTS

These studies were supported by USPHS grant EY05404. We thank Dr. A. Tyl Hewitt for his comments on the manuscript, Dr. M. Lehar for technical assistance, and Mrs. Doris Golembieski for secretarial work.

REFERENCES

Abrams L, Politi LE, Adler R (1988). Differential susceptibility of isolated mouse retinal neurons and photoreceptors to kainic acid toxicity: in vitro analysis of cell survival and neurotrans-mitter-related activities. Submitted.

Adler R (1988). Trophic factors in neuronal development. In Meisami E, Timiras P (eds): "Handbook of Human Growth and Developmental Biology," Boca Raton: CRC Press, pp 67-74.

Adler R (1987a). In vitro techniques for the investigation of trophic factors active on CNS neurons. In Perez-Polo JR (ed): "Handbook of Nervous System and Muscle Factors," Boca Raton: CRC Press, pp 151-173.

Adler R (1987b). The differentiation of retinal photoreceptors and neurons in vitro. In Osborne N, Chader G (eds): "Progress in Retinal Research," London: Pergamon Press, pp 1-27.

Adler R (1986). Trophic interactions in retinal development and in retinal degenerations. In vivo and in vitro studies. In: Adler R, Farber D (eds): "The Retina: A Model for Cell Biology Studies, Part I," Orlando: Academic Press, pp. 112-150.

Adler R, Lindsey JD, Elsner CL (1984). Expression of cone-like properties by chick embryo neural retina cells in glial-free monolayer cultures. J Cell Biol 99:1173-1178.

Adler R (1982). Regulation of neurite growth in purified retina cultures. Effects of PNPF, a substratum-bound neurite-promoting factor. J Neurosci Res 8:165-177.

Adler R, Varon S (1982). Neuronal survival in intact ciliary ganglia in vivo and in vitro: CNTF as a target surrogate. Dev Biol 92:470-475.

Akagawa K, Barnstable CJ (1986). Identification and characterization of cell types in monolayer cultures of rat retina using monoclonal antibodies. Brain Res 383:110-120.

Appel S (1981). A unifying hypothesis for the course of amyotrophic lateral sclerosis, Parkinsonism and Alzheimer's disease. Ann Neurol 10:499-505.

Araki M (1984). Immunocytochemical study on photoreceptor cell differentiation in the cultured retina of the chick. Dev Biol 103:313-318.

Barnstable LJ (1980). Monoclonal antibodies which recognize different cell types in the rat retina. Nature (London) 286:231.

Berg DK (1982). Cell death in neuronal development. In Spitzer NC (ed): "Neuronal Development," New York: Plenum Press, pp. 297-331.

Blanks JC, Adinolfi AM, Lolley RN (1974). Photoreceptor degeneration and synaptogenesis in retinal degeneration (rd) mice. J Comp Neur 156:95-106.

Bottenstein JE, Sato G (1979). Growth of a rat neuroblastoma cell line in serum-free supplemented medium. Proc Natl Acad Sci 76:514-517.

Farber DB, Lolley RN (1974). Cyclic guanosine monophosphate: elevation in degenerating photoreceptor cells of the C3H mouse retina. Science 186:449-451.

Farber DB, Shuster TA (1986). Cyclic nucleotides in retinal function and degeneration. In Adler R, Farber D (eds): "The Retina: A Model for Cell Biology Studies, Part I," Orlando: Academic Press, pp 239-296.

Fletcher RT, Sanyal S, Krishna G, Aguirre G, Chader G (1986). Genetic expression of cyclic GMP phosphodiesterase activity defines abnormal photoreceptor differentiation in neurological mutants of inherited retinal degeneration. J Neurochem 46:1240-1245.

Hamburger V, Oppenheim RW (1982). Naturally occurring neuronal death in vertebrates. Neurosci Comment 1:39-55.

Hughes WF, McLoon SC (1979). Ganglionic cell death during normal retinal development in the chick: comparisons with cell death induced by early target field destruction. Exp

Neurol 66:587-601.
Johnson JE, Barde YA, Schwab M, Thoenen H (1986). Brain-derived neurotrophic factor supports the survival of cultured rat retinal ganglion cells. J Neurosci 6:3031-3038.
Korsching S (1986). The role of nerve growth factor in the CNS. Trends in Neuroscience 9:570-573.
Lasanski A, DeRobertis E (1960). Submicroscopic analysis of the genetic dystrophy of visual cells in C_3H mice. J Biophys Biochem Cytol 7:679-683.
Levi Montalcini R (1987). The nerve growth factor 35 years later. Science 237:1154-1162.
Lindsey JD, Hewitt AT, Adler R (1988). Photoreceptor survival-promoting activity in interphotoreceptor matrix preparations: characterization and partial purification. Invest Ophthalmol Vis Sci (Suppl) 29:242.
Lolley RN (1982). Cyclic GMP synthesis and hydrolysis in visual cells of rd mouse retina. In Clayton RM (ed): "Problems of Normal and Genetically Abnormal Retinas," London: Academic Press, pp 201-206.
Manthorpe M, Varon S, Adler R (1981). Cholinergic neuronotrophic factors. VI: age-dependent requirements by chick embryo ciliary ganglionic neurons. Dev Biol 85:156-163.
Morgan IG (1983). Kainic acid as a tool in retinal research. Prog in Retinal Research 2:249-266.
Papermaster DS, Schneider BG (1982). Biosynthesis and morphogenesis of outer segment membranes in vertebrate photoreceptor cells. In McDevitt D (ed): "Cell Biology of the Eye," New York: Academic Press, pp 475-531.
Politi LE, Adler R (1988). Selective failure oflong term survival of isolated photoreceptors from both homozygous and heterozygous rd (retinal degeneration) mice. Exp Eye Res 47:269-282.
Politi LE, Lehar M, Adler R (1988). Development of neonatal mouse retinal neurons and photoreceptors in low density cell cultures. Invest Ophthalmol Vis Sci 29:534-543.
Sanyal S (1982). A survey of cytomorphological changes during expression of the retinal degeneration (rd) gene in the mouse. In Clayton RM (ed): "Problems of Normal and Genetically Abnormal Retinas," London: Academic Press, pp 223-232.
Sengelaub DR, Finlay BI (1981). Early removal of one eye reduces normally occurring cell death in the remaining eye. Science 213:573-574.
Varon S, Adler R (1981). Trophic and specifying factors directed to neuronal cells. Adv. Cell Neurobiol. 2:115-163.
Vidal-Sanz M, Bray GM, Villegas-Perez MP, Thanos S, Aguayo

AJ (1987). Axonal regeneration and synapse formation in the superior colliculus by retinal ganglion cells in the adult rat. J Neurosci 7:2894-2909.

Wiggert B, Ling L, Rodrigues M, Hess H, Redmond TM, Chader GJ (1986). Immunochemical distribution of interphotoreceptor retinoid-binding protein in selected species. Inv Ophthalmol Vis Sci 27:1041-1049.

Young RW (1984). Cell death during differentiation of the retina in the mouse. J Comp Neurol 229:362-373.

Inherited and Environmentally Induced
Retinal Degenerations, pages 183–189

FUNCTIONAL AND BIOCHEMICAL ABNORMALITIES IN THE RETINAS OF MICE HETEROZYGOUS FOR THE RD GENE

Mary J.Voaden, Nicholas J.Willmott, Ali A.Hussain and Sahar Al-Mahdawi.
Departments of Visual Science (MJV, NJW) and Pathology (SA-M), Institute of Ophthalmology, London WC1H 9QS, England, U.K. and Comparative Ophthalmology Unit (AAH, NJW-present) The Animal Health Trust, Kennett CB8 7PN, England, U.K.

rd/rd Mice. Several causes have been proposed for the rod dysplasia that occurs in mice homozygous for the photoreceptor degeneration gene designated as rd, but the primary genetic defect has not been established. Rod breakdown is preceeded, at about 8 days of age, by a large increase in the concentration of cyclic guanosine monophosphate (cGMP), resulting from a reduction in the activity of the rod specific cGMP phosphodiesterase (PDE; Farber & Lolley, 1976), and this may lead to cell death.

Although PDE-immunoreactive polypeptides are present they fail to form a normal complex (Lee et al., 1988). Thus, cross-reactivity with a specific anti-PDE monoclonal antibody is disrupted (Lee et al., 1985) and the enzyme cannot be activated by histone (Lolley et al., 1987). However, doubt remains as to the primary genetic defect, since rhodopsin phosphorylation and antibody interaction are also abnormal (Shuster & Farber, 1986; Takemoto et al., 1985) and changes in membrane fatty acids have been detected as early as 5-6 days of age (Marcheselli et al, 1988).

+/rd Mice. Mice heterozygous for the rd gene (+/rd) have normal retinal structure and a full complement of rhodopsin (Doshi et al., 1985). Moreover, rhodopsin phosphorylation and, by implication, the carboxy terminal, are also reported as normal (Shuster & Farber, 1986). However, the retinal concentration of cGMP is reduced by 40% (Ferrendelli & Cohen, 1976; Doshi et al., 1985) and the K_m of the PDE, as measured in a homogenate of bleached retina, is increased (Doshi et al., 1985). The Vmax of the PDE and the activity of guanylate cyclase appear normal.

Cyclic GMP. About 90% of the cGMP present in normal photoreceptors is bound (Pugh & Cobbs, 1986), in part to non-catalytic binding sites on the PDE (Yamazaki et al., 1980; Fesenko & Krapivinsky, 1986; Shinozawa et al., 1987). Since the overall concentration of cGMP is reduced by 40% in +/rd retinas, there must be a reduction in cGMP binding.

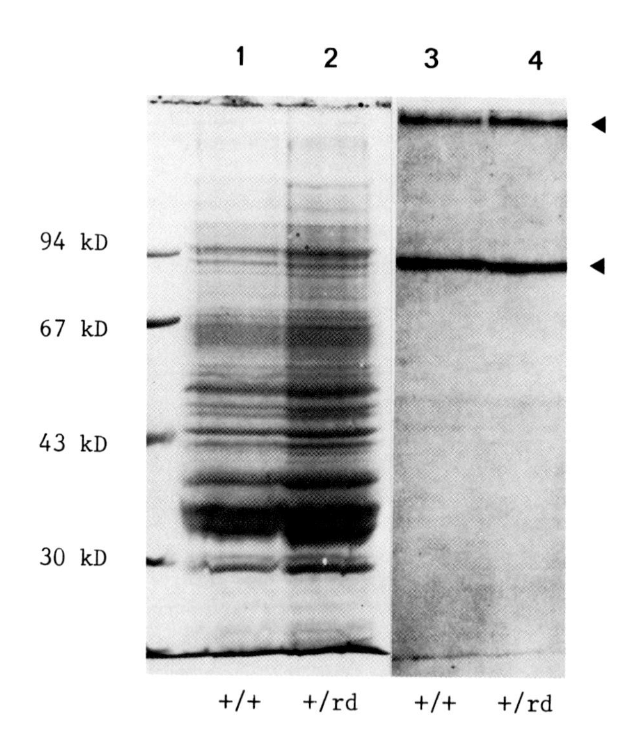

Figure 1. Cyclic GMP binding sites in +/+ and +/rd rod outer segments. Rod outer segments, isolated from bleached retinas by flotation on 43% sucrose, were incubated under UV light with 90μM tritiated cGMP and the cGMP-protein complexes thus formed, separated by electrophoresis on a 10% polyacrylamide gel and detected by fluorography (Shinozawa et al., 1987). 17 μg +/+ and 22 μg +/rd ROS protein were loaded. There is no obvious difference between +/+ and +/rd retina protein profiles (lanes 1&2) or extent of cGMP binding (lanes 3&4).

Two major binding proteins have been detected in bleached ROS of +/+ and +/rd mice (arrows, Fig. 1). One at about 90 kDa most probably corresponds to the subunits of PDE and the other is likely to be equivalent to the 250 kDa, cGMP-binding protein detected by Shinozawa et al., (1987) in frog ROS. Binding capacities, as estimated by fluorography, appear normal in the +/rd animal (Fig. 1), suggesting that the defect that leads to a reduction in the endogenous level in +/rd mice lies in binding affinity.

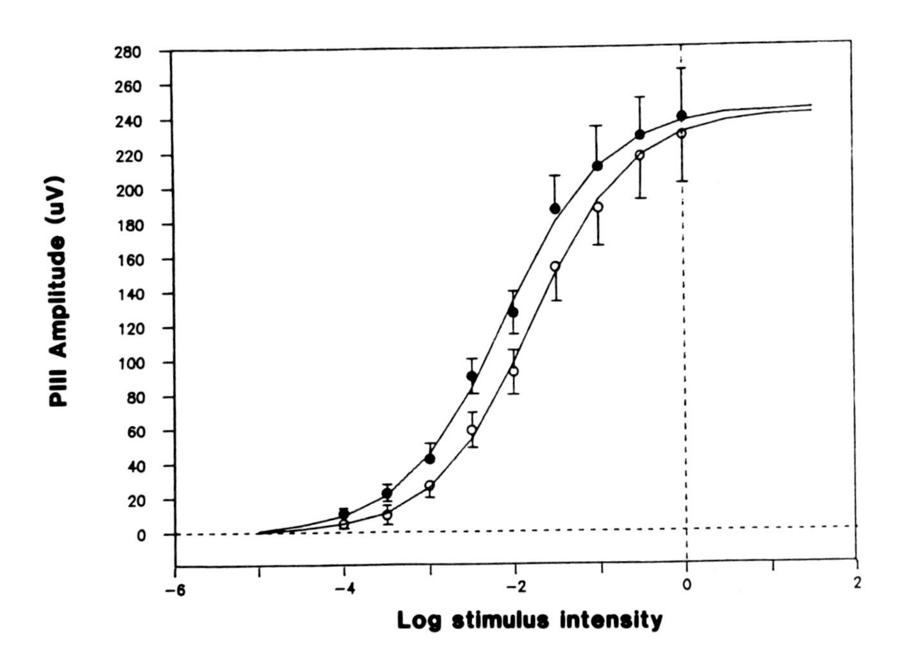

Figure 2. Amplitude of the isolated trans-retinal photoresponse (PIII) as a function of stimulus intensity for +/+ and +/rd mouse retinas. Each point represents the mean ± S.E.M. of at least 17 separate estimations. Curves drawn through the PIII measurements represent sigmoidal distributions of the type V = Vmax / exp(B(I-Io)) + 1, where V is the response amplitude in µV, Vmax is the maximum response amplitude, B is the slope parameter of response at Io, I is log light intensity, and Io is the log light intensity required to elicit a half maximal response.
o, retinas from +/+ mice; ●, retinas from +/rd mice.

Photoresponses and Calcium. Photoresponses in +/rd retinas are abnormal. In particular, the PIII implicit time is increased (Arden & Low, 1980) and, at low light intensities, sensitivity is increased (Fig. 2; Low, 1987). An increase in PDE K_M should lead to an increase in free cGMP in rod outer segments and this, in turn, would increase the rod dark current. Corollaries are an increase in intracellular calcium and, possibly, increased sensitivity (Pugh & Cobbs, 1986).

Rod guanylate cyclase is progressively inhibited by increasing concentrations of calcium up to a level of 10^{-5} M (Pepe et al., 1986). Thus a transient increase in the concentration of intracellular cGMP is noted when a retina is superfused with the calcium chelator, EGTA (Fig. 3A). The rise is due to cyclase activation and the subsequent decline to PDE activity. We have found that the rise in cGMP after EGTA is significantly delayed in intact +/rd retinas (Fig. 3A). The kinetics of guanylate cyclase and its response to calcium withdrawal, as assessed in a retinal homogenate in vitro, are normal (Doshi et al., 1985). Therefore, it is possible that an increase in endogenous calcium is causing the delay.

Present evidence suggests that the photoreceptor membrane current is principally determined by the concentration of free cGMP, and that photic responses are due to a reduction in its concentration through hydrolysis (Pugh & Cobbs, 1986) and, perhaps, binding (Fesenko & Krapivinsky, 1986). Assuming equilibrium between free and bound cGMP, it should be possible to correlate the total retinal content (95% of which may be present in photoreceptor cells; Orr et al., 1976) with the photoreceptor response waveform (PIII). However, there is no correlation in the +/rd retinas exposed to EGTA (Fig. 3). For, whereas the peak in PIII amplitude corresponds with the normal retina and occurs within 2.0 minutes, the cGMP peak is delayed to 4.0 minutes. It is possible that the 3-fold increase in PIII amplitude represents saturation and is achieved when total intracellular cGMP has increased about 4-fold - the level reached in +/rd retinas 2.0 minutes after exposure to EGTA. However, results from other species suggest that this is unlikely since, in the dark, less than 10% of the cGMP-sensitive pores in ROS are open (relative to the maximal amount that can be opened when cGMP is elevated) (Pugh & Cobbs, 1986). Alternatively, factors other

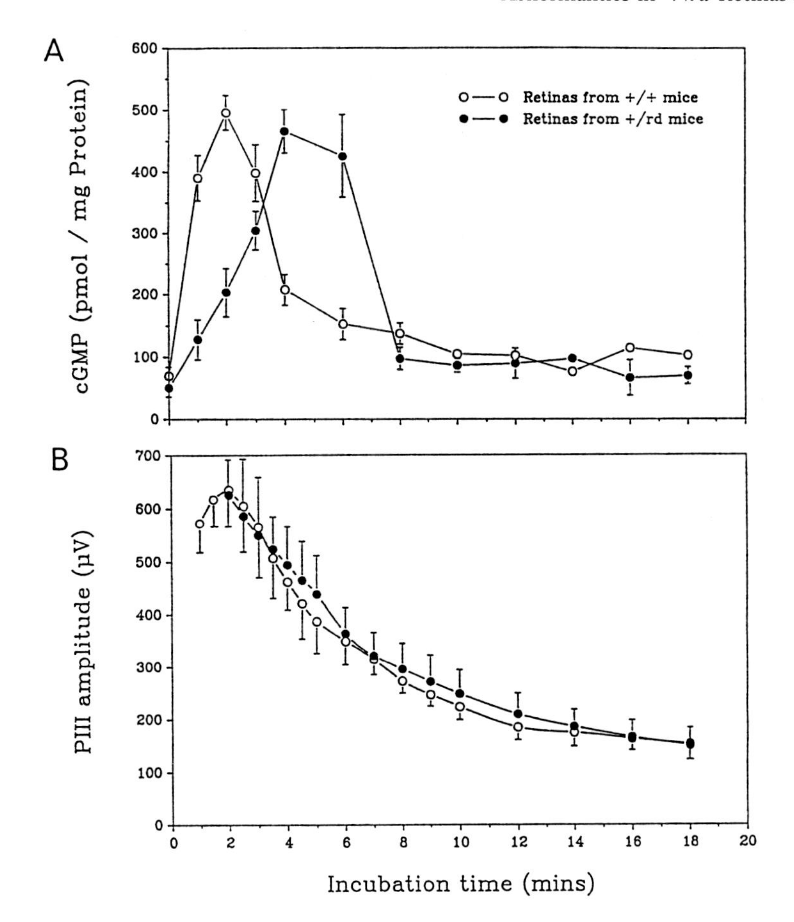

Figure 3. Effect of 3.0 mM EGTA on Retinal cGMP and PIII Amplitude in +/+ and +/rd mice. Cyclic GMP was measured by radioimmunoassay as described by Doshi et al., (1985). Glutamate-isolated, trans-retinal responses to light flashes (4 m.sec. duration) were obtained on a Medelec Mainframe system. Dark-adapted mouse retinas were incubated in normal Earle's medium until the response stabilized. The medium was then changed to calcium-free Earle's medium, containing 3.0 mM EGTA, and photoresponses elicited every 30 seconds over the following 20 minute period. Each point represents the mean of at least four separate estimations ± S.E.M.. A, Retinal cGMP; B, PIII Amplitude. (Taken with modification from Willmott et al., 1988).

than cGMP might be influencing the +/rd response. For example, it is known that the plasma membrane of the normal photoreceptor contains pores responsive to calcium and magnesium, and that there are complex interactions between the divalent cations and cGMP (Pugh & Cobbs, 1986; Stern et al., 1987; Macleish & Stern, 1987). Perhaps the genetic defect, partially expressed in +/rd retinas, is perturbing phototransduction sufficiently for separate processess, normally coincident, to be discerned. The model might, therefore, be informative as regards the processes involved in normal phototransduction and adaptation in rods.

ACKNOWLEDGEMENTS

The expert technical assistance of Mr David Moore B.Sc. is gratefully acknowledged. We thank Prof GB Arden for helpful discussions, Dr R Curtis for providing ERG facilities, The American Retinitis Pigmentosa Foundation for financial support and the British Retinitis Pigmentosa Society for a studentship for NJW. Electrophoresis equipment was a gift in memory of Mary Jane Voller.

REFERENCES

Arden GB, Low JC (1980). Altered kinetics of the photoresponse from retinas of mice heterozygous for the retinal degeneration gene. J Physiol 308:80p.

Doshi M, Voaden MJ, Arden GB (1985). Cyclic GMP in the retinas of normal mice and those heterozygous for early-onset photoreceptor dystrophy. Exp Eye Res 41:61-65.

Farber DB, Lolley RN (1976). Enzymic basis for cyclic GMP accumulation in degenerative photoreceptor cells of mouse retina. J Cyclic Nucleotide Res 2:139-148.

Ferrendelli JA, Cohen AI (1976). The effects of light and dark adaptation on the levels of cyclic nucleotides in retinas of mice heterozygous for a gene for photoreceptor dystrophy. Biochem Biophys Res Commun 73:421-427.

Fesenko EE, Krapivinsky GB (1986). Cyclic GMP binding sites and light control of free cGMP concentration in vertebrate rod photoreceptors. Photobiochem Photobiophys 13:345-358.

Lee RH, Lieberman BS, Hurwitz RL, Lolley RN (1985). Phosphodiesterase-probes show distinct defects in rd mice and Irish setter dog disorders. Invest Ophthalmol Vis Sci 26:1569-1579.

Lee RH, Navon SE, Brown BM, Fung BK-K, Lolley RN (1988). Characterization of a phosphodiesterase-immunoreactive polypeptide from rod photoreceptors of developing rd mouse retinas. Invest Ophthal Vis Sci 29:1021-1027.

Lolley RN, Navon SE, Fung BK-K, Lee RH (1987). Inherited disorders of rd mice and affected Irish setter dogs. In Hollyfield JG, Anderson RE, LaVail MM (eds): "Degenerative retinal disorders: clinical and laboratory investigations" New York: Alan R. Liss, pp 269-287.

Low JC (1987). The electroretinogram of heterozygous +/rd mice. J Physiol 382:116P.

Macleish PR, Stern JH (1987). Direct effects of divalent cations and cyclic GMP on excised patches of rod outer segment membrane. Neurosci Res Suppl.6:S67-S74

Marcheselli VL, Scott BL, Racz E, Lolley R, Bazan NG (1988). Early changes in membrane fatty acids of developing photoreceptor cells of RD mice. Invest Ophthal Suppl 29:383.

Orr HT, Lowry OH, Cohen AI, Ferrendelli JA (1976). Distribution of 3',5'-cyclic AMP and 3',5'-cyclic GMP in rabbit retina _in vivo_: selective effects of dark and light adaptation and ischemia. Proc Natl Acad Sci U.S.A. 73:4442-4445.

Pepe IM, Boero A, Vergani L, Panfoli I, Cugnoli C (1986). Effect of light and calcium on cyclic GMP synthesis in rod outer segments of toad retina. Biochim Biophys Acta 889:271-276.

Pugh EN, Cobbs WH (1986). Visual transduction in vertebrate rods and cones: a tale of two transmitters, calcium and cyclic GMP. Vision Res 26:1613-1643.

Shinozawa T, Sokabe M, Terada S, Matsuka H, Yoshizawa T(1987) Detection of cyclic GMP binding protein and ion channel activity in frog rod outer segments. J Biochem 102:281-290

Shuster TA, Farber DB (1986). Rhodopsin phosphorylation in developing normal and degenerative mouse retinas. Invest Ophthalmol Vis Sci 27:264-268.

Stern JH, Knutsson H, MacLeish PR (1987). Divalent cations directly affect the conductance of excised patches of rod photoreceptor membrane. Science 236:1674-1677.

Takemoto DJ, Hansen J, Takemoto LJ (1985). Altered rhodopsin accessibility in the retinal dystrophic mouse retina. Biochem Biophys Res Comm 132:804-810.

Willmott NJ, Hussain AA, Voaden MJ (1988). Biochemical and electrophysiological abnormalities in the photoreceptors of mice heterozygous for the rd gene. Biochem Soc Trans 16:1074-1075

Yamazaki A, Sen I, Bitensky MW, Casnellie JE, Greengard P (1980). Cyclic GMP-specific, high-affinity, noncatalytic binding sites on light-activated phosphodiesterase. J Biol Chem 255: 11619-11624.

Inherited and Environmentally Induced
Retinal Degenerations, pages 191–215

THE IDENTIFICATION OF A NEW BIOCHEMICAL ALTERATION EARLY IN THE DIFFERENTIATION OF VISUAL CELLS IN INHERITED RETINAL DEGENERATION

Nicolas G. Bazan

Louisiana State University Medical Center, LSU Eye Center, and The Eye, Ear, Nose and Throat Hospital, New Orleans, LA*

INTRODUCTION

Phospholipids create specific domains within cellular membranes for ionic channels, receptors for light, neurotransmitters, growth factors and hormones, and for other components of cell signal transduction (e.g. G-proteins). Phospholipids also serve as storage sinks for second messengers released upon cell activation. These messengers or mediators, in turn, elicit events within the cell that lead to physiological responses. In the retina a number of intracellular and extracellular messengers arise from membrane lipids (Bazan, 1988). For example, inositol trisphosphate (IP_3), derived from the phosphodiesteratic cleavage by phospholipase C of phosphatidylinositol 4,5-bisphosphate upon receptor occupancy, promotes the ionization of intracellular calcium. Calcium in turn elicits actions leading to physiological responses. Oxygenated metabolites of arachidonic acid (e.g., prostaglandins or lipoxygenase reaction products) often leave the cell after their formation, exerting effects on surface receptors of the same cell or on other cells. Platelet activating factor (PAF), another mediator derived from

*Mailing address: LSU Eye Center, 2020 Gravier Street, Suite B, New Orleans, LA 70112

membrane phospholipids, is involved in several biological effects (Braquet et al., 1987) as well as pathophysiological actions such as ischemia-reperfusion injury of neural cells (Panetta et al.,1987).

Docosahexaenoic acid is highly concentrated in the phospholipids of photoreceptor cells (Aveldano de Caldironi et al., 1981; Bazan and Reddy, 1985). Although the physiological significance of this fatty acid is not clearly understood, primates deprived of its dietary precursor, linolenic acid, show impaired visual acuity and electroretinographic alterations (Neuringer and Conner, 1984, 1985, 1986 and 1988). The retina synthesizes oxygenated metabolites, docosanoids (Bazan et al., 1984a), analogous to those made through the arachidonic acid cascade. Whether these metabolites play a role as messenger remains to be ascertained. In the model of docosahexaenoic acid metabolism discussed below, possible effects of docosanoids in the intercellular communication between photoreceptors and retinal pigment epithelium are indicated. Moreover, lipoxygenase reaction products derived from arachidonic acid, including leukotrienes, have been found to be actively synthesized in retinas after K+ depolarization (Birkle and Bazan, 1984) and in retinal pigment epithelium of frogs during light-darkness environmental changes (Bazan et al., 1987a; Bazan et al., 1987b). This indicates that the release to the interphotoreceptor matrix of these lipoxygenase products may play a role in the cross talk between the cells conforming to the limits of the matrix (Bazan et al., 1987a, 1987b). In canine ceroid lipofuscinosis, a model of retinal degeneration and aging, changes were found in docosanoids of the retina and retinal pigment epithelium (Reddy et al., 1985). Decreased amounts of docosahexaenoic acid have been found in patients with retinitis pigmentosa (Converse et al., 1983; Anderson et al., 1987) and Usher's syndrome (Bazan et al., 1986a; Bazan and Scott, 1987). It should be noted, however, that in this dual neurosensory degeneration arachidonic acid is also decreased.

Another feature of docosahexaenoic acid in the retina is that some phospholipids contain two such fatty acids per molecule; these unique molecular species were called supraenoic (Aveldano de Caldironi and Bazan, 1977). The presence of these lipids (Aveldano de Caldironi and Bazan, 1977; Miljanich et al., 1979; Aveldano and Bazan, 1983) and their metabolism (Aveldano and Bazan, 1980; Aveldano et al., 1983; Wiegand and Anderson, 1983) have both been studied, and as yet there is no clear understanding of their functional role. Recently a novel group of docosahexaenoic acid elongation products, very long chain polyenoic fatty acids (Aveldano, 1987; Aveldano and Sprecher, 1987), were found in the photoreceptors. Hexaenoic molecular species (one docosahexaenoate per molecule) as well as supraenoic molecular species (two docosahexaenoates per molecule) turnover much faster than other molecular species of phospholipids in the retina (Aveldano de Caldironi and Bazan, 1980; Aveldano et al., 1983). Furthermore, radiolabeling studies with (2-^{3}H)glycerol of hexaenoic and supraenoic molecular species observed in the retina (Aveldano de Caldironi and Bazan, 1980) as well as in photoreceptor membranes (Aveldano et al., 1983; Wiegand and Anderson, 1983) suggested that a very rapid de novo biosynthesis of docosahexaenoate-containing phospholipids takes place in the photoreceptors. The precursor of phospholipids in the de novo pathway, phosphatidic acid, has been found to be highly enriched in docosahexaenoate, both in photoreceptor membranes (Bazan et al., 1982b) and in retinal microsomal membranes (Giusto and Bazan, 1979). The retina of the toad is endowed with a high content of diacylglycerol containing about 40% docosahexaenoate to possibly support the large requirements of docosahexaenoyl phospholipids of rod photoreceptors (Aveldano and Bazan, 1972, 1973). These observations lead to the hypothesis that a sizable proportion of docosahexaenoyl-phospholipids in the retina is synthesized through the de novo pathway (Giusto and Bazan, 1979; Bazan and Giusto, 1980; Bazan et al., 1982b; Bazan, 1982a,b,c). This pathway was subsequently demonstrated (Bazan et al., 1984b), and contradicts the notion that polyunsaturated fatty acyl groups are introduced only by deacylation-reacylation reactions. Hexaenoic molecular species of phospholipids in early developing embryos also display very rapid turnover as assessed by ^{32}P, indicating that active membrane biogenesis in cells other than the photoreceptors share similar metabolic features (Barassi and Bazan, 1974).

We summarize here recent studies designed to explore for alterations in the fatty acyl content and composition of phospholipids in rod photoreceptor cells of the mutant rd mouse as compared with controls during development and differentiation.

We approached this problem by determining the endogenous fatty acyl groups of individual phospholipids from rod photoreceptor cells dissociated from developing mouse retinas. Our results show selective accumulation of certain fatty acids and a decrease in others. Below is an outline of the striking differences we have found, as well as of recent metabolic experiments using radiolabeled linolenic acid, the precursor of docosahexaenoic acid. Moreover, a model that includes various events that lead to the avid retention of docosahexaenoic acid during outer segment renewal and photoreceptor biogenesis is discussed. An understanding of the biosynthetic pathways of docosahexaenoyl-phospholipids underlying membrane biogenesis-related events leading to the construction of the heterogeneous visual cell will lead to bridging the gap between gene expression and cell differentiation. To begin to test these ideas, the fatty acid compositions of the individual phospholipids in rod photoreceptor cells isolated from the mutant rd mouse were compared with controls.

SELECTIVE UTILIZATION OF POLYUNSATURATED FATTY ACIDS FOR PHOSPHOLIPID BIOSYNTHESIS DURING THE DEVELOPMENT AND DIFFERENTIATION OF ROD PHOTORECEPTOR CELLS

The phospholipid classes, phosphatidylcholine, phosphatidylethanolamine, phosphatidylinositol and phosphatidylserine, were studied in isolated rod photoreceptor cells during postnatal development in mice. The aim of this study was to quantitatively assess the endogenous fatty acyl chains of each of the phospholipids during the period in which retinal rod photoreceptors develop and differentiate, generating a rod outer segment, an inner segment containing an ellipsoid (rich in mitochondria), and a myoid (rich in endoplasmic reticulum), a nucleus, an axon and a synaptic

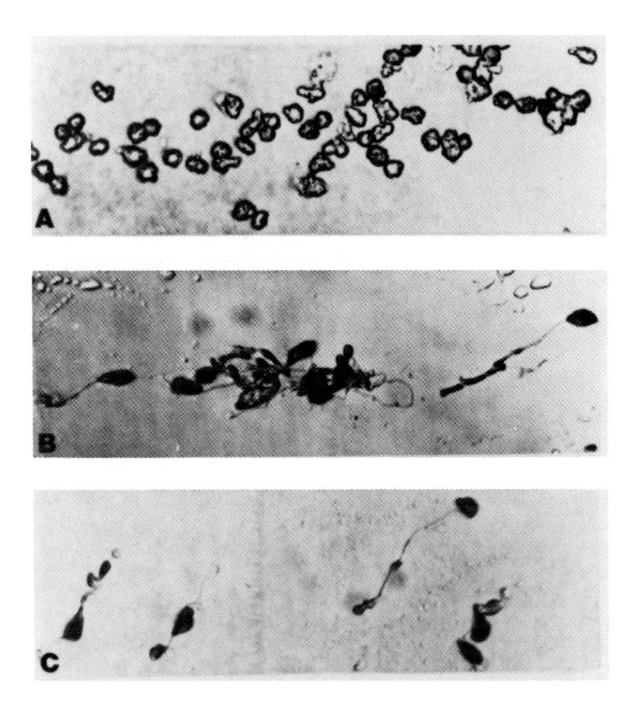

Figure 1. Isolated rod photoreceptor cells from normal mouse retinas at postnatal day 5 (A) and postnatal day 11 (B). C depicts a cell preparation from rd mouse at postnatal day 11. Magnification x 875. (Reprinted with publisher's permission from Scott et al., 1988.)

terminal. Because rod outer segments are enriched in phospholipids containing docosahexaenoyl chains, we compared cells that were dissociated at 5-6 days of age, when no rod outer segment is yet present, with those dissociated at 11-13 days of age, when rod outer segments have already begun to be differentiated.

Figure 1 shows the dissociated cells under light microscopy. Several morphological features were used to characterize the dissociated rod photoreceptor cells. Details are given elsewhere

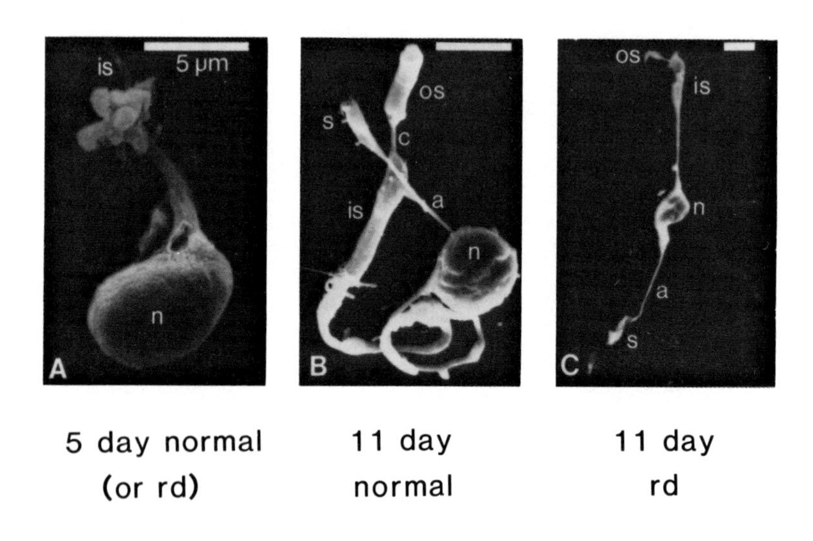

Figure 2. Scanning electron micrograph of isolated rod photoreceptors. The morphological appearance of photoreceptors at postnatal day 5 is similar in normal and rd mice. Nucleus (n) and short inner segment (is) with ruffled projections and single cilium. An axonal extension may be hidden beneath the cell. In B and C, the following additional structures are seen: os, outer segment; c, connecting cilium; a, axon; s, synaptic spherule. (Reprinted with publisher's permission from Scott et al., 1988)

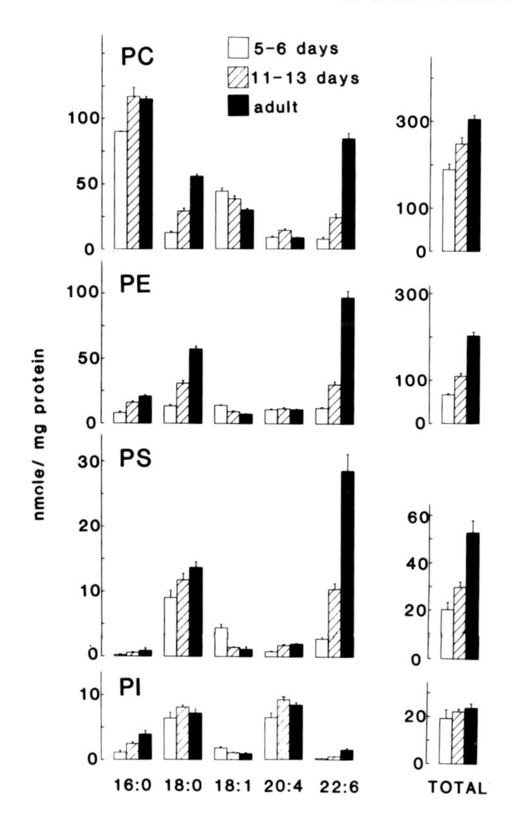

Figure 3. Changes in the phospholipid content and acyl chain composition during the postnatal development and differentiation of rod photoreceptor cells in the mouse. White bars (5-6 days), notched bars (11-13 days) and black bars (adults). PC, phosphatidylcholine; PE, phosphatidylethanolamine; PS, phosphatidylserine; and PI, phosphatidyl inositol. (Reprinted with publisher's permission from Scott et al., 1988.)

(Scott et al., 1988). Figure 2 depicts structural aspects of the dissociated cells as seen by scanning electron microscopy.

Figure 3 illustrates the biochemical differentiation of membrane phospholipids in rod photoreceptor cells. The most striking modification is the enrichment in docosahexaenoate in phosphatidylcholine, phosphatidylethanolamine and phosphatidylserine. Other fatty acyl groups are also increased; stearate increases in all three phospholipids, and palmitate increases in phosphatidyl-

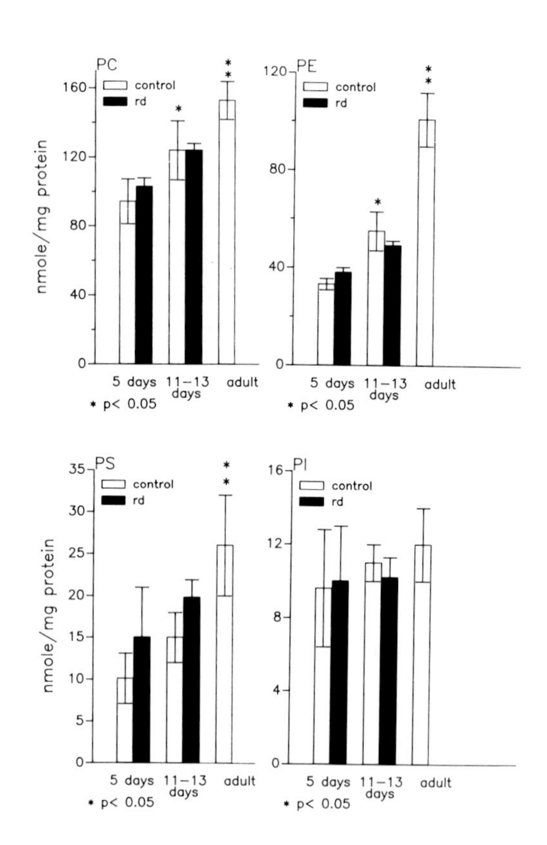

Figure 4. Endogenous fatty acyl chain content in individual phospholipids from rod photoreceptor cells isolated from the mouse retina at the indicated ages. Phosphatidylcholine and plasmalogens (PC), phosphatidylethanolamine and plasmalogens (PE), phosphatidylserine (PS) and phosphatidylinositol (PI).

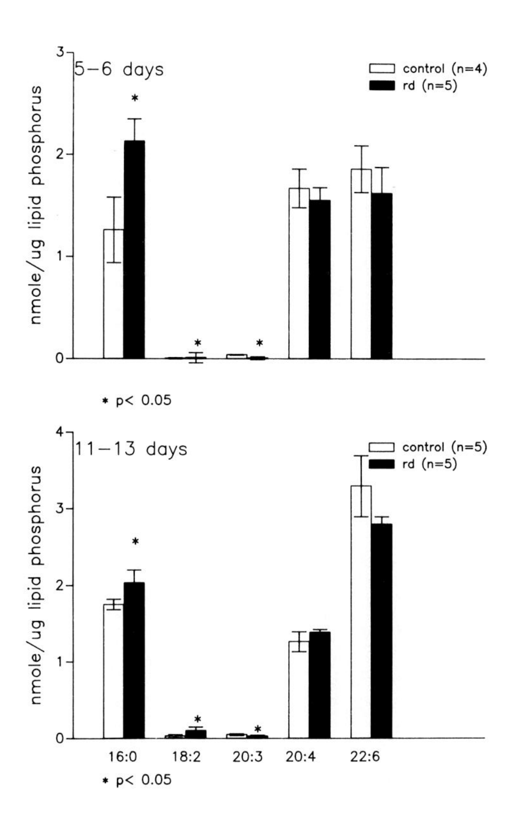

Figure 5. Content of fatty acids in phosphatidylcholine from control and rd mouse.

ethanolamine. It is of interest that while all phospholipids increased, oleate tended to decrease, particularly in phosphatidylcholine. We found major differences between the two polyunsaturated fatty acids; docosahexaenoic acid greatly increased, whereas arachidonic acid did not change in phosphatidylethanolamine, and only minor changes were observed in other phospholipids. Phosphatidylinositol is the precursor of the polyphosphoinositides and is part of the inositol lipid cycle engaged in cell signal transduction. One of the features of the inositol lipids is their relatively high proportion of the molecular species stearoyl-arachidonyl. In rod photoreceptors, most of this molecular species is already present at 5 days of postnatal development (Fig. 3), implying that the supply and utilization of arachidonic acid in phospholipid biosynthesis precedes that of docosahexaenoic acid.

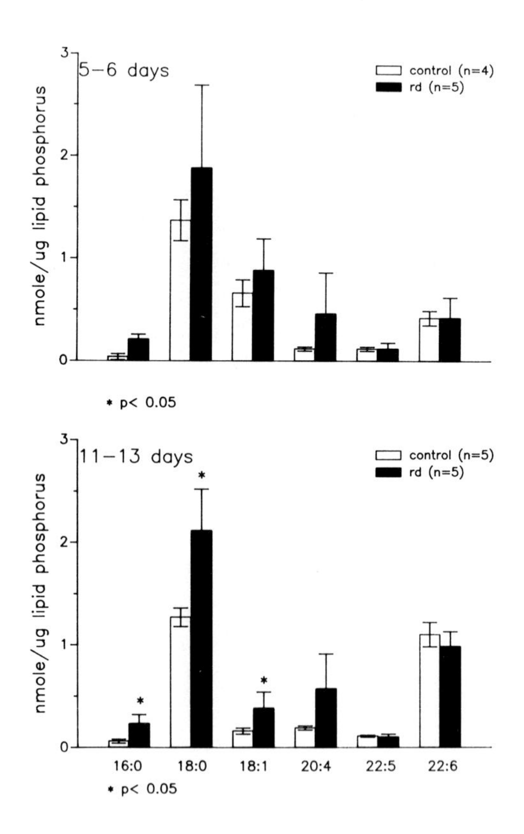

Figure 6. Content of fatty acids in phosphatidylethanolamine from control and rd mouse.

ALTERATIONS IN ENDOGENOUS FATTY ACID CONTENT AND COMPOSITION IN INDIVIDUAL PHOSPHOLIPIDS FROM ROD PHOTORECEPTOR CELLS ISOLATED FROM THE MUTANT RD MOUSE AT 5-6 DAYS OF POSTNATAL DEVELOPMENT

Striking changes were found in the endogenous fatty acids of individual phospholipids from rod photoreceptor cells isolated from rd mutant mice (Scott and Bazan, 1988). Fig. 4 illustrates the content of individual phospholipids. It does not include data for adult rd mice because the loss of rod cells led to extensive retinal atrophy and secondary loss of membrane lipids. These studies focused on developing cells, particularly on the issue of whether changes can be detected in endogenous phospholipids at 11-13, or at 5-6 days of postnatal development. At 11-13 days, there are already marked morphological alterations in the cells (Fig. 2) and biochemical modifications may already be a consequence of these abnormalities. These findings raised the question of whether membrane lipid alterations can be detected prior to morphological changes. In cells isolated at 5-6 days of postnatal development, selective alterations were found in the fatty acid content and composition of certain phospholipids. Phosphatidylcholine showed an increased content of 14:0 and of 22:5ω6, and a tendency to increase in palmitate (16:0). At the same time, there is a tendency towards a decrease in docosahexaenoic acid and in 22:5ω3. These changes are also seen later, at 11-13 days of age. It is very interesting that arachidonate (20:4) remains unaltered at both ages, strongly suggesting selective alterations in fatty acids other than members of the linoleic-arachidonic fatty acid family (Fig. 5).

In phosphatidylethanolamine, an enhancement in palmitate and in linoleate was found at 5-6 days of age, with a decreased content of a minor acyl group identified as 20:3. In this phospholipid, neither arachidonic nor docosahexaenoic acid content was changed. At 11-13 days of age, only a tendency to decrease was seen in docosahexaenoic acid; no change in arachidonate was observed (Fig. 6).

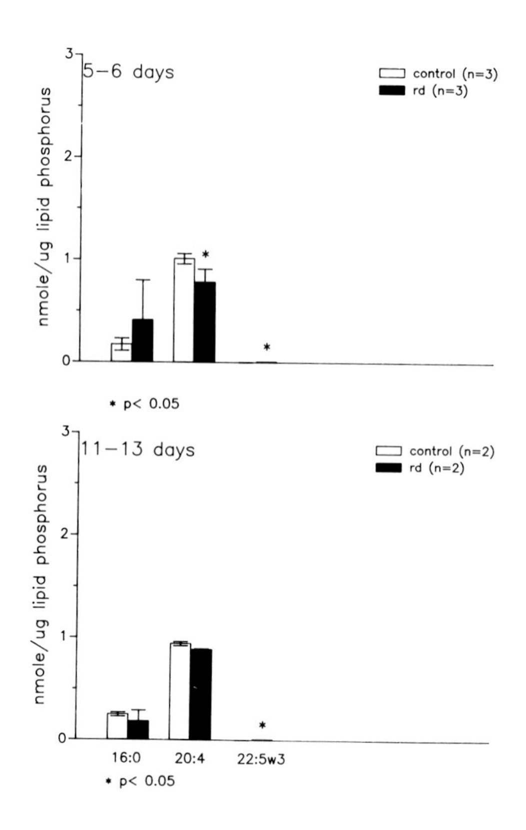

Figure 7. Content of fatty acids in phosphatidylserine from control and rd mouse.

In phosphatidylserine, at 5-6 days of age, only palmitate changed significantly (increased). At 11-13 days of age, palmitate, stearate and oleate were seen to increase. However, docosahexaenoate was not changed at any of these ages (Fig. 7).

Phosphatidylinositol displayed only minor changes when control and rd mice are compared (Fig. 8).

PHOSPHOLIPID SYNTHESIS AND SUPPLY OF POLYUNSATURATED FATTY ACIDS TO PHOTORECEPTOR CELLS

Photoreceptor Shedding and Phagocytosis by Retinal Pigment Epithelium Cells.

The visual cells are renewed daily by a shedding of the tips of the outer segments followed by engulfment in the retinal pigment epithelial cells. Subsequently the phagolysosomal system of these cells digests the photoreceptor membranes

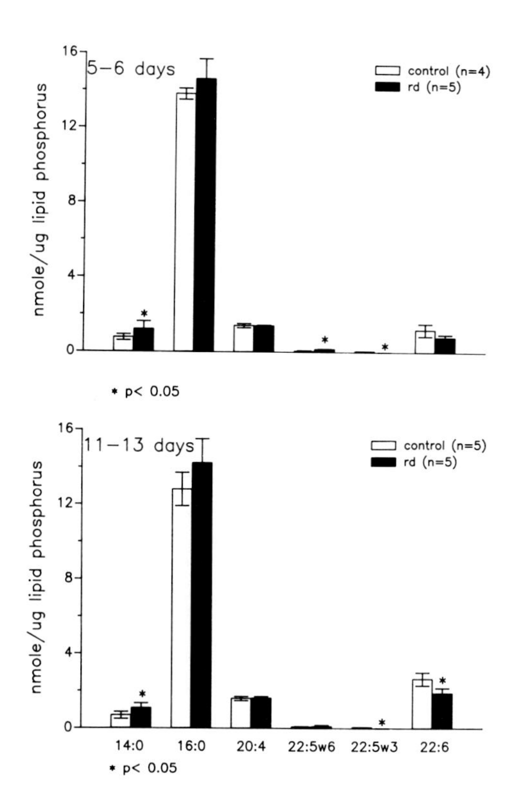

Figure 8. Content of fatty acids in phosphatidylinositol from control and rd mouse.

(Bok, 1985). The fate of docosahexaenoate of phospholipids from photoreceptor membranes during and after each renewal cycle is not well understood. It has been well established for several years that retinal docosahexaenoate remains unchanged in quantity even after prolonged dietary deprivation of its precursors. To achieve significant depletion, deprivation was extended in several studies over one generation (Salem et al., 1986). Observations from these studies indicate that the retina has the intrinsic capacity to avidly retain docosahexaenoate in spite of its daily renewal. Figure 9 outlines a model of the events that the fatty acid may undergo. After shedding and phagocytosis docosahexaenoic acid may be retrieved by a short loop to the inner segment of the visual cell through the interphotoreceptor matrix. Interphotoreceptor retinal binding protein as well as other as yet unidentified proteins of the interphotoreceptor matrix from the monkey eye contain sizable amounts of non-covalently bound docosahexaenoic acid (Bazan et al., 1985a). In addition, the interphotoreceptor matrix may also play the role of a retrieval route for docosahexaenoic acid arising from the metabolism of visual cells should it involve the release of docosahexaenoate. In fact active release of free docosahexaenoic acid (and release to the media) was found in the retina implying the presence of phospholipase A_2 (Aveldano and Bazan, 1974; Aveldano de Caldironi et al., 1981; Giusto and Bazan, 1983). In rod outer segments a light sensitive, G-protein mediated phospholipase A_2 activity acting on docosahexaenoyl-phospholipids has been reported (Bazan et al., 1988). Moreover, acylation and deacylation reactions also are active in outer segments (Zimmerman and Keys,1988). An unexplored additional route of retention may involve the blood stream, the long loop of Figure 9. After digestion of photoreceptor membranes in the retinal pigment epithelium, docosahexaenoic acid as such or as an acyl chain of a glycerolipid may be released to the choriocapillaris. The liver will take up these lipids and release them back to the blood stream when the inner segments of the visual cells set in motion the biogenesis of photoreceptor membranes.

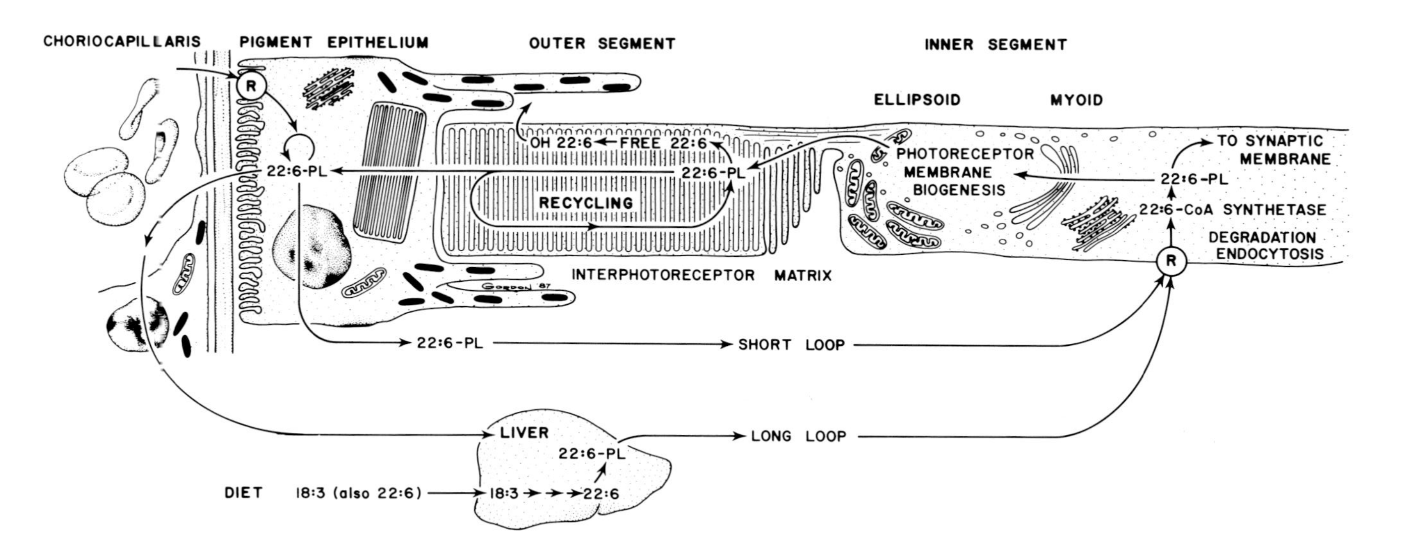

Figure 9. Model depicting metabolic events of docosahexaenoic acid during the life cycle of visual cells. Most of the dietary linolenic acid (18:3) is indicated to be metabolized by the liver to docosahexaenoic acid and then released to the blood stream as a 22:6 - PL (a glycerolipid in a lipoprotein). The long loop supplies this 22:6 to the visual cell through a receptor-mediated event at the inner segment and/or retinal pigment epithelium. A short loop and a recycling event are also depicted. The generation of free 22:6 and of oxygenated metabolites that may contribute as intercellular signaling messengers is also indicated.

Brain cells as well as neural retinal cells may also have similar requirements for docosahexaenoic acid through their life cycle, since they also contain large amounts of this fatty acid (Scott and Bazan, 1987). The recapture of docosahexaenoic acid prevalently by the eye (and brain), and to a much lesser extent by non-neural tissues, may involve a selective recognition and uptake by the microvasculature at the retina (and brain) and/or choriocapillaris. Since the latter are the main suppliers of nourishment for the photoreceptor cells, they may actively take up and deliver docosahexaenoic acid to the retinal pigment epithelial cells. This fatty acid is likely part of a lipoprotein secreted by the liver and also comprises either a neutral lipid or a phospholipid. Metabolic processing may take place during its transit through the retinal pigment epithelium. An additional feature of the model is that receptors are present either/or in both retinal pigment epithelial cells and inner segments of visual cells. Once docosahexaenoic acid is taken up by the cell, a low Km docosahexaenoyl Coenzyme A synthetase (Reddy et al.,1984; Reddy and Bazan, 1984a and 1984b) gives rise to an activated fatty acid that does not accumulate, and that is subsequently acylated into glycerolipids.

A third alternative may be that the renewal of outer segments involves the sorting out of docosahexaenoyl-phospholipids in the distal tips of photoreceptors and that they are conserved by the visual cells. Such selectivity implies a high degree of rearrangement close to the tips of photoreceptors and therefore the lipid arriving during phagocytosis will be docosahexaenoate-poor. In fact the content of this fatty acid in the retinal pigment epithelium is much lower than that of the photoreceptor cells (Handelman and Dratz, 1986). The short loop, the long loop and the selective preservation of docosahexaenoate in the outer segment are not mutually exclusive. Since docosahexaenoic acid is such a valuable currency for visual cells several convergent events may lead to tenacious retention of the fatty acid. Additional protective features may be also expressed in photoreceptors and related cells, such as the ability to maintain high levels of antioxidants (e.g. vitamin E) and antioxidant enzymes.

Photoreceptor Cell Differentiation.

To probe components of the model depicted in Figure 9 some of the studies described here were performed during the postnatal development and differentiation of photoreceptor cells. We have described above that there is in fact a selective accumulation of docosahexaenoate in differentiating photoreceptor cells. Since the construction of the outer segment has to be accomplished during a few days, labeling experiments using (1-^{14}C) linolenic acid in newborn mice were performed (Scott et al, 1987b; Cai et al, 1988). The very rapid uptake by the liver of this fatty acid, the subsequent elongation and desaturation to docosahexaenoic acid in the liver, and the relatively low uptake by brain and retina of the labeled precursor lead to the suggestion that the liver plays a central role in supplying docosahexaenoic acid to the differentiating photoreceptor cell and to the synapses (Scott and Bazan, 1987; Cai et al., 1988). This support function of the liver may also take place during the entire life cycle of the visual cell to assure the maintenance of a high content of docosahexaenoic acid. The visual cells and the retina, although also able to elongate and desaturate precursors of docosahexaenoic acid, are not as efficient as the liver is in this regard. In fact 1-^{14}C eicosapentaenoic acid (20:5) intravitreally injected in the rat eye is elongated and desaturated to docosahexaenoic acid rather rapidly (Bazan et al., 1982a).

CONCLUSIONS

The alterations in fatty acyl chains of some phospholipids in dissociated rod photoreceptors cells from rd mouse at 5-6 days of postnatal age imply that the expression of the mutation gives rise to early changes in the chemical compostion of membranes. These changes and the abnormalities in the cyclic GMP metabolism (Farber and Lolley, 1974) are the only known alterations preceding the failure to differentiate outer segments. Whether or not these changes are unique to the rd model or extend to other inherited retinal degenerations remains to be elucidated. However, the compositional and metabolic studies here summarized have given rise to a model that may explain the events and pathways that lead to the uniquely enriched photoreceptor membranes in docosahexaenoyl-phospholipids. These new observations and ideas will hopefully contribute to the understanding of photoreceptor cell degenerations.

In inherited retinal degenerations, membrane compositional alterations may reflect defects in the enzyme/enzymes responsible for the elongation and/or desaturation of essential fatty acids or for the synthesis of the unique hexaenoic or supraenoic molecular species of phospholipids. In view of the model given in Figure 9, abnormal genetic expression may result in impaired uptake or transport of docosahexaenoic acid (or of docosahexaenoyl glycerolipids). Moreover, alterations in the signals required to turn on the interorgan cycles of polyunsaturated fatty acids (retina, retinal pigment epithelium, liver)may occur. In age-related retinal diseases such as senile macular degenerations, oxidative stress involving one or more of the following factors may lead to loss of function: antioxidant deficiency (vitamin E, selenium or antioxidant enzymes), light damage, free radical-induced injury, lipid peroxidation and impairments in the nourishment of the photoreceptor cells. These impairments may lead to shortage of the supply of docosahexaenoic acid through the interorgan cycle. In light-induced retinal regenerations, perturbation of the photoreceptors may be triggered by lipid peroxidation and by alterations in the ability to take up precursors and to synthesize appropriate molecular species of phospholipids and other membrane components. In cystoid macular edema, lipid oxidation products may alter the cycle supporting the renewal of photoreceptors and at the same time, either directly or through the generation of other lipid mediators, may in turn produce microvascular injury.

In sum, the impairment of the unique structural features of visual cells' phospholipids, the hexaenoic and supraenoic molecular species, and the selective pathways and interorgan cycles supporting membrane biogenesis during the differentiation and renewal of visual cells provide a new target towords understanding retinal degenerations.

ACKNOWLEDGEMENTS

This work was supported in part by NIH grant EY04428 and the Edward Schlieder Foundation. Figures 4 through 8 are plots of data from Scott et al., 1988.

REFERENCES

Anderson RE, Maude MB, Lewis RA, Newsome DA, Fishman GA (1987). Abnormal plasma levels of polyunsaturated fatty acid in autosomal dominant retinitis pigmentosa. Exp Eye Res 44:155-159.

Aveldano de Caldironi MI, Bazan NG (1977). Acyl groups, molecular species and labeling by ^{14}C-glycerol and ^{3}H-arachidonic acid of vertebrate retina glycerolipids. Adv Exp Med Biol 83:397-404.

Aveldano de Caldironi MI, Bazan NG (1980). Composition and biosynthesis of molecular species of retinal phosphoglycerides. Neurochem Internat 1:381-392.

Aveldano de Caldironi MI, Giusto NM, Bazan NG (1981). Polyunsaturated fatty acids of the retina. Prog in Lipid Res 20:49-57.

Aveldano MI (1987). A novel group of very long chain polyenoic fatty acids in dipolyunsaturated phosphatidylcholines from vertebrate retina. J Biol Chem 262:1172-1179.

Aveldano MI, Bazan NG (1972). High content of docosahexanoate and of total diacylglycerol in retina. Biochem Biophys Res Comm 48:689-693.

Aveldano MI, Bazan NG (1973). Fatty acid

composition and level of diacylglycerols and phosphoglycerides in brain and retina. Biochim Biophys Acta 296:1-9.

Aveldano MI, Bazan NG (1974). Displacement into incubation medium by albumin of highly unsaturated retina free fatty acids arising from membrane lipids. Febs Letters 40:53-56.

Aveldano MI, Bazan NG (1983). Molecular species of phosphatidylcholine, -ethanolamine, -serine, and inositol in microsomal and photoreceptor membranes of bovine retina. J Lipid Res 24:620-627.

Aveldano MI, Pasquare de Garcia SJ, Bazan NG (1983). Biosynthesis of molecular species of inositol, choline, serine, and ethanolamine glycerophospholipids in the bovine retina. J Lipid Res 24:628-638.

Aveldano MI, Sprecher H (1987). Very long chain (C_{24} to C_{36}) polyenoic fatty acids of the n-3 and n-6 series in dipolunsaturated phosphatidylcholines from bovine retina. J Biol Chem 262:1180-1186.

Barassi CA, Bazan NG (1974). Metabolic heterogeneity of phosphoglyceride classes and subfractions during cell cleavage and early embryogenesis: Model for cell membrane biogenesis. J Cell Physiol 84:101-114.

Bazan HEP, Careaga MM, Sprecher H, Bazan NG (1982a). Chain elongation and desaturation of eicosapentaenoate to docosahexaenoate and phospholipid labeling in the rat retina in vivo. Biochim Biophys Acta 712:123-128.

Bazan HEP, Ridenour B, Birkle DL, Bazan NG (1986b). Unique metabolic features of docosahexaenoate metabolism, related to functional roles in brain and retina. In Horrocks L, Freysz L, Toffano G (eds): "Phospholipid Research and the Nervous System. Biochemical and Molecular Pharmacology," Padova: Liviana Press, pp 67-78.

Bazan HEP, Sprecher H, Bazan NG (1984a). De novo biosynthesis of docosahexaenoyl phosphatidic acid in bovine retinal microsomes. Biochim Biophys Acta 796:11-19.

Bazan NG (1982a). Biosynthesis of phosphatidic acid and polenoic phospholipids in the central nervous system. In Horrocks LA, Ansell GB,

Porcellatti G (eds): "Phospholipids in the Nervous System, Vol 1, Metabolism," New York: Raven Press, pp49-62.

Bazan NG (1982b). Metabolism of phosphatidic acid. In Lajtha F. (ed): "Handbook of Neurochemistry", Vol. 3, New York: Plenum Press, pp 17-39.

Bazan NG (1982c). Metabolism of phospholipids in the retina. Vision Res 22:1539-1548.

Bazan NG (1988). Lipid-derived metabolites as possible retina messengers: Arachidonic acid, leukotrienes, docosanoids, and platelet activating factor. In Redburn D, Pasantes-Morales H (eds): "Extracellular and Intracellular Messengers in the Vertebrate Retins", New York: Alan R. Liss, pp 269-300.

Bazan NG, Bazan HEP, Birkle DL, Rossowska M (1987b). Synthesis of leukotrienes in the frog retina and retinal pigment epithelium. J Neurosci Res 18:591-596.

Bazan NG, Birkle DL (1987). Polyunsaturated fatty acids and inositol phospholipids at the synapse in neuronal responsiveness. In Ehrlich Y, et al (eds): "Molecular Mechanisms of Neuronal Responsiveness," New York: Plenum Press, pp 45-68.

Bazan NG, Birkle DL, Reddy TS (1984b). Docosahexaenoic acid (22:6, n-3) is metabolized to lipoxygenase reaction products in the retina. Biochem Biophys Res Comm 126:741-747.

Bazan NG, Birkle DL, Reddy TS (1985). Biochemical and nutritional aspects of the metabolism of polyunsaturated fatty acids and phospholipids in experimental models of retinal degeneration. In LaVail MM, Anderson G, Hollyfield J (eds): "Retinal Degeneration: Contemporary Experimental and Clinical Studies," New York: Alan R. Liss, pp 159-187.

Bazan NG, di Fazio de Escalante MS, Careaga MM, Bazan HEP, Giusto NM (1982b). High content of 22:6 (docosahexaenoate) active [2-^{3}H]glycerol metabolism of phosphatidic acid from photoreceptor membranes. Biochim Biophys Acta 712:702-706.

Bazan NG, Giusto NM (1980). Docosahexaenoyl chains are introduced in phosphatidic acid during de novo synthesis in retinal microsomes. In Kates

M, Kuksis A (eds): "Control of Membrane Fluidity," New Jersey: Humana Press, pp 223-236.
Bazan NG, Marcheselli VL, Ma AD, Jelsema C (1988). The role of GTP binding proteins (G-proteins) in the hydrolysis of docosahexaenoyl (22:6)-phospholipids in photoreceptors. [Suppl] Invest Ophthalmol Vis Sci, 29:83.
Bazan NG, Rossowska M, Woodland JM, Bazan HEP, Birkle DL (1987a). Changes in peptide and non-peptide leukotrienes (LT) of the retinal pigment epithelium correlated with photoreceptor shedding and phagocytosis in xenopus laevis. [Suppl] Invest Opthalmol Vis Sci, 28:185.
Bazan NG, Reddy TS (1985). Retina. In Lajtha A, (eds): "Handbook of Neurochemistry, Vol. 8," New York: Plenum Press, pp 507-575.
Bazan NG, Reddy TS, Bazan HEP, Birkle DL (1986a). Metabolism of arachidonic and docosahexaenoic acids in the retina. Prog Lipid Res 25:595-606.
Bazan NG, Reddy TS, Redmond TM, Wiggert B, Chader GJ (1985a). Endogenous fatty acids are covalently and non covalently bound to interphotoreceptor retinoid-binding protein in the monkey retina. J Biol Chem 260:13677-13680.
Bazan NG, Scott BL (1987b). Docosahexaenoate acid metabolism and inherited retinal degeneration. In Hollyfield JG, Anderson RE, LaVail MM (eds): "Degeneration Retinal Disorders: Clinical and Laboratory Investigations," New York: Alan R. Liss, pp 103-118.
Birkle DL, Bazan NG (1984). Effects of K^+ depolarization on the synthesis of prostaglandins and hydroxyeicosatetra(5,8,11,14)enoic acids (HETE) in the rat retina. Evidence for esterification of 12-HETE in lipids. Biochim Biophys Acta 795:564-573, 1984.
Bok D (1985). Retinal photoreceptor pigment epithelium interactions. Invest Ophthalmol Vis Sci 26:1659-1693.
Braquet P, Touqui L, Shen TY, Vargaftig BB (1987). Perspectives in platelet activating factor research. Pharmacol Rev 39:97-145.
Cai F, Scott BL, Bazan NG (1988). Delivery of omega-3 fatty acids to developing photoreceptor cells. Invest Ophthalmol Vis Sci [Suppl] 29:245.

Converse CA, Hammer HM, Packard CJ, Shepherd J (1983). Plasma lipid abnormalities in retinitis pigmentosa and related conditions. Trans Ophthalmol Soc UK 103:508-512.
Farber DB, Lolley RN (1974). Cyclic guanosine monophosphate: Elevation in degenerating photoreceptor cells of the C3H mouse retina. Science 186:449-451.
Giusto NM, Bazan NG (1979). Phosphatidic acid in retinal microsomes contains a high proportion of docosahexaenoate. Biochem Biophys Res Comm 91:791-794.
Giusto NM, Bazan NG (1983). Anoxia-induced production of methylated and free fatty acids in retina, cerebral cortex and white matter. Comparison with triglycerides and with other tissues. Neurochem Pathol 1:17-41.
Handelman GJ, Dratz EA (1986). The of role of antioxidants in the retina and retinal pigment epithelium and the nature of prooxidant-induce damage. Adv in Free Radical Biol and Med 2:1-89.
Neuringer M, Conner WE (1986). N-3 fatty acids in the brain and retina: Evidence for their essentiality. Nutrition Rev 44:285-294.
Neuringer M, Conner WE, Daigle D, Barstad L (1988). Electroretinogram abnormalities in young infant rhesus monkeys deprived of omega-3 fatty acids during gestation and postnatal development or only postnatally. Invest Ophthalmol Vis Sci [Suppl] 29:145.
Neuringer M, Conner WE, Luck SJ (1985). Suppression of ERG amplitude by repetitive stimulation in rhesus monkeys deficient in retinal docosahexaenoic acid. Invest Ophthalmol Vis Sci [Suppl] 54:31.
Neuringer M, Conner WE, Van Petten C, Barstad L (1984). Dietary omega-3 fatty acid deficiency and visual loss in infant rhesus monkeys. J Clin Invest 73:272-276.
Panetta T, Marcheselli VL, Braquet P, Spinnewyn B, Bazan NG (1987). Effects of a platelet activating factor antagonist (BN 52021) on free fatty acids, diacylglycerols, polyphosphoinositides and blood flow in the gerbil brain: Inhibition of ischemia-reperfusion induced cerebral injury. Biochem Biophys Res Comm

149:580-587.
Reddy TS, Bazan NG (1984a). Activation of polyunsaturated fatty acids by rat tissues in vitro. Lipids 19:987-989.
Reddy TS, Bazan NG (1984b). Synthesis of arachidonoyl coenzyme A and docosahexaenoyl coenzyme A in retina. Curr Eye Res 3:1225-1232.
Reddy TS, Bazan NG (1985a). Synthesis docosahexaenoyl-, arachidonoyl- and palmitoyl-coenzyme A in the ocular tissues. Exp Eye Res 41:87-95.
Reddy TS, Birkle DL, Armstrong D, Bazan NG (1985b). Change in content, incorporation and lipoxygenation of docosahexaenoate acid in retina and retinal pigment epithelium in canine ceroid lipofuscinosis. Neurosci Lett 59:67-72.
Reddy TS, Sprecher H, Bazan NG (1984). Long-chain acyl coenzyme A synthetase from rat brain microsomes: Kinetic studies using [1 ^{14}C]docosahexaenoic acid substrate. Eur J. Biochem 145:21-29.
Salem N Jr, Kim H-Y, Yergey JA (1986). Docosahexaenoic acid: Membrane function and metabolism. "Health Effects of Polyunsaturated Fatty Acids in Seafood, Vol. 15," London, England: Academic Press, pp 263-317.
Scott BL, Bazan NG (1988). Developing retinal photoreceptor cells accumulate polyunsaturated fatty acids. Amer Soc Neurochem 79:108.
Scott BL, Bazan NG (1987). Docosahexaenoate synthesis in the developing mouse brain. J Neurochem [Suppl] 48:S80C.
Scott BL, Lolley RN, Bazan NG (1988). Developing rod photoreceptors from normal and mutant rd mouse retinas: Altered fatty acid composition early in development of the mutant. J Neurosci Res 20:202-211.
Scott BL, Moises J, Bazan NG (1987d). Maternal supply of n-3 essential fatty acids to the developing mouse retina. Soc Neurosci 13:239.
Scott BL, Moises J, Lolley RN, Bazan NG (1987c). Selective accumulation of docosahexaenoic acid (DHA) in dissociated rod photoreceptor cells during mouse postnatal development. Invest Ophthalmol Vis Sci [Suppl] 28:340.
Scott BL, Reddy TS, Bazan NG (1987).

Docosahexaenoate metabolism and fatty acid composition in developing retinas and rd mutant mice. Exp Eye Res 44:101-113.
Wiegand RD, Anderson RE (1983). Phospholipid molecular species of frog rod outer segment membranes. Exp Eye Res 37:159-173.
Zimmerman WF, Keys S (1988). Acylation and deacylation of phospholipids in isolated bovine rod outer segments. Exp Eye Res 47:247-260.

Inherited and Environmentally Induced
Retinal Degenerations, pages 217–232

EFFECTS OF RETINAL DEGENERATIONS ON THE CONE MATRIX SHEATH

Lincoln V. Johnson[1], Janet C. Blanks[1,2,3]
and Gregory S. Hageman[4]

Departments of [1]Anatomy & Cell Biology and [2]Ophthalmology, University of Southern California, School of Medicine, Los Angeles, California; [3]Estelle Doheny Eye Foundation, Los Angeles, California; and [4]Department of Ophthalmology, Bethesda Eye Institute, St. Louis University, St. Louis, Missouri, USA

INTRODUCTION

A number of animal models exhibiting hereditary degeneration of the neural retina have been employed in studies of the cell biology of degenerative processes. Our knowledge of the pathogenetic mechanisms of hereditary retinal degenerations, especially those which involve the light sensitive photoreceptor cells, has been enhanced by such studies. Animals with various forms of photoreceptor degeneration have been utilized, these include: rd (LaVail and Sidman, 1974), rds (Van Nie et al., 1978; Sanyal et al., 1980), and pcd (Mullen and LaVail, 1975; LaVail et al., 1982) mice, RCS rats (Bourne et al., 1938; Dowling and Sidman, 1962; Bok and Hall, 1971; Mullen and LaVail, 1976), a variety of dogs (see Aguirre and Rubin, 1979), cats (Narfstrom, 1985), chickens (Ulshafer et al., 1984; Ulshafer and Allen, 1985) and primates (El-Mofty et al., 1980; Vainisi et al., 1974). No known animal model mimics exactly the retinal degenerations observed in human retinitis pigmentosa and it appears that the variety of forms of retinal degeneration observed in humans and animal models may be the result of a variety of genetic disorders with varied biochemical manifestations.

However, it is hoped that the investigation of photoreceptor degeneration in such animal models will provide a basis for understanding the pathogenesis of diseases such as retinitis pigmentosa.

The common progression of most cases of human and animal photoreceptor degenerations is the initial loss of outer segment membrane, that portion of the photoreceptor cell which contains the photosensitive pigment(s), followed by cell body degeneration and death. In a number of types of photoreceptor degenerations cone photoreceptor cells are markedly less susceptible to degeneration than are rod photoreceptor cells (see LaVail, 1981; Schmidt, 1985). In the rd mutant mouse, as early as 18 days after birth almost all surviving photoreceptor cells are cones (Noell, 1958; Carter-Dawson et al., 1978). Similarly, in the pcd mouse it has been shown that rods degenerate faster than cones so that, at 6.5 months, only 20% of rods remain compared to 50% of cones (LaVail, 1981; LaVail et al., 1982). Rod degeneration also precedes cone degeneration in the RCS rat (LaVail, 1981) and in degenerative retinae of Irish setters, collies, and miniature French poodles (Buyukmihoi et al., 1980; Aguirre et al., 1978, 1982a,b; Woodford et al., 1982). In addition, a number of cases of human retinitis pigmentosa have been described in which a single layer of cone photoreceptor cell bodies and no rod photoreceptors remain in the peripheral retina; cone degeneration is even less advanced in the fovea (Kolb and Gouras, 1974; Szamier and Berson, 1977, 1982). In animal models in which photoreceptor degeneration is initiated by depletion of dietary vitamin A (Carter-Dawson et al., 1979) or by exposure to iodoacetate (Noell, 1952, 1965) rod degeneration precedes that of cones.

Understanding the reason for the increased survivability of cones as compared to rods in degenerative retinae may provide us with an important key with which to help unravel the mysteries of photoreceptor degenerations. Because there are relatively few documented differences between these two cell types, one can only speculate on the possible basis for the differential susceptibilities of rod and cone photoreceptors to degeneration. Three likely alternatives are (1) differences in their biochemical composition and metabolism, (2) differential associations of the two types of photoreceptors with the retinal pigmented epithelium, and (3) differences in the interaction of the two cell types with the surrounding interphotoreceptor matrix.

In recent studies of a variety of vertebrate retinae we have made observations of potential relevance to the last of the above alternatives. We have demonstrated a significant difference in the composition of the interphotoreceptor matrix (IPM) surrounding cone photoreceptor inner and outer segments as compared to the IPM surrounding rod photoreceptors in a variety of species from birds to primates. In all species examined, peanut agglutinin (PNA), a lectin with high binding affinity for galactose-galactosamine disaccharide linkages, binds selectively to cone cell membranes and to discrete, cylindrical domains of interphotoreceptor matrix which ensheath cone, but not rod, inner and outer segments (Johnson et al., 1986; Hageman and Johnson, 1986; Sameshima et al., 1987; Blanks et al., 1988). These interphotoreceptor matrix domains, termed "cone matrix sheaths" appear to be proteoglycan-like and contain chondroitin 6-sulfate glycosaminoglycan as a major constituent (Hageman and Johnson, 1987; Varner et al., 1987). In the studies described here, we have examined the fate of the cone matrix sheaths in a variety of animal models exhibiting inherited and experimentally induced photoreceptor degeneration.

METHODS

At various postnatal developmental time points eyes were dissected from mice homozygous for the rd (retinal degeneration) or the rds (retinal degeneration slow) gene, or from homozygous mutant RCS rats and fixed by immersion in 4% paraformaldehyde in 100 mM sodium cacodylate buffer (pH 7.2). Following fixation and subsequent rinsing, retinas were dissected free from the sclera, embedded in acrylamide and sectioned as described previously (Johnson and Blanks, 1984). Retinal sections were exposed to fluorescein conjugated peanut agglutinin (FITC-PNA, Vector Laboratories, Burlingame, CA) to label the cone-associated extracellular matrix using our published procedures

(Johnson and Blanks, 1984; Johnson et al., 1986). Eyes from rats with experimentally induced taurine deficiency (Hageman and Schmidt, 1987) were treated similarly, as were age-matched normal control animals for all groups.

RESULTS

A. rd Mouse

At postnatal day 12 a slight reduction in the thickness of the outer nuclear layer (ONL) is noted reflecting degeneration and death of photoreceptor cells. Cone matrix sheaths labeled by PNA show signs of shortening and slight disruption in the distal (outer segment-associated) regions while inner segment-associated matrix remains heavily labeled and relatively intact (Fig. 1A). By 16 days postnatally, a marked reduction in the thickness of the outer nuclear layer is noted, reflective of a major loss of photoreceptors. Despite this PNA-labeled cone matrix sheath components remain between the ONL and the retinal pigmented epithelium (Fig. 2A). Similar observations are made in retinas of animals sixty days postnatally (Fig. 3A). Even at this late point in the degeneratiave process, when few to no photoreceptors remain, and those that do appear to be cones (Carter-Dawson et al., 1978), there is evidence of remaining cone matrix sheath material. Over this same period, little change is noted in the cone matrix sheaths in control retinas except for their elongation concomitant with outer segment development (Figs. 1B, 2B, 3B).

B. rds Mouse

Homozygous rds mice do not form outer segments (Sanyal and Jansen, 1981). Thus at 60 days postnatally, when approximately 50% of photoreceptor cells have degenerated (Sanyal et al., 1980), cone matrix sheaths in these animals are comprised of only inner segment-associated material and appear shorter than in controls in which outer segment-associated matrix is present (Figs. 4A,B). Late in the degenerataive process when almost all photoreceptors have degenerated (10-12 months postnatally), remnants of PNA-binding cone matrix sheath components remain detectable (not shown).

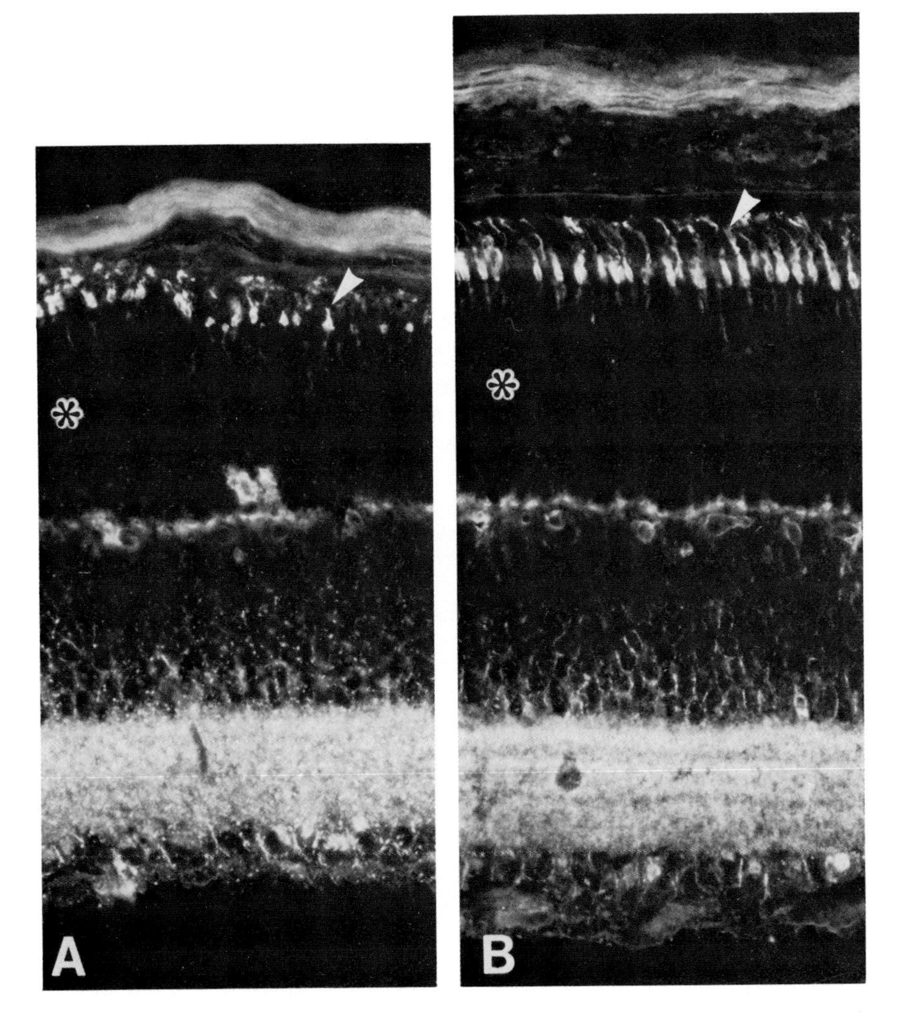

Figure 1. Fluorescence micrographs of retinal sections from postnatal day 12 rd (A) and normal control (B) mice exposed to fluorescein-conjugated PNA. A slight reduction in the thickness of the outer nuclear layer (*) is apparent in the mutant retina as is shortening and disruption of the PNA-labeled cone matrix sheaths (arrowheads).

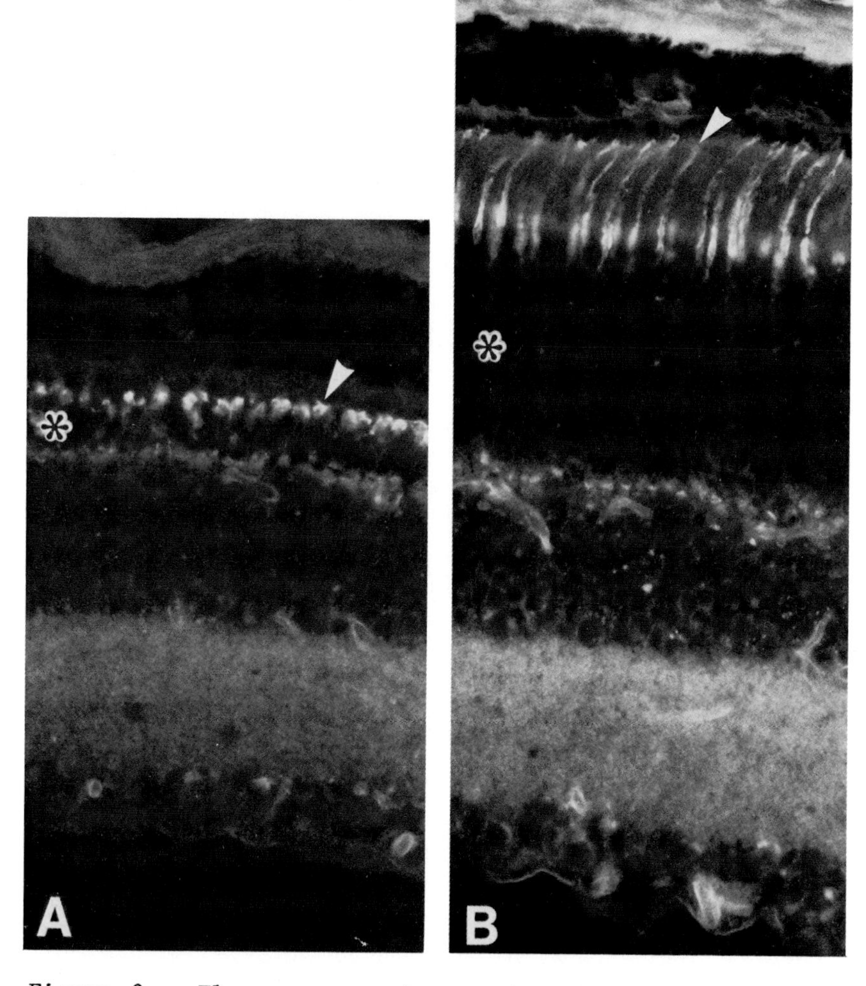

Figure 2. Fluorescence micrographs of retinal sections from postnatal day 16 rd (A) and normal control (B) mice exposed to fluorescein-conjugated PNA. A marked reduction in the thickness of the outer nuclear layer (*) has occurred as the result of massive photoreceptor cell death. PNA-labeled cone matrix sheath material (arrowhead in A) remains but is condensed and poorly organized compared to that of the control (arrowhead in B).

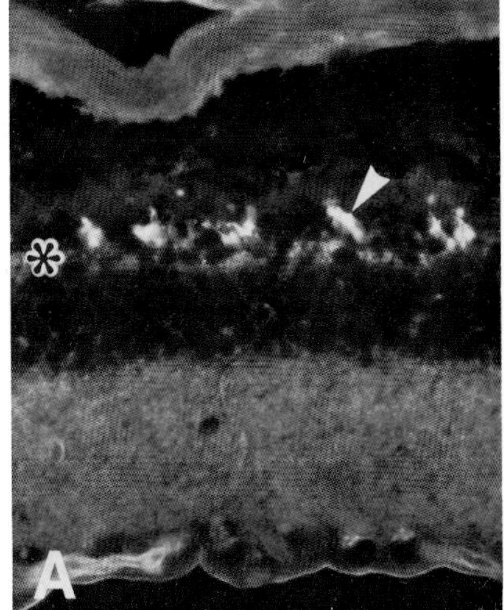

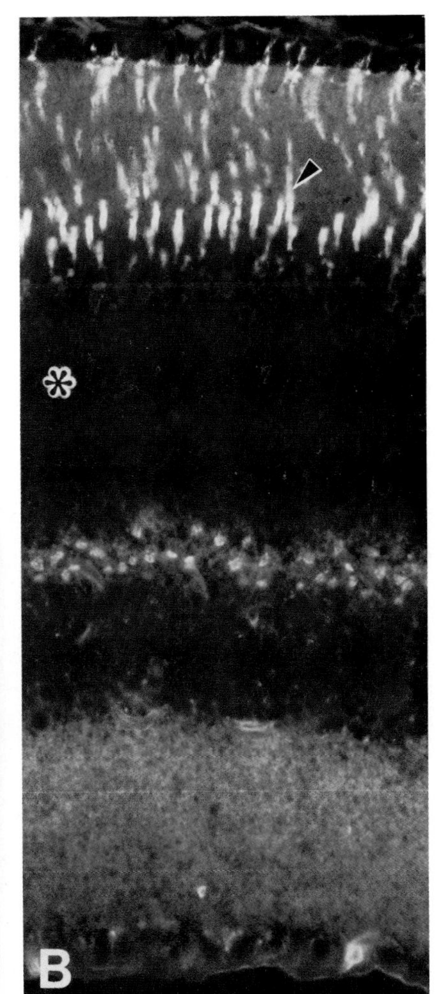

Figure 3. Fluorescence micrographs of retinal sections from postnatal day 60 rd (A) and normal control (B) mice exposed to fluorescein-conjugated PNA. Only few photoreceptor cell nuclei remain in the outer nuclear layer (*) of the mutant retina (A). However, even at this late point in the degenerative process, significant amounts of PNA-labeled material are present in adjacent areas (arrowhead in A). The cone matrix sheaths in the normal retina (arrowheads in B) appear discontinuous due to the tangential nature of the section.

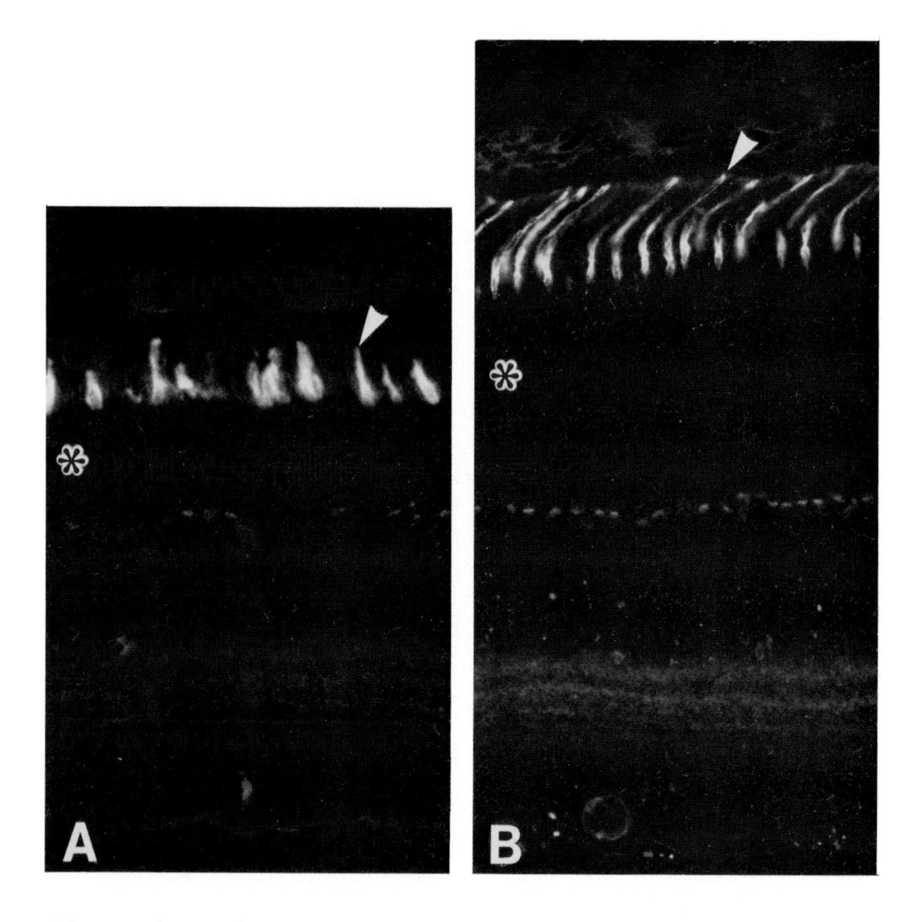

Figure 4. Fluorescence micrographs of retinal sections from postnatal day 60 rds (A) and normal control (B) mice exposed to fluorescein-conjugated PNA. The cone matrix sheath in the rds retina (arrowhead A) is shorter (inner segment-associated) and more diffusely stained than in the normal control (arrowhead in B) most likely due to the lack of outer segment formation by the mutant photoreceptor cells. Some decrease in the thickness of the outer nuclear layer (*) is also noted in the mutant retina, indicative of degenerative cell loss.

C. RCS Rat

The RCS rat, which suffers photoreceptor cell degeneration as a result of abnormal phagocytic activity by the adjacent pigmented epithelium (Bok and Hall, 1971; Mullen and LaVail, 1976), has relatively normal appearing cone matrix sheaths at postnatal day 20 (Fig. 5A) despite the fact that significant amounts of outer segment debris have accumulated in the subretinal space. At 40 days postnatally, the cone sheaths appear shortened, perhaps condensed, in the inner segment region and disrupted distally or obscured by impinging debris material (Fig. 5B). By 55 days of age in the RCS rat, PNA-binding components of the cone matrix sheath are still detectable in the inner segment region (Fig. 5C). However, by postnatal day 90 (Fig. 5D) only occasional examples of foci of PNA binding material are noted adjacent to the few remaining cone nuclei in the outer nuclear layer.

D. Taurine-Deficient Rat

Rats with experimentally induced taurine deficiency initiated at the time of weaning exhibit ultrastructural abnormalities in photoreceptor inner and outer segments with cones being affected earlier than rods (Hageman and Schmidt, 1987). Following 40 days of taurine deficiency, disruption of the distal (outer segment-associated) regions of cone matrix sheaths is observed (Fig. 6A). Large unstained areas in the outer segment region that appear to correspond to regions previously occupied by cone matrix sheaths are often observed. By 60 days of taurine deficiency, cone matrix sheaths are not detectable (Fig. 6B), suggesting a progression of the sheath degeneration/dissolution observed earlier (Fig. 6A). Instead, a more uniform fluorescence resulting from the binding of PNA is observed in the outer segment region (Fig. 6B).

DISCUSSION

The observation that cone photoreceptor cells appear to be less susceptible to the adverse effects of hereditary retinal degenerative diseases has been made frequently. The _rd_ mouse is an extreme example in which late in the degenerative process all surviving

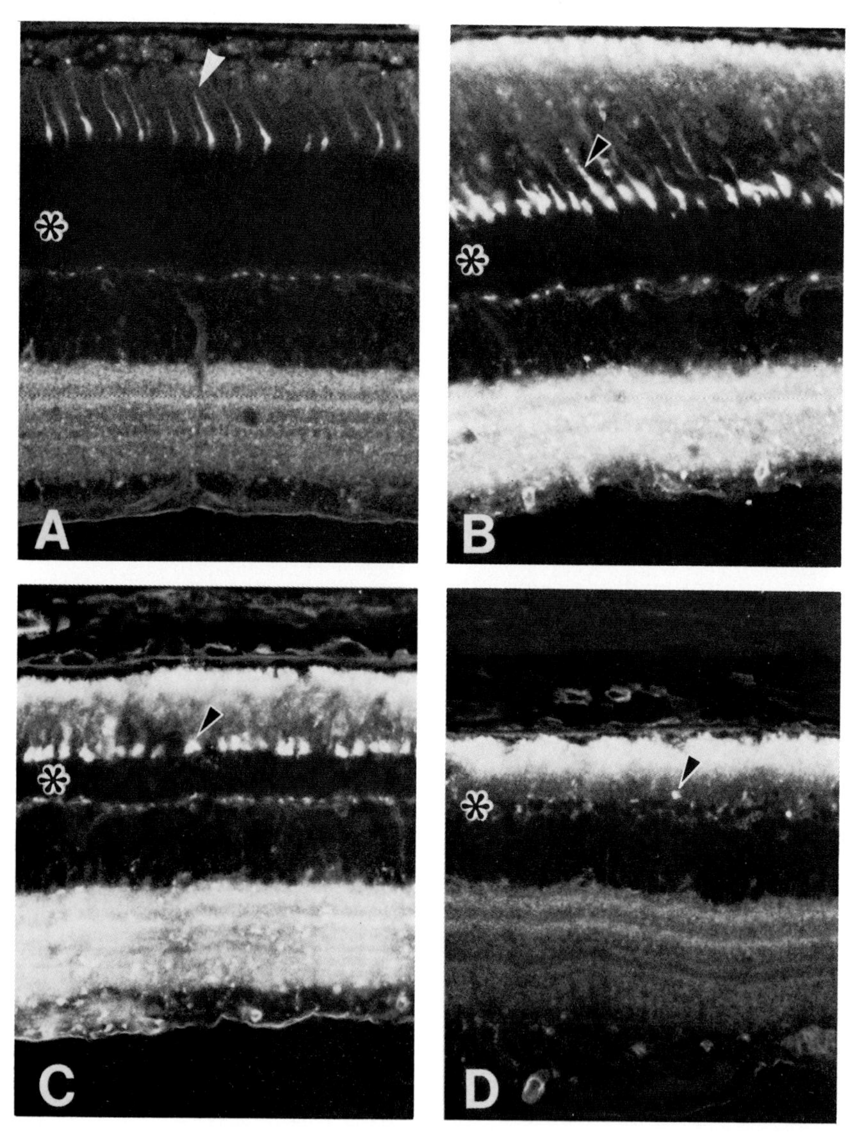

Figure 5. Fluorescence micrographs of retinal sections from postnatal day 20 (A), 40 (B), 55 (C) or 90 (D) RCS rats exposed to fluorescein-conjugated PNA. At day 20 (A) cone matrix sheaths (arrowhead) appear relatively normal, except for being a bit shortened probably as a result of the accumulation of outer segment debris in the subretinal

photoreceptors are cones (Carter-Dawson et al., 1978; LaVail, 1981). The Irish setter retina exhibits a similar progression of photoreceptor degeneration (Aguirre et al., 1978). In other mutant strains, including the RCS rat (LaVail, 1981) and the pcd mouse (LaVail, 1981; LaVail et al., 1982), rod cell loss is accelerated as compared to cones but in neither does a 100% cone retina result. In contrast, experimentally induced taurine deficiency in rats appears to result in cone photoreceptors being affected earlier and more severely than rods.

It is of interest to note that in each of the mutant strains in which we have monitored the fate of the cone matrix sheath (rd, rds, RCS), PNA-detectable sheath material remains present until relatively late in the degenerative process. This observation is consistent with the aforementioned observations that cones show increased survivability in these mutants. Whether the presence of the cone matrix sheath confers some survival advantage to cone photoreceptor cells or its presence is merely indicative of surviving cones remains to be determined. However, in the taurine deficiency model in which cones are more severely affected, there are observable effects on the cone matrix sheaths that occur relatively early, concomitant with ultrastructural alterations in photoreceptor inner and outer segments (Hageman and Schmidt, 1987).

Assuming the cone matrix sheath does confer some survival advantage to cone photoreceptor cells, how might this effect be mediated? It is possible that the sheath could prolong the maintenance of outer segment-RPE interaction, although it is not clear why this would be advantageous in the RCS rat. Alternatively, sheath

---

space. Later in the degenerative process (B,C) PNA-labeled cone matrix sheath material (arrowheads) becomes primarily inner segment-associated and at times is difficult to distinguish from the auto-fluorescent debris material. At very late time points (D), only very few nuclei remain in the outer nuclear layer (*), and only occasional foci of PNA-labeled material (arrowhead) are noted.

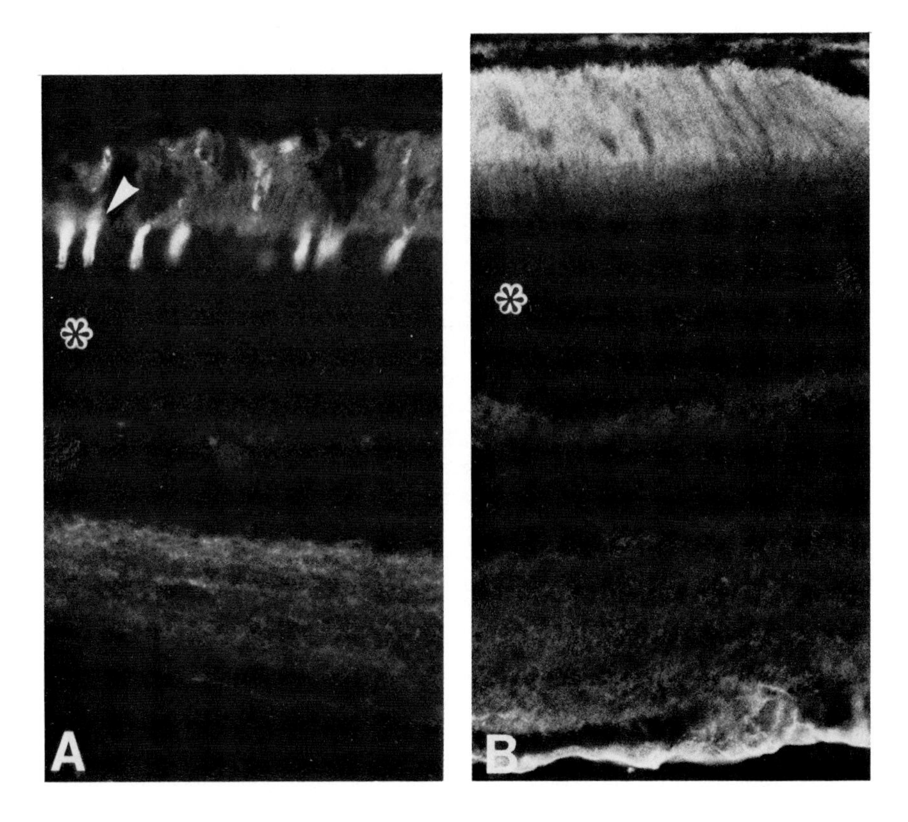

Figure 6. Fluorescence micrographs of retinal sections from rats with experimentally induced tuarine deficiency for 40 (A) or 60 (B) days exposed to fluorescein-conjugated PNA. Early effects on the cone matrix sheaths (arrowhead in A) are noted with apparent degeneration of the distal (outer-segment associated) regions. Taurine deficiency ultimately leads to the disappearance of PNA-detectable cone matrix sheaths and an elevation of uniformly distributed PNA-specific fluorescence in the outer segment region (B). Little change in outer nuclear layer (*) thickness is noted under these experimental conditions.

material could serve as a selective shield, inhibiting the diffusion of toxic substances toward the cone inner and outer segments or facilitating the diffusion of toxic substances away from them. The sheath might also act to stabilize outer segment membrane in some way, making it less susceptible to the structural disruption typically observed in rod outer segments. Ongoing studies of the function of the cone matrix sheath in the normal retina should provide some insight into how it might (or might not) be involved in enhanced cone survival in hereditary retinal degenerataive diseases.

ACKNOWLEDGEMENTS

The authors gratefully acknowledge the technical assistance of Mr. Kaj Anderson, Mr. Robert Claycomb, Ms. Maria Estevez, Ms. Juana Roy, Ms. Chris Spee, Ms. Lois Samuels, Ms. Joyce Tombran-Tink, and Ms. Susan Way and the clerical assistance of Ms. Nancy Polito and Ms. Wanda Hill. Parts of this work were supported by grants from the National Institues of Health to LVJ (EY 04741), JCB (EY 03042) and GSH (EY 06463). Animals employed in these studies were maintained in facilities fully accredited by the American Association for Laboratory Animal Science and treated according to the Guiding Principles in the Care and Use of Animals (DHEW Publication #NIH 80-23).

REFERENCES

Aguirre G, Alligood J, O'Brien P, Buyukmiohi N (1982a). Pathogenesis of progressive rod-cone degeneration in miniature poodles. Invest Ophthalmol Vis Sci 23:610-630.

Aguirre G, Farber DB, Lolley RN, Fletcher RT, Chader GJ (1978). Rod-cone dysplasia in Irish setters: A defect in cyclic GMP metabolism in visual cells. Science 201:1133-1134.

Aguirre G, Farber DB, Lolley RN, O'Brien P, Alligood J, Fletcher RT, Chader GJ (1982b). Retinal degenerations in the dog. III. Abnormal cyclic nucleotide metabolism in rod-cone dysplasia. Exp Eye Res 35:625-642.

Aguirre G, Rubin L (1979). Diseases of the retinal pigment epithelium in animals. In Zinn KM, Marmor MF (eds.): "The Retinal Pigment Epithelium", Harvard University Press, Cambridge, pp 334-356.

Blanks JC, Hageman GS, Johnson LV, Spee C (1988). Ultrastructural visualization of primate cone photoreceptor matrix sheaths. J Comp Neurol 270:288-300.

Bok D, Hall MO (1971). The role of the pigment epithelium in the etiology of inherited retinal dystrophy in the rat. J Cell Biol 49:664-682.

Bourne MC, Campbell DA, Tansley K (1938). Hereditary degeneration of the rat retina. Br J Ophthalmol 22:613-623.

Buyukmiohi N, Aguirre G, Marshall J (1980). Retinal degenerations in the dog. II. Development of the retina in rod-cone dysplasia. Exp Eye Res 30:575-591.

Carter-Dawson L, Kuwabara T, O'Brien PJ, Bieri JG (1979). Structural and biochemical changes in vitamin A-deficient rat retinas. Invest Ophthalmol Vis Sci 18:437-446.

Carter-Dawson L, LaVail MM, Sidman RL (1978). Differential effect of the rd mutation on rods and cones in the mouse retina. Invest Ophthalmol Vis Sci 17:489-498.

Dowling JE, Sidman RL (1962). Inherited retinal dystrophy in the rat. J Cell Biol 14:73-109.

El-Mofty AAM, Eisner G, Balazds EA (1980). Retinal degeneration in rhesus monkeys, Macaca mulatto: Survey of three seminatural free-breeding colonies. Exp Eye Res 31:147-166.

Hageman GS, Johnson LV (1986). Biochemical characterization of the major peanut agglutinin-binding glycoproteins in vertebrate retinae. J Comp Neurol 249:499-510.

Hageman GS, Johnson LV (1987). Chondroitin 6-sulfate glycosaminoglycan is a major constituent of primate cone photoreceptor matrix sheaths. Curr Eye Res 6:639-646.

Hageman GS, Schmidt SY (1987). Taurine-deficient pigmented and albino rats: Early retinal abnormalities and differential rates of photoreceptor degeneration. In LaVail MM, Hollyfield JG, Anderson RE (eds): "Degenerative Retinal Disorders: Clinical and Laboratory Investigations", A.R. Liss, pp 497-515.

Johnson LV, Blanks JC (1984). Application of acrylamide as an embedding medium in studies of lectin and antibody binding in the vertebrate retina. Curr Eye Res 3:969-974.

Johnson LV, Hageman GS, Blanks JC (1986). Interphotoreceptor matrix domains ensheath vertebrate cone photoreceptors. Invest Ophthalmol Vis Sci 27:129-135.

Kolb H, Gouras P (1974). Electron microscopic observations of human retinitis pigmentosa, dominantly inherited. Invest Ophthalmol Vis Sci 13:487-498.

LaVail MM (1981). Analysis of neuronal mutants with inherited retinal degeneration. Invest Ophthalmol Vis Sci 21:638-657.

LaVail MM, Blanks JC, Mullen RJ (1982). Retinal degeneration in the pcd cerebellar mutant mouse. I. Light microscopic and autoradiographic analysis. J Comp Neurol 212:217-230.

LaVail MM, Sidman RL (1974). C57BL/6J mice with inherited retinal degeneration. Arch Ophthalmol 91:394-400.

Mullen RJ, LaVail MM (1975). Two new types of retinal degeneration in cerebellar mutant mice. Nature 258:528-530.

Mullen RJ, LaVail MM (1976). Inherited retinal dystrophy: Primary defect in pigment epithelium determined with experimental rat chimeras. Science 192:799-801.

Narfstrom K (1985). Progressive retinal atrophy in the abyssinian cat. Invest Ophthalmol Vis Sci 26:193-200.

Noell WK (1952). The impairment of visual cell structure by iodoacetate. J Cell Comp Physiol 40:25-28.

Noell WK (1958). Differentiation, metabolic organization, and viability of the visual cell. Arch Ophthalmol 60:702-733.

Noell WK (1965). Aspects of experimental and hereditary retinal degeneration. In Graymore CN (ed.): "Biochemistry of the Retina", Academic Press, London, pp. 51-72.

Sameshima M, Uehara F, Ohba N (1987). Specialization of the interphotoreceptor matricies around cone and rod photoreceptor cells in the monkey retina, as revealed by lectin cytochemistry. Exp Eye Res 45:845-863.

Sanyal S, DeRuiter A, Hawkins RK (1980). Development and degeneration of retina in rds mutant mice: Light microscopy. J Comp Neurol 194:193-207.

Sanyal S, Jansen HG (1981). Absence of receptor outer segments in the retina of rds mutant mice. Neurosci Lett 21:23-26.

Schmidt SY (1985). Retinal degenerations. In Lajtha A (ed.): "Handbook of Neurochemistry", Vol 10, Plenum Publishing Corp, pp. 461-507.

Szamier RB, Berson EL (1977). Retinal ultrastructure in advanced retinitis pigmentosa. Invest Ophthalmol Vis Sci 16:947-962.

Szamier RB, Berson EL (1982). Histopathologic study of an unusual form of retinitis pigmentosa. Invest Ophthalmol Vis Sci 22:559-570.

Ulshafer RJ, Allen CB (1985). Hereditary retinal degenerations in the Rhode Island red chicken: Ultrastructural analysis. Exp Eye Res 40:865-877.

Ulshafer RJ, Allen CB, Dawson WW, Wolf ED (1984). Hereditary retinal degeneration in the Rhode Island red chicken. I. Histology and ERG. Exp Eye Res 39:125-135.

Vainisi SG, Beck BB, Apple DG (1974). Retinal degenerations in a baboon. Am J Ophthalmol 78:279-284.

Van Nie R, Ivanyi D, Demant P (1978). A new H-2-linked mutation, rds, causing retinal degeneration in the mouse. Tissue Antigens 12:106-108.

Varner HH, Rayborn ME, Osterfeld AM, Hollyfield JG (1987). Localization of proteoglycan within the extracellular matrix sheath of cone photoreceptors. Exp Eye Res 44:633-642.

Woodford BJ, Liu Y, Fletcher RT, Chader CJ, Farber DB, Santos-Anderson R, Tso MOM (1982). Cyclic nucleotide metabolism in inherited retinopathy in collies: A biochemical and histochemical study. Exp Eye Res 34:703-714.

Inherited and Environmentally Induced
Retinal Degenerations, pages 233–250

A COMPARATIVE SURVEY OF SYNAPTIC CHANGES IN THE ROD PHOTORECEPTOR TERMINALS OF rd, rds AND DOUBLE HOMOZYGOUS MUTANT MICE

Somes Sanyal and Harry Jansen

Department of Anatomy, Faculty of Medicine,
Erasmus University, Rotterdam, The Netherlands

INTRODUCTION

Histogenesis of retina in the mouse proceeds over a period of several weeks, starting from the mid gestation stage and continuing until about 14 days postnatal. As the neuroblast cells of the embryonic optic cup continue to divide by mitosis, a proportion of the cells successively withdraw from the mitotic cycle and differentiate into one or the other of the glial, neuronal or the photoreceptor cell-types. A definite sequence of cellular origin (Sidman, 1961, 1970; Young 1985a, b) and morphological differentiation (Hinds, Hinds 1974; 1978; Bhattacharjee 1976; 1977; Sanyal, Bhattacharjee 1979) has been described in the developing retina of mice. Generally speaking, a relatively higher proportion of the cells originating at earlier stages of development contribute to the pool of glial and neuronal cells while an increasing proportion of cells, originating at later stages, contribute to the receptor cell population.

The photoreceptor cells in the mouse retina, comprising of 97% rod cells and 3% cone cells (Carter-Dawson, LaVail 1979a), originate and differentiate in a sequential manner (Carter-Dawson, LaVail 1979b; Hinds, Hinds 1979). At the ultrastructural level, the receptor inner segments are first seen to appear in the perinatal retina; the outer segments develop between postnatal day 5 - 14 and the synaptic contacts are established between postnatal day 7 - 14 (Olney 1968; Blanks et al., 1974a), the individual cells showing a high degree of asynchrony in development of the receptor as well as the synaptic structures.

Of the various mutant genes affecting the retina in the mouse (LaVail 1981), the genes rd (retinal degeneration, Sidman, Green 1965) and rds (retinal degeneration slow, van Nie et al., 1978) cause specific loss of photoreceptor cells and the lesions have been listed as early onset photoreceptor dysplasia (Chader et al., 1988). In the homozygous rd/rd mutant retina the first indication of gene expression is seen in retarded development of the receptor inner and outer segments followed by rapid loss of photoreceptor cells starting from around postnatal day 10 (Caley et al., 1972; Sanyal, Bal 1973). Initially, the rods are affected preferentially so that almost all rods disappear by day 21, whereas the cones degenerate at a slower rate over the next two months (Carter-Dawson et al., 1978). Synaptogenesis in the receptor terminals of this mutant, particularly within the rods, has been shown to be retarded, and few, if any rod synapses are seen at day 14 (Blanks et al., 1974b). Since the period of rapid loss of rods overlaps with the period of synaptogenesis the real effect of the mutant gene on this aspect of photoreceptor differentiation remains unclear.

In the developing homozygous rds/rds mutant retina the action of the gene is also recorded early in the complete failure of receptor outer segment disc formation (Sanyal, Jansen 1981; Cohen 1983). Loss of photoreceptor cells, both rods and cones, starts around day 14 and progresses very slowly (Sanyal et al., 1980). The process of synaptogenesis in the photoreceptor terminals appears to follow in normal sequence although some definite alterations are recorded at later stages of degeneration (Jansen, Sanyal 1984).

In the retina of double homozygous rd/rd,rds/rds mutant mice the receptor outer segments fail to develop due to the action of the rds gene and degenerative changes are already present at day 11. However, the rate of photoreceptor cell loss is slower than in the rd retina and as a result, many rod cells are still present at day 21 (Sanyal, Hawkins 1981). Initial observation has suggested that the terminals of many of these cells contain normal or rds type synapses.

In this chapter we compare the structural changes in the synapses within the rod terminals of the three different mutant genotypes mentioned above and discuss the possible relation between the expression of the mutant genes and the observed synaptic changes.

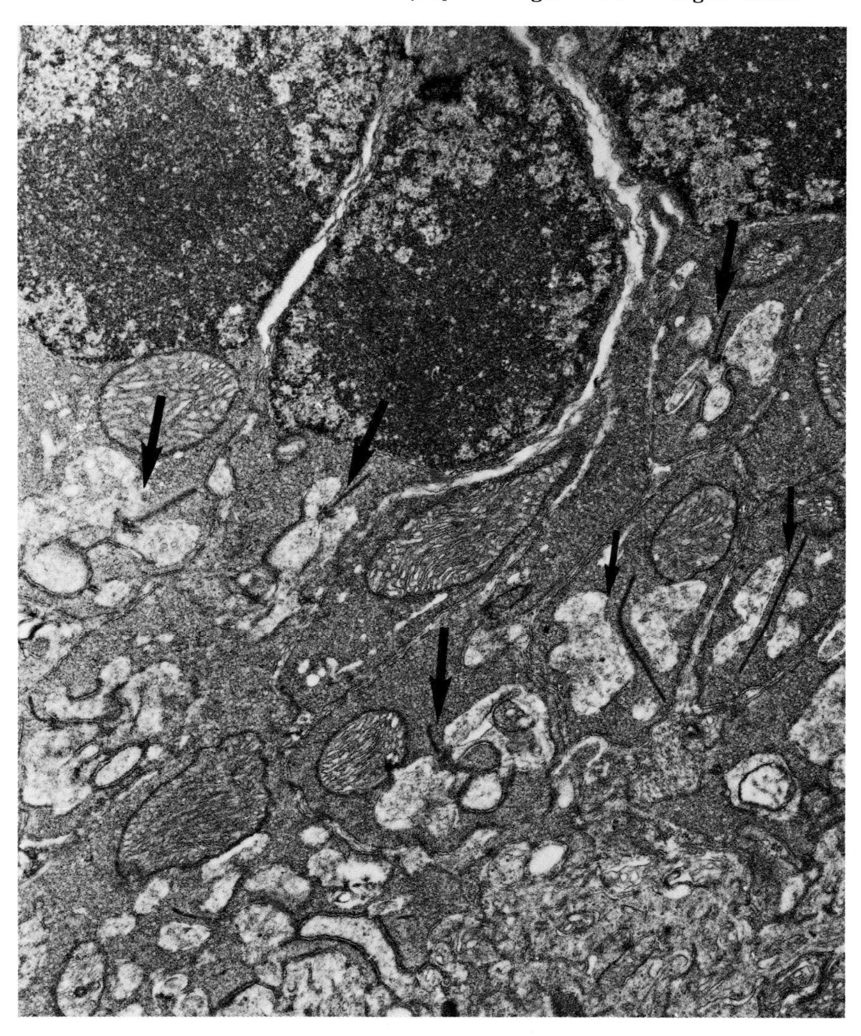

Fig. 1. Electron micrograph showing part of the outer plexiform layer of the normal retina of the mouse at the age of 21 days. Three rod nuclei marked by large masses of heterochromatin are located in the upper part of the picture. Several rod terminals with profiles of typical synaptic configuration - one ribbon with two lateral horizontal cell processes and one opposing medial bipolar element (large arrow) are seen. Some other profiles (small arrow) on a different sectional plane show only the ribbon and the lateral processes. x15000

MATERIALS AND METHODS

Four different congenic lines of C3H mice, originally derived from C3HfHeA strain, with different allelic combination at the rd and rds loci - (1) rd/rd, (2) rds/rds, (3) rd/rd,rds/rds and (4) +/+,+/+ - provided the materials used in this study. Therefore, the animals have uniform genetic background and differ only in the locus of interest. Animals were maintained in cyclic light under standard conditions of husbandry.

Details of preparative and histological procedures have been described earlier (Jansen, Sanyal 1984). Briefly, eyes were fixed in aldehyde mixture, embedded in epoxy resin, and ultrathin sections were contrasted with uranyl acetate and lead citrate. Successive 10 µm long stretches in a section were used for quantitative analysis in the electron microscope and averages (n=40 to 80) were used for comparison. The number of terminals with profile of a complete synapse (three elements), as opposed to the ones with less than three elements (Fig. 1) were counted in these samples and expressed as percentage of total. The number of synapses with more than one ribbon, as opposed to synapses with one single ribbon (Figs. 8, 9) was similarly counted and expressed as percentage of total.

RESULTS

In the following description and quantitative data on the synaptogenesis in the photoreceptor terminals in the retina of various mutant and normal mice, only the terminals of the rods, both spherules and paranuclear types, have been included and the pedicle terminals belonging to the cones, also termed beta terminals, which are generally not very frequently encountered in the normal retina, have not been included. Typically, one rod terminal contains one single synaptic complex. Within this complex one synaptic ribbon (also called synaptic lamella) is surrounded laterally by two processes of horizontal cell and one medially located bipolar cell dendrite (Fig. 1)

Synaptogenesis in the rod terminals of the normal retina

In course of development in the normal retina, synaptic contacts between the rod terminals and the second order

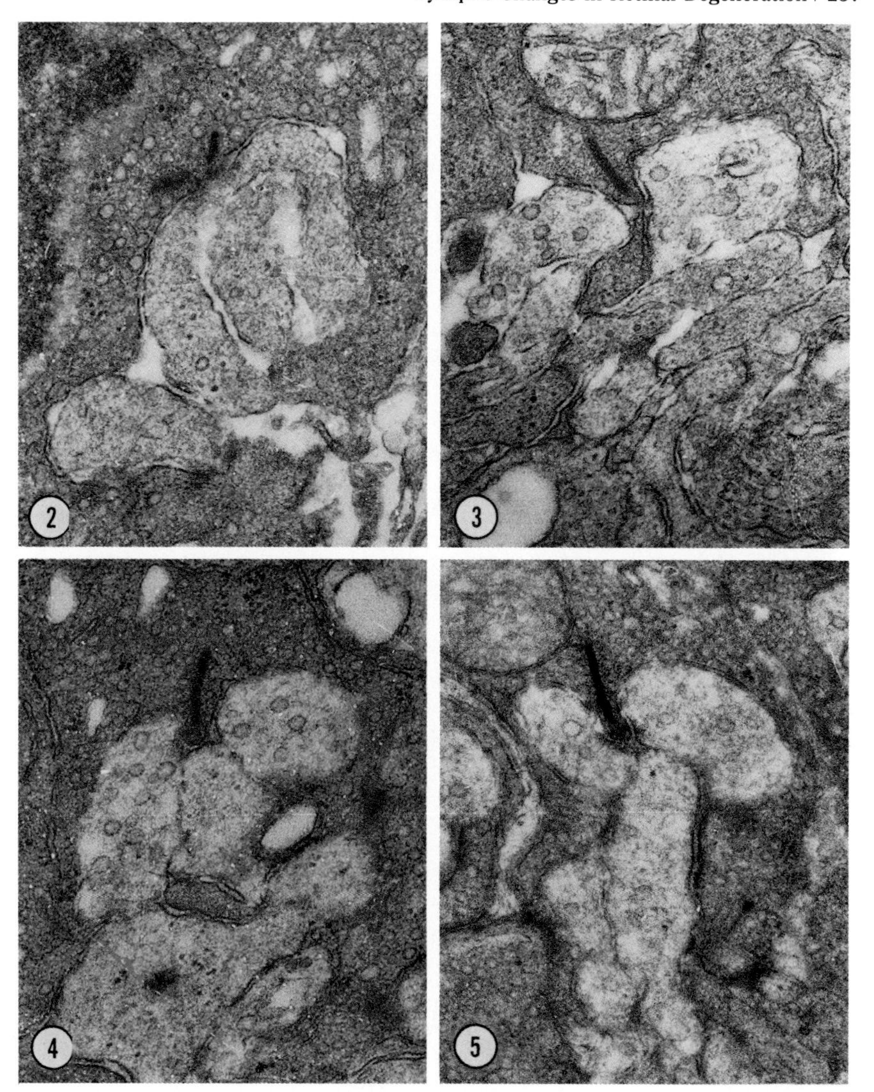

Figs. 2 - 5. Stages in the synaptogenesis within the rod terminals of normal mouse. x41000
Fig. 2. Rudimentary ribbons adjoining horizontal cell processes at 4 days after birth. Fig. 3. At 7 days, note lateral alignment of the horizontal cell processes.
Fig. 4. At 11 days, note the presence of the medially located bipolar component. Fig. 5. A typical profile of a normal rod synapse is frequently observed at 14 days.

neurons are established through a series of stages. In mice, as described by Blanks et al., (1974a), earliest stages are recognized by the presence of a synaptic ribbon and appositional contact with processes from horizontal cells (Fig. 2). These processes invaginate deeper within the terminal (Fig. 3); such stages are more frequently encountered at 4 - 7 days after birth. The bipolar cell dendrite appears at a later stage within this complex and assumes a median position facing the ribbon (Fig. 4). Synaptogenesis within the rod terminals proceeds in a highly asynchronous manner so that profiles of complete synaptic structures which are indistinguishable from that of the adult retina are frequently encountered in the retina of 14 day old mouse (Fig. 5) but many other terminals have synapses which are still at a formative stage.

Synaptic changes in the rod terminals of rd/rd mice

In the retina of homozygous rd/rd mutant mice loss of photoreceptor cells starts from after day 10. At day 11 most rod cells are still present, however, many of these cells have started to degenerate and the general ultrastructure at the receptor end of the cells has started to deteriorate. In the electronmicrographs of the outer plexiform layer at this stage (Fig. 6) only a few terminals are observed in which both horizontal and bipolar cell elements are present in the synapse; while many terminals with profiles of fewer processes are encountered adjoining a normal appearing ribbon. These observations confirm the previous finding of Blanks et al., (1974b) that synaptogenesis in the rod terminals of rd/rd mice is retarded in comparison to the normal. However, in the retina of 17 day old mutant mice the central retina is seen to contain about two rows of photoreceptor cell perikarya, and some rod cells are still present which show a normal synaptic complex (Fig. 7). Even at day 21 rod cells are very rarely seen but some of the terminals contain a normal appearing synaptic complex.

Synaptic changes in the rod terminals of rds/rds mice

Synaptic changes within photoreceptor terminals of these mutants have been described earlier (Jansen, Sanyal 1984). As already mentioned the development of the receptor outer segments in this mutant is completely blocked. Hence, the gene can be considered as acting rather early in development, but loss of photoreceptor cells starts only

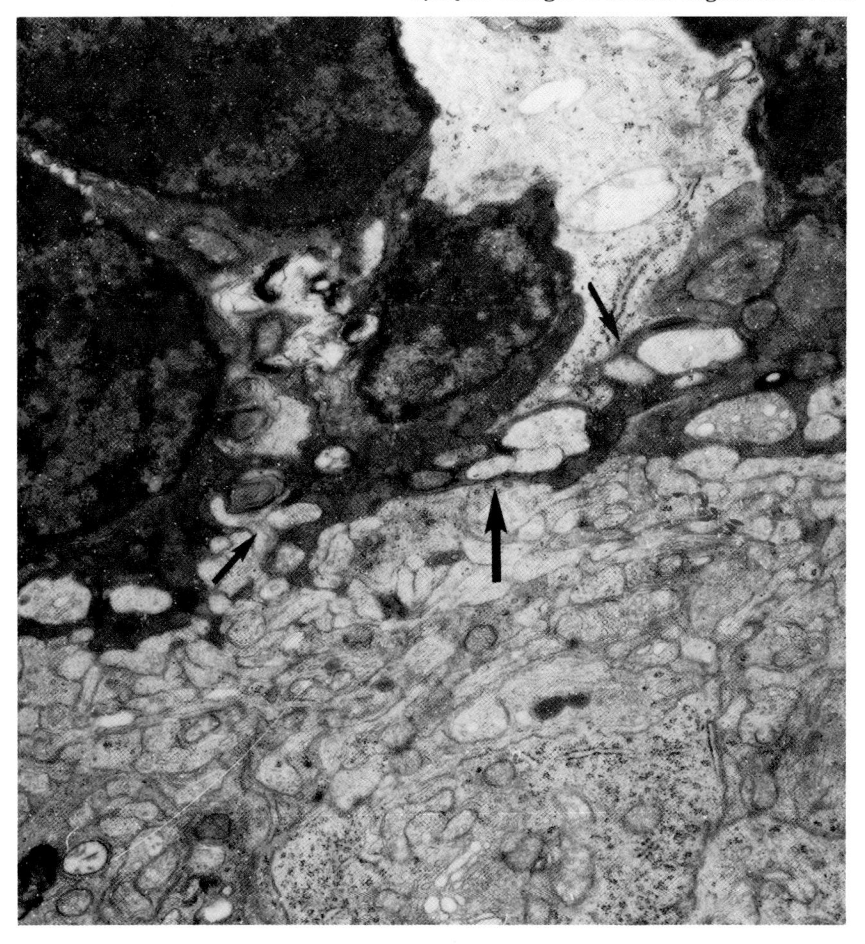

Fig. 6. Part of the outer plexiform layer of retina from a homozygous rd/rd mutant mouse at 11 days. A profile of a typical synaptic configuration (large arrow) belonging to the rod perikarya located above is seen along with a number of atypical synaptic structures (small arrow). x6300

after day 14 and progresses very slowly. As a result the cell population at day 17 and 21 is only slightly reduced. At the initial stages, the ultrastructural changes in the course of synaptogenesis in the rds/rds retina is indistinguishable from the normal. However, with loss of cells, some of the terminals of the surviving rod cells show an increase in the number of synaptic ribbons along with

Fig. 7. Part of homozygous rd/rd mutant retina at 17 days. Although most of the photoreceptor cells have disappeared, some of the rod perikarya, still present, show persistent synaptic contact (arrow) including a typical ribbon and profiles of horizontal and bipolar cell elements. x24000

enlargement or increase in the number of profiles belonging to the horizontal cell processes and also possibly to the bipolar cell dendrites. With progressive depletion of the photoreceptor cell population, the number of rod terminals present in the outer plexiform layer is gradually reduced but some of these terminals continue to show increased number of ribbons and enlarged or increased number of neuronal profiles.

Synaptic changes in the rod terminals of rd/rd,rds/rds mice

In the retina of these mutant mice, which are homozygous for both genes, the rate of photoreceptor cell loss

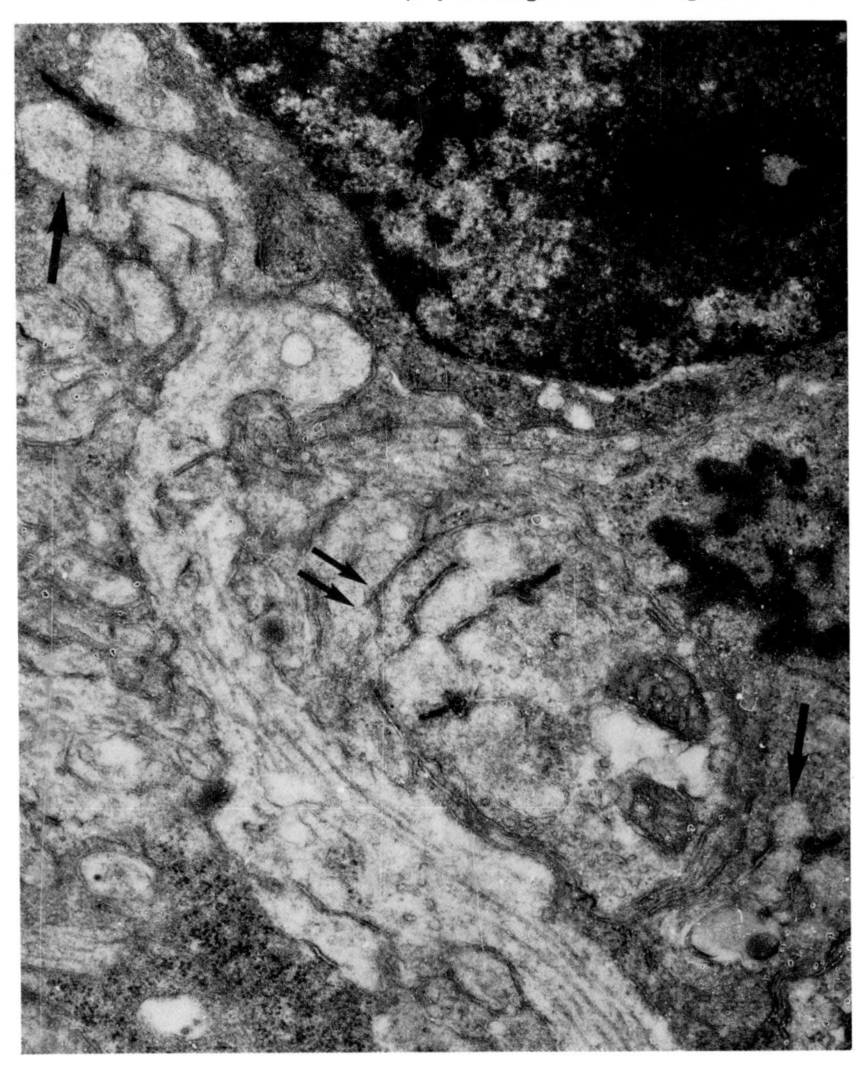

Fig. 8. Part of the outer plexiform layer from the retina of a double homozygous rd/rd,rds/rds mouse at 21 days. Since degeneration is slower in this genotype than in the single homozygous rd/rd mouse, more rod cells are present at this stage. Some of these contain a normal appearing synapse (large arrow) while some rod terminals show a synapse with extra ribbons and altered configuration (see also Fig. 9) often present in the single mutant rds/rds retina of this age. x16300

is significantly slower than in the rd/rd mice (Sanyal, Hawkins 1981). Although no clear difference is recorded in the early stages of synaptogenesis between these mutants and the rd/rd mice, the prolonged survival of the photoreceptor cells is found to have clear effects on the synapses in the rod terminals at later stages. This is seen in the continued presence of a normal synaptic configuration within some of the rod terminals at day 21 (Fig. 8). At the same time, some of these persistent synapses show presence of extra ribbons and features of synaptic growth, observed in the rod terminals of rds/rds mutant mice. Fig. 9 shows two such synapses located within paranuclear terminals belonging to rod perikarya. The multiple synaptic ribbons in these terminals often appear as centers of discrete synaptic complexes. The profiles of the second order neuronal processes that participate in these multiple sites appear to have enlarged in size. Although the exact structural relationship within such complex synapses can only be properly visualized in three dimensional reconstructions, which is being currently undertaken, the general impression is that the observed changes have resulted from an overall nelargement of the synaptic components within these terminals.

Quantitative data

Table 1 summarizes the data on synaptic changes in the rod terminals of the various mutants and normal mice at 17 and 21 days of age. The effect of differential loss of photoreceptor cells in the three different mutant retinas, in comparison to the normal, is seen to have proportionately reduced the number of rod terminals. Whereas a slight reduction in the number of terminals is also seen in the normal retina between day 17 and 21, this is likely to be due to the growth of the eyeball in course of which the existent terminals are spread out.

Increase in the frequency of the terminals with complete synaptic profile in the normal retinas between day 17 and 21 indicates that some new synapses are still being formed during this period. In the homozygous rd/rd retina the frequency of rod terminals with complete synaptic profile is similar to normal at 17 days, and at 21 days the number of rods still present is too few for comparison. In the rds/rds retina the frequency of rods terminals with complete synapse is higher than normal at day 17 and shows

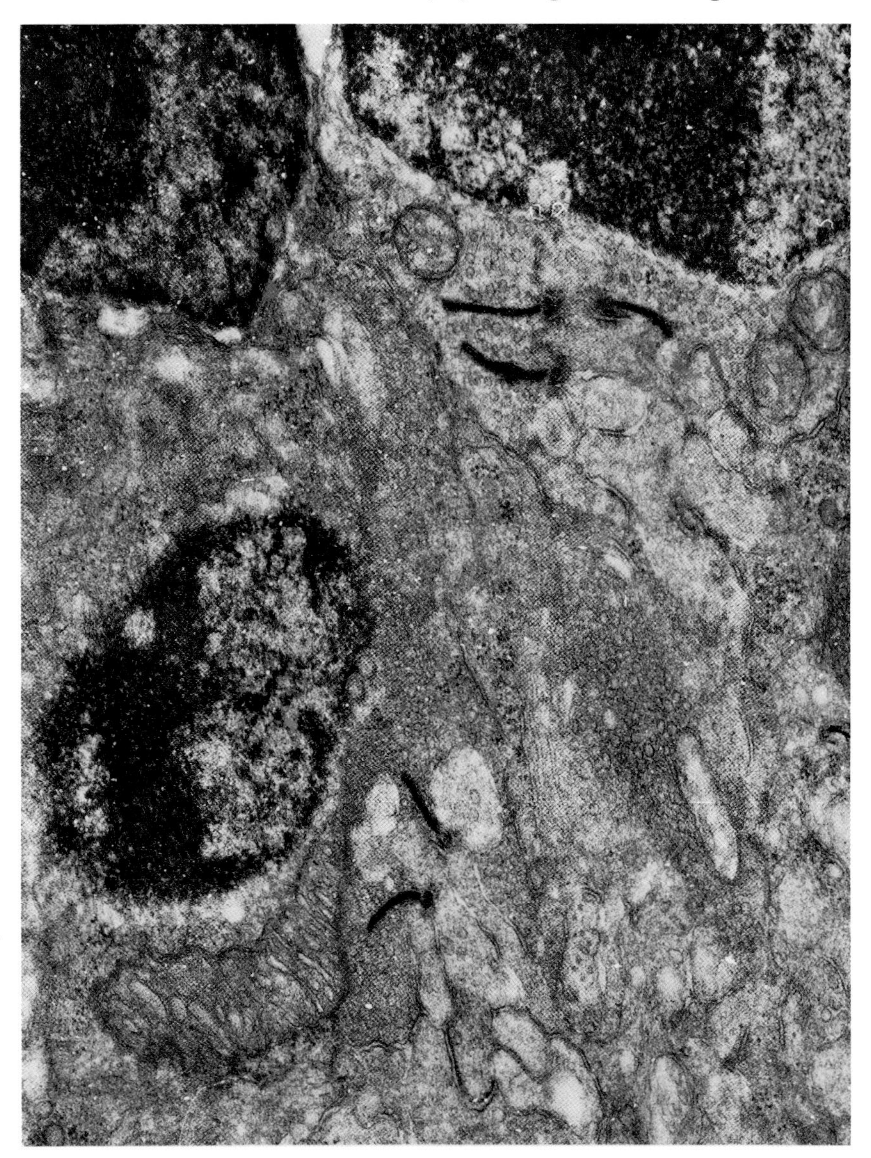

Fig. 9. Retina of double homozygous rd/rd,rds/rds mutant mouse at 21 days. Some of the surviving rod cells, identified by massive heterochromatin containing nuclei, show in their terminals extra ribbons and multiple profiles of processes from the second order neurons. Such changes are also seen in the rds/rds retina. x48000

TABLE 1. Data on the frequency of rod terminals (all pedicles i.e., cone terminals are excluded from these counts), synapses and synaptic ribbons in the retina of normal and mutant mice bearing different genotypic combination at the *rd* and *rds* loci. All figures are mean of three different individuals ± S.E.M, obtained from successive 10 µm length of outer plexiform layer (OPL).

Genotype	Age in days	Number of terminals per 10 µm OPL	% Terminals with complete synaptic profile	% Synapse with more than one ribbon
+/+,+/+	17	4.8 ± 1.4	22.8 ± 1.3	0.6 ± 0.5
	21	3.4 ± 0.4	36.4 ± 4.0	2.9 ± 0.8
rd/*rd*	17	0.7 ± 0.1	25.1 ± 8.2	7.7 ± 1.2
	21	rods are rarely present at this age		
rds/*rds*	17	4.7 ± 1.4	30.8 ± 1.6	6.5 ± 3.4
	21	1.4 ± 0.3	54.4 ± 5.3	24.7 ± 2.5
rd/*rd*,*rds*/*rds*	17	1.3 ± 0.2	17.6 ± 7.9	4.0 ± 3.0
	21	0.8 ± 0.2	16.8 ± 1.1	9.8 ± 5.2

a considerably further increase at day 21. This increase in the frequency of terminals with synaptic profile over the normal frequency is surprising and appears likely to be due to the enlargement of the synaptic complex occurring in this mutant during this time. In the double homozygous *rd/rd,rds/rds* mutants, however, the frequency of rod terminals with complete synapse is considerably less than in the other groups. The thickness of the outer nuclear in this mutant at this period is higher than in the *rd/rd* mice but shows considerable regional and some individual variation which is seen in a higher standard error.

As mentioned earlier, the increase in the number of synaptic ribbons present in a rod synapse is considered to result from the growth of the synaptic structure. The number of synapses with extra i.e., more than one ribbon, is seen to be considerably higher in the different mutant groups than in the normal retina. However, the *rds/rds* retina shows the highest increase, and in the double homozygous *rd/rd,rds/rds* and the *rd/rd* mutant retinas this increase is considerably less but still significantly higher than in the normal.

DISCUSSION

In an earlier report, arguments have been presented (Sanyal 1987) that both *rd* and *rds* genes act within the photoreceptor cells. The *rd* gene prior to causing cell death retards the differentiation of the photoreceptor cells (Sanyal 1982). When loss of photoreceptor cells starts at around 10-11 days, the cells are not fully differentiated, particularly the bipolar contact within the rod terminals is retarded (Blanks et al., 1974b) even though the bipolar cells are not affected (Blanks, Bok 1977). However, some progress towards the maturation of the surviving photoreceptor cells take place and the presence of rod terminals with complete synaptic profiles in these mutants at day 17 presumably results from this phase of development. The mutants show some detectable ERG (Noell 1958, Karli 1965) at the time of eye opening, and continued functional capa bility dependent upon the cones (Drager, Hubel 1978). The cones have been shown to survive longer than the rods (Carter-Dawson et al., 1978). Given the state of asynchrony, prevailing among the differentiating photoreceptor cell population at the time of rapid cell death it is interesting to ask if some of the surviving cells complete their synaptic development or if the cells which have already established synaptic contacts can survive better, as has been speculated in some recent studies (Adler 1986, Politi, Adler 1988).

The primary biochemical lesion in the retina of rd/rd mice has been identified as reduced phosphodiesterase level resulting in high accumulation of cyclic GMP (Farber, Lolley 1974; 1976). Experimentally increased cyclic GMP level has also been shown to cause photoreceptor cell loss (Lolley et al., 1977). In the double homozygous rd/rd,rds/rds mice, some of the photoreceptor cells including rods survive longer than in the rd/rd mice (Sanyal, Hawkins 1981) but the levels of phosphodiesterase and cyclic GMP in both genotypes have been found to be exactly identical (Fletcher et al., 1986; Chader et al., 1987). Since some of these surviving rod cells show normal synaptic structures or even enlarged ones as in the rds/rds retina, it appears likely that the rd gene does not specifically prevent synaptic development.

The present findings in the rds/rds retina confirm the previous report that this gene does not affect synaptic development; on the contrary surviving cells show enlargement of their synaptic components (Jansen, Sanyal 1984). Furthermore, similar enlargement of the rod synapses has also been shown to occur in the retina of normal albino mice, following partial photoreceptor cell loss induced by constant light (Jansen, Sanyal 1987). It has been speculated that the observed synaptic changes constitute compensatory growth following partial deafferentation, caused by slow photoreceptor cell loss.

SUMMARY AND CONCLUSIONS

In the developing retina of homozygous rd/rd mutant mice the time of onset of degenerative changes and the period of rapid photoreceptor cell loss overlaps the later phase of differentiation during which the maturation of the receptors and the completion of the synaptic connections take place. It remains unresolved if the retarded synaptogenesis within the photoreceptor terminals of the rod cells is a direct effect of the mutant gene or an indirect consequence of premature cell death.

In the retina of homozygous _rds/rds_ mutant mice early indication of gene expression within the photoreceptor cells is also recorded as photoreceptor outer segments fail to develop. However, during the very slow rate of degeneration, the surviving cells develop normal synaptic contacts within their terminals. Furthermore, some of the rod cells, though not the cones, go on to enlarge their synaptic structures as more and more photoreceptor cells are lost.

In the retina of double homozygous mutant _rd/rd,rds/rds_ mice the photoreceptor cells remain lacking in outer segments, as would be expected, but curiously enough, survive longer than in the _rd/rd_ retina. Among this population of photoreceptor cells with extended life-span profiles of rod terminals are frequently encountered which contain a normal synaptic structure - one synaptic ribbon with two laterally placed processes of horizontal cells and one medially facing process of a bipolar cell. A number of these terminals also show signs of synaptic enlargement. Thus it can be concluded that some of the terminals of the rod photoreceptor cells of the double homozygous _rd/rd,rds/rds_ mice, which survive longer than in the _rd/rd_ mice, develop synapses that are either comparable to normal or resemble those of the _rds/rds_ retina. These findings suggest that retarded synaptogenesis within the rod terminals of the _rd/rd_ retina is likely to result from a pathogenic defect affecting the whole cell and is not due to a specific or exclusive action of the mutant gene on the synaptic components involved.

Acknowledgements

Authors' thanks are due to RK Hawkins, H Visser and Ellen Voigt for technical assistance and efficient care of the mouse colony, and to Edith Klink for secretarial help. This work was partly supported by grant no. EY 06841 from the National Eye Institute of the National Institutes of Health, Bethesda, MD., U.S.A.

REFERENCES

Adler R (1986). Trophic interactions in retinal development and in retinal degeneration. In _vivo_ and in _vitro_ studies. In Adler R, Farber D (eds): "The Retina: A Model for Cell Biological Studies", Part I, New York: Academic Press, p 111-150.

Bhattacharjee J (1976). Developmental changes of carbonic anhydrase in the retina of the mouse: Histochemical study. Histochem J 8:63-70.

Bhattacharjee J (1977). Sequential differentiation of retinal cells in the mouse studied by diaphorase staining. J Anat 123:273-282.

Blanks JC, Adinolfi AM, Lolley RN (1974a). Synaptogenesis in the photoreceptor terminal of the mouse retina. J Comp Neurol 156:81-94.

Blanks JC, Adinolfi AM, Lolley RN (1974b). Photoreceptor degeneration and synaptogenesis in retinal degenerative (rd) mice. J Comp Neurol 174:95-106.

Blanks JC, Bok D (1977). An autoradiographic analysis of postnatal cell proliferation in the normal and degenerative mouse retina. J Comp Neurol 156:317-328.

Caley DW, Johnson C, Liebelt RA (1972). The postnatal development of the retina in the normal and rodless CBA mouse: a light and electron microscopic study. Am J Anat 133:179-212.

Carter-Dawson LD, LaVail MM (1979a). Rods and cones in the mouse retina I. Structural analysis using light and electron microscopy. J Comp Neurol 188:245-262.

Carter-Dawson LD, LaVail MM (1979b). Rods and cones in the mouse retina II. Autoradiographic analysis of cell generation using tritiated thymidine. J Comp Neurol 188: 263-272.

Carter-Dawson LD, LaVail MM, Sidman RL (1978). Differential effect of the rd mutation on rods and cones in the mouse retina. Invest Ophthalmol Vis Sci 17:489-498.

Chader GJ, Aguirre GD, Sanyal S (1988). Studies on animal models of retinal degeneration. In Tso MOM (ed): "Retinal Diseases: Biomedical Foundations and Clinical Management", Philadelphia: JB Lippincott Co. p. 80-99.

Chader GJ, Fletcher RT, Barbehenn E, Aguirre G, Sanyal S (1987). Studies on abnormal cyclic GMP metabolism in animal models of retinal degeneration: Genetic relationships and cellular compartmentalization. In Hollyfield JG, Anderson RE, LaVail MM (eds): "Degenerative Retinal Disorders: Clinical and Laboratory Investigations", New York: Alan R Liss, Inc., p. 289-307.

Cohen AI (1983). Some cytological and initial biochemical observations on photoreceptors in retinas of rds mice. Invest Ophthalmol Vis Sci 24:832-843.

Drager U, Hubel D (1978). Studies of visual function and its decay in mice with hereditary retinal degeneration. J Comp Neurol 180:85-114.

Farber D, Lolley R (1974). Cyclic guanosine monophosphate: elevation in degenerating photoreceptor cells of the C3H mouse retina. Science 186:449-451.

Farber D, Lolley R (1976). Enzymatic basis for cyclic GMP accumulation in degenerative photoreceptor cells of mouse retina. J Cyclic Nucleotide Res 2:139-148.

Fletcher RT, Sanyal S, Krishna G, Aguirre G, Chader GJ (1986). Genetic expression of cyclic GMP phosphodiesterase activity defines abnormal photoreceptor differentiation in neurological mutants of inherited retinal degeneration. J Neurochem 46:1240-1245.

Hinds JW, Hinds Pl (1974). Early ganglion cell differentiation in the mouse retina: An electron microscopic analysis utilizing serial sections. Dev Biol 37: 381-416.

Hinds JW, Hinds Pl (1978). Early development of amacrine cells in the mouse retina: An electron microscopic, serial section analysis. J Comp Neurol 179:277-300.

Hinds JW, Hinds PL (1979). Differentiation of photoreceptors and horizontal cells in the embryonic mouse retina: An electron microscopic, serial section analysis. J Comp Neurol 187:495-512.

Jansen HG, Sanyal S (1984). Development and degeneration of retina in rds mutant mice: Electron microscopy. J Comp Neurol 224:71-84.

Jansen HG, Sanyal S (1987). Synaptic changes in the terminals of rod photoreceptors of albino mice after partial visual cell loss induced by brief exposure to constant light. Cell Tissue Res 250:43-52.

Karli P, Stoeckel MD, Porte A (1965). Dégénérescence des cellules visuelles photo-réceptrices et persistence d'un sensibilité de la retine à la stimulation photique. Observations au microscopie electronique. Z Zellforsch Mikrosk Anat 65:238-252.

LaVail MM (1981). Analysis of neurological mutants with inherited retinal degeneration. Invest Ophthalmol Vis Sci 21:638-657.

Lolley R, Farber D, Rayborn M, Hollyfield J (1977). Cyclic GMP accumulation causes degeneration of photoreceptor cells: simulation of an inherited disease. Science 196: 664-666.

Noell WK (1958). Studies on visual cell viability and differentiation. Ann NY Acad Sci 74:337-361.

Olney JW (1968). An electron microscopic study of synapse formation, receptor outer segment development, and other aspects of developing mouse retina. Invest Ophthalmol 7:

250-268.
Politi L, Adler R (1988). Selective failure of long-term survival of isolated photoreceptors from both homozygous and heterozygous rd (retinal degeneration) mice. Exp Eye Res 47:269-282.
Sanyal S (1982). A survey of cytomorphological changes during expression of the retinal degeneration (rd) gene in the mouse. In Clayton RM et al. (eds): "Problems of Normal and Genetically Abnormal Retinas", London: Academic Press, p. 223-232.
Sanyal S (1987). Cellular site of expression and genetic interaction of the rd and the rds loci in the retina of the mouse. In Hollyfield JG, Anderson RE, LaVail MM (eds): "Degenerative Retinal Disorders: Clinical and Laboratory Investigations", New York: Alan R Liss, Inc., p. 175-194.
Sanyal S, Bal AK (1973). Comparative light and electron microscopic study of retinal histogenesis in normal and rd mutant mice. Z Anat EntwGesch 142:219-238.
Sanyal S, Hawkins RK (1981). Genetic interaction in the retinal degeneration of mice. Exp Eye Res 33:213-222.
Sanyal S, Jansen HG (1981). Absence of receptor outer segments in the retina of rds mutant mice. Neurosci Lett 21:23-26.
Sanyal S, De Ruiter A, Hawkins RK (1980). Development and degeneration of retina in rds mutant mice: Light microscopy. J Comp Neurol 194:193-207.
Sidman RL (1961). Histogenesis of mouse retina studied with thymidine-H. In Smelser GK (ed) "The Structure of the Eye", New York: Academic Press, p 487-505.
Sidman RL (1970). Autoradiographic methods and principles for study of the nervous system with ^{3}H-thymidine. In Ebbesson SOE, Nauta WJ (eds): "Contemporary Research Techniques in Neuroanatomy", New York: Springer Verlag, p 252-274.
Sidman Rl, Green MC (1965). Retinal degeneration in the mouse; Location of the rd locus in linkage group XVII. J Heredity 56:23-29.
Van Nie R, Ivanyi D, Demant P (1978). A new H-2 linked mutation, rds causing retinal degeneration in the mouse. Tissue Antigens 12:106-108.
Young RW (1985a). Cell proliferation during postnatal development of the retina in the mouse. Develop Brain Res 21:229-239.
Young RW (1985b). Cell differentiation in the retina of the mouse. Anat Rec 212:199-205.

Inherited and Environmentally Induced
Retinal Degenerations, pages 251–264

IMMUNOCYTOCHEMICAL LOCALIZATION OF OPSIN IN DEGENERATING PHOTORECEPTORS OF RCS RATS AND *rd* AND *rds* MICE

Izhak Nir* and David S. Papermaster

Department of Pathology, University of Texas Health Science Center at San Antonio, San Antonio, TX and *Technion Faculty of Medicine, Haifa, Israel

INTRODUCTION

Retinal dystrophies of several species greatly affect retinal photoreceptor cell integrity (LaVail, 1981). Despite substantial photoreceptor cell damage however, light responses could be observed in the absence of outer segments. Morphological analysis of a human retina from a 24 year old patient with sex-linked RP revealed substantial abnormalities in both rods in cones (Szamier et al, 1979). There was a virtual absence of organized cone outer segments from the parafovea through the midperiphery. This pattern was of particular interest since the patient had full visual fields, with a large test light spot, three weeks prior to death. Similar capacity to respond to light has been observed in mutant mice and rats with inherited retinal dystrophies. Since rhodopsin is the only light-sensitive molecule so far identified in the photoreceptor cell, this residual light sensitivity suggested that rhodopsin, at sites in the rod other than the outer segment, might be capable of transduction of light to a neural response. In order to evaluate this question, we explored opsin distribution in rodents bearing various forms of retinal dystrophy to determine if there were any general features of opsin localization which could contribute to persisting visual function. Dystrophic RCS rats with the *rdy* mutation, and dystrophic mice with the *rd* (C57BL/6J strain) and *rds* (O20/A strain) mutations were studied.

The RCS rat develops photoreceptors normally. Failure of the pigment epithelium to ingest shed rod outer segment tips, however, results in the accumulation of disk debris in the subretinal space as the rats mature (Bok and Hall, 1971; Mullen and LaVail, 1976). This indirectly affects photoreceptor viability and by the 60th postnatal day there are no intact outer or inner segments (Dowling and Sidman, 1962).

The rods and cones of mice with the *rd* dystrophy develop nearly to the normal stage, achieved by postnatal day 10 (P10). Thereafter, rapid degeneration occurs. ERG responses cannot be recorded in the *rd* mouse after P20. By 30 days only a few photoreceptor nuclei remain (Caley et al, 1972; Carter-Dawson et al, 1978). The defect in the *rd* mouse appears to be a consequence of abnormal accumulation of cyclic nucleotides (especially cGMP) in the photoreceptors; accordingly the mutation directly affects the photoreceptor cells and does not appear to be a secondary defect arising from another cell (Farber and Lolley, 1974).

Outer segments fail to develop altogether in mice with the *rds* (retinal degeneration slow) mutation. The opsin content in the retina of this mutant is about 3% of normal (Schalken et al, 1985). Only a cilium projects from the inner segment (Jansen and Sanyal, 1984). Often, several lamellar-shaped membranes are formed on the distal end of the cilium and opsin-laden vesicles accumulate in the interphotoreceptor space (Nir and Papermaster, 1986; Jansen et al, 1987; Usukura and Bok, 1987). Photoreceptor cell loss is gradual and completed within one year. It is not yet proven that the retinal dystrophy in the *rds* mouse is a result of a gene defect in the photoreceptor cells or a secondary result of abnormalities in other cells. Studies with tetraparental chimaeras showed that the defect was expressed in the photoreceptors also in regions of interactions with normal pigment epithelium (Sanyal and Zeilmaker, 1984).

In all these mutants, light responses were measured at stages of total absence of outer segments (and even inner segments) in the affected retinas. For example, in the *rds* mouse, in the absence of outer segments ERG'S with lower than normal amplitude were recorded (Reuter and Sanyal 1984). Limited light evoked decline in cyclic nucleotides was also found (Cohen 1983). In view of the observed visual responses, in dystrophic retinas, we conducted immunocytochemical studies at the electron microscopic level to explore further the capacity of these retinas to express opsin on the cell membrane of the damaged rods.

IMMUNOCYTOCHEMICAL PROCEDURES

Opsin's distribution in photoreceptor domains which were easily accessible to externally applied antibodies were studied by techniques of immersion immunocytochemistry of fixed retinas (pre-embedding technique). Pre-embedding immunocytochemical procedures are especially useful for detection of exposed antigenic determinants in the plasma membrane of outer and inner segments. After separation of the neural retina from the pigment epithelium, the intact retina is immersed in the reagents sequentially. Since the photoreceptors project into the incubation media, their plasma membranes become accessible to the reagents in the incubation media. Sheep anti-bovine opsin (Papermaster et al, 1978) was used as the first-stage reagent because it reacts mainly with opsin's N-terminal domain (Hargrave et al, 1986) which is exposed extracellularly (and intradiscally). When conditions for pre-embedding procedures are adequate this technique is highly sensitive and readily detects cell surface antigens.

Since the outer limiting membrane (which consists of adherent junctions at the inner segment level) is not permeable to antibodies, photoreceptor plasma membranes in the outer nuclear and outer plexiform layers cannot be reached by antibodies when the intact retina is incubated (except on the cut edge of the tissue). For detection of plasma membrane antigens in the deeper retinal layers as well as the visualization of intracellular antigens, post-embedding immunocytochemical procedures were employed (Schneider amd Papermaster, 1982). For that purpose, the fixed retina was embedded in a hydrophilic resin (LR Gold, Polysciences, Warrington, PA) which was polymerized by UV light at 4°. Opsin antigenicity is retained during this preparation and antibodies which are directed both against the N- and C-terminals of opsin can be used. For post-embedding labeling, sheep anti-bovine opsin (mainly anti-N-terminal) and a monoclonal antibody (mAb) 1D4 (generously provided by Dr. R. Molday) which is directed against the C-terminal of opsin (MacKenzie et al, 1984) were used. Thin sections were incubated on drops of the immunocytochemical reagents. Bound antibodies are visualized with colloidal gold conjugates.

POLARIZED DISTRIBUTION OF OPSIN IN NORMAL RETINAS

Adult normal photoreceptor are highly polarized cells. Opsin is largely sequestered in the outer segment. Earlier studies using pre-embedding procedures demonstrated that the photoreceptor plasma membranes enveloping the outer segment contained abundant opsin (Jan and Revel, 1974; Nir and Papermaster, 1983). The inner segment plasma membrane normally contains very little immunocytochemically detectable opsin. The boundary between high and low density of labeling is the connecting cilium (see Nir, et al., 1984, Figure 8). The cilium was postulated to function as an unidirectional gate which prevents backflow of opsin to the inner segment (Peters et al, 1983).

The polarized distribution of opsin in the photoreceptor's plasma membrane was initially described in the amphibian retina (Nir and Papermaster, 1983). Subsequently, mammalian retinas were also shown to be highly polarized in their opsin distribution in inner and outer segments (Nir et al, 1984, Besharse, 1986). However, other studies (Fekete and Barnstable, 1983) have shown the presence of opsin in the outer nuclear and outer plexiform layer (see below).

Studies of developing rat retinas showed that the segregation of opsin to the outer segment plasma membrane was characteristic only of the mature retina. During early stages of photoreceptor development, prior to formation of outer segments, opsin was readily detected in the inner segment plasma membrane. In the developing rat rod, the rod inner segment lost its opsin only after outer segments were formed (see Nir et al, 1984, figures 3 and 4; Hicks and Barnstable, 1986). Thereafter, opsin could not be detected in significant levels in the inner segment plasma membrane. The ability of opsin to accumulate in the developing photoreceptor's inner segment plasma membrane, in the absence of outer segments, led us to speculate that opsin might accumulate in the inner segment's plasma membrane even in adult dystrophic retinas that have lost or damaged their outer segments.

OPSIN ACCUMULATION IN THE INNER SEGMENT PLASMA MEMBRANE

***rds* mouse:** Outer segments do not form in this mutant. Thus the photoreceptors of the mature *rds* retina resemble normal immature photoreceptors prior to outer segment differentiation. It was expected, therefore, that as in immature retinas, opsin will be present in the inner segment plasma membrane. Indeed, pre-embedding labeling with sheep antiopsin clearly showed the presence of opsin in the inner segment plasma membrane of these retinas (See Nir and Papermaster, 1986, Figure 1).

RCS rat: Photoreceptors develop normally in this animal, and fully differentiated outer segments are formed. Cells begin to degenerate only after significant amounts of undigested debris accumulates in the subretinal space. When analyzed for opsin distribution, prior to the onset of outer segment degeneration, the normal polarized distribution of opsin in the plasma membrane was observed. Thereafter, as outer segments deteriorated, opsin accumulated in the inner segment plasma membrane (Figure 1).

***rd* mouse:** Loss of polarity in this mutant resembled the pattern which was observed in the RCS rat. By P10, prior to the onset of outer segment degeneration, the rods were nearly fully polarized. Thereafter, rapid loss of outer segments and cell degeneration was accompanied by an increase in opsin content in the inner segment plasma membrane (Figure 2). As a result of these studies, it was clear that opsin is retained in the inner segment plasma membrane of photoreceptors that lack outer segments.

OPSIN ACCUMULATION IN THE OUTER NUCLEAR AND OUTER PLEXIFORM LAYERS

As photoreceptor degeneration advances, inner segments are eventually lost. Light perception could still be detected despite the complete absence of both outer and inner segments. Behavioral studies with one year old RCS rats revealed discrimination between light and dark at a time that only photoreceptor nuclei remained (LaVail et al,

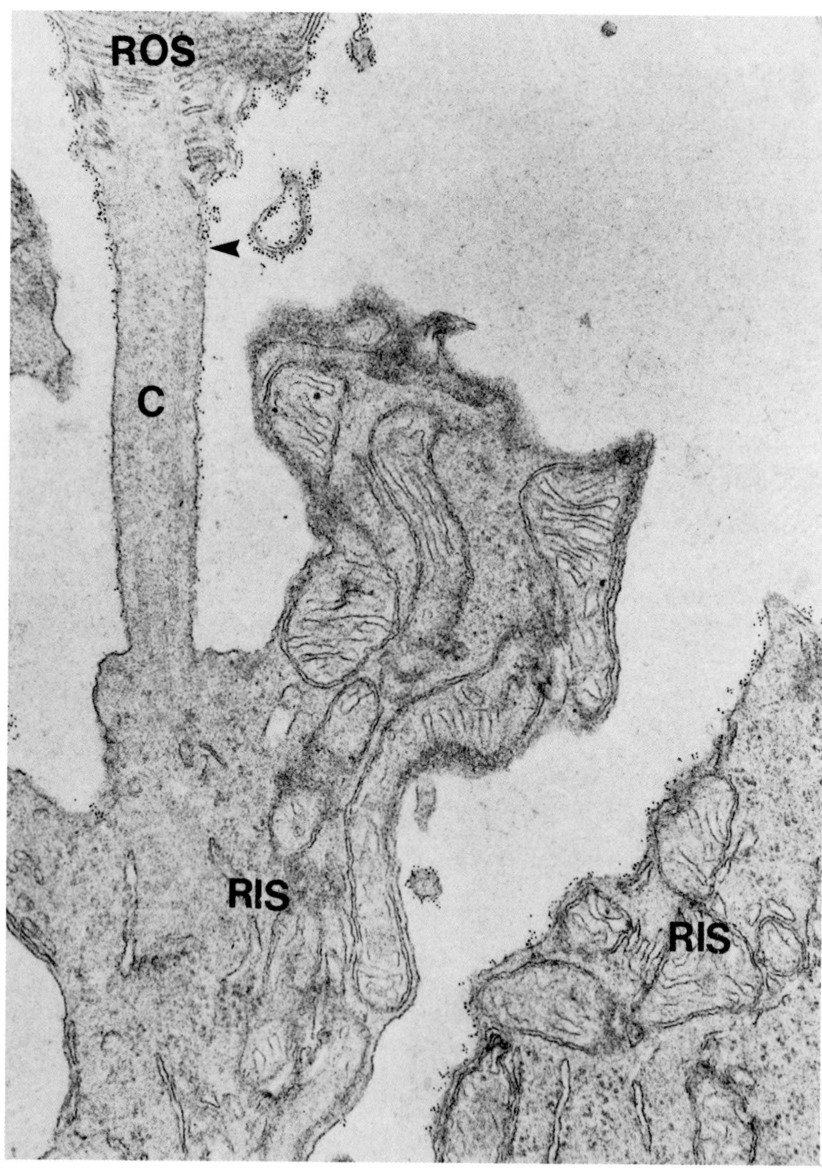

Figure 1. RCS rat, 36 days old. Pre-embedding labeling with biotinyl-sheep anti-opsin and avidin-ferritin. The inner segment plasma membrane of the cell on the right side is densely labeled. The cell on the left is still polarized: ferritin particles are seen on the outer segment (ROS) but not on the inner segment plasma membrane (RIS). Note the transition (arrow) in labeling density along the cilium (C). X 40.300.

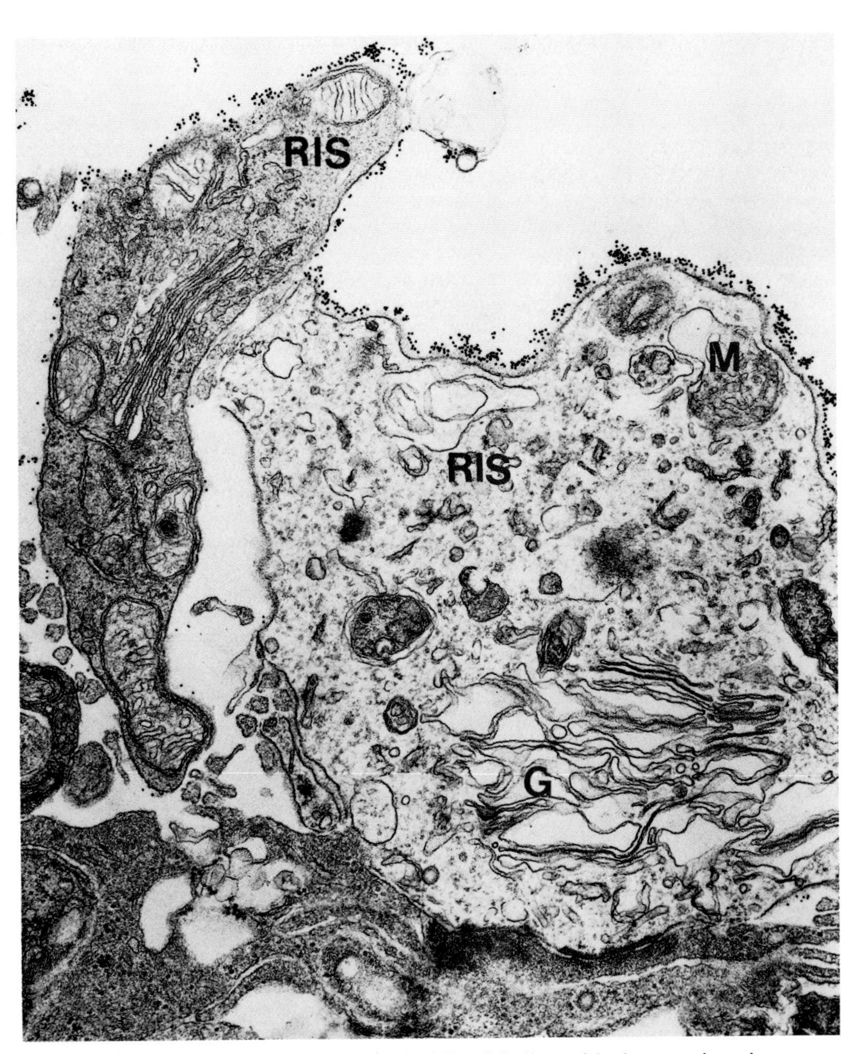

Figure 2. *rd* mouse 14 days old. Pre-embedding labeling with sheep anti-opsin, visualized with 10 nm gold. The inner segment (RIS) plasma membranes are densely labeled. The inner segment on the right is damaged. The Golgi (G) and mitochondria (M) are dilated and disorganized. The inner segment on the left is relatively undamaged and has intact Golgi and mitochondria. X 30.000.

1974; Kaitz, 1976). Visually evoked cortical responses were also recorded in advanced stages of retinal degeneration (Noell and Salinsky, 1985). Approximately normal visual responses were recorded from parts of the tectum under photopic conditions in 24 day old *rd* mice, a stage when only residual photoreceptor nuclei and synaptic terminals persist (Drager and Hubel, 1978). Persistence of brightness discrimination was revealed by behavioral studies even in 100 day old *rd* mice (Nagy and Misanin, 1970). The next step in our study was, therefore, to determine the distribution of opsin in photoreceptors which had lost both outer and inner segments. Since the photoreceptor nuclei lie on the vitreal side of the outer limiting membrane, post-embedding cytochemical procedures were used for detection of opsin in these regions.

Normal retinas: Opsin was detected at low density in the plasma membrane which envelopes the photoreceptor perikaya of the normal adult retina even when relatively high concentrations of antibody (100 ug/ml) were used in post-embedding immunocytochemical studies (Figure 3A). By contrast, in immature retinas, opsin was detected in the same domain at a relatively high density (Figure 3B). With a lower antibody concentration (25 ug/ml), while opsin was still clearly detectable in immature retinal perikaya, its labeling around perikarya was drastically reduced in the normal adult retina. Thus, although opsin content in the adult perikaya is higher then the adjacent inner segment plasma membrane, its labeling density is substantially lower than the labeling of perikaya in immature retinas.

RCS rat: Opsin was readily detected in the perikaryal plasma membrane. Even in 8 month old rats, the plasma membranes of the few remaining pyknotic nuclei contained opsin (Figure 4A). In addition to perinuclear plasma membranes, opsin was also detected in plasma membranes which enclose some of the synaptic terminals (Figure 4B).

***rd* mouse:** By P10, opsin at low density was noted in the outer nuclear layer. The labeling density in this domain increased considerably and rapidly as outer segment degeneration progressed. After P14, high labeling density was observed in the perikaryal plasma membrane (Figure 5). This finding was also reported by others (Ishiguro et al, 1987). High labeling density was also observed in the outer plexiform layer (Figure 6).

***rds* mouse:** Both young (3 weeks old) and older mice displayed opsin in the outer nuclear and outer plexiform layers (Figure 7). Similar observations were reported by Jansen et al (1987).

DELAYED DEGENERATION OF CONES

Most forms of human RP present clinically with the early onset of night blindness, indicating that rod degeneration preceeds that of cones. Early studies of the RCS rat and *rd* mouse suggested that in these animals, cone nuclei also survived longer than rods (LaVail et al, 1974; Carter-Dawson et al, 1978). When we applied antiopsin, by the post-embedding procedure, to the sections of the *rd* mouse retina, opsin was detected up to the age of P30 in the plasma membrane which enveloped rod nuclei and synaptic terminals but unlabeled perikarya were also observed (Figure 5). Thereafter, only unlabeled photoreceptor nuclei, apparently from cones, remained. Although we do not have, at present, an antibody which will detect cone photopigment(s) in these surviving nuclei, we postulate that if they are cones, the cone photopigments will be retained around them the same way opsin molecules persist in surviving rod nuclei of younger mice.

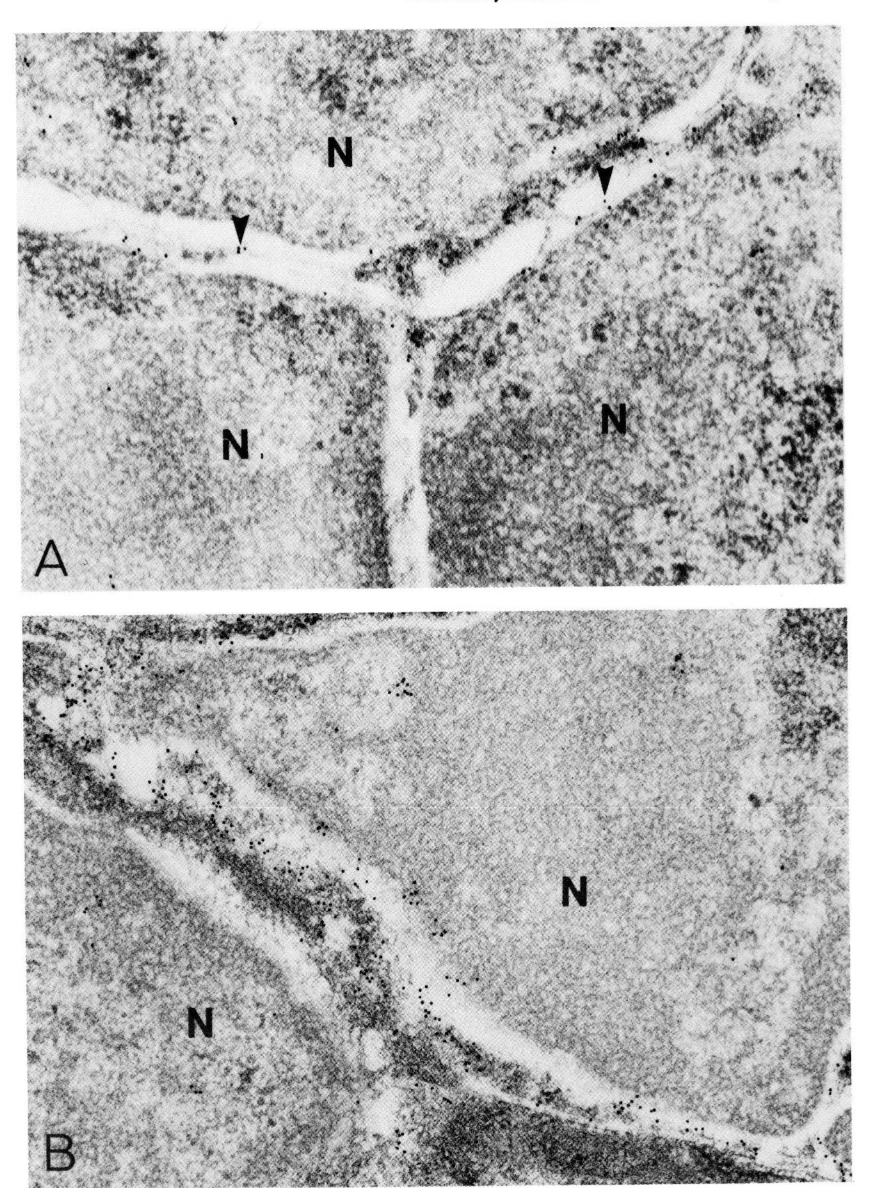

Figure 3. Normal Long Evans rats, 30 days old (A) and 7 days old (B). Post-embedding labeling with sheep anti-opsin (100 μg/ml), visualized with 10 nm gold. In the 30 day old rat (A) antibody binds sparsely along the plasma membranes (arrows) which envelopes the nucleus (N). In the 7 day old rat (B) greater labeling density is observed along the plasma membrane. (A)X 54.000 (B)X 49.500.

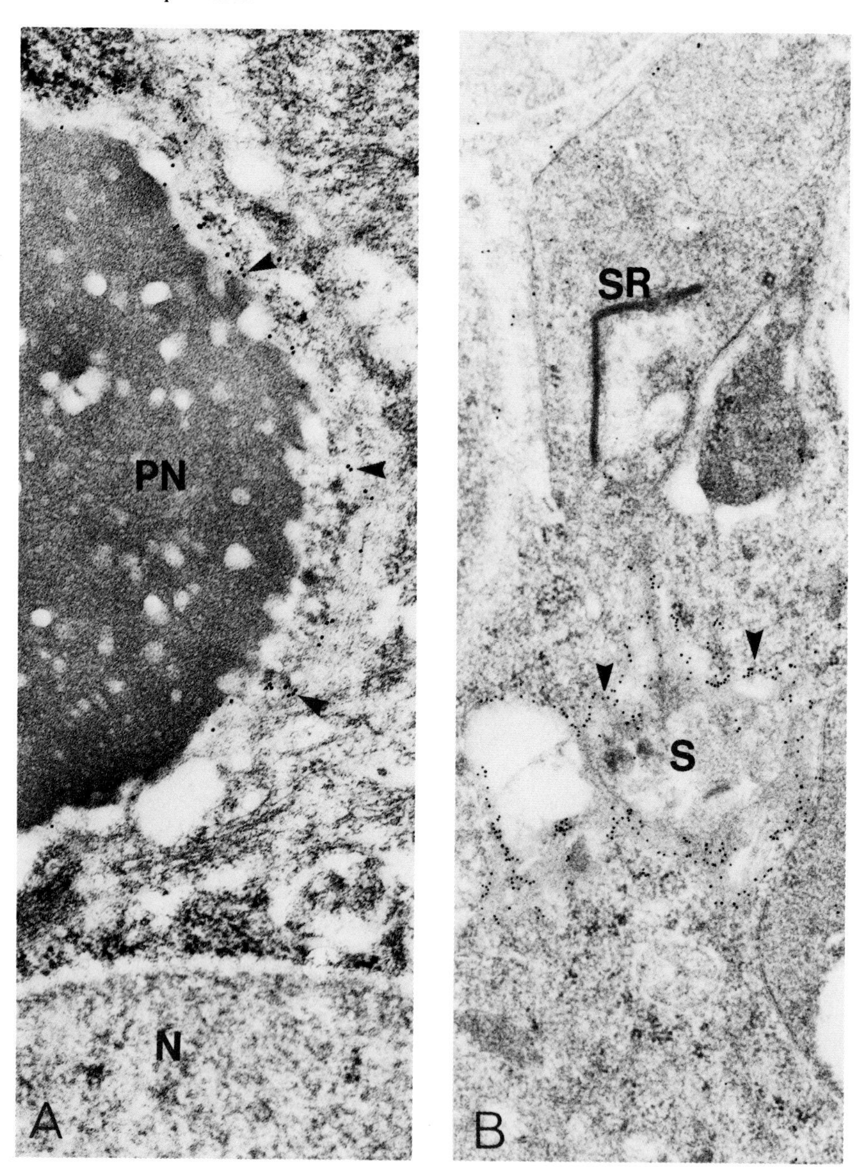

Figure 4. RCS rat. Post-embedding labeling with sheep anti-opsin visualized with 10nm gold. (A) The membrane surrounding a pyknotic nucleus (PN) in an 8 month old retina is labeled (arrows). The membrane which envelopes an adjacent, non-pyknotic nucleus (N) is not labeled. X 57.000. (B) The plasma membrane of the synaptic terminal region (S) is labeled (arrows) in a 6 weeks old retina (SR, synaptic ribbon). X 39.000.

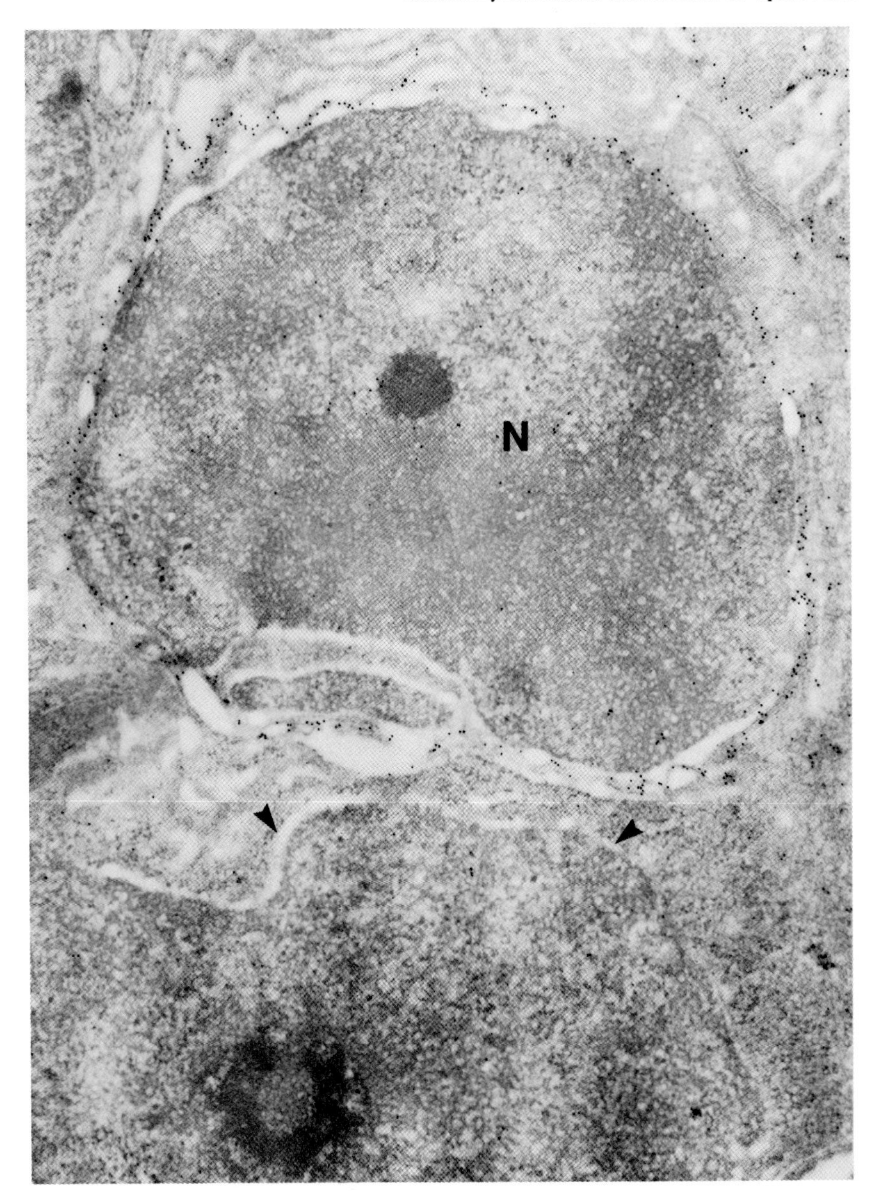

Figure 5. *rd* mouse 17 days old. Post-embedding labeling with mAb 1D4 visualized with 10 nm gold. The plasma membrane which envelops the rod nucleus (N) is densely labeled. Another nucleus (arrows) possibly of a cone, is unlabeled. X 33.000.

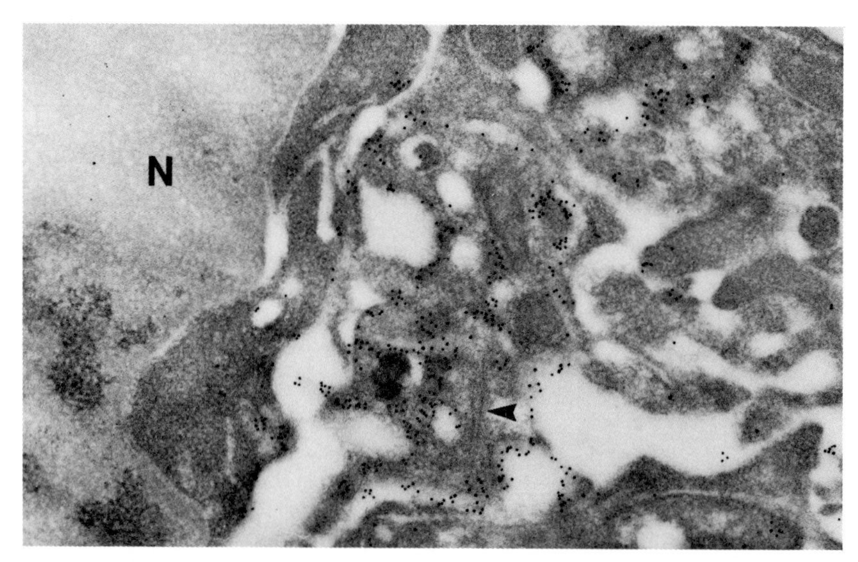

Figure 6. *rd* mouse 14 days old. Post-embedding labeling with mAb 1D4, visualized with 10 nm gold. Labeling in the synaptic terminal region is seen. Synaptic ribbons are identified with arrowheads (N-nucleus). X 45.000.

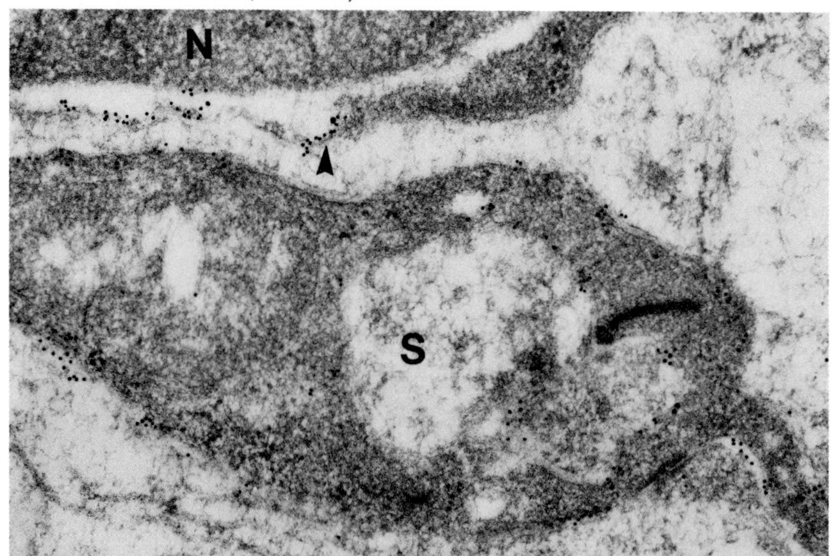

Figure 7. *rds* mouse 21 days old. Post-embedding labeling with sheep anti-opsin, visualized with 10 nm gold. The plasma membrane which envelops a synaptic terminal (S) is labeled at high density. The perikaryal plasma membrane (arrow) is also labeled (N-nucleus). X 52.000.

EVENTS THAT UNDERMINE THE POLARIZED DISTRIBUTION OF OPSIN

A remaining question concerns the mechanism by which opsin accumulates in the inner segment plasma membrane of these mutant rodent retinas. In RCS rat and *rd* mouse retinas, polarized distribution of opsin predominantly to the outer segment, is lost as outer segments degenerate. It is possible that accumulation of opsin in the inner segment plasma membranes of the RCS rats and *rd* mice might result from back diffusion of opsin from the outer segment following a breakdown of the postulated unidirectional gate in the connecting cilia of rods. The gate may fail to form properly in the *rds* mouse altogether so that its inner segments are never adequately cleared of opsin. Alternatively, in the absence of an outer segment, which never develops in the *rds* mouse and is lost in *rd* mice and RCS rats, newly synthesized opsin may be inserted directly into the lateral plasma membrane and accumulate there.

In order to evaluate this alternative explanation, we studied opsin's synthesis in both the *rd* and *rds* mice (Agarwal et al, ms in preparation). Retinas were isolated from dystrophic mice at various ages and incubated for two hours in a medium containing [^{35}S]-methionine. In the *rd* mouse, opsin was detected in 15 day old retinas immunochemically by Western blots of SDS-PAGE gels of retinal homogenates. That part of this opsin was newly synthesized was demonstrated by autoradiography of the gels. Since most of the outer segments had already disappeared at 15 days, opsin was apparently inserted into the only remaining photoreceptor plasma membranes.

At postnatal day 11, the opsin content of the *rds* mouse retina was comparable to normal. In one month old and three month old *rds* mice, opsin content was considerably lower when compared to normal mice. This observation is in agreement with the previously described low level of opsin in *rds* retina (Schalken et al, 1985, Usukura and Bok, 1987). Analysis of the autoradiographs of newly synthesized proteins separated by SDS-PAGE revealed, remarkably, that opsin synthesis in the *rds* retina, *when corrected for cell loss,* proceeded at an apparently normal rate per cell. Thus, only a small part of the newly synthesized opsin was retained in the plasma membrane enclosing the residual rod cell structures in these mice. This domain is obviously limited in its capacity to store opsin. The remainder of the newly synthesized opsin is either transported to the interphotoreceptor space and destroyed there or it is destroyed inside the cell by autophagocytosis. Most detectable opsin molecules in the *rds* retina are packaged in small vesicles which are shed into the extracellular space between photoreceptors rather than being assembled into an outer segment (Nir and Papermaster, 1986; Usukura and Bok, 1987; Jansen et al, 1987).

Intracellular opsin-containing post-Golgi vesicles were shown to fuse with the apical inner segment plasma membrane of frog rods, possibly at the grooves of the periciliary ridge complex (Peters et al, 1983). These opsin molecules subsequently translocate predominantly to the outer segment (Papermaster et al, 1985). If a similar mechanism exists in the mammalian retina, the presence of opsin at significant levels in the perikarya and synaptic terminal of dystrophic rat and mouse retinas indicates that, in the absence of outer segments, some of the opsin molecules remain in the inner segment plasma membrane while others apparently redistribute over its surface.

SUMMARY AND CONCLUSIONS

Opsin is normally sequestered predominantly in the outer segment disc and plasma membranes of adult photoreceptors. Absence of opsin from the inner segment plasma membrane in normal photoreceptors is probably not due to the inability of the inner segment plasma membrane to retain opsin. Rather, in the adult mammalian retina, if opsin is inserted at sites in the apical inner segment plasma membrane, in a fashion comparable to the pathway in amphibians, it is rapidly transported predominantly to the outer segment by unknown mechanisms.

Dystrophic *rds* retinas, lacking an outer segment, display newly synthesized opsin throughout the plasma membrane. If opsin is transported to the inner segment plasma membrane as a specific insertional site, diffusion in the plane of the membrane may redistribute opsin throughout the plasma membrane which encloses the nucleus and the synaptic terminal. Alternatively, opsin may be inserted randomly throughout the entire cell's plasmalemma beneath the cilium. Selective transport to the outer segment may preferentially clear the inner segment of most of its opsin and nearly clear the perikaryal and synaptic terminal's plasmalemma in normal cells. In dystrophic retinas, however, as outer segments degenerate or fail to form, opsin is detected readily in the remaining plasma membrane sites. In the *rd* mouse, some of the opsin molecules in the inner segment plasma membrane might be newly synthesized while others may arise from molecules which reached the inner segment by back-diffusion from the outer segment at least at early stages in the degeneration while outer segments survive.

The opsin in the plasma membrane which envelopes the residual rod nuclei and synaptic terminals in dystrophic retinas may account for the persisting light perception in retinas which have lost both the rod outer and inner segments. Dystrophic retinas, such as the *rd* mouse and RCS rats and possibly human RP retinas, in which cone nuclei survive long after rods disappear, might retain light perception because of cone photo-pigments in the outer nuclear and outer plexiform layers. To explore these questions further, the localization of other components of the transduction cascade and the determination of the efficiency of their coupling in dystrophic cells is necessary. We need to know where the cyclic GMP-sensitive sodium channels lie in these dystrophic cells and the cellular requirements for proximity of these components to generate a signal.

Outer segment-free photoreceptors, bearing opsin in their plasma membranes, resemble other cells which have receptor-mediated alterations in membrane permeability to ions. Indeed, primitive photoreceptors may have begun in this form long ago and only subsequently amplified their sensitivity to light by elaboration of outer segments. Thus, modern retinas with inherited retinal dystrophies may provide a glimpse of the past in the phylogeny and ontogeny of these remarkable cells.

Acknowlegements: We thank Nancy Ransom for her technical assistance, Sandy Gomez for typing the manuscript, and Dr. Neeraj Agarwal, Dusanka Deretic and Barbara Schneider for helpful discussions. The research was supported, in part by the Veterans Administration, by grant of Israel ministry of health and NIH grants EY-845 and EY-6891.

REFERENCES

Besharse JC (1986). Photosensitive membrane turnover: Differentiated membrane domains and cell-cell interactions. In Adler R and Farber D (eds). "The Retina, A model for Cell Biological Studies". New York: Academic Press, Vol. 1 pp 297-352.
Bok D, Hall MO (1971). The role of the pigment epithelium in the etiology of inherited retinal dystrophy in the rat. J Cell Biol. 49:664-682.
Caley DS, Johnson C, Liebelt RA (1972). The postnatal development of the retina in the normal and rodless CBA mouse: A light and electron microscopic study. Am J Anat 133:179-212.
Carter-Dawson LD, LaVail MM, Sidman, RL (1978). Differential effect of the *rd* mutation on rods and cones in the mouse retina. Invest Ophthalmol Vis Sci 17:489-498.
Cohen AI (1983). Some cytological and initial biochemical observations on photoreceptors in retinas of *rds* mice. Invest Ophthalmol Vis Sci 24:832-843.
Dowling JE, Sidman RL (1962). Inherited retinal dystrophy in the rat. J Cell Biol 14:73-109.
Drager UC, Hubel DH (1978). Studies of visual function and its decay in mice with hereditary retinal degeneration. J Comp Neurol 180:85-114.
Farber DB, Lolley RN (1974). Cyclic guanosine monophosphate: elevation in degenerating photoreceptor cells of the C3H mouse retina. Science 186:449-451.
Fekete DM, Barnstable CJ (1983). The subcellular localization of rat photoreceptor-specific antigens. J Neurocytol 12:785-803.
Hargrave PA, Adamus G, Arendt A, McDowell JH, Wang J, Szary A, Curtis D, Jackson RW (1986). Rhodopsin's amino terminus is a principal antigenic site. Exp Eye Res 42:363-373.
Hicks D, Barnstable CJ (1986). Lectin and antibody labelling of developing rat photoreceptor cells: an electron microscope immunocytochemical study. J Neurocytol 15:219-230.
Ishiguro S, Fubuda K, Kanno C, Mizuno K (1987). Accumulation of immunoreactive opsin on plasma membranes in degenerating rod cells of *rd/rd* mutant mice. Cell Structure and Function 12:141-155.
Jan LY, Revel JP (1974). Ultrastructural localization of rhodopsin on the vertebrate retina. J Cell Biol 62:257-273.
Jansen HG, Sanyal S, DeGrip WJ, Shalken JJ (1987). Development and degeneration of retina in *rds* mutant mice: ultraimmunohistochemical localization of opsin. Exp Eye Res 44:347-361.
Jansen HG, Sanyal S (1984). Development and degeneration of retina in *rds* mutant mice: Electron microscopy. J Comp Neurol 224:71-84.
Kaitz M (1976). The effect of light on brightness perception in rats with retinal dystrophy. Vis Res 16:141-148.
LaVail MM (1981). Analysis of neurological mutants with inherited retinal degeneration. Invest Opthalmol Vis Sci 21:638-657.
LaVail MM, Sidman M, Rausin R, Sidman RL (1974). Discrimination of light intensity by rats with inherited retinal degeneration: a behavioral and cytological study. Vision Res 14:693-702.

MacKenzie D, Arendt A, Hargrave P, McDowell JH, Molday RS (1984). Localization of binding sites for carboxyl terminal specific anti-rhodopsin monoclonal antibodies using synthetic peptides. Biochemistry 23:6544-6549.

Mullen RJ, LaVail MM (1976). Inherited retinal dystrophy: Primary defect in pigment epithelium determined with experimental rat chimeras. Science 192:799-801.

Nagy ZM, Misanin JR (1970). Visual perception in the retinal degenerate C3H mouse. J Comp Physiol Psych 72:306-310.

Nir I, Cohen D, Papermaster DS (1984). Immunocytochemical localization of opsin in the cell membrane of developing rat photoreceptors. J Cell Biol 98:1788-1795.

Nir I, Papermaster DS (1986). Immunocytochemical localization of opsin in the inner segment and ciliary plasma membrane of photoreceptors in retinas of rds mutant mice. Invest Ophthalmol Vis Sci 27:83-843.

Nir I, Papermaster DS (1983). Differential distribution of opsin in the plasma membrane of frog photoreceptors: an immunocytochemical study. Invest Ophthalmol Vis Sci 24:868-878.

Nir I, Sagie G, Papermaster DS (1987). Opsin accumulation in photoreceptor inner segment plasma membrane of dystrophic RCS rats. Invest Ophthalmol Vis Sci 28:62-69.

Noell WK, Salinsky MC (1985). The preservation of visual evoked cortical responses at an advanced stage of retinal degeneration in the *rdy* rat in retinal degeneration: In LaVail, MM, Hollyfield JG, RE Anderson (eds): "Retinal degeneration: Experimental and Clinical Studies." New York. Alan R. Liss pp. 301-320.

Papermaster DS, Schneider BG, Zorn MA, Kraehenbuhl JP (1978). Immunocytochemical localization of opsin in outer segments and Golgi zones of frog photoreceptors. J Cell Biol 77:196-210.

Papermaster DS, Schneider BG, Besharse JC (1985). Vesicular transport of newly synthesized opsin from the golgi apparatus toward the rod outer segment. Invest Ophthalmal Vis Sci 26:1386-1404.

Peters KR, Palade GE, Schneider BG, Papermaster DS (1983). Fine structure of a periciliary ridge complex of frog retinal rod cells revealed by ultrahigh resolution scanning electron microscopy. J Cell Biol 96:265-276.

Reuter JH, Sanyal S (1984). Development and degeneration of retina in *rds* mutant mice: the electroretinogram. Neurosci Lett 48:231-237.

Sanyal S, Zeilmaker GH (1984). Development and degeneration of retina in *rds* mutant mice: Light and electron microscopic observations in experimental chimeras. Exp. Eye Res 39:231-246.

Schalken JJ, Jansen JJM, deGrip WJ, Hawkins RK, Sanyal S (1985). Immunoassay of rod visual pigment (opsin) in the eyes of *rds* mutant mice lacking receptor outer segments. Biochim Biophys Acta 839:122-126.

Schneider BG, Papermaster DS (1982). Immunocytochemistry of retinal membrane protein biosynthesis at the electron microscope level by the albumin embedding technique. Meth Enzymol 96:485-495.

Szamier RB, Berson EL, Klein R, Meyers S (1979). Sex-linked retinitis pigmentosa: ultrastructure of photoreceptors and pigment epithelium. Invest Ophththal Visual Sci 18:145-160.

Usukura J, Bok D (1987). Changes in the localization and content of opsin during retinal development in the *rds* mutant mouse: Immunocytochemistry and immunoassay. Exp Eye Res 45:501-515.

Inherited and Environmentally Induced
Retinal Degenerations, pages 265–274

IMMUNOGOLD LOCALIZATION OF ACTIN AND OPSIN IN RDS MOUSE PHOTORECEPTORS

Michael H. Chaitin

Bascom Palmer Eye Institute and the Department of Anatomy and Cell Biology, University of Miami School of Medicine, Miami, Florida 33136

INTRODUCTION

Rod outer segments (ROS) comprise a plasma membrane enclosed stack of discs that is continually renewed through the competing processes of disc morphogenesis at the ROS base (Young, 1967; Steinberg et al., 1980) and disc shedding from the tip (Young and Bok, 1969; LaVail, 1976). Although the mechanisms responsible for these two phenomena are not yet understood, important insight into the morphogenic aspect of ROS renewal has recently been gained. Immunoelectron microscopy studies have localized an actin-rich domain to the distal portion of the connecting cilium in photoreceptors from several vertebrate species (Chaitin et al., 1984; Chaitin and Bok, 1986). Since the distal cilium is also the site of new disc formation via ciliary membrane evaginations (Steinberg et al., 1980), it has been suggested that an actin-mediated contractile mechanism or cytoskeletal network may regulate some aspect of ROS disc morphogenesis (Chaitin et al., 1984; Chaitin et al., 1988).

A rhodamine-phalloidin binding study has demonstrated that at least a portion of this distal ciliary actin is in the filamentous form (Vaughan and Fisher, 1987). The polarity and distribution of these filaments has recently been observed at the electron microscope level by using myosin subfragment-1 binding to form characteristic arrowhead complexes on the actin filaments (Chaitin and Burnside, 1988). When isolated retinas are maintained in an in vitro culture system, the inclusion of cytochalasin D reportedly depolymerizes these ciliary actin filaments and disrupts normal disc morphogenesis (Williams et al., 1988). Previously formed membrane evaginations continue to grow beyond the normal disc dimensions, however new evaginations do not occur. Thus, it is becoming increasingly apparent that the mechanism which regulates ROS disc morphogenesis involves an actin filament network.

In rodent retinas, ROS begin to differentiate at about five to seven days postnatal. Cilia project from the visual cell bodies prior to this event and opsin can be immunocytochemically localized to the inner segment and ciliary plasma membranes at this time (Nir et al., 1984; Besharse et al., 1985; Hicks and Barnstable, 1986; Chaitin, 1988). Upon the differentiation of an ROS, the opsin labeling density on the inner segment membrane becomes dramatically reduced until almost none remains at maturity (Nir et al., 1984; Hicks and Barnstable, 1986). The proximal cilium is essentially free of label, however the distal ciliary and ROS plasma membranes show dense labeling for opsin (Nir et al., 1984; Besharse et al., 1985; Hicks and Barnstable, 1986). In a separate study, it has also been demonstrated that the actin-rich ciliary domain is established within the distal end of elongated cilia prior to the differentiation of an ROS (Chaitin et al., 1988).

In the homozygous rds (retinal degeneration slow) mouse, photoreceptors appear to develop normally up until the time when ROS differentiation should begin (Sanyal et al., 1980; Sanyal and Jansen, 1981). Cilia project into the subretinal space (Cohen, 1983; Sanyal and Jansen, 1981; Jansen and Sanyal, 1984) and opsin is situated on the ciliary plasma membrane (Nir and Papermaster, 1986; Usukura and Bok, 1987a; Jansen et al., 1987), however ROS development is absent (Sanyal and Jansen, 1981). Opsin distribution is never fully polarized to the distal cilium, and opsin of a lower labeling density can be detected on the inner segment plasma membrane using immunocytochemical techniques. This inner segment and ciliary membrane opsin is evidently responsible for the ERG responses (Reuter and Sanyal, 1984) and light-modulated cyclic nucleotide levels (Cohen, 1983) which are exhibited by rds retinas. Opsin has also been detected on vesicular structures which fill the subretinal space (Cohen, 1983; Jansen and Sanyal, 1984; Nir and Papermaster, 1986; Jansen et al., 1987; Usukura and Bok, 1987a). The source of these vesicles may be the distal cilium where an abortive attempt at disc morphogenesis is apparently taking place. In some instances, opsin-rich lamellar structures are observed on the ends of these rds cilia.

Recent studies have also shown that actin is localized to its normal ciliary domain at early ages in rds photoreceptors (Chaitin et al, 1988). It is not yet known, however, whether the actin filament distribution or the regulation of this actin are normal. Additional work is needed to answer these questions.

In the current study, we have used double immunogold labeling to localize actin and opsin in rds photoreceptor cilia of various ages. The purpose of this work was to determine if these two proteins maintain their distribution in older rds retinas, even though ROS formation is absent.

MATERIALS AND METHODS

Animals

Homozygous rds mutant mice were selected from a colony maintained on a 12 hr light: 12 hr dark cycle.

Antibodies

Actin was purified from chicken gizzard and injected into rabbits for antibody production (Chaitin and Hall, 1983). Specific anti-actin was affinity purified and used at a concentration of 0.02 mg/ml.

Sheep anti-opsin antiserum was obtained from Dr. David Papermaster and an IgG fraction was prepared by Dr. Steven Fliesler using an Affi-Gel blue column. Upon receipt of this preparation, the anti-opsin was biotinylated and used at a concentration of 0.30 mg/ml.

Streptavidin-colloidal gold (5 nm) and goat anti-rabbit IgG-colloidal gold (15 nm) were obtained from Janssen Life Sciences Products and used at a dilution of 1:35 with Tris-BSA buffer (0.02M Tris, pH 8.2, 0.15M NaCl, 1% BSA, 0.05% sodium azide).

Immunogold Labeling

After animal sacrifice and enucleation of the eyes, neural retinas were isolated and immersion labeled with anti-opsin using a modified version of the procedure described by Nir et al (1984). Following a PBS (0.01M sodium phosphate, 0.85% NaCl, 0.02% sodium azide, pH 7.4) rinse, the retinas were fixed in 1% glutaraldehyde in 0.15M sodium phosphate, pH 7.4 for 1 hr at room temperature (R.T.). During a PBS rinse, the retinas were cut into strips and these were immersion labeled as follows; 0.05M glycine in PBS (10 min), 2% BSA in PBS (10 min), biotinylated anti-opsin (90 min), PBS (twice for 20 min each), Tris-BSA (20 min), streptavidin-gold (90 min), Tris-BSA (20 min), PBS (twice for 20 min), 1% glutaraldehyde in 0.15M sodium phosphate (20 min), PBS rinse. The tissue was then stored in 2% paraformaldehyde in 0.1M sodium phosphate, pH 7.4 before being embedded in Lowicryl K_4M (Chaitin and Bok, 1986).

Ultrathin tissue sections were collected onto formvar coated nickel grids and anti-actin labeling was done at R.T. by transferring the grids sequentially onto drops of 0.1M sodium phosphate, pH 7.4 (PB, 5 min), 4% BSA in PB (10 min), 0.02M glycine in PB (15 min), anti-actin (2 hr), PB (3 X 1 min), Tris-BSA (3 X 1 min), IgG-gold (30 min), Tris-BSA (3 X 1 min), PB (3 X 1 min), 2% glutaraldehyde in PB (20 min), water (8 X 1 min). The sections were then stained wih 5% aqueous uranyl acetate for 15 min at 60°C and viewed on a JEOL 100CX transmission electron microscope.

RESULTS AND DISCUSSION

In this immunogold labeling study, we have used antibodies specific for opsin and actin in order to localize these two proteins in rds mouse photoreceptors. Immersion labeling of neural retinas with anti-opsin was chosen so as to detect membrane opsin with high sensitivity. Anti-actin labeling, on the other hand, was done on thin sections of Lowicryl embedded tissue so as to detect actin within the photoreceptor cilia. In order to distinguish the localization of these two proteins, a 5 nm colloidal gold probe was used to tag opsin and a 15 nm gold probe was used for actin.

Our results agreed with those of several other investigators who have studied opsin localization in young rds mouse photoreceptors (Nir and Papermaster, 1986; Usukura and Bok, 1987a; Jansen et al., 1987). Opsin was present in the ciliary plasma membrane, as was the case for normal cilia prior to ROS differentiation (Nir et al, 1984; Besharse et al., 1985; Hicks and Barnstable, 1986; Chaitin, 1988). At the time when ROS formation should begin, opsin label was most dense over the distal cilium (Fig. 1a,b). Labeling over the inner segment plasma membrane was lighter, and the proximal cilium was relatively unlabeled. Opsin-rich lamellar structures were sometimes observed on the ends of these cilia (Fig. 1b). In older animals, this labeling pattern was maintained, however the amount of inner segment label was more variable (Fig. 2a,b). The continued presence of inner segment opsin is probably due to defective disc morphogenesis which is occurring at the distal cilium and to a lower than normal incorporation of opsin into new disc membranes. These results agree with those of a recent study in which ciliary opsin label was observed in surviving rds photoreceptors up to at least two months of age (Jansen et al., 1987). Neither this study nor the one reported in the current paper support the contention of Usukura and Bok (1987a) that the amcunt of opsin in rds photoreceptors peaks at about 15 days postnatal, thereafter decreasing to background levels by 30 days of age. Their study used both Western blot analysis and immunoelectron microscopy to reach its conclusions. An important difference in the specificity of their anti-opsin as compared to those used in the other two studies cannot be discounted at this time.

Studies on chimaeric mice have demonstrated that the rds gene acts within the neural retina (Sanyal and Zeilmaker, 1984) and expresses itself as an abortive attempt at ROS disc morphogenesis. This disorder may be due to a defect in the morphogenic mechanism, a possibility which led us to investigate actin localization within rds photoreceptor cilia. Using immunogold labeling techniques, actin was found to be situated within its normal distal ciliary domain in young and old rds mice (Fig. 1,2). However, it is not yet known if the actin filament network or its regulation are normal. Additional work is currently in progress to answer these questions.

An alternative explanation for the rds disorder is that there is an abnormality in one of the membrane proteins,

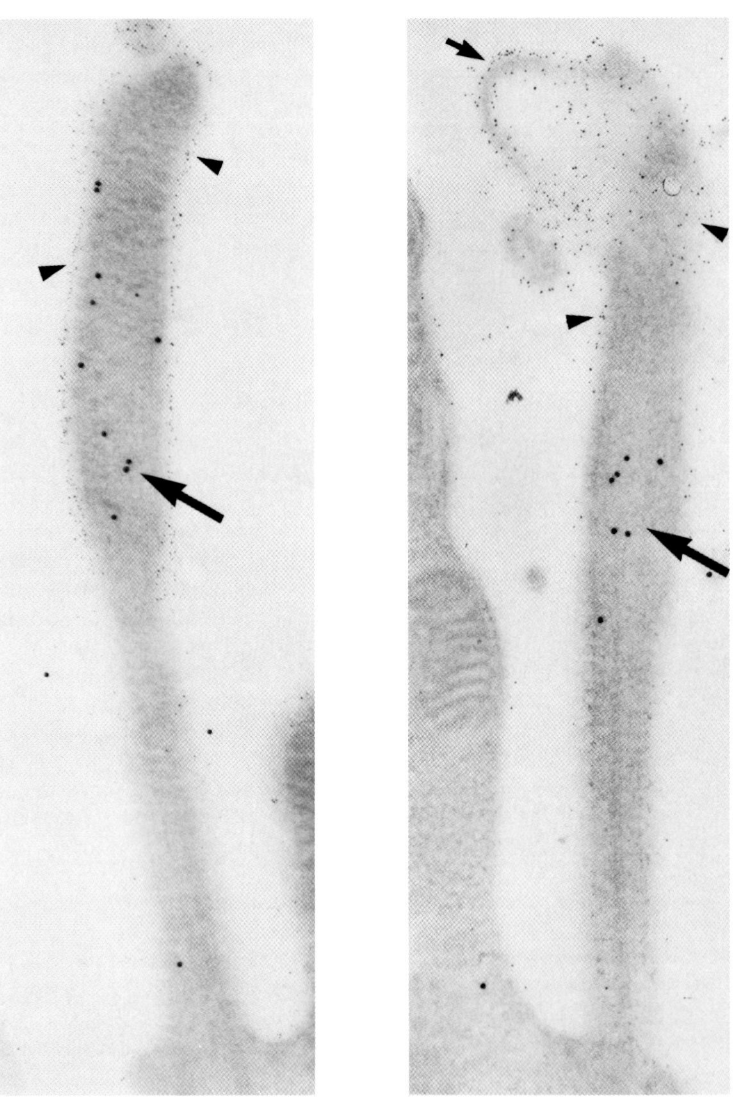

Fig. 1. (a,b) Cilia from 10 day old rds photoreceptors. Anti-actin binds to the distal cilium and is detected using 15 nm gold (large arrows). Anti-opsin densely labels the distal ciliary plasma membrane and is detected with 5 nm gold (arrowheads). Inner segment label is light and the proximal cilium is free of label. In (b) note the opsin-rich lamellar structure (small arrow). x51,300 (a); x53,200 (b).

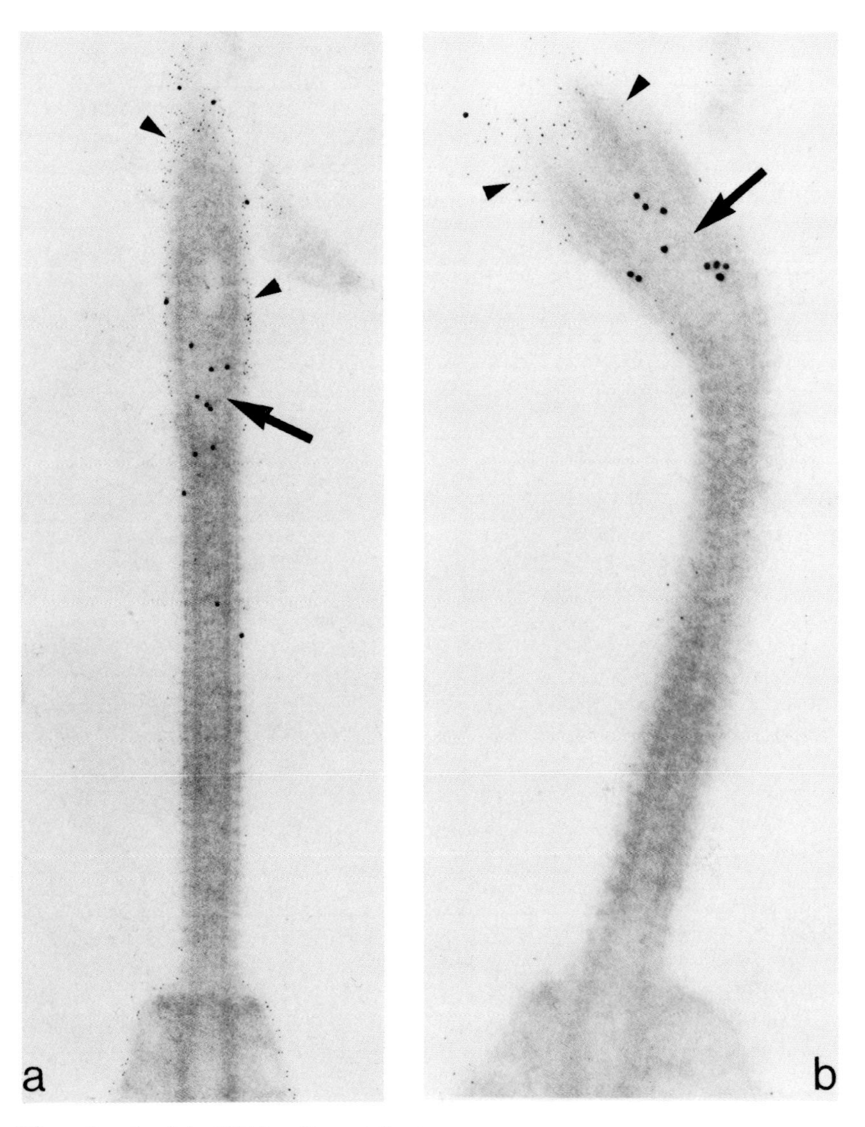

Fig. 2. (a,b) Cilia from 50 day old rds photoreceptors. Anti-actin binds to the distal cilium and is detected using 15 nm gold (large arrows). Anti-opsin densely labels the distal ciliary plasma membrane and is detected using 5 nm gold (arrowheads). The proximal cilium is relatively free of label, whereas label on the inner segment plasma membrane is variable. x39,200 (a); x57,000 (b).

such as opsin, which becomes incorporated into new ROS discs. In this regard, a suggestion was recently made that incomplete glycosylation of rds mouse opsin might result in the vesicular structures which fill the subretinal space (Usukura and Bok, 1987b). This would be similar to the situation observed when amphibian retinas are treated with tunicamycin, a substance which blocks the glycosylation of glycoproteins, including opsin (Fliesler et al., 1985). In a study of rds opsin, however, it was shown that rather than being underglycosylated, the opsin molecule may be slightly hyperglycosylated (Usukura and Bok, 1987b). There is no current evidence to suggest that this could led to a breakdown in disc morphogenesis. Additional work is necessary before we can determine whether opsin or other ROS proteins are truly abnormal in the rds photoreceptor.

Despite numerous studies, our understanding of the rds retinal disorder is incomplete. It can be demonstrated, however, that actin is localized to its normal ciliary domain and opsin is present in the ciliary plasma membrane at appropriate developmental stages. The actin-rich domain persists in older rds mice, as does membrane opsin, even though ROS formation is absent. As studies of this animal model continue, we should be able to pinpoint the defect at the molecular level and, at the same time, achieve a better understanding of the mechanism which regulates ROS disc morphogenesis.

ACKNOWLEDGEMENTS

The author thanks Mr. Richard Carlsen for his expert technical assistance and Dr. Steven Fliesler for providing the IgG fraction of anti-opsin. This work was supported by NIH Grants R29 EY06590 to Dr. Chaitin and P30 EY02180 to the Department of Ophthalmology.

REFERENCES

Besharse JC, Forestner DM, Defoe DM (1985). Membrane assembly in retinal photoreceptors. J Neurosci 5:1035-1048.

Chaitin MH (1988). Double immunogold localization of actin and opsin in developing mouse photoreceptors. Invest Ophthal Vis Sci Supple 29:107.

Chaitin MH, Bok D (1986). Immunoferritin localization of actin in retinal photoreceptors. Invest Ophthal Vis Sci 27:1764-1767.
Chaitin MH, Burnside B (1988). Actin filament polarity at the site of outer segment disc morphogenesis. Proc Int'l Soc Eye Res V:120.
Chaitin MH, Carlsen RB, Samara GJ (1988). Immunogold localization of actin in developing photoreceptor cilia of normal and rds mutant mice. Exp Eye Res 47:437-446.
Chaitin MH, Hall, MO (1983). The distribution of actin in cultured normal and dystrophic rat pigment epithelial cells during the phagocytosis of rod outer segments. Invest Ophthal Vis Sci 24:821-831.
Chaitin MH, Schneider BG, Hall MO, Papermaster DS (1984). Actin in the photoreceptor connecting cilium: immunocytochemical localization to the site of outer segment disk formation. J Cell Biol 99:239-247.
Cohen AI (1983). Some cytological and initial biochemical observations on photoreceptors in retinas of rds mice. Invest Ophthal Vis Sci 24:832-843.
Hicks D, Barnstable CJ (1986). Lectin and antibody labelling of developing rat photoreceptor cells: an electron microscope immunocytochemical study. J Neurocytol 15:219-230.
Jansen HG, Sanyal S (1984). Development and degeneration of retina in rds mutant mice: electron microscopy. J Comp Neurol 224:71-84.
Jansen HG, Sanyal S, de Grip WJ, Schalken JJ (1987). Development and degeneration of retina in rds mutant mice: ultraimmunohistochemical localization of opsin. Exp Eye Res 44:347-361.
LaVail MM (1976). Rod outer segment disk shedding in rat retina: relationship to cyclic lighting. Science 194:1071-1074.
Nir I, Cohen D, Papermaster DS (1984). Immunocytochemical localization of opsin in the cell membrane of developing rat retinal photoreceptors. J Cell Biol 98:1788-1795.
Nir I, Papermaster DS (1986). Immunocytochemical localization of opsin in the inner segment and ciliary plasma membrane of photoreceptors in retinas of rds mutant mice. Invest Ophtholmol Vis Sci 27:836-840.
Reuter JH, Sanyal S (1984). Development and degeneration of retina in rds mutant mice: the electroretinogram. Neurosci Lett 48:231-237.
Sanyal S, de Ruiter A, Hawkins RK (1980). Development and degeneration of retina in rds mutant mice: light micros-

copy. J Comp Neurol 194:193-207.
Sanyal S, Jansen HG (1981). Absence of receptor outer segments in the retina of rds mutant mice. Neurosci Lett 21:23-26.
Sanyal S, Zeilmaker GH (1984). Development and degeneration of retina in rds mutant mice: light and electron microscopic observations in experimental chimaeras. Exp Eye Res 39:231-246.
Steinberg RH, Fisher SK, Anderson, DH (1980). Disc morphogenesis in vertebrate photoreceptors. J Comp Neurol 190: 501-518.
Usukura J, Bok D (1987a). Changes in the localization and content of opsin during retinal development in the rds mutant mouse: immunocytochemistry and immunoassay. Exp Eye Res 45:501-515.
Usukura J, Bok D (1987b). Opsin localization and glycosylation in the developing Balb/c and rds mouse retina. In Hollyfield JG, Anderson RE, LaVail MM (eds): "Degenerative Retinal Disorders: Clinical and Laboratory Investigations" New York: Alan R. Liss, pp 195-207.
Vaughan DK, Fisher SK (1987). The distribution of F-actin in cells isolated from vertebrate retinas. Exp Eye Res 44: 393-406.
Williams DS, Linberg KA, Vaughan DK, Fariss RN, Fisher SK (1988). Disruption of microfilament organization and deregulation of disk membrane morphogenesis by cytochalasin D in rod and cone photoreceptors. J Comp Neurol 272:161-176.
Young RW (1967). The renewal of photoreceptor cell outer segments. J Cell Biol 33:61-72.
Young RW, Bok D (1969). Participation of the retinal pigment epithelium in the rod outer segment renewal process. J Cell Biol 42:392-403.

Inherited and Environmentally Induced
Retinal Degenerations, pages 275–289

Postnatal development of photoreceptor proteins in mutant mice and Abyssinian cats with retinal degeneration.

[1]Theo van Veen,[1]Rafael Cantera, [2]Kristina Narfström, [2]Sven Erik Nilsson,[3]Somes Sanyal [4]Barbara Wiggert, [4]Gerald J Chader,
[1]Laboratoy of Molecular Neuroanatomy, Dept of Zoology, University of Lund, S-22362 Lund, Sweden
[2]Dept of Ophthalmology, University of Linköping, Linköping, Sweden.
[3]Dept of Anatomy, Erasmus University, Rotterdam, The Netherlands
[4] Laboratory of Retinal Cell and Molecular Biology, National Eye Institute, National Institutes of Health, Bethesda MD, USA

Introduction

Mutant mice, double homozygous for the retinal degeneration (*rd*) and retinal degeneration slow (*rds*) genes, fail to develop their photoreceptor outer segments (Sanyal *et al.*1980, Sanyal 1988). Although other retinal layers appear to develop normally, a progressive degeneration is observed in the photoreceptor cell layer such that, the outer nuclear layer is reduced somewhat slower than in the single homozogous *rd,* and considerably faster than in the single homozogous *rds* mutants (Sanyal 1988, van Veen *et al.* 1988). In contrast, mutant mice homozygous for the retinal degeneration (*rd*) gene develop photoreceptor outer segments, but degeneration starts around the time of outer segment development and proceeds very rapidly so that only a single row of outer nuclei is present at the 21 days age of (Tansley 1951). The development of photoreceptor specific proteins and their mRNAs in the *rd* mutant has the same onset of expression as in the normal control (Bowes *et al.*1988); the development of IRBP also occurs simultaneously in the mutant and normal retina (van Veen *et al.* 1988, Carter-Dawson *et al.* 1986). Immunochemical and immuno-cytochemical techniques demonstrated that the IRBP is present intra-cellularly in abnormally high concentrations in mutants where the *rd* gene is expressed, i.e. in both homozygous *rd* and double homozygous *rd/rds* mice. There is also a significant increase in the cyclic GMP content of affected retinae along with a decrease in cyclic GMP phosphodiesterase (PDE) activity in the *rd* mutant, and a decrease in both substances in the *rds* mutant (Farber and Lolley 1976, Cohen 1983, Sanyal *et al.* 1984, Fletcher *et al.* 1986). Similarly, rhodopsin content in these retinae has also been shown to be markedly reduced in the *rd* (Caravaggio and Bonting

1963, Yanasaki and Mizuno, 1970), as well as in the *rds* mutant retina (Schalken *et al.* 1985, Jansen *et al.* 1987).

The Abyssinian cat, homozygous for a progressive retinal atrophy gene, develops retinal lesions comparable to retinitis pigmentosa in man (Narfström 1983). The lesions, which start to occur at approximately two years of age and involve the photoreceptor population of the retina, are slowly progressive and lead to a generalized retinal atrophy in another 2-4 years (Narfström 1985 b). Although retinal differentiation seems to be normal with light microscopy, a higher frequency of disorganized rod outer segments are found in affected kittens prior to the time of retinal maturation (Narfström and Nilsson 1987 a,b).The IRBP content of the retina is markedly decreased at early stage 2 of the disease and at stage 3 (a stage when a marked cell loss is apparent) almost no IRBP immunoreactivity can be demonstrated (Narfström *et al.*1988). In both the mutant mice and the affected Abyssinian cat, GFAP is expressed in the Müller cells of the retina at an early time of retinal degeneration (Ekström *et al.*1988).

Materials and methods

Two strains of mice -pigmented C3H and albino Balb/c bearing congenic lines with the following genotypic combinations: +/+,+/+ (control); *rd/rd,* +/+; +/+, *rds/rds*; +/+, *rds*/+; and *rd/rd, rds/rds* had been produced and provided the material used in this study. Most of the immunocytochemical study was performed on materials of C3H background. In some cases, Albino Balb/c materials were used for the observation of the RPE. All animals were reared in cyclic light and maintained in normal housing conditions with unlimited access to food and water. At least two eyes of different animals of each investigated stage were used.

Abyssinian cats, homozygous for the progressive retinal atrophy gene, and as controls, healthy household mixed-breed cats were studied. The degree of retinal atrophy was assessed with ophthalmoscopy and staged according to Narfström (1985 a,b). One eye of each stage was used for immunocytochemistry and the other eye for immunochemistry. For critical stages i.e. (stage 1-2) two eyes of different animals were investigated.

Immunocytochemistry: Light-adapted eye-cups, (mice and cats) sampled at the same time of day, were fixed in 4% paraformaldehyde dissolved in 0.1M Sorensen phosphate buffer (pH 7.2), washed in Tyrode buffer to which successively 25% sucrose was added and serially sections (6-10 m) were obtained using a cryostat. Phosphate buffered saline (PBS) containing 0.25% bovine serum albumin (BSA) and 0,25% Triton X-100 was used to dilute the antisera. The sections were incubated with rabbit anti bovine transducin-α(1.1000-1:3000) (Dr A Spiegel, NIH) rabbit anti bovine S-antigen (1:1000-1:3000) (Dr I.Gery, NIH) sheep anti bovine opsin (1:10.000-1:30.000) (Dr Papermaster, University of Texas) and goat anti bovine IRBP (1:1000-1:3000) for 12 hours at room temperature. Secondary antibodies were swine anti-rabbit IgG and rabbit anti-goat IgG (crossreacts with sheep opsin antibodies). The secondary antibodies were purchased from DAKOPATTS (Copenhagen) and used at a dilution 1:50 for 30 min. Incubation with rabbit- or goat peroxidase anti-peroxidase complex (1:50 for 30 min.) was performed as the last step in the immunoreactions. The slides were rinsed with PBS containing 0.25% Triton X-100 between each step of the immunoreactions. The immunoreactions were visualized by using 3,3-diamino benzidine tetrahydrochloride 0.03% and hydrogen peroxide (0.0013%) in Tris HCl buffer (pH 7.6) for 5-15 min.

Results

Mouse models for retinal degeneration.

*Opsin immunoreactions:*In the normal mouse retina (*+/+,+/+*), an opsin-immunoreaction (IR) is first detected in the inner segments (IS) of the photoreceptor cells at postnatal day 7. From day 11, a distinct immunoreaction (Fig.1, +11 and +28) is only present in the outer segments (OS). In the single homozygous *rd/rd,+/+* retina, a distinct immunolabeling of the inner segments (IS) appears also at postnatal day 7, and at day 11 also a weak immunolabeling of the photoreceptor cell somata is seen (Fig.1, rd11). The photoreceptor cell somata become more intensely labeled at 14 days (Fig.1, rd14), when a considerable reduction in the number of photoreceptor cells has occurred. At day 21 a single row of photoreceptor cells remain in the outer nuclear layer (ONL); These are structurally markedly changed and homogeneously labeled (Fig.1, rd21). Only a very few, scattered and homogeneously labeled cells are found at 2 months.

In the single homozygous*+/+,rds/rds* retina, a weak immunoreaction is apparent in the IS at day 7. The same immunoreactive pattern is seen at 14 and 28 days (Fig.1, rds14) and at 2 months. Later (9 and 18 months) the

---

Figs.1-3 Differential Interference Contrast micrographs of mouse retinae processed for immunocytochemistry with antibodies rainsed against opsin, S-antigen or transducin-α with the peroxidase anti-peroxidase (PAP)-method. Mice were of the following allelic combinations at the *rd* and *rds* loci: (+)*+/+, +/+*; (**rd**)*rd/rd, +/+*; (**rds**)*+/+, rds/rds*; (**rd/rds**)*rd/rd, rds/rds*; and (**rds/+**)*+/+, rds/+*. Scale bars = 100 μm. The numbers indicate post-natal age in days or months (**m**). Abbreviations: **RPE**, retinal pigment epithelium; **OS**, photoreceptor outer segments; **IS**, photoreceptor inner segments; **ON**, outer nuclear layer and **IN**, inner nuclear layer.

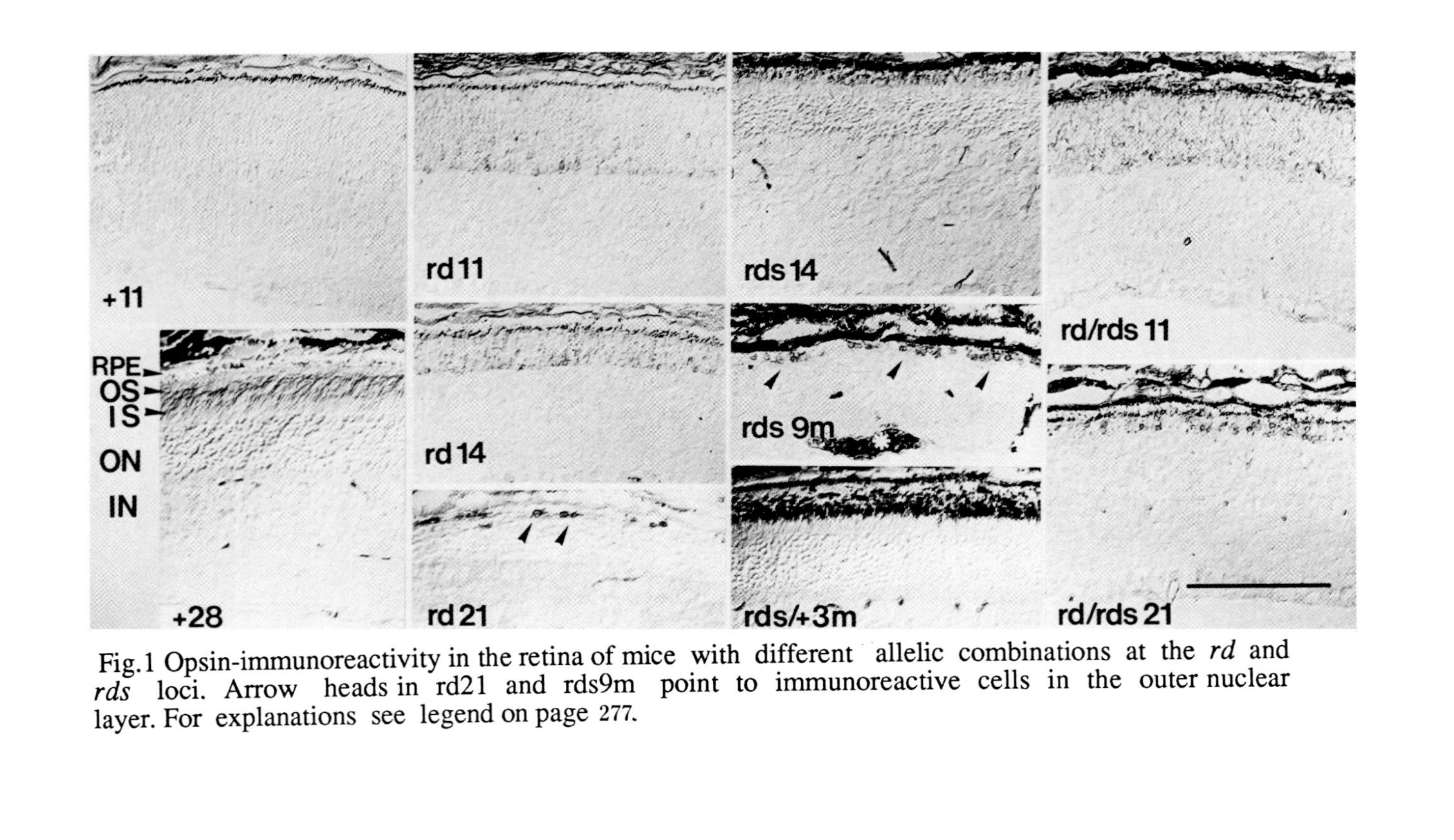

Fig.1 Opsin-immunoreactivity in the retina of mice with different allelic combinations at the *rd* and *rds* loci. Arrow heads in rd21 and rds9m point to immunoreactive cells in the outer nuclear layer. For explanations see legend on page 277.

whole remaining photoreceptor cell (Fig.1, rds9m) is immuno-labeled. In the double homozygous *rd/rd,rds/rds* retina, a weak immunoreaction is apparent at 7 days in a thin band corresponding to the IS. and, at day 11, even the outer nuclear layer (ONL) appears to be immunoreactive (Fig.1, rd/rds11). This pattern, with a strong immunolabeling of the IS and a weaker immunolabeling of the ONL is also found at 14 days. At 21 days, the immunolabeling of the ONL is more distinct but not all the perikarya are labeled with the same intensity. At this age, numerous photoreceptor cell have died but the degree of cell death is not as marked as in the single homozygous *rd* mutant (Fig.1, rd/rds21). In the single heterozygous *+/+,rds/+* retina, a weak immunoreaction in the IS is seen at 7 days. The same pattern as in the control mice i.e.a well defined immunoreaction in the OS and all other layers negative, is present at 3 and 16 months.

Alpha subunit of Transducin (TD-α)immunoreactions:
retina (+/+,+/+), the antibodies against TD-αreact strongly with the outer segments (OS), and weaker with the inner segments (IS) and perikarya of the photoreceptor cells Fig.2). The development of TD-α in the normal phenotype proceeds as follows: No immunoreactive elements are seen during the first weak of postnatal development, a weak immunoreaction appears at postnatal day 7, this reaction in the IS is still rather week at day 11 (Fig.2, +11), but becomes stronger by the second week (Fig.2, +14) By one month of age, the immunoreaction is very distinct (Fig.2, +28) and follows the immunoreactivity of opsin. In the single homozygous *rd/rd,+/+* retina, a positive immunoreaction is seen at 7 days. It appears somewhat stronger than in the normal retina, because here the cell somata are immunopositive from the first week. No differences in the intensity of the immunolabeling of IS and OS can be resolved, but it appears that both these structures and the receptor terminals are more heavily labeled than the perikarya at 11 days (Fig.2, rd11). Furthermore, as the degeneration of the distal portions of the photoreceptors advances, the immunoreaction becomes stronger in the perikarya (Fig.2, rd14); by day 21, only scattered immuno-positive cells are seen (Fig.2, rd21). At this stage, both the perikarya and their processes are homogeneously immunolabeled. No immunopositive elements are observed at, and after two months.

In the single homozygous *+/+,rds/rds* retina, no immunoreaction is detected during the first postnatal week; at day 11 (Fig.2, rds11) the reaction is still rather weak. From this stage, the immunolabeling is located in both perikarya (weakly) and inner segments (Fig.2, rds28). The terminals are not more strongly immunolabeled than the perikarya. Finally, by 9 and 18 months, the whole receptor cell appears to be homogeneously labeled.

In the *rd/rd,rds/rds* retina, the course of early post natal development is simalar to the other mouse models studied. At 11 days (Fig.2, rd/rds11) a weak labeling appears at the level of the terminals and a stronger labeling in

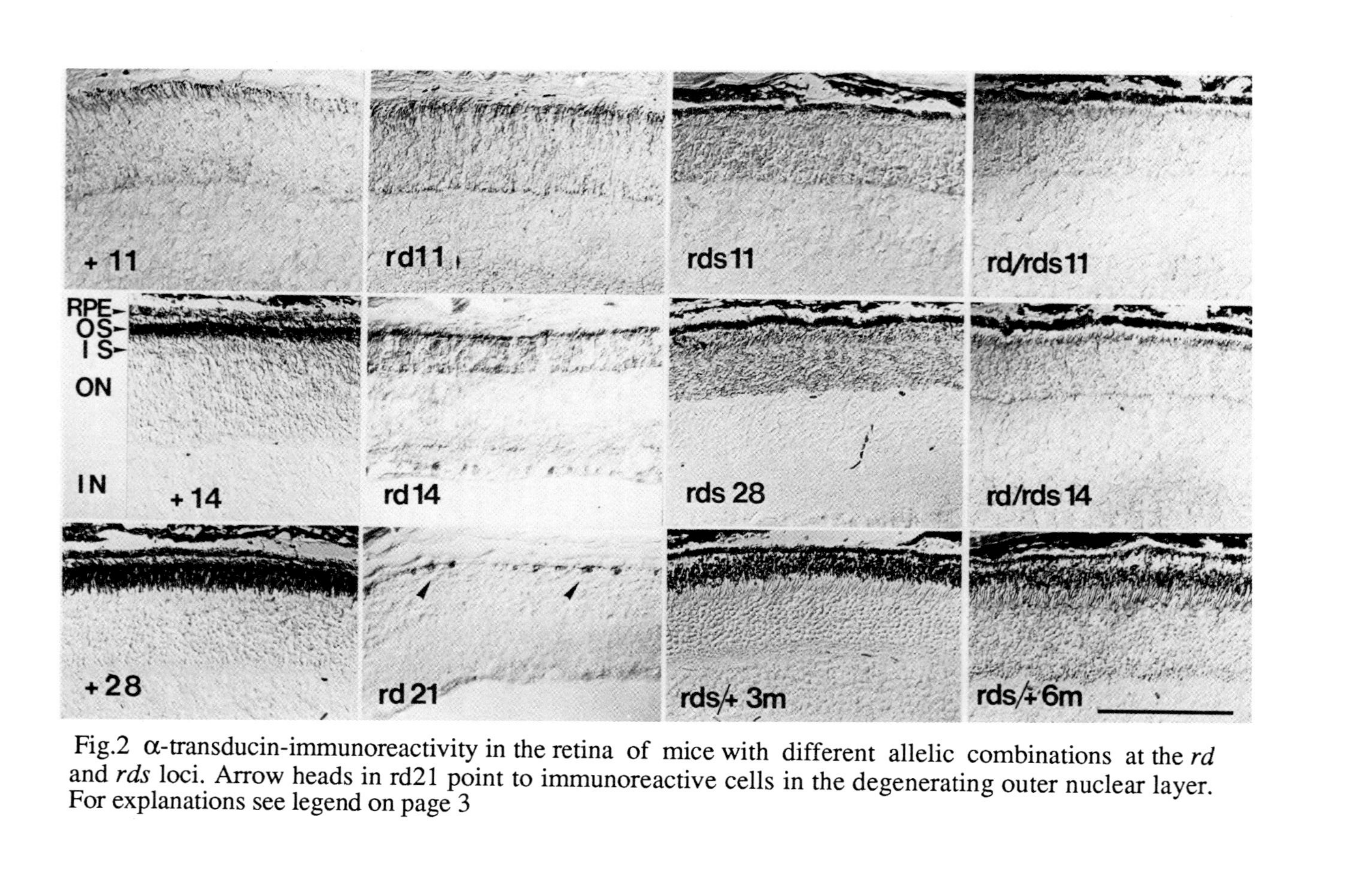

Fig.2 α-transducin-immunoreactivity in the retina of mice with different allelic combinations at the *rd* and *rds* loci. Arrow heads in rd21 point to immunoreactive cells in the degenerating outer nuclear layer. For explanations see legend on page 3

the IS. This pattern is even more defined at 14 days, when the terminals are weakly but distinctly labeled and the inner segments show a stronger labeling (Fig.2, rd/rds14). At 21 days, a weak labeling is apparent in the ONL. A stronger labeling persists in the band corresponding to the inner segments. At this age about 4 rows of photoreceptor cells remain in the ONL. In the *rds*/+ retina, a very weak immunoreaction is seen at 7 days. In the other stages studied (3, 6, and 16 months), the pattern of the immunoreaction is relatively similar to that in the normal retina: very intensive labeling of the outer segments, less intensive labeling of the inner segments and little or no labeling of the ONL (Fig.2, rds/+ 3m and 6m).

S-antigen immunoreactivity: Directly after birth (P0), S-antigen IR is present in presumptive differentiated photoreceptors of the retina in all mouse models studied (control and mutants). Many of these migrating IR cells have not reached their final position yet and are thus found over the whole retina. At day 4, in the normal control retina, the immunolabeling is most intensive in the developing IS. Outer segments are yet not developed at this stage. At day 11, the most intensive immunolabeling is seen at the OS, while a weaker immunoreaction is seen in the ONL and in displaced cells in INL. (Fig.3, +11). The number of displaced photoreceptor cells falls rapidly after 14 days. At day 14, the developing OS, are heavily immunolabeled while the IS, the ONL, as well as the displaced photoreceptor cells are weaker labeled. At day 28, the pattern of immunolabeling is essentially the same as described for earlier stages. However, a differentiation in the labelling intensity is seen at the distal level of the photoreceptors: the IS are weakly labeled, the proximal portion of the OS is very intensively labelled and the distal portion is weakly labeled (Fig.3, +28). From 4 months and thereafter, the immunolabeling of the perikarya in the ONL is weaker and two strongly labeled bands are very apparent: at the level of the terminals and at the level of the proximal 2/3 of the OS.

In the homozygous *rd* mutant mouse retina, a very similar development of S-antigen-IR is present until day 11 (Fig.3, rd11). After this stage, an increase in intracellular labelling seems is observed. At day 14, the reduction of photoreceptor cells is marked and numerous displaced cells are present. At 21 days, only a single row of photoreceptors are present; these cells have short processes, and are strongly and homogeneously labelled. At 2 months of age, this mutant yet has numerous immunopositive cells in the remainder of the ONL. In the homozygous *rds* retina, the early development of S-antigen IR is more similar to that of the control mice. The number of displaced immunopositive cells found in the INL appears to increase from 11 to 14 days and decreases again thereafter: only a few cells of this type can be seen at 28 days and 2 months. At day 28, the immunoreaction seems to be more enhanced at the level of the IS and terminals. At 9 months of age, the ONL is drastically reduced, and displays a homogeneously strong S-antigen-IR. Photoreceptor death progresses

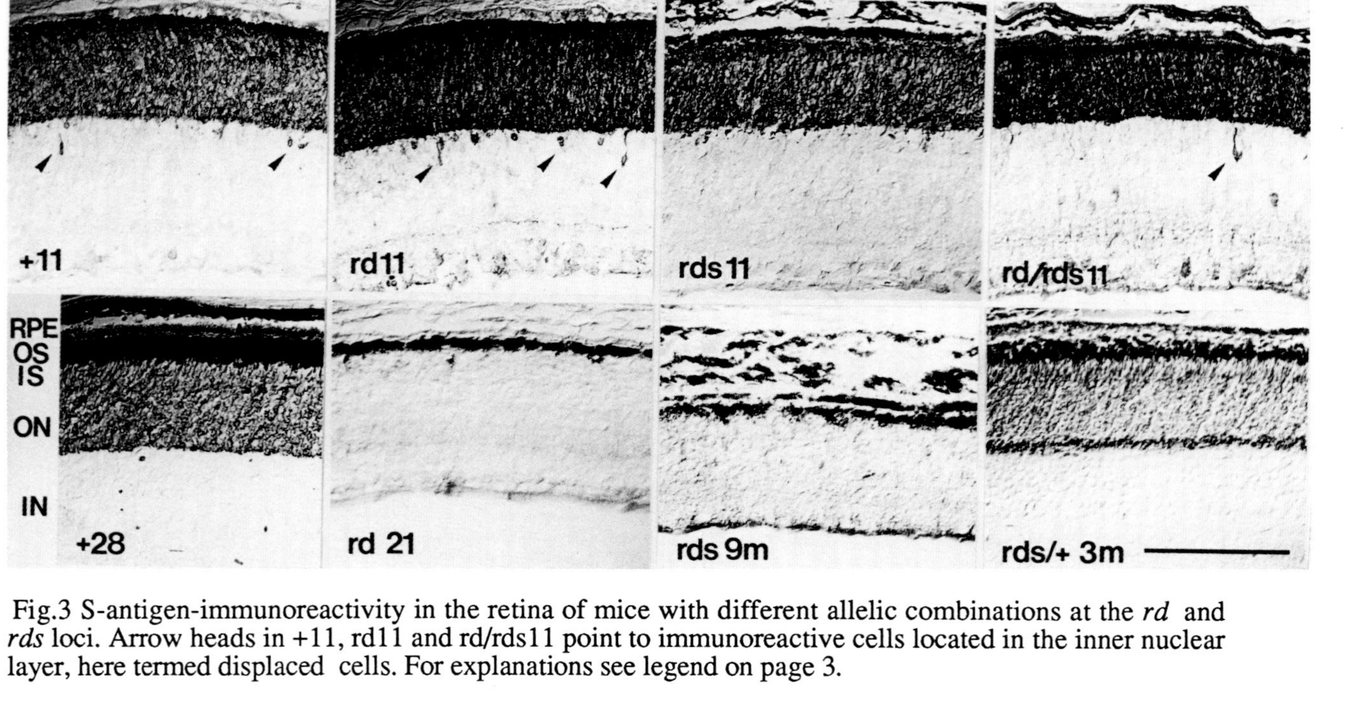

Fig.3 S-antigen-immunoreactivity in the retina of mice with different allelic combinations at the *rd* and *rds* loci. Arrow heads in +11, rd11 and rd/rds11 point to immunoreactive cells located in the inner nuclear layer, here termed displaced cells. For explanations see legend on page 3.

successively and, at 18 months of age, only one row of strongly labelled photoreceptors is seen in the ONL.

In the double homozygous *rd/rd,rds/rds* retina, an homogeneously labeled band is seen at 7 days that includes the terminals, the perikarya and the inner segments. A few immunopositive displaced cells are present. The immunoreaction seems more enhanced in the photoreceptor cell somata than in the controls of the same age. The number of immunopositive, displaced cells increases until day 14 and decreases thereafter. The pattern of immunolabeling is essentially the same at 14, 21, and 28 days, i.e. a strong IR is present in the entire photoreceptor cell. At 21 days, only four rows of photoreceptor cells, and at 28 days the four rows are now even more reduced, and only one row of photoreceptor cells is left.

In the single heterozygous *+/+,rds/+* retina, the OS are strongly immunolabeled while the IS and the perikarya display a much weaker immunoreaction. There are also a few, weakly immunoreactive, displaced cells. At 3 and 6 months, the bands corresponding to the terminals and the OS are strongly labeled, while the ONL and the IS are less intensively labeled (Fig.2, rds/+3m), this is comparable with the normal control retina.

Abyssinian cat model for progressive retinal atrophy.

Immunocytochemical and immunochemical evidence demonstrates that the course of IRBP development in diseased Abyssinian cat retinae and control cat retinae is comparable until stage 2 of the disease after which a

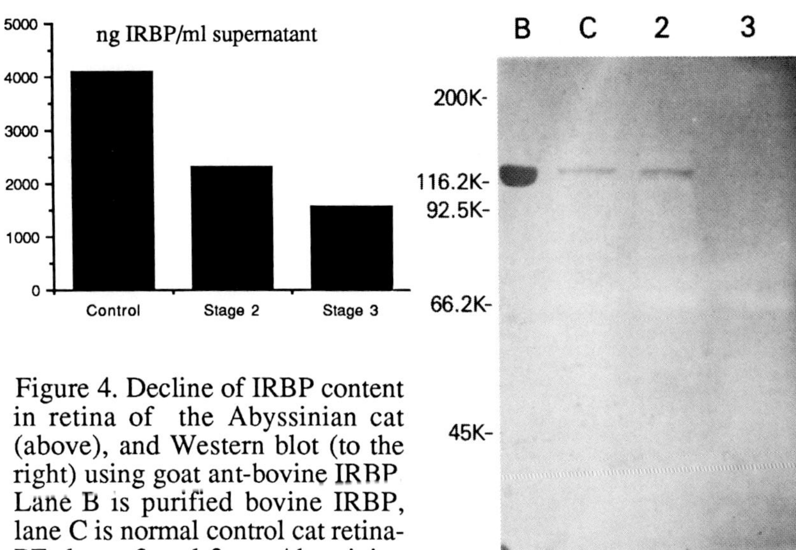

Figure 4. Decline of IRBP content in retina of the Abyssinian cat (above), and Western blot (to the right) using goat ant-bovine IRBP Lane B is purified bovine IRBP, lane C is normal control cat retina-PE, lanes 2 and 3 are Abyssinian cat retinae and PE from stages 2 and 3 respectively.

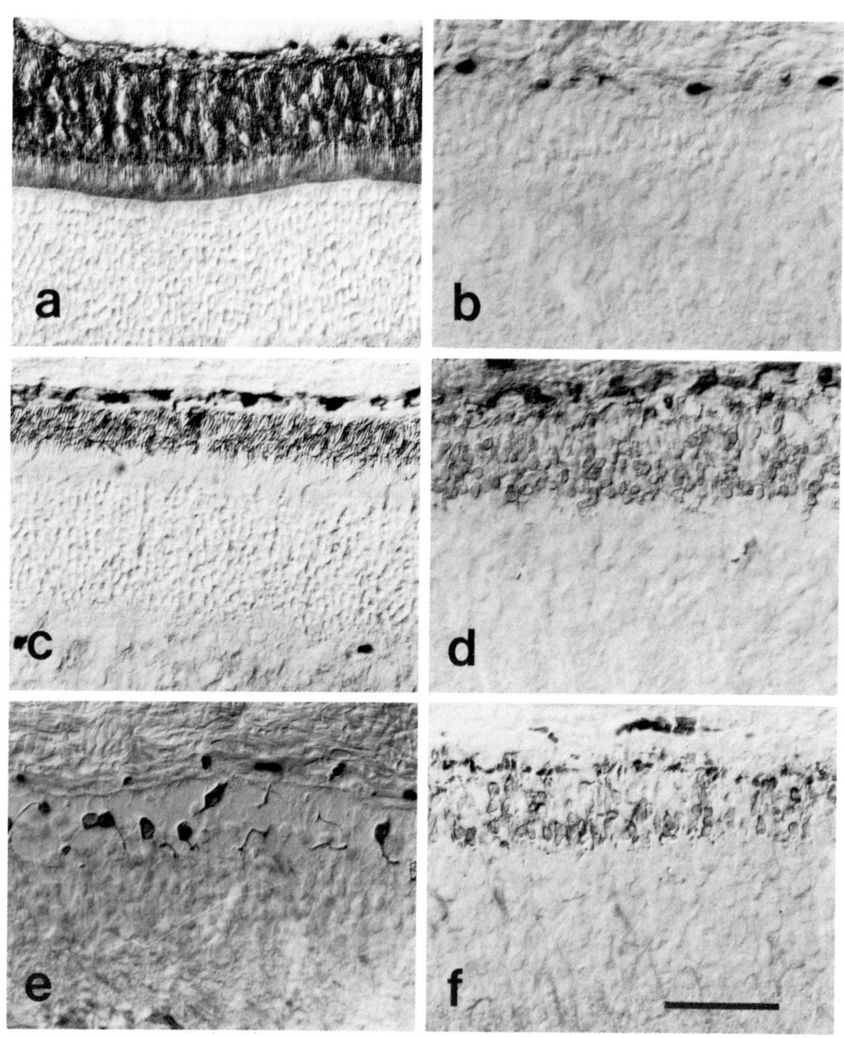

Figure 5a-f. Differential Interference Contrast microscopy of 10 μm cryostate sections of retinae from Abyssinian cats homozygous for the Progressive Retinal Atrophy gene. a. IRBP immunoreaction (IR) in the retina of stage 1. Note the strong immunoreaction between the OLM and the RPE. b. IRBP-IR in the retina of stage 3. No IR is present in this more centrally located region of the retina. c. Opsin-IR in the retina of stage 1. d. Opsin-IR in the retina of stage 3. Note the intracellular location of IR and the absence of ROS. e. S-antigen-IR at stage 4. Only few, morphologically changed photoreceptors remain at this stage. f. TD-α-IR in the retina of stage 3. Note the intracellular distribution of the immunoreaction. Scale bar=50μm

considerable reduction of IRBP is observed (Figs.4 and 5). This reduction continues, and in stage 3 of the disease, hardly any immunoreaction is detectable using immunocytochemistry.
Immunochemically, however, measuring the IRBP content of the whole retina, IRBP is stil present but in a much lower concentration than in the control animals. At stage 4, no immunoreactivity is detectable using immunocytochemistry. At this stage, almost all photoreceptor elements of the retina have died. It is important to note that the decrease in IRBP content precedes the dramatic photoreceptor cell loss found at later stages of the disease. As with IRBP, the other proteins tested (Fig. 5 opsin, TD-α, and S antigen) display a seemingly normal development in the diseased retinae until stage 2, although a slight reduction in opsin and TD-α is noticeable at stage 2. At stage 3 of the disease, the ROS are morphologically strongly changed or absent. At this stage, immunoreactivity for opsin, and TD-α is present not only in the remaining outer segments, but also in the photoreceptor cell somata. A strong S-antigen IR is present in the photoreceptors in all stages of the disease. For example, the majority of remaining IR elements of the diseased retina even at stage 4 are strongly S-antigen-IR. (Fig.5e,f).

Discussion

The photoreceptor elements in models for retinal degeneration used in this study, seem to differentiate normally with respect to early occurrence and development of photoreceptor-specific proteins. Differences in immunoreactivity of these proteins compared to controls generally occur in connection with ROS development or ROS degeneration. Per se, this is not remarkable because in the normal retina, the immunoreactivity is situated in ROS (opsin), in ROS/RIS (TD-α), or in the whole photoreceptor cell but strongest in the ROS (S-antigen). When photoreceptor outer segments do not develop as in the *rds* or *rd*/*rds* mice, or degenerate as in the *rd* mouse and Abyssinian cat, the IR is found in the photoreceptor soma. This is also true for IRBP even though it is a secreted protein in the normal case (van Veen *et al.*1986). When biochemical (Schalken *et al.*1985) measurements or fundus reflectometry (Narfström *et al.*1987) of opsin content of the retina show a decline, this is not necessarily due to a decline in opsin content in the individual remaining photoreceptors, but could be due to the general loss or non formation of photoreceptor elements. However, the major site of the photoreceptor proteins is in the ROS and thus the lack of this structure will influence the cellular distribution and amount of these proteins per individual photoreceptor. The synthesis of photoreceptor-specific proteins in the individual photoreceptors is regulated by messenger RNA's for the different proteins. In *rd* mutant mice, mRNA's for the different photoreceptor proteins are still present in retinae at an age where high photoreceptor cell death is present (Bowes *et al.*1988). In the case of *rds* mutant mouse retinae, the photoreceptors survive for a considerably time (many survive during the whole life span of the animals). In this model, the IR for opsin, TD-α, and S-antigen is remarkably enhanced intracellularly compared to normal control animals of the same age and to heterozygotes, *+/rds*. The IR of the heterozygous *+/rds* more closely resembles that in normal control animals even though the outer segments of these animals are disorganized and photoreceptor celldeath is present in mature animals (Sanyal 1988). It would appear that a simple lack of synthesis of one or more of these photoreceptor specific proteins is not the primary defect that leads to photoreceptor cell death in these mutants.

The Abyssinian cat homozygous for the retinal atrophy gene does not exhibit an intracellular increase of IR for the different photoreceptor specific proteins in early stages of the disease. An increase of intra-cellular staining is directly coupled to the degenerative process and loss of ROS. It therefore seems that the synthesis of photoreceptor proteins, their incorporation in membranes and/or transport is impaired at the time of photoreceptor degeneration. Importantly, it seems that the photoreceptor cell loses polarity at the time of OS loss. Causal relationships in this situation merit further investigation.

In contrast to the proteins directly involved in phototransduction, IRBP is secreted into the extracellular matrix, between the outer limiting membrane and the pigment epithelium (Pfeffer *et al.*1983). In the Abyssinian cat, this protein declines long before the massive ROS degeneration and cell death occurs. In this model, no increase in intracellular IRBP occurs after ROS degeneration as in the different mutant mice models (van Veen *et al.*1988). However, in the double homozygous *rd/rd,rds/rds* mouse, a decline in IRBP content in the retina is present well before extensive cell death occurs (van Veen *et al.* 1988). IRBP is an important component of the interphotoreceptor matrix, binding retinoids. It may thus function to protect the photoreceptor outer and inner segments from deleterious effects of retinoids on membranes (Meeks *et al.*1983). Loss of extracellular IRBP therefore may cause or greatly accelerate ROS degeneration.

Summary

Using immunocytochemical techniques, development of opsin, transducin alpha and S-antigen in photoreceptor cells of the mice homozygous or heterozygous for the *rd* or *rds* genes has been found to be similar to that of control animals during the first postnatal week. Even though the absolute amounts of these proteins are low (eg. opsin in the *rds* retina) or decrease in the postnatal period (as in the *rd* retina), we can demonstrate their persistence during the entire degeneration process. In fact, the content of the proteins in the photoreceptor perikarya actually appear to be higher after postnatal day 11 in all mutants studied. Thus, one of the major manifestations of the mutant retinae is a loss of polarity of the photoreceptor cells at the time of ROS degeneration without a loss in the ability to synthesize these important proteins of the visual cycle. This correlates well with the apparent defect in IRBP secretion and its intracellular accumulation in mutant photoreceptor cells as previously observed (van Veen *et al.*1986). In the Abyssinian cat model for progressive retinal atrophy, the development an cellular distribution of all the proteins studied are similar in affected and control retinae until the beginning of stage 2 of the disease. At this time, outer segments begin to degenerate and immunoreactivity increases in the photoreceptor perikarya. The IRBP content of the retina declines markedly at stage 2, preceding extensive loss of photoreceptors.

Acknoweledgements: The authors are indebted to Carina Rasmussen for skillful technical assistance Mr Henk Visser for taking care of the mouse colony. We are especially grateful to Dr Papermaster, Dr Gery and Dr Spiegel for their kind gifts of antibodies. This study was supported by The Swedish Natural Science Research Council (44 46-110) The Medical Research Council (MFR 12X-734), the Retinitis Pigmentosa Foundation for Fighting Blindness, USA, and the Crown Princess Margareta's Committee for the visually handicapped.

References

Bowes C, Veen Th van, Farber DB (1988) opsin, G-protein and 48k-Da protein in normal and rd mouse retinas: Developmental expression of mRNAs and proteins, and light/dark cycling of mRNAs. Exp Eye Res.47, 369-390

Caravaggio L and Bonting S (1963) The rhodopsin cycle in the developing vertebrate retina. II. Correlative study in normal mice and mice with hereditary retinal degeneration. Exp Eye Res 2:12-19.

Carter- Dawson L, Alvarez R, Fong S-I, Liou G, Sperling H, Bridges C (1986) Rhodopsin, 11-*cis* Vitamin A and interstitial retinol-binding protein (IRBP) during retinal development in normal and *rds* mutant mice. Dev Biol 116: 431-438

Ekström P, Sanyal S, Narfström K, Chader GJ, Veen Th van (1988) Development of glial fibrillary acidic protein immunoreactivity in Müller radial glia during hereditary retinal degeneration. Invest Ophthalm. Vis. Sci. 29/9, 1363-1371

Farber D and Lolley R (1976) Enzymatic basis for cyclic GMP accumulation in degenerative photoreceptor cells of mouse retina. J Cyclic Nucleotide Res 2:139-148.

Fletcher RT, Sanyal S, Krishna G, Aguirre G, Chader GJ (1986) Genetic expression of cyclic GMP phodphodiesterase activity difines abnormal photoreceptor differentiation in Neurological mutants of inherited retinal degeneration. J Neurochem 46: 1240-1245.

Jansen H, Sanyal S, de Grip W, and Shalcken J (1987) Development and degeneration of retina in rds mutant mice: ultra-immunohistochemical localization of opsin. Exp Eye Res 44:347-361.

Meeks RG, Zaharevitz D, Chen F (1981) Membrane effects of retinoids: Possible Correlation with toxicity. Arch.Bioch. Biophys.207: 141-147

Narfström K (1983) Hereditary progressive retinal atrophy in the Abyssinian cat. J Heredity 74: 273-276

Narfström K (1985a) Progressive retinal atrophy in the Abyssinian cat: Clinical characteristics. Invest Ophthalm Vis Sci. 26: 193-200

Narfström K (1985b) Retinal degeneration in a strain of Abyssinian cats: A hereditary, clinical, electrophysiological and morphological study. Linköping University Medical Dissertations No.208

Narfström K, Nilsson SEG (1986) Progressive retinal atrophy in the Abyssinian cat: Electron microscopy. Invest Ophthalm. Vis. Sci.27: 1569-1576

Narfström K, Nilsson SEG (1987a) Hereditary retinal degeneration in a strain of Abyssinian cats: Developmental studies. Invest Ophthalm Vis. Sci.(Suppl) 28: 141

Narfström K, Nilsson SEG (1987b) Hereditary rod-cone degenerayion in a strain of Abyssinian cats. Prog. Clin. Biol. Res. 247: 349-368

Narfström K, Nilson SEG, Wiggert B, Lee L, Chader GJ, Veen T van (1988) Reduced IRBP level, a pausible cause for retinal degeneration in the Abyssinian cat. Cell and Tiss. Res. submitted.

Pfeffer B, Wiggert B, Lee L, Zonnenberg B, Newsome D, and Chader G J (1983) The presence of a soluble Interphotoreceptor Retinoid-Binding Protein in the retinal interphotoreceptor space. J Cell Physiol 117:333

Sanyal S, de Ruiter A, Hawkins RK (1980) Development and degeneration of retina in rds mutant mice. Light microscopy. J.Comp Neurol. 194: 193-207.

Sanyal S, Fletcher R, Liu Y, Aguirre G, and Chader G (1984) Cyclic nucleotide content and phosphordiesterase activity in the<< mouse (020/A) retina. Exp Eye Res 38:247-256.

Sanyal S (1988) Cellular site of expression and genetic interaction of the *rd* and the *rds* loci in the retina of the mouse. In: Degenerative Retinal Disorders: Clinical and Laboratory Investigations (Eds. Hollyfield JG, Anderson RE, LaVail MM) Alan R Liss, Inc. New York pp

Schalken J, Janssen J, de Grip W, Hawkins R and Sanyal S (1985) Immunoassay of rod visual pigment (opsin) in the eyes of rds mutant mice lacking receptor outer segments. Biochim Biophys Acta 839:122-126.

Veen Th van, Katial A, Shinohara T, Barrett DJ, Wiggert B, Chader GJ, Nickerson JM (1986 a) Retinal photoreceptor neurons and pinealocytes accumulate mRNA for interphotoreceptor retinoid-binding ptotein (IRBP). FEBS lett. 208: 133-137

Tansley K (1951) Hereditary degeneration of the mouse retina. Br J Ophthalmol 35:573-582.

Veen T van, Katial A, Shinohara T, Barrett D, Wiggert B, Chader G, Nickerson J (1986) Retinal photoreceptor neurons and pinealocytes accumulate mRNA for interphotoreceptor retinoid-binding protein (IRBP). FEBS Lett. 208, 133-137

Veen T van, Ekström P, Wiggert B, Lee L, Hirose Y, Sanyal S, Chader G (1988) A developmental study of interphotoreceptor retinoid-binding protein (IRBP) in the homozygous or hetrozygous rd and rds mutant mouse retina. Exp. Eye Res 47, 291-305

Yamasaki I and Mizuno K (1970) Rhodopsin and ERG of hereditary dystrophic mice and experimental retinitis pigmentosa. Jap J Ophthalmol 14:151-158.

Inherited and Environmentally Induced
Retinal Degenerations, pages 291–300

INTERPHOTORECEPTOR RETINOID-BINDING PROTEIN (IRBP) AND OPSIN IN THE RDS MUTANT MOUSE: EM IMMUNOCYTOCHEMICAL ANALYSIS

Louvenia Carter-Dawson and
Michael Burroughs

Sensory Sciences Center, Graduate
School of Biomedical Sciences,
University of Texas Health Science
Center at Houston, Texas 77030

INTRODUCTION

Photoreceptor cells in mice homozygous for the rds (retinal degeneration slow) gene fail to develop outer segments and degenerate over the course of one year. Although no intact outer segments are formed, some opsin positive cytoplasmic masses and membranous lamellae develop (Jansen, Sanyal, de Grip and Schalken, 1986; Usukura and Bok, 1988). A reduction in amount and/or translation of mRNA for opsin probably accounts for the small amount of opsin synthesized in rds retinas (McGinnis, Burga, Whelan and Triantafyllas, 1988; Neeraj, Nir and Papermaster, 1988). The light activated cGMP phosphodiesterase activity is present in rds retinas (Cohen, 1983) as well as an electroretinogram (ERG) that is reduced in amplitude (Reuter and Sanyal, 1984). The presence of a light activated cGMP phosphodiesterase and a detectable ERG suggest that all components of the visual cycle are present although some in reduced amounts.

IRBP (interphotoreceptor retinoid-binding protein) is thought to be another important component of the visual cycle. It has been implicated in the transfer of retinoids between the RPE and the photoreceptors (Adler and

Martin, 1982; Chader, Wiggert, Lai, Lee and Fletcher, 1983; Fong, Liou, Landers, Alveraz, Gonzales-Fernandez, Glazebrook, Lam and Bridges, 1984). However, other data suggest that IRBP may not be a specific transporter of retinoids, but perhaps is an important buffer protein for retinoids traversing the IPM (Ho, Massey, Pownall, Anderson and Hollyfield, 1988). IRBP is localized to the interphotoreceptor space with an apparent higher density at the apical region of the RPE (Bunt-Milam and Saari, 1983). To further characterize the rds mutant, the distribution of IRBP was examined by EM immunocytochemistry in postnatal developing retinas. The localization of opsin also was examined at selected ages.

Mice homozygous for the rds gene and BALB/c controls were sacrificed at postnatal days 9 (P9), 14, 18 and 21. The eyes were fixed and processed for embedding in LR white, IRBP was localized on ultra-thin sections using the two stage labeling method (primary antibodies: rabbit anti-monkey IRBP or rabbit anti-bovine IRBP; secondary antibody: goat anti-rabbit IgG conjugated to 5 nm gold). Opsin was localized by the same method on adjacent sections at P18 and P21 (primary antibodies: monoclonal antibodies specific for the carboxyl and amino termini of rhodopsin; secondary antibody: goat anti-mouse IgG conjugated to 5nm gold).

RESULTS

A small amount of IRBP was detected at P9 in BALB/c and rds retinas. Occasionally intracellular labeling was observed, but most of the immunoreactivity for IRBP was located at the apical region of the RPE. No distinguishing differences were observed in the distribution of IRBP between the rds and the BALB/c retinas at this age.

By P14, differences were detected in the

distribution of IRBP in the rds and the BALB/c photoreceptors. The golgi apparatus of some rds photoreceptors contained a much higher density of label for IRBP than neighboring cells as well as those of BALB/c controls (Fig. 1). However,

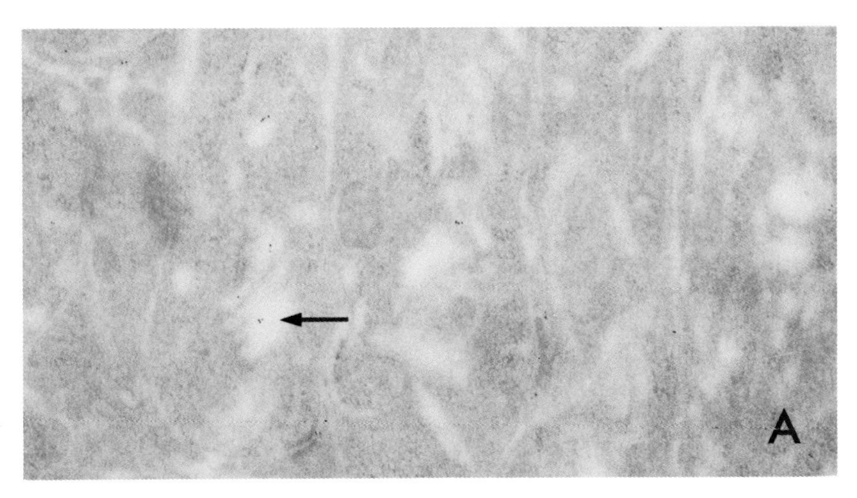

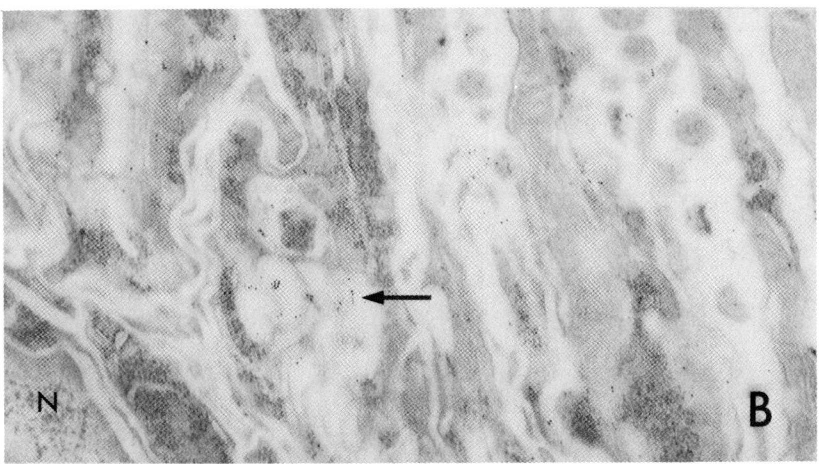

Fig. 1. At P14 the golgi apparatus of BALB/c photoreceptors is lightly labeled for IRBP (A, arrow), but the golgi apparatus of some rds photoreceptors is heavily labeled for IRBP (B, arrow). N, nucleus. X 40,000.

the density of label for IRBP at the apical RPE region appeared similar to controls.

At P18 and P21, the density of label for IRBP was obviously abnormal in rds retinas. The apical RPE region contained a lower density of label for IRBP than the controls. (Fig. 2). The golgi regions of most photoreceptor cells at these ages were lightly labeled similarly to controls at P14, but the golgi regions of some cells were enlarged or dilated and were labeled heavily for IRBP (Figs. 3 and 4). These cells represented approximately 12% of the photoreceptor population.

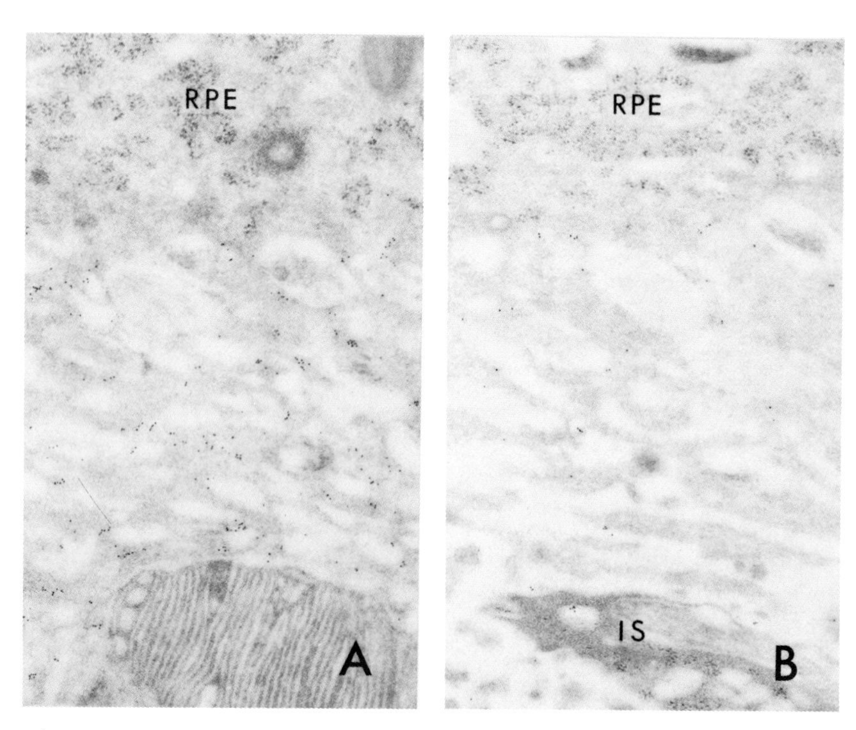

Fig. 2. The apical RPE region of P18 BALB/c retinas (A) is labeled more densely for IRBP than that of the rds retina of the same age (B). RPE, retinal pigment epithelium; IS, inner segment. X 32,000.

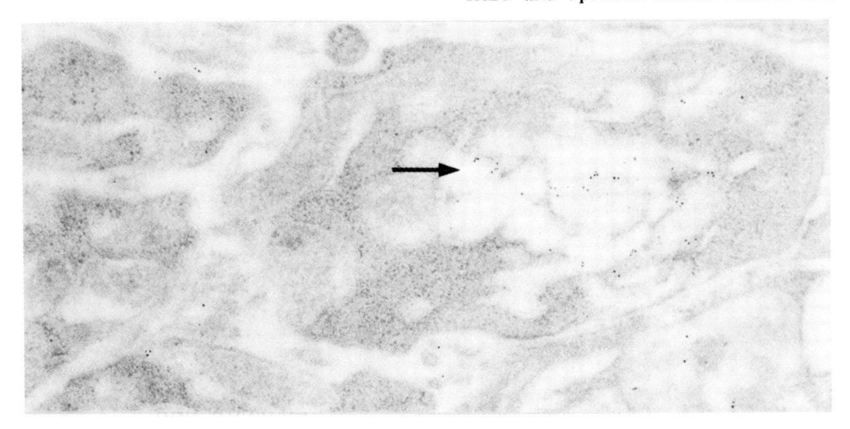

Fig. 3. This P18 rds photoreceptor shows an enlarged golgi (arrow) labeled heavily for IRBP. N, nucleus. X 40,000.

As previously described by Jansen, Sanyal, de Grip and Schalken (1987) and Usukura and Bok (1988), label for opsin was found associated with membranous vesicles, the cilium and membranous lamellae. In the current study, many, but not all of the extracellular vesicles were labeled for opsin (Fig. 5). IRBP was occasionally associated with some of the extracellular vesicles (Fig. 6).

The golgi of some photoreceptor cells in the rds retina were densely labeled for IRBP as well as opsin. Adjacent sections incubated with IRBP or opsin antibodies showed a small number of cells that were labeled for both proteins (Fig.7). Not every cell labeled with IRBP antibodies was labeled with opsin antibodies and vice versa. Only light labeling as seen in photoreceptors of BALB/c as well as the majority of photoreceptors in the rds mutant retina. It is difficult to determine whether the cells were labeled above background levels for each protein on adjacent sections. Thus, similar co-localization of IRBP and opsin in controls is inconclusive.

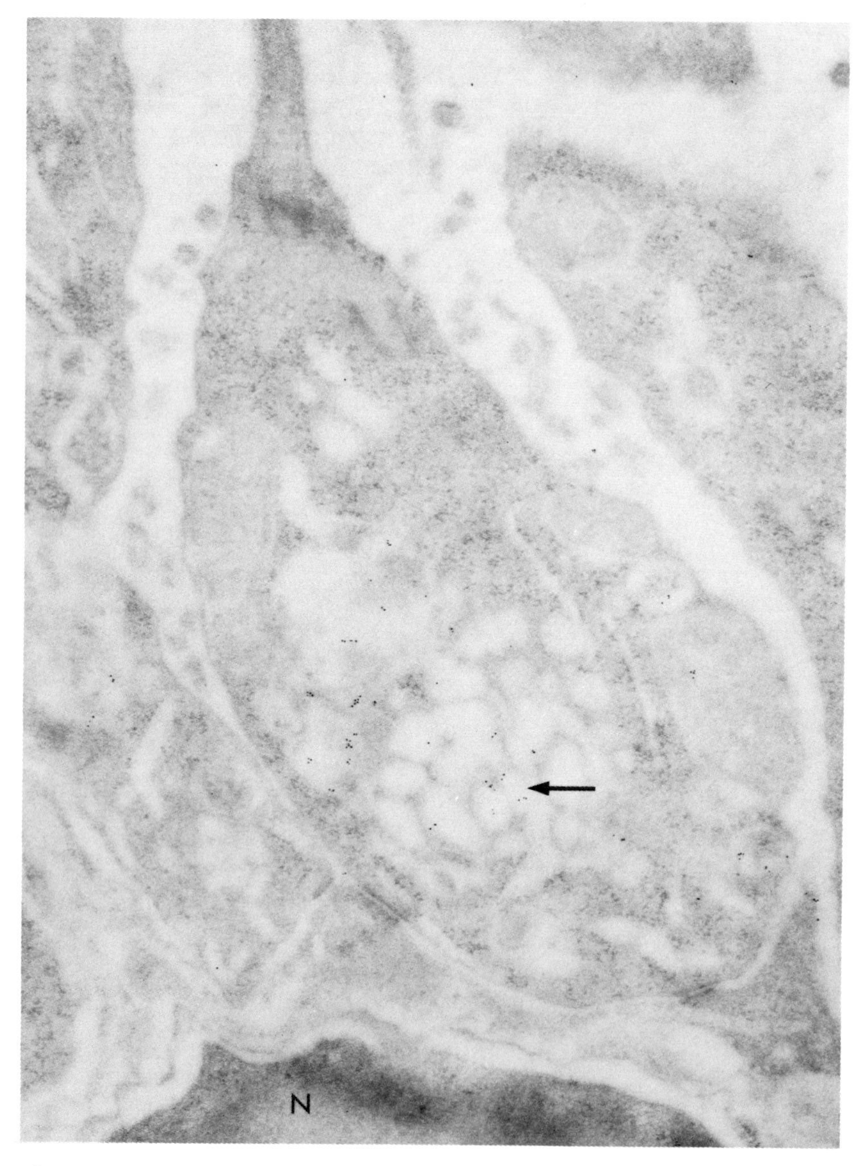

Fig. 4. The golgi of this P21 rds photoreceptor is dilated (arrow) and is heavily labeled for IRBP. N, nucleus. X 40,000.

It is unclear from the present data whether the accumulation of IRBP or opsin in the golgi apparatus occurs as a primary or secondary effect of the rds mutation. There is a possibility that these changes occur during the early stages of deterioration in all degenerating photoreceptor cells. Analyses of photoreceptors in retinas degenerating from lesions that are not intrinsic to photoreceptors (i.e. RCS rat, light damage etc.) may prove useful in determining whether these type changes occur in any deteriorating photoreceptor or as a direct result of the rds mutation.

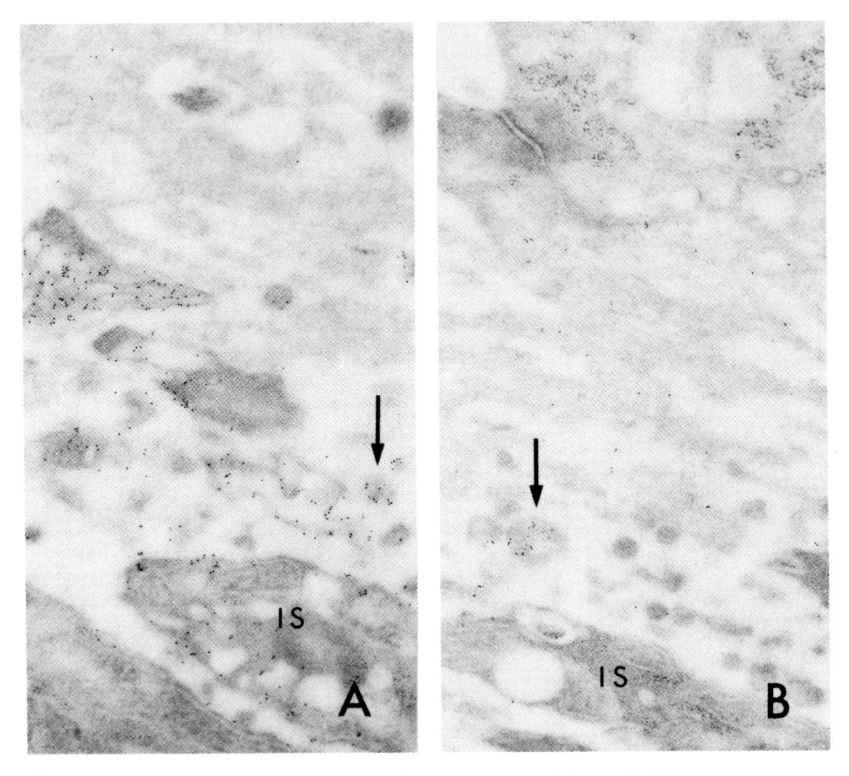

Fig. 5. At P18, most, but not all of the extracellular vesicles are labeled for opsin (A, arrow). Occasionally vesicles are labeled for IRBP (B, arrow). IS, inner segment. X 32,000.

It is also unclear whether the co-localization of IRBP and opsin within the golgi apparatus is an artifact of the degenerative process or an amplification of a normal occurrence. Further studies are needed to determine whether IRBP and opsin are normally processed simultaneously in the golgi.

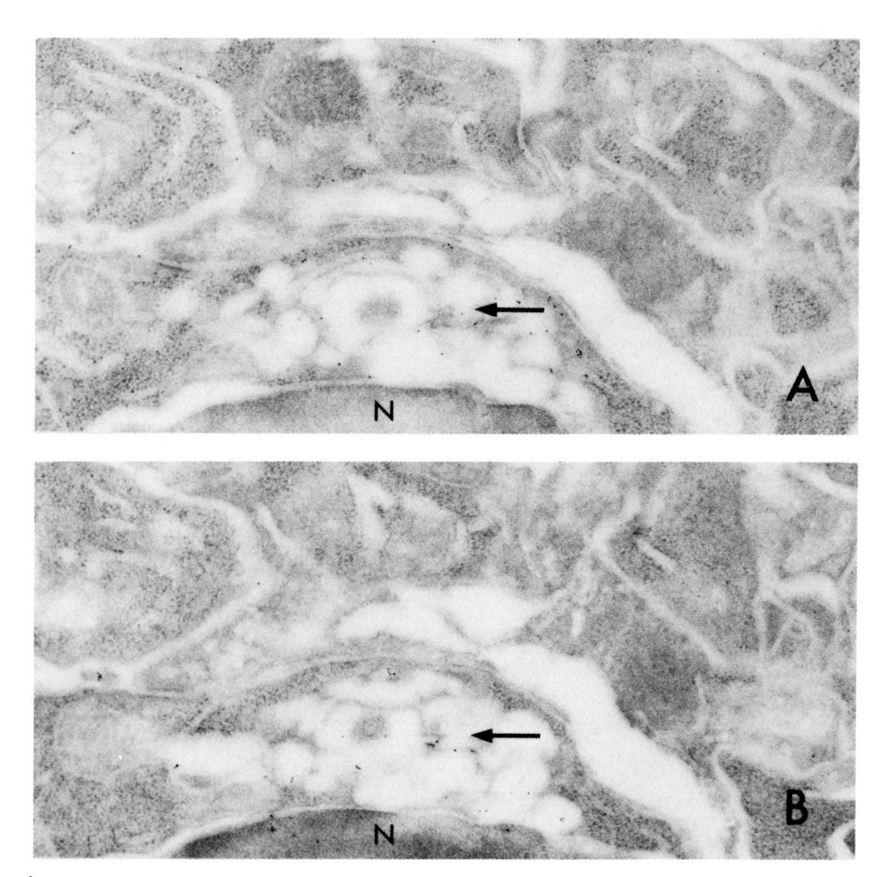

Fig. 6. Labeling of the golgi for opsin (A, arrow) and IRBP (B, arrow) on adjacent sections of the same rds photoreceptors at P18. N, nucleus. X 40,000

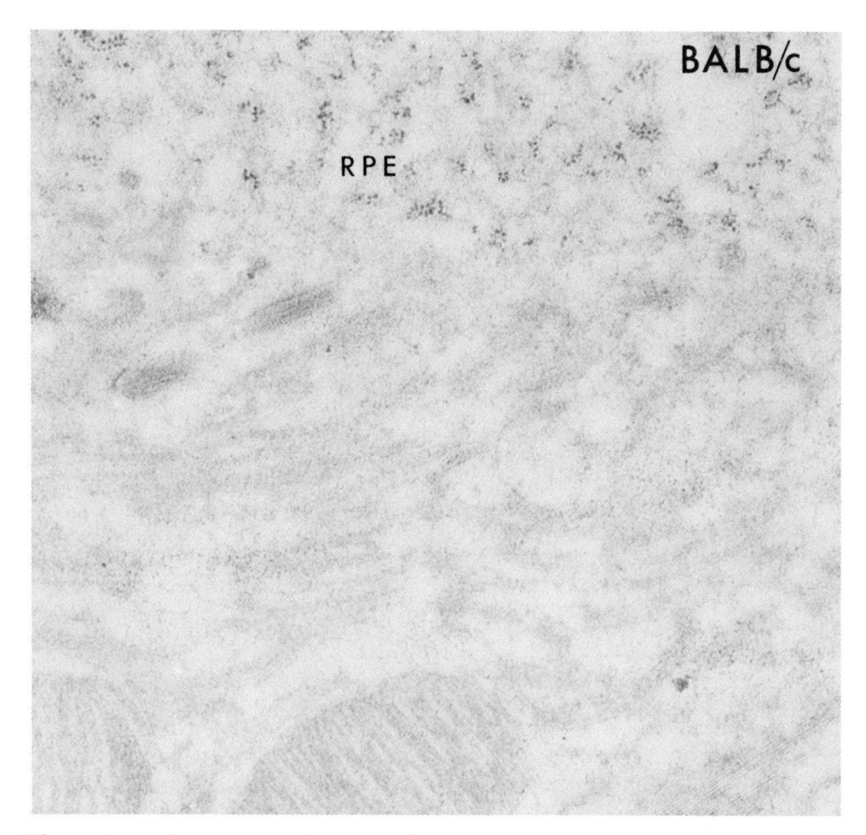

Fig. 7. Immunocytochemical control section of P21 BALB/c apical RPE. The section was treated the same as those with anti-IRBP IgG, except pre-immune IgG was substituted for immune IgG. RPE, retinal pigment epithelium. X 40,000.

REFERENCES

Adler AJ and Martin KJ (1982). Retinal-binding proteins in bovine interphotoreceptor. Biochem Biophys Res Commun 108: 1601-1608.
Bunt-Milam AH and Saari JC (1983). Immunocytochemical localization of two retinoid binding proteins in vertebrate retina. J Cell Biol 97: 703-712.

Chader GJ, Wiggert B, Lai Y-L, Lee L and Fletcher R (1983). Interphotoreceptor retinoid-binding protein: a possible role in retinoid transport to the retina. In Progress in Retinal Research. Osberne N and Chader GJ editors. Oxford Pergamon Press pp. 162-189.
Cohen AI (1983). Some cytological and initial biochemical observations on photoreceptors in retinas of rds mice. Invest Opthalmol Vis Sci 24: 832-843.
Fong S-L, Liou G, Landers RA, Alvarez RA, Gonzalez-Fernandez F, Glazebrook PA, Lam DMK and Bridges CDB (1984). The characterization, localization and biosynthesis of interstitial retinol binding glycoprotein in the human eye. J Neurochem 42 1667-1676.
Ho PM-T, Massey J, Pownall H, Anderson RE and Hollyfield JG (1988). Mechanism of vitamin A transfer between liposomes and IRBP. Invest Opthalmol Vis Sci (suppl) 29: 122.
Jansen HG, Sanyal S, de Grip WJ and Schalken JJ (1987). Development and degeneration of retina in rds mutant mice: Ultraimmunohistochemical localization of opsin. Exp Eye Res 44: 347-361.
McGinnis JF, Bugra K, Whelan JP and Triantayllos J (1988). Comparative analysis of gene expression in photoreceptor cells in normal and mutant mice. Invest Opthalmol Vis Sci (suppl) 29: 168.
Neeraj A, Nir I and Papermaster DS (1988). Opsin mRNA levels in rds mutant mice. Invest Opthalmol Vis Sci (suppl) 29: 169.
Reuter JH and Sanyal S (1984). Development and degeneration of retina in rds mutant mice: the electroretinogram. Neurosci Lett 48: 231-237.
Usukura J and Bok D (1987). Changes in the localization and content of opsin during development in the rds mutant mouse: Immunocytochemistry and immunoassay. Exp Eye Res 45: 501-515.

Inherited and Environmentally Induced
Retinal Degenerations, pages 301–314

LOCALIZATION OF 5′-NUCLEOTIDASE ACTIVITY IN RCS RAT RETINAS

Margaret J. Irons

Division of Anatomy and Experimental Morphology, Department of Biomedical Sciences, Faculty of Health Sciences, McMaster University, Hamilton, Ontario, Canada L8N 3Z5

INTRODUCTION

In the course of an ultrastructural cytochemical study on acid phosphatase localization in RPE cells, a previously undetected enzyme activity, i.e. manganese-dependent pyrimidine 5′-nucleotidase (MDPNase) activity was identified in the rod outer segments (ROS) and RPE cells in the normal rat retina (Irons, 1987a). Analysis of the distribution of this enzyme activity with respect to cyclic light revealed that the MDPNase activity was concentrated in the tips of the ROS and in phagosomes during the 2–3 hr period after light onset when shedding and phagocytic activity are maximal. Prior to the shedding peak, heavy cytochemical staining was observed over the apical processes (AP) of the RPE cells. This reaction later disappeared as the tips of the ROS and phagosomes became stained. Following the shedding peak, staining of the AP was greatly reduced and the ROS were uniformly weakly labelled from base to tip (see Fig. 6 of Irons, 1987a). This striking redistribution of cytochemical reaction product led to the hypothesis that the MDPNase might be a marker for the portion of the ROS to be shed and phagocytosed (Irons, 1987a). A preliminary study comparing the distribution of this enzyme activity at various times of day in the outer retinas of one-month old normal rats and Royal College of Surgeons (RCS) rats with inherited retinal degeneration revealed marked staining differences between the two strains at this age (Irons MJ, ARVO Abstracts 1985). This paper presents our observations on the RCS rat from an historical perspective, starting with the initial findings in the one-month old animals, followed by recent work on developing retinas which has aided our understanding of the complex staining patterns seen in the older RCS rats.

MATERIALS AND METHODS

Tan-hooded RCS rats were raised from birth in a 12L:12D cycle. One group of rats, 28–30 days of age, was perfused under anaesthesia with cacodylate-buffered 2.5% glutaraldehyde, pH 7.4, containing 0.1% $CaCl_2$ and 5% sucrose. Two rats per time point were fixed at 1 or 3 hr before light onset and 0, 1, 2, 3, 4, 6 and 8 hr after light onset; those killed during the dark part of the cycle were handled under dim red light. A second group of control and RCS rats was perfused 2 hr after light onset on postnatal days 4, 8, 10, 12, 14, 16, and 20. The eyes were enucleated and the posterior eyecups were subsequently processed for light and electron microscopic cytochemical localization of MDPNase activity as described previously (Irons, 1987a). Briefly, 75 µm-thick retinal sections were incubated in the acid nucleotidase medium of Novikoff (1963), modified by the addition of manganese ions. Final concentrations of reagents in this medium, designated MDPNase medium, were 5.4 mM cytidine-5′-monophosphate (5′-CMP), 5 mM $MnCl_2$, 4 mM lead acetate and 120 mM sucrose in 40 mM acetate buffer, pH 5.2. Control sections were incubated in Mn^{++}-free medium and in MDPNase medium containing 20 mM sodium fluoride. To facilitate comparison of the cytochemical reactions in retinas fixed at various ages and times of day, a standard incubation time of 1 hr at 37°C was used throughout.

RESULTS AND DISCUSSION

The retina of the one-month old RCS rat is characterized by the presence of a substantial layer of membranous debris in the subretinal space. Close to the apical surfaces of the RPE cells, this lamellar material is commonly arranged in large whorls. In incubated sections from animals of this age, a very similar pattern of cytochemical staining was observed in retinas fixed from 3 hr before to 8 hr after light onset (Fig. 1). The AP of the RPE cells were prominently stained and numerous

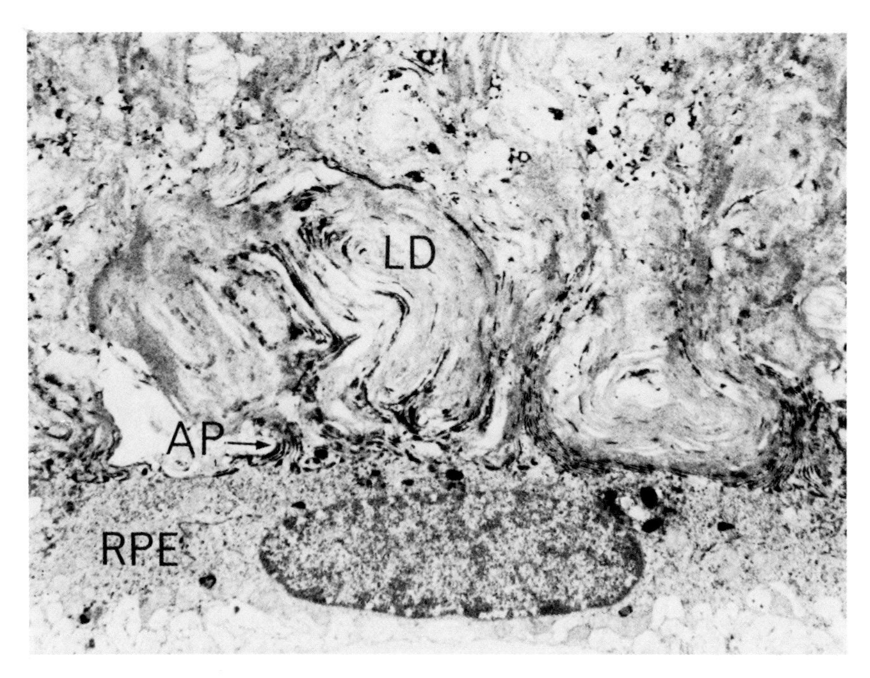

Fig. 1. Typical distribution of MDPNase activity in the outer retina of an RCS rat fixed on P28. LD- lamellar debris. (x 7,160)

electron dense deposits of cytochemical reaction product were scattered throughout the debris layer (Fig. 1). Reaction product associated with the AP was on the ecto surfaces of these membranes, often appearing to fill the narrow extracellular spaces between adjacent processes (Fig. 2). In addition, the cytochemical reaction outlined unstained rounded structures, often present in chains within the stained layers (arrows, Fig. 2). The whorls of lamellar debris surrounded by the AP were for the most part unreactive, although distinct streaks of reaction product were interspersed amongst the unstained layers (arrowheads, Fig. 2). The swirling patterns formed by the stained AP and the stained layers within the membranous whorls were remarkably complex, making it impossible at this stage to determine where the AP ended and the lamellar debris began.

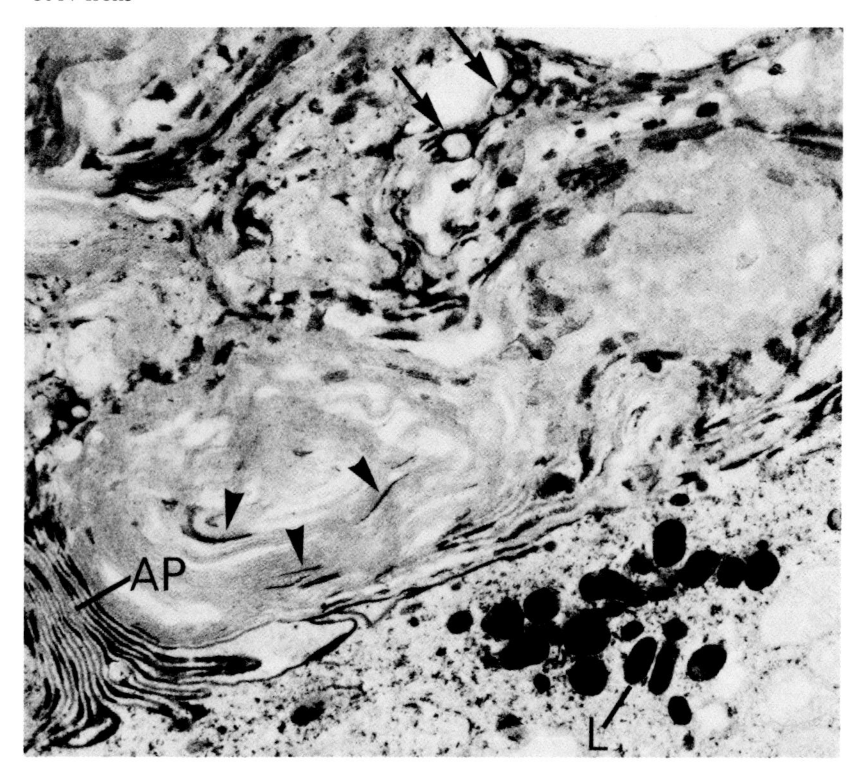

Fig. 2. Higher power view of the cytochemical reaction in the P28 RCS retina. L– lysosomes in RPE cell. (x 18,400)

Within the RPE cell bodies, the numerous lysosomes, which typically formed small clusters, were heavily stained at each time point examined (Fig. 2). Cytochemical staining of the Golgi apparatus was also observed. Additionally, much smaller precipitates were scattered throughout the cytoplasm (Fig. 2).

In one-month old RCS rats, the discs in the distal ends of the outer segments are highly disorganized and in apparent continuity with the array of membranes forming the debris layer. In contrast, the proximal ends of many of the ROS are surprisingly well organized, closely resembling those seen in the normal retina. Following incubation in the MDPNase medium, most of the discs within the outer segments at this level were unstained. The reaction product was localized to discrete areas at the rims and in the interiors of a few of the discs. This distribution of reaction product closely resembled that observed in control rats of the same age. (See Fig. 8 of Irons, 1987b.)

The specificity of the cytochemical reaction in the RCS rats was analyzed in sections from each retina incubated in MDPNase medium containing 20 mM NaF. Fluoride at this concentration was found to inhibit enzyme activity over the ROS and all sites in the RPE cells although in some sections a few discrete foci of lead salts remained in the debris layer (Fig. 3a). Following incubation in Mn^{++}-

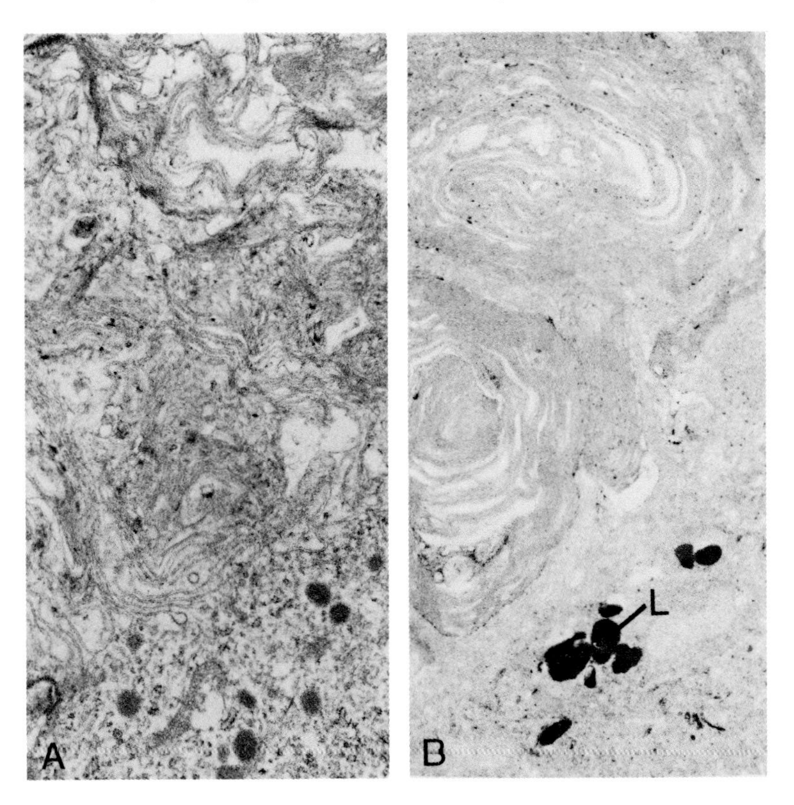

Fig. 3. Control sections from P28 RCS retinas incubated in MDPNase medium containing 20 mM NaF (a) or in Mn^{++}-free medium (b). L- lysosome (x 12,530)

free medium, staining of the ROS, AP and all structures in the debris layer was greatly diminished, although the lysosomes in the RPE cells remained well stained (Fig. 3b). This correlated with our previous observation in normal rats (Sprague-Dawley and RCS rdy^{+}) that cytochemical staining of the AP and outer segment membranes is Mn^{++}-dependent (Irons, 1987a,b). Thus the same enzyme activity, i.e. MDPNase, was no doubt visualized in both the control and dystrophic RCS retinas.

Comparison of the reaction patterns in the one-month old dystrophic and control rats fixed at the same nine time points in the diurnal cycle revealed major differences between the two strains. First, the staining of the AP differed considerably. Whereas in the normal RPE cells the reaction over the AP was very prominent in the dark and was greatly diminished during and after the shedding peak, a very heavy reaction was observed in the dystrophic cells at all time points examined (Compare Figs. 2,4). Secondly, the distribution of reaction product within the

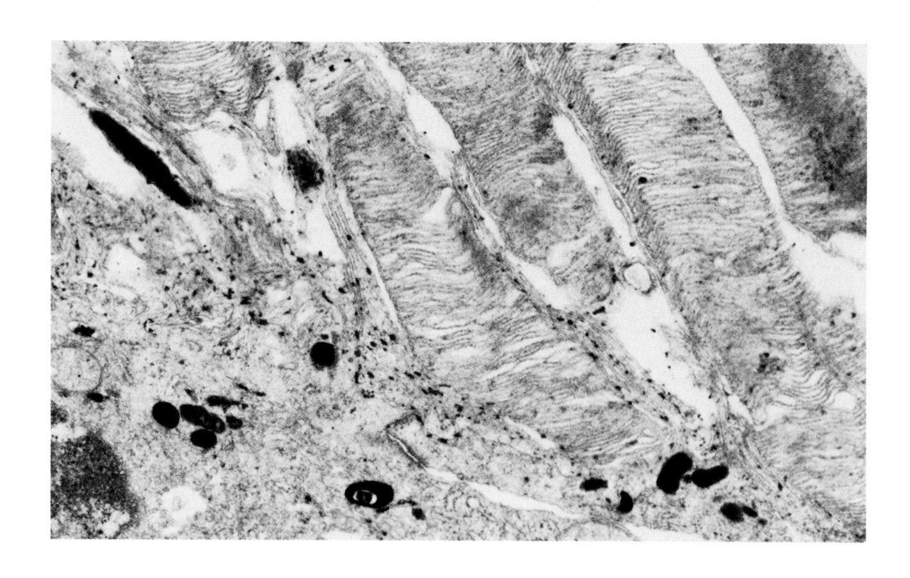

Fig. 4. Normal rat retina fixed 6 hr after light onset on P28. Note the relatively weak MDPNase activity associated with the AP at this time of day. (x 18,400)

disorganized membranes adjacent to the RPE cell surfaces in the RCS rats was clearly very different from that observed in the tips of the ROS in the normal retinas at various times of day. It then became important to determine how these differences came about, and whether they were significant with respect to the phagocytic defect in the RCS rat (Herron, Riegel, Myers and Rubin, 1969; Bok and Hall, 1971; Mullen and LaVail, 1976). For this purpose, morphological features and distribution of MDPNase activity were compared in the differentiating outer retinas of control and RCS rats fixed on P4– P20.

The primitive ROS, which develops as an extension of the future connecting cilium around P6–8, is a whisklike structure in which disc-like flattened saccules of various sizes extend parallel to the long axis of the cilium. Subsequent elongation of the developing outer segments continues in a similar fashion in both normal and RCS rats until at least P12. During the first day or two of growth, extensive sheets of lamellar material are produced which lay parallel to the RPE cell surfaces, surrounded by the immature AP. Sometime between P8 and P12, discs of uniform size begin to be produced and are aligned in the usual manner, perpendicular to the cilium. Thus within the forming ROS on P12, the highly disarrayed apical discs (or membranes) appear to abruptly give way to a well-organized stack of basal discs (*, Fig. 5). A point to be stressed from the morphological part of the study is that the ROS of both normal and dystrophic rats appear highly disorganized during early development. In fact, the normal retina at this stage could easily be mistaken for a dystrophic retina (Fig. 5).

In the normal retina, the ROS acquire a more or less mature appearance by P16–20. This seems to result both from overall growth of the stack of mature-type discs and from the disappearance of the disorganized apical discs. Phagosomes resembling the latter membranes appear in the normal RPE cells as early as P12; it thus seems likely that these primitive discs removed from the apical ends of the ROS by phagocytosis during P12–16. In contrast, phagocytosis is not apparent in the dystrophic RPE cells during this (or any other) period. Hence it appears that these early membranes remain in the subretinal space, contributing to the so-called membranous "debris".

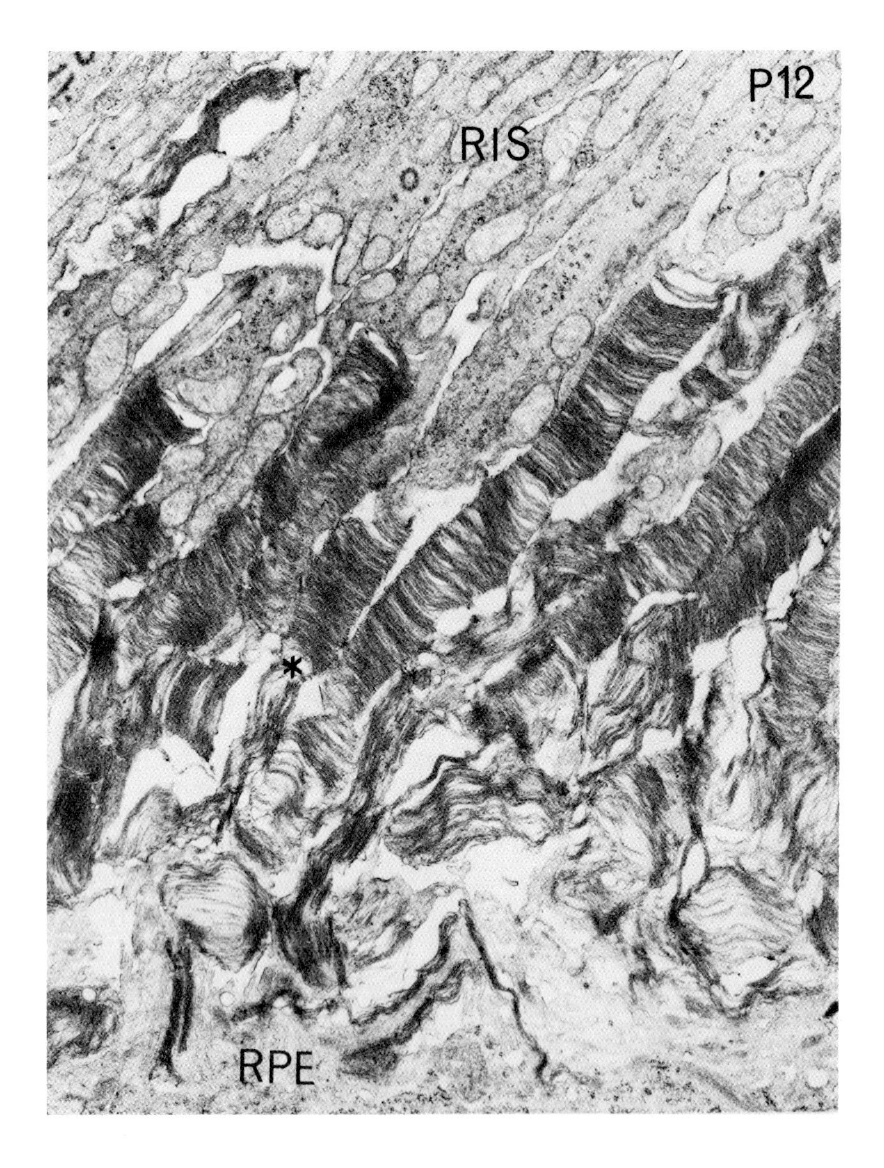

Fig. 5. Normal rat retina fixed on P12. The apical ends of the forming ROS, adjacent to the RPE, are comprised of sheets of disorganized lamellae whereas the basal discs are organized into neat stacks. RIS- rod inner segment. (x12,530)

It is known from the results of in vitro studies (Edwards and Szamier, 1977; Chaitin and Hall, 1983) that the phagocytic defect is expressed in cultured RPE cells obtained from immature RCS rats (P12 and younger). It appears from the current study that this defect begins to exert its influence in vivo around P12 as well. Thus elucidation of the cause of the phagocytic defect will require understanding of a normal photoreceptor-RPE interaction that becomes established while the outer segments are still at a very early stage of development.

What have we learned about this interaction by examining MDPNase activity in the developing retinas? Certainly that the situation appears to be even more complex than in the older retinas! A complete account of the cytochemical reaction patterns at various stages of retinal development and their significance with regard to the normal phagocytic mechanism will appear elsewhere. This article will address the issue of the possible involvement of the MDPNase in the phagocytic defect in the RCS rat.

Briefly, comparable MDPNase activity was observed in the developing photoreceptors of both normal and dystrophic rats at least until P12. In the early membranes produced between P8-12, prominent deposits of reaction product appeared within certain vesicles and discs (Fig. 6). These heavily-stained structures stood out as isolated dark streaks when viewed at low power (arrows, Fig. 7). In addition, a much finer punctate reaction was associated with many of the lamellar membranes, giving them a stippled appearance at high power (Fig. 6). By P16 in the normal animal, the outer retina had undergone striking maturational changes and the corresponding cytochemical reaction closely resembled that of one-month old normal animals. In the dystrophic animals, the basal portions of the ROS appeared to elongate normally and the cytochemical reaction at this level looked completely normal at P16 (Fig. 8). In contrast, the features of the outer segment-RPE interface changed very little between P12 and P16 and the cytochemical reaction at this level remained essentially unchanged from P12 (Figs. 7,9). In fact, the reaction seen in the so-called "debris" adjacent to the RPE cell surface in a one-month old RCS rat is very similar to that observed in the 12-day-old retina of either normal or dystrophic rats (Figs. 1, 7).

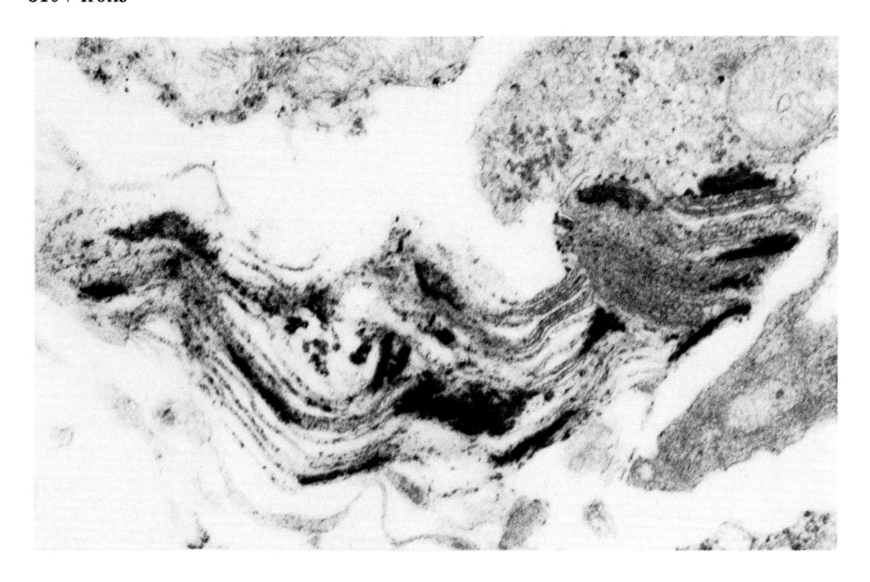

Fig. 6. MDPNase activity in the membranes of a developing outer segment in an 8-day old RCS rat retina. (x 55,900)

The observed cytochemical reaction patterns are compatible with the contention that the debris layer in the one-month old dystrophic animal contains most, if not all of the outer segment membranes produced during the life of the photoreceptors. The distribution of MDPNase activity within these membranes appears to be normal. Our conclusion from this work is that what originally appeared to be a very abnormal pattern of reaction product in the debris layer of the P28 RCS rat can be explained simply as a consequence of the phagocytic defect starting to be expressed around P12, at which time phagocytosis begins in the normal animal.

What do we conclude regarding the previously observed abnormal staining of the AP of the RCS RPE cells on P28? In the developing retinas, no staining differences in the AP of control and dystrophic cells were observed. In the differentiating outer retinas of both strains, the AP of the RPE cells were generally weakly stained (Figs. 7,9) Prominent staining of the AP in the RCS retinas was first apparent around P16 at the earliest, by which time

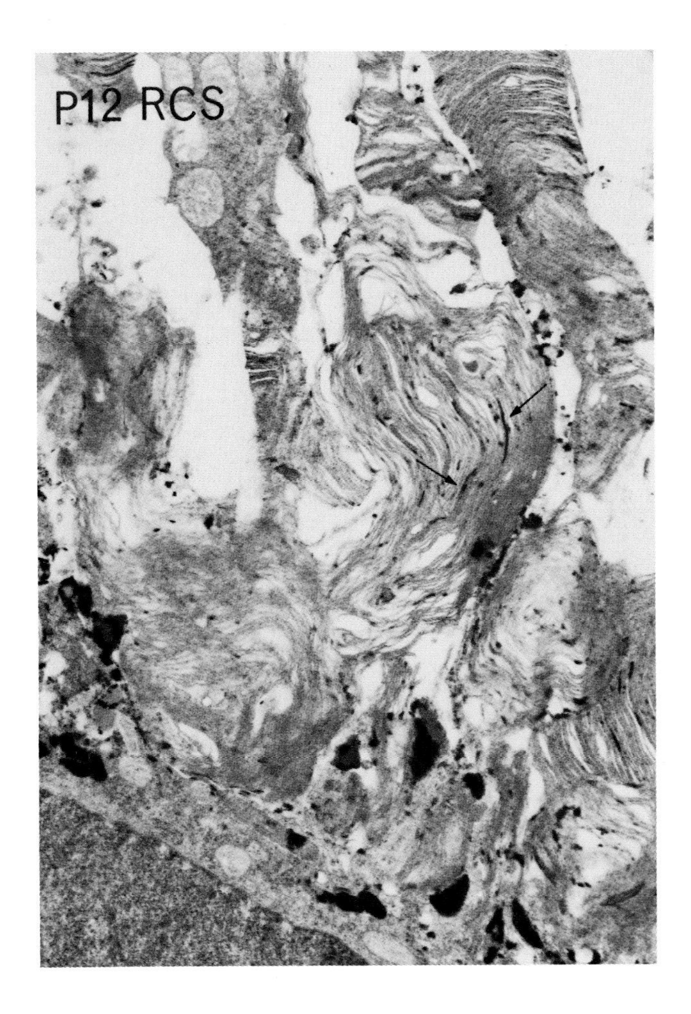

Fig. 7. Distribution of MDPNase activity in the RCS retina on P12. Very similar staining is observed in retinas of normal rats of this age. (x 12,530)

photoreceptor degeneration has begun. Thus there is no evidence from this study to implicate the MDPNase enzyme in the etiology of the phagocytic defect in the RCS rat. The persistence of intense staining of the AP throughout the day in the phagocytosis-defective RPE cells of the one-month old RCS rats does however strengthen our hypothesis

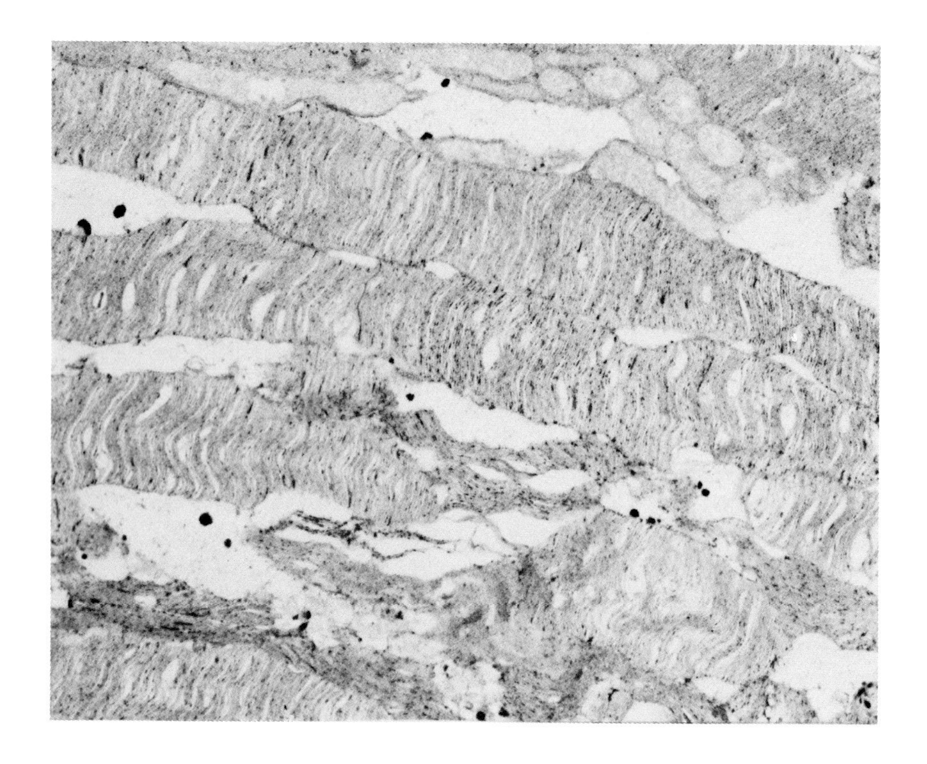

Fig. 8. Distribution of MDPNase activity in the basal portions of the ROS in an RCS rat retina fixed on P16. A similar distribution of reaction product is observed in normal animals of this age. (x 12,530)

that in the normal retina the enzyme may play a role in maintaining the mechanism of rhythmic shedding and phagocytosis, which becomes established after P15 (Tamai and Chader, 1979).

ACKNOWLEDGMENT

This work was supported by the Medical Research Council of Canada and the RP Eye Research Foundation of Canada.

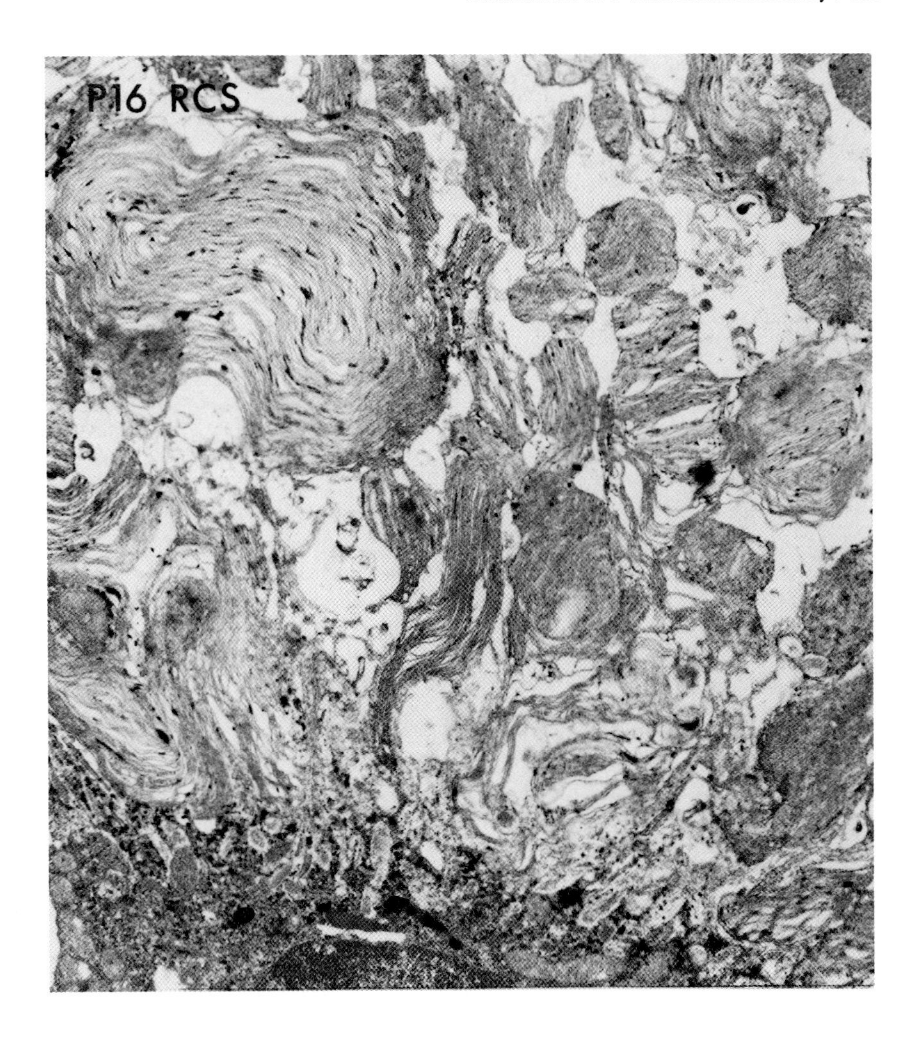

Fig. 9. Localization of MDPNase activity in the RCS retina fixed on P16. The distribution of reaction product in the disorganized membranes adjacent to the RPE cell surface is essentially the same as that observed in both normal and dystrophic animals on P12. (x 12,530)

REFERENCES

Bok D and Hall MO (1971). The role of the pigment epithelium in the etiology of inherited retinal dystrophy in the rat. J Cell Biol 49:664-682.

Chaitin MH and Hall MO (1983). Defective ingestion of rod outer segments by cultured dystrophic rat pigment epithelial cells. Invest Ophthalmol Vis Sci 24: 821-831.

Edwards RB and Szamier RB (1977). Defective phagocytosis of isolated rod outer segments by RCS rat retinal pigment epithelium in culture. Science 197: 1001-1003.

Herron WL, Riegel BW, Myers OE and Rubin ML (1969). Retinal dystrophy in the rat- a pigment epithelial disease. Invest Ophthalmol Vis Sci 8:595-604.

Irons MJ (1987a). Redistribution of Mn^{++}-dependent pyrimidine 5'-nucleotidase (MDPNase) activity during shedding and phagocytosis. Invest Ophthalmol Vis Sci 28: 83-91.

Irons MJ (1987b) Cytochemical localization of Mn^{++}-dependent pyrimidine 5'-nucleotidase activity in isolated rod outer segments. Exp. Eye Res. 44:789-803.

Mullen RJ and Lavail MM (1976). Inherited retinal dystrophy: primary defect in pigment epithelium determined with experimental rat chimeras. Science 192:799-801.

Tamai M and Chader GJ (1979). The early appearance of disk shedding in the rat retina. Invest Ophthalmol Vis Sci 18: 913-917.

Inherited and Environmentally Induced
Retinal Degenerations, pages 315–330

INHERITED RETINAL DYSTROPHY IN THE RCS RAT: COMPOSITION OF THE OUTER SEGMENT DEBRIS ZONE

Michael T. Matthes and Matthew M. LaVail

Departments of Anatomy and Ophthalmology
Beckman Vision Center, University of California
San Francisco, CA 94143-0730

The hallmark of inherited retinal dystrophy in the RCS rat is the accumulation of rhodopsin-containing debris membranes in the outer segment zone before and during the degenerative period (Dowling, Sidman 1962; Noell 1963). The debris accumulation results from the failure of the retinal pigment epithelium (RPE) to phagocytize rod outer segment (ROS) membranes as it does in normal retinas (Herron et al. 1969; Bok, Hall 1971; LaVail et al. 1972; Goldman, O'Brien 1978). Thus, with continued ROS disc synthesis, the ROS become longer than normal, and some disorganized debris membranes are interposed between the apical tips of the ROS and the RPE even before photoreceptor cell death begins in this mutant (LaVail et al. 1972). As photoreceptor cells degenerate and disappear, mostly between postnatal day (P) 20 and 75, the outer segment zone becomes progressively more disorganized and thickens, reaching its maximum at about P35. Thereafter, the combined outer segment zone and disorganized membranes appear by light microscopy as a single debris zone that remains thicker than the normal ROS zone until about P75 (LaVail, Battelle 1975). Much of the debris is lost by about P100 (LaVail, Battelle 1975). When the pink-eyed rats are reared in the dark, or if their eyes are pigmented and the animals are reared either in cyclic light or in the dark, the debris persists for a much longer time (LaVail, Battelle 1975). Even in pink-eyed rats reared in cyclic light, however, small pockets of debris membranes may survive for the life of the animal (LaVail et al. 1974).

The debris membranes with their excessive rhodopsin content have been mentioned in more than a hundred research publications on retinal dystrophy in the rat, many of which have ascribed an important role to them. For example, the debris has been implicated in abnormal photoreceptor-RPE interactions in the mutant retina, i.e., a slowed transfer of retinol from the ROS to the RPE (Delmelle et al. 1975). Moreover, the debris accumulation has been involved in virtually every hypothetical cause of photoreceptor cell death in this form of retinal degeneration. These suggestions include 1) the debris as a diffusion barrier to metabolites (Herron et al. 1969); 2) extracellular release of acid hydrolases from the RPE due to the excessive rhodopsin content of the debris (Burden et al. 1971; Dewar, Reading 1975; Reading 1980); 3) the presence of a substance in the debris that induces an abnormality of cyclic nucleotide metabolism in photoreceptors (Lolley, Farber 1976); and 4) exclusion of the interphotoreceptor matrix by the debris membranes (LaVail et al. 1981).

In the past several years, we have made numerous ultrastructural observations of the outer segment debris zone in RCS rats that demonstrate a much larger number of cellular elements in the debris than has generally been appreciated. Since these morphological observations have a bearing on the interpretation of previous biochemical (Delmelle et al. 1975; Organisciak, Noell 1976; Organisciak et al. 1982) and other (Katz et al. 1986) studies of the debris zone, as well as on future investigations of this mutant, we present them here.

MATERIALS AND METHODS

Rats of different ages and genotypes were examined cytologically to determine the contents of the debris zone. Pink-eyed, retinal dystrophic RCS rats were examined at P20, 23, 45, 55, 66 and 75. Congenic, black-eyed, dystrophic RCS-p^+ rats (LaVail 1981) were examined at P32 and P96. Congenic, pink-eyed RCS-rdy^+ rats (LaVail 1981) were used as controls and were examined at representative ages. In most cases, two or more of the retinal dystrophic animals were examined at each of the ages. The animals were maintained under our standard conditions with a normal 12 hr light-12 hr dark lighting cycle. Under ether anesthesia the rats were killed by vascular perfusion with mixed aldehydes and their eyes were prepared for light and electron microscopic examination as described elsewhere (LaVail, Battelle 1975). All work with the animals was done in full compliance with the ARVO Resolution on the Use of Animals in Research and the UCSF Committee on Animal Research.

The components of the debris zone in selected animals were analyzed stereologically using the intersection point counting method (Weibel et al. 1966). Electron microscope photomontages were made of the entire outer segment debris zone at a magnification of 20,000X. The outer segment debris zone in each photomontage had an area of approximately 160 μm^2. Point counts were made separately for the apical and basal halves of the outer segment debris zone by dividing the zone halfway between the apical surface of the RPE and the apical surface of the photoreceptor inner segments, if present, or if not, the apical surface of the cells of the residual outer nuclear layer. Micrographs were collected from the posterior to equatorial regions of the retina. The volume density of each component (e.g., debris membranes or interphotoreceptor space) was expressed as a percentage of the total volume by the formula $[V_v=P_p=P_a/P_T]$, where P_a is the number of points for a given component and P_T is the number of total points in a given montage (Weibel 1979).

RESULTS

The region between the RPE and outer limiting membrane of a normal, pink-eyed RCS-*rdy*⁺ rat is illustrated in Figure 1a to allow comparison with various stages of degeneration in dystrophic RCS rats. This zone in normal rats consists primarily of photoreceptor inner and outer segments and the interphotoreceptor space surrounding them, which is filled with the interphotoreceptor matrix (IPM) (Sidman 1958; Zimmerman, Eastham 1959; Fine, Zimmerman 1963; Röhlich 1970). Other components of this zone are processes of cells that border the interphotoreceptor space, the Müller cells, photoreceptor cells and RPE cells. Microvillous processes of Müller cells extend outward from the outer limiting membrane between the inner segments usually about half the inner segment length (ca. 7 μm), although some can be found that are longer. Calycal processes extending outward from the inner segments are infrequently seen.

Processes of the RPE are usually thin, microvillous-like structures that surround the apical tips of the ROS (Fig. 1b) and extend inward for variable lengths, some as far as one-third or one-half the ROS length, but most are appreciably shorter. The RPE cell processes surrounding presumptive phagosomes (LaVail 1976) are occasionally expanded somewhat in diameter and have pale cytoplasm (Fig. 1b). The RPE cells in the pink-eyed RCS-*rdy*⁺ rats and the pink-eyed dystrophic RCS rats frequently contain premelanosomes typical of pink-eyed animals, which are usually non-melanized or, occasionally, very lightly pigmented (Sidman, Pearlstein 1965). These premelanosomes are often found in the somewhat expanded processes of the RPE of normal RCS-*rdy*⁺ rats (Fig. 1b).

In dystrophic RCS rats at P20 and P23, when photoreceptor cells have just begun to disappear, the outer segment debris zone is thicker than normal (cf. Figs. 1a and 1c), as noted above. Although the zone can be distinguished into a basal outer segment region and an apical region with abundant debris membranes, the basal zone has already undergone some changes from normal. The outer segments are often more disorganized than normal, and there is a significantly greater than normal amount of extracellular space in this zone (LaVail et al. 1982), which is filled with a greater than normal amount of IPM (LaVail et al. 1981; Porrello et al. 1986).

In the apical debris zone at P20 and P23, whorls of debris membranes are the most obvious feature; these are interspersed with fragments of ROS (Fig. 1c). In some regions (not illustrated), larger than normal extracellular spaces are also present among the membranous components of the apical part of the debris zone, as shown in micrographs of previous studies (Bok, Hall 1971; LaVail et al. 1972; Goldman, O'Brien 1978). The apical processes of the RPE cells (Fig. 1c) appear similar to those in the normal RCS-*rdy*⁺ retinas, except they mostly surround lamellar whorls of membranes at the apical surface of the RPE (Fig. 1c) rather than intact ROS or relatively intact fragments of ROS that have recently been shed.

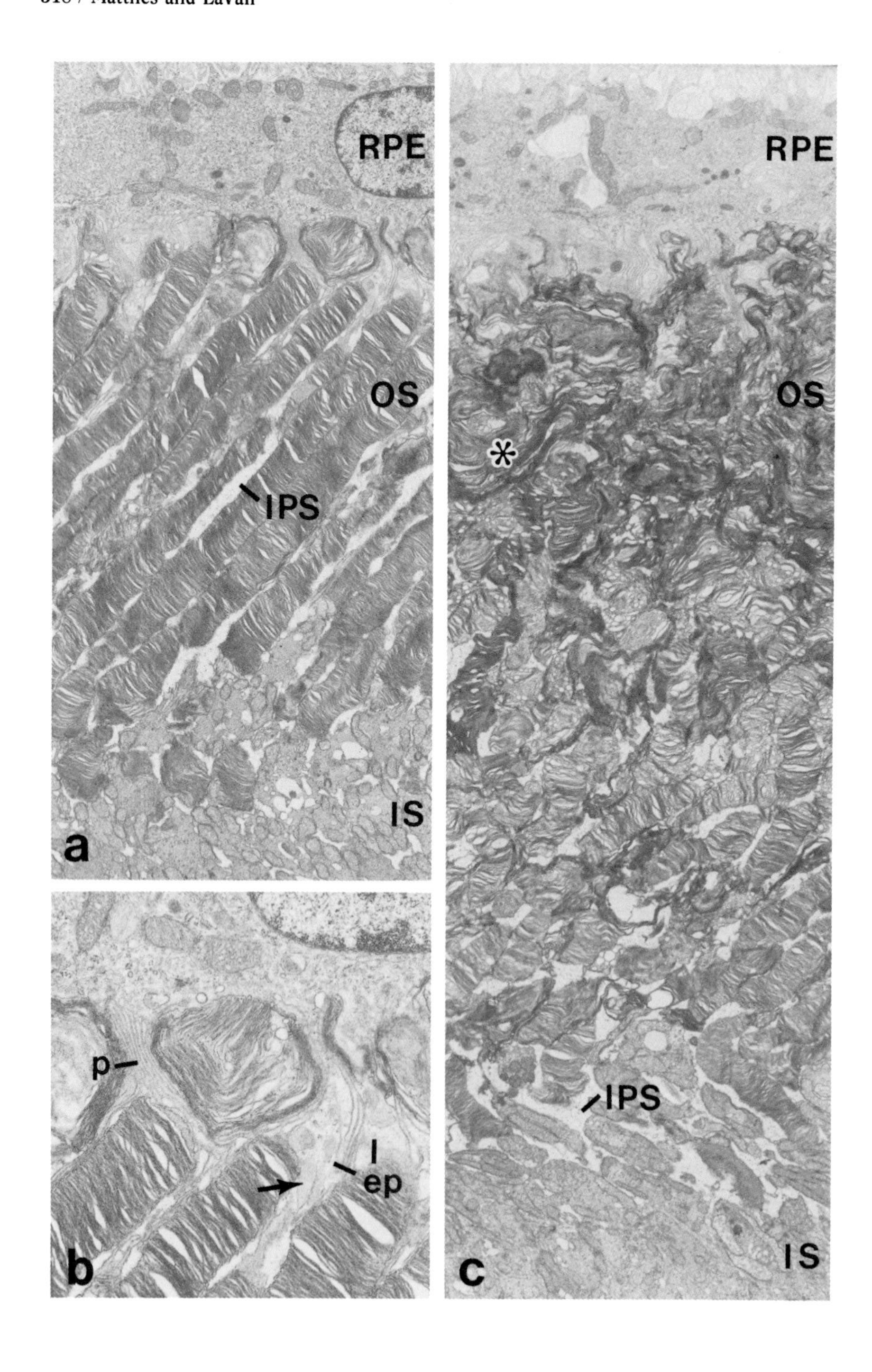
RPE
OS
IPS
IS
a
p
l
ep
b
RPE
OS
*
IPS
IS
c

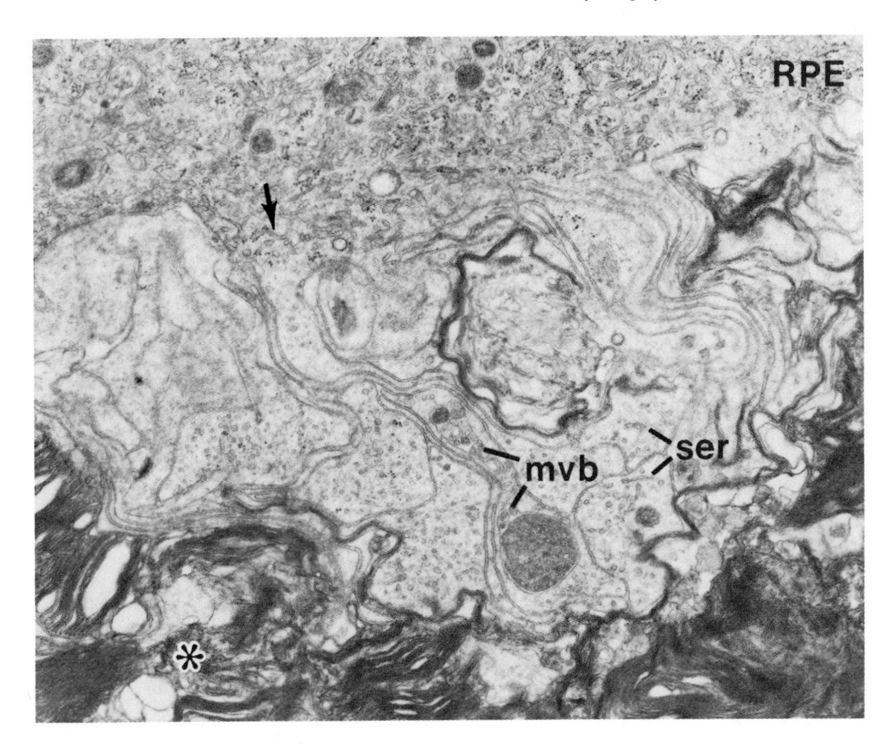

Figure 2. RCS rat retina at P32. Many of the apical processes of the RPE cell are expanded and contain vesicular and tubular profiles of the smooth endoplasmic reticulum (ser) and multivesicular bodies (mvb), but little or no rough endoplasmic reticulum or polyribosomes. The arrow points to the continuity of one expanded process with an RPE cell body. *, debris membranes. (X16,150)

Figure 1. (Opposite page) Electron micrographs of normal RCS-*rdy*$^{+}$ (a and b) and dystrophic RCS (c) retinas at P23. In the normal retina (1a), the outer segment zone consists primarily of photoreceptor outer segments (OS), whereas the zone in the mutant retina (1c) is significantly thicker than normal and consists of whorls of membranous debris (*) in the apical part and somewhat disorganized outer segments in the basal part. Fig. 1b is an enlarged portion of Fig. 1a; it illustrates the slender, microvillous like apical processes of the RPE (p) as well as pale, expanded processes (ep) that contain premelanosomes (arrow). IPS, interphotoreceptor space; IS, inner segment; RPE, retinal pigment epithelium. (a and c, X4,380; b, X9,600).

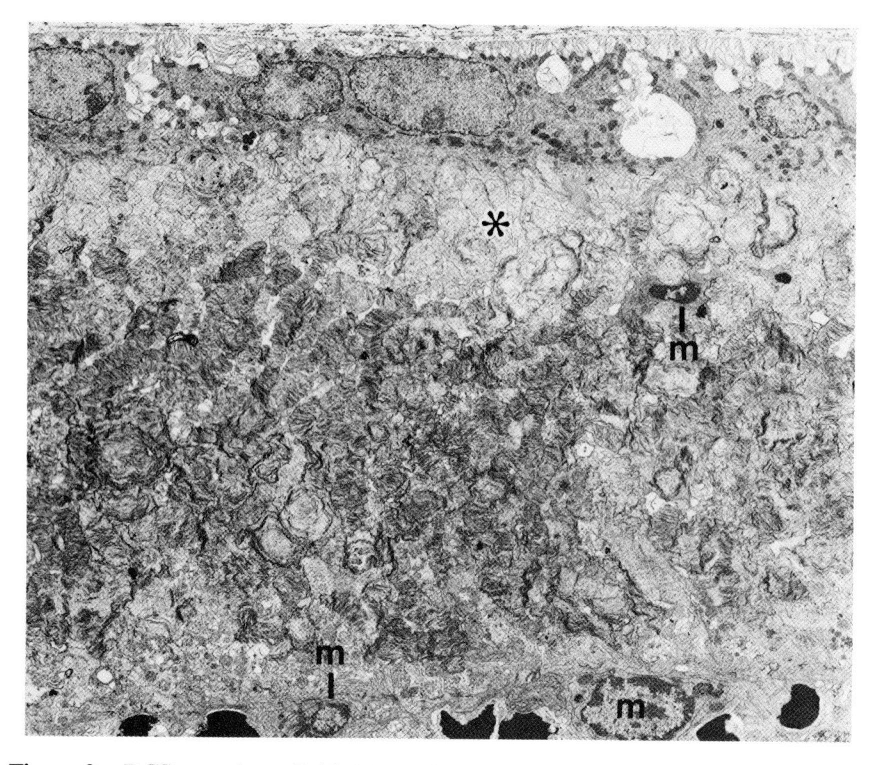

Figure 3. RCS rat retina at P45. Most of the outer segment zone consists of small fragments of outer segments and debris membranes. Some macrophages (m) are present in the outer segment debris zone and in the outer nuclear layer among the degenerating photoreceptor nuclei. The apical third of the debris zone is more electron lucent than the remainder of the zone (*). (X2,580)

About one week later, at P32, massive cell death is evident in the outer nuclear layer and a much greater disruption of the ROS zone is obvious. Photoreceptor inner segments are decreased in length at this age, but Müller cell processes are not. Thus, the Müller cell processes usually extend outward to the basal part of the outer segment debris zone. At many points along the length of the RPE, the apical processes are changed from their normal structure. Many are significantly expanded in diameter (Fig. 2). As they extend from the RPE cell body, they become more electron lucent than the cytoplasmic matrix of the RPE cell body. The expanded processes contain abundant profiles of smooth endoplasmic reticulum, many of which are cut in cross-section and appear as small vesicles (Fig. 2). Multivesicular bodies are also common in the processes, but little or no rough endoplasmic reticulum or polyribosomes are present (Fig. 2). As described previously (LaVail 1979), macrophages begin to invade the outer segment debris zone in significant

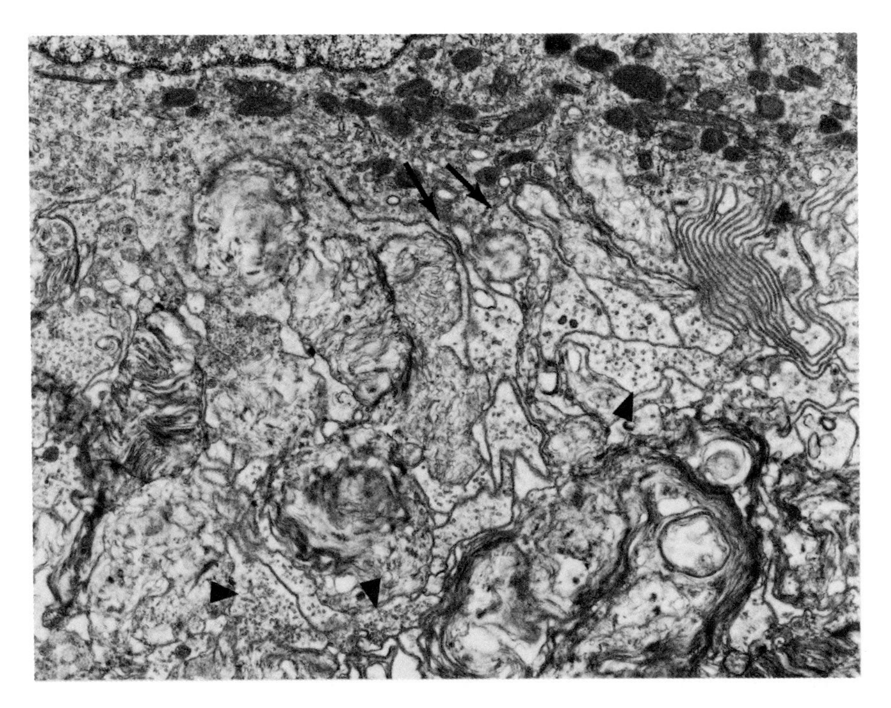

Figure 4. Higher magnification of Figure 3. The pale apical zone consists of some membrane whorls and outer segment fragments, but primarily of expanded RPE cell processes, many of which contain abundant vesicular profiles (arrowheads). The arrows point to the continuity of two expanded processes with an RPE cell body. (X9,830)

numbers by about P30. Their processes apparently do not pervade the debris zone far from their cell bodies at this age, since electron micrographs of the debris zone from RCS rats at P32 which do not contain macrophage cell bodies also show few cellular elements other than identifiable RPE cell processes.

By P45, the debris zone of RCS rat retinas shows a much greater loss of ROS integrity and much greater cytoplasmic (cellular) content (Figs. 3 and 4) than at earlier ages. One of the most conspicuous changes from earlier ages is the abundance of expanded apical processes of the RPE first seen at P32 (Fig. 2). These electron lucent processes extend inward from the apical surface of the RPE as far as one-third to one-half the thickness of the debris zone (Figs. 3 and 4). The pale processes occasionally contain premelanosomes of the pink-eyed variety. Most often, when the expanded processes extend inward, away from the RPE cell body, they are filled with small vesicles and tubules presumably of smooth endoplasmic reticulum. However, the vesicles are usually so uniform in size that when expanded

RPE cell processes are cut in cross-section, they often appear similar to profiles of synaptic terminals (Fig. 4), albeit without the other features of synaptic boutons. The abundance of expanded RPE cell processes give the apical part of the debris zone a much more electron lucent appearance than the basal part when viewed at low magnification (Fig. 3).

At this age, photoreceptor inner segments are grossly reduced in length (Fig. 3), and fragments of inner segments in various stages of degeneration and Müller cell processes of almost normal length are found in the basal portion of the debris zone. Macrophages are abundant in the outer nuclear layer and debris zone at P45. In most cases, they can be distinguished from RPE cells and photoreceptor cells by their characteristic nuclear heterochromatin pattern and cytoplasmic content, including a paucity of smooth endoplasmic reticulum, sharply defined rough endoplasmic reticulum, and a dense cytoplasmic matrix when fixed with mixed aldehydes (Essner, Gorrin 1979). Using these criteria, processes of macrophages can be identified with certainty in many cases, even when the cell body is not in the section, and many such processes are found at P45 and at older ages.

By P55 and P66, and particularly by P75, when most photoreceptor cell nuclei have disappeared, the appearance of the RPE and debris zone is more variable than at earlier ages. At some foci, the debris zone is actually missing and the RPE cells abut the former outer limiting membrane. In most regions of the retina, however, the debris zone is still as thick or thicker than a normal outer segment zone. In these regions, the debris zone consists mostly of outer segment membranes in various stages of degeneration. However, the cytoplasmic or cellular elements described for the P45 retinas make up an even greater part of the debris zone at P55-P75. Examples of the debris zone at the older ages are shown in figures 5-7. In addition to the defined cytoplasmic elements (processes of macrophages, Müller cells and RPE cells, as well as fragments of photoreceptor inner segments), a large number of unidentifiable cellular elements are present (Figs. 5 and 6a). These, presumably, are of the same cellular types as those that are identifiable, but which have become less characteristic in staining properties and in cytoplasmic contents due to the massive cell degeneration in this region.

In order to confirm the apparent extension of RPE cell processes far into the debris zone, we examined the retinas of pigmented RCS-p^+ dystrophic rats with the hope that the melanosomes in the processes of the RPE cells would provide a cytological marker for this cell type. The age of P96 was chosen to compare with the P75 RCS pink-eyed retinas because of the 10- to 30-day slower rate of photoreceptor degeneration in the pigmented eyes (depending on the position in the retina) and the similar debris thickness of the retinas at these ages (LaVail, Battelle 1975). Examination of the debris zone did, in fact, reveal broad, melanosome-containing processes at many different regions of the zone, even deep in the basal region near the residual outer nuclear layer (Figs. 6b and 7). Although it is possible for macrophages to ingest melanosomes from dead or dying (or even normal) RPE cells (Eckmiller, Steinberg 1981), the processes containing melanosomes at P96

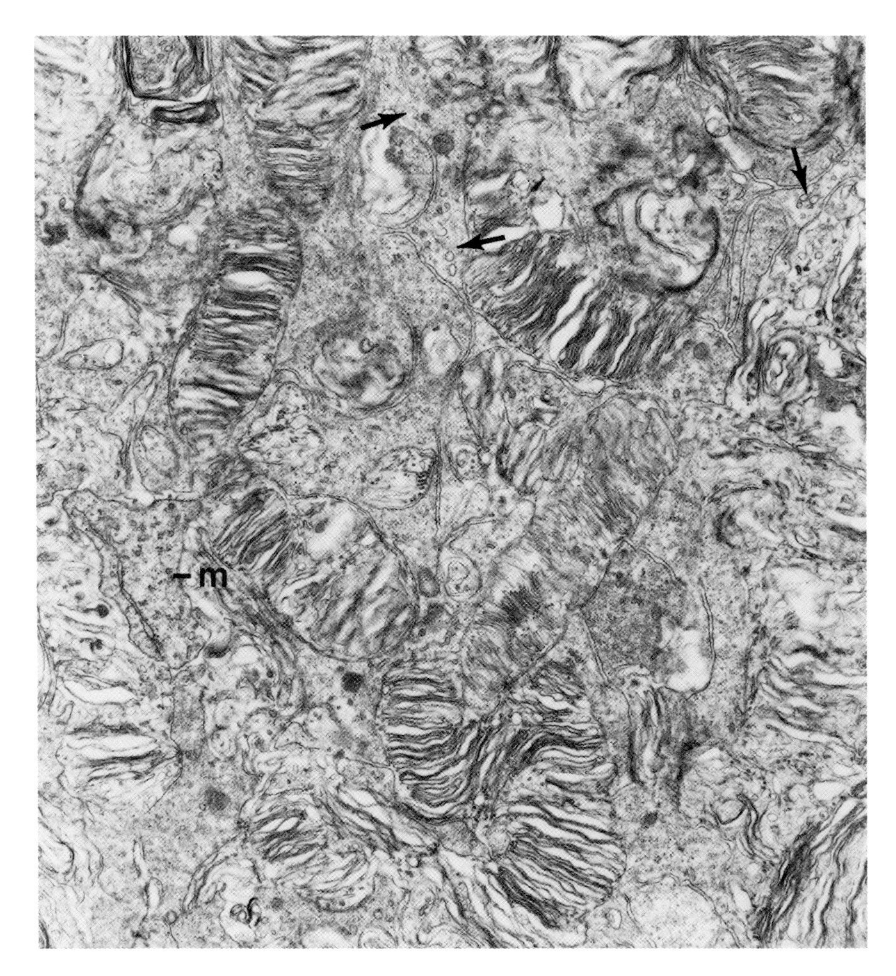

Figure 5. RCS rat retina at P55. This figure is taken from the central part of the outer segment debris zone and illustrates the high cellular (cytoplasmic) content of the debris zone at this and older ages. One macrophage process (m) can be identified, and the arrows indicate probable RPE cell processes, but most of the processes cannot be identified with certainty. (X14,900)

show cytoplasmic features characteristic of RPE cells, such as an abundance of smooth endoplasmic reticulum, absence of abundant rough endoplasmic reticulum and a relatively pale cytoplasmic matrix compared to that of macrophages (Fig. 7). Consistent with the identification of the melanosome-containing processes as those of RPE cells, Essner and Gorrin (1979) found that in pigmented RCS-p^+ rats the invading macrophages never contain melanosomes or premelanosomes.

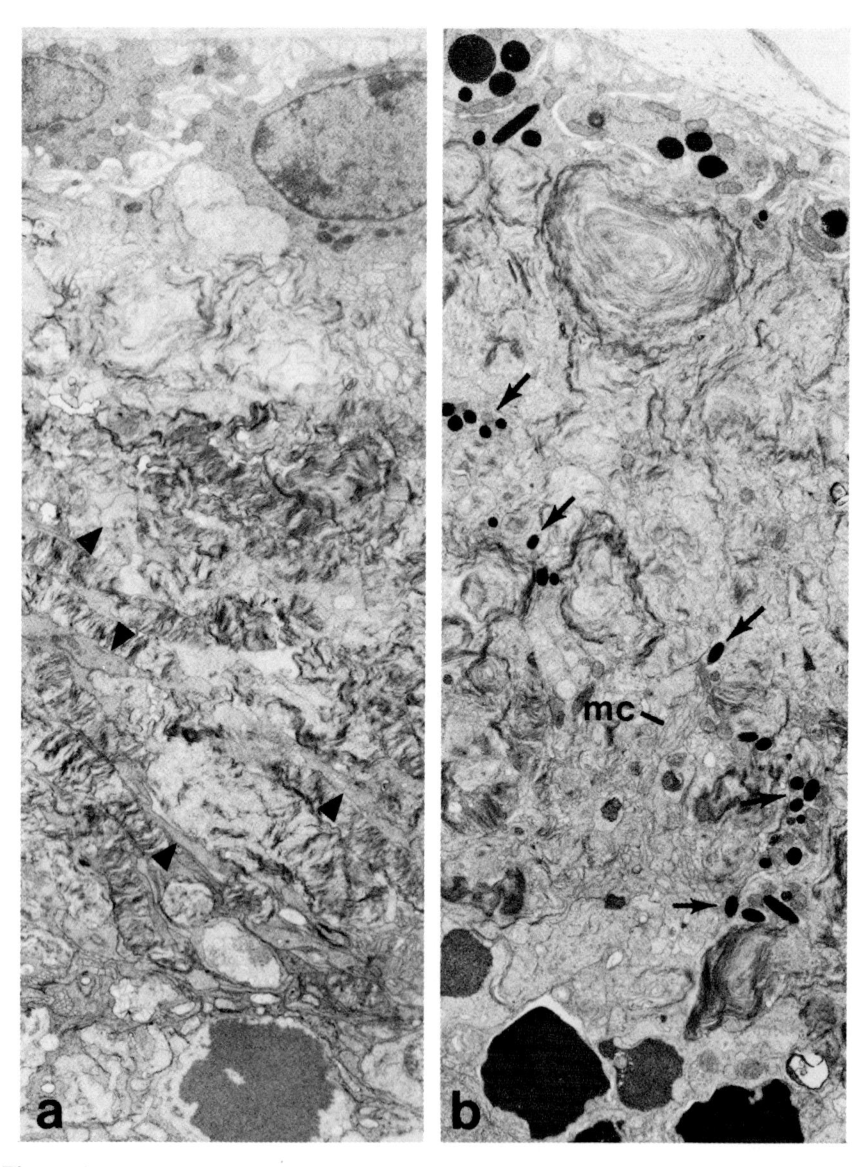

Figure 6. RCS rat retina at P75 (a) and pigmented RCS-p^+ rat retina at P96 (b). In the P75 retina (a), the outer segment debris zone shows a high degree of cellularity with many cytoplasmic processes of unknown origin (arrowheads). The P96 retina (b) demonstrates that in pigmented animals many of the processes at all levels of the debris zone contain melanosomes (arrows). mc, Müller cell microvillous processes. (a, X4,950; b, X4,750)

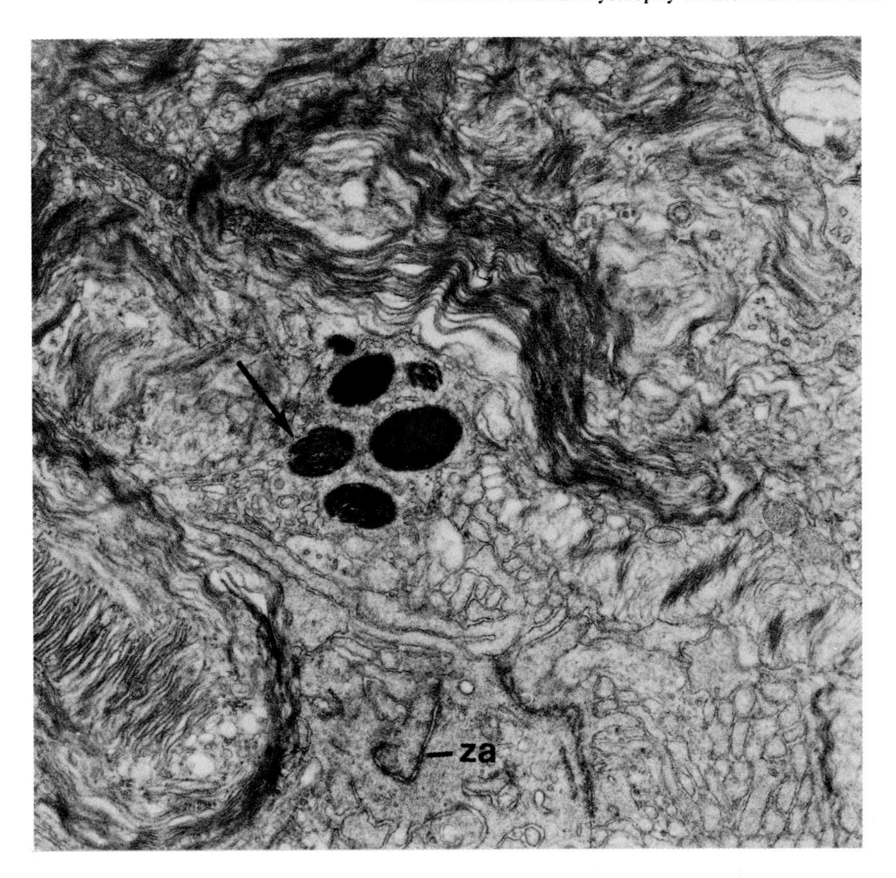

Figure 7. Pigmented RCS-p^+ rat retina at P96. Higher magnification of a region comparable to that shown in Fig. 6b. A cell process containing melanosomes can be identified as that of an RPE cell on the basis of the abundant smooth endoplasmic reticulum and absence of rough endoplasmic reticulum (arrow). This process is in the most basal region of the outer segment debris zone, as shown by the zonulae adherentes (za) that make up the outer limiting membrane and the abundant microvillous processes of Müller cells (mc). (X19,870)

In order to gain some quantitative measure of the degree of cellularity of the debris zone, we have carried out a morphometric analysis of this region in RCS rats, primarily at P75, although some retinal sections were examined at earlier ages. In micrographs used for the analysis, we avoided fields that included macrophage cell bodies, so our estimates of the cytoplasmic content of the debris zone are conservative. We initially attempted to categorize each cytoplasmic element in the debris zone. We consistently find 8 distinctly different cytoplasmic profiles in the debris

Table 1. Volume Density (%) of Different Components of ROS Debris Zone

Region Component	Normal RCS-*rdy*⁺ *(n=3)	P45 RCS (n=6)	P75 RCS (n=6)	P75 RCS **(max)
Apical				
Membrane	81.7	72.0	74.7	67.0
Cytoplasm	11.5	21.7	22.7	30.5
IPS	6.8	6.3	2.6	2.5
Basal				
Membrane	90.2	82.5	79.8	57.5
Cytoplasm	1.5	11.2	18.4	41.0
IPS	8.3	6.3	1.8	1.5
Total (Apical + Basal)				
Membrane	86.0	77.3	77.2	62.3
Cytoplasm	6.5	16.4	20.6	35.6
IPS	7.5	6.3	2.2	2.1

*One at P76 and two at P180; these two ages did not differ significantly from each other, so the data were pooled.
**Photomontage (of 6 examined) that exhibits the greatest cytoplasmic content.
IPS, interphotoreceptor space.

zone. Although some can be categorized as processes of RPE cells, macrophages or Müller cells, many of the 8 types of processes cannot be defined with certainty. Therefore, we have pooled all 8 types into a single "cytoplasm" category. Likewise, 4 distinct membranous types are regularly seen, all presumably derived from ROS or attenuated RPE cell processes (LaVail, Battelle 1975; LaVail et al. 1972), but these have been pooled into a single "membrane" category. The other category is interphotoreceptor space, which includes any extracellular space either with or without obvious interphotoreceptor matrix filling it.

The data in Table 1 reflect the visual impression that a substantial portion of the outer segment debris zone by P75 consists of cytoplasmic elements. The total mean of 6 photomontages at this age shows that 20.6% of the area is occupied by cytoplasmic elements (Table 1). This varied from region to region, and the photomontage that illustrated the greatest degree of cellularity measured 35.6%. In general, RPE cell processes make up the major cellular component of the apical half of the zone and macrophage processes constitute the major cellular component of the basal half of the zone, but exceptions are common. The point counts from the P45 RCS photomontage illustrate a lower overall cellular percentage (16.4%) than at P75 (20.6%), as would be expected from the appearance of the retinas at the different ages.

DISCUSSION

The debris zone of the degenerating retina of the RCS rat is generally assumed to consist entirely of degenerating ROS membranes, although it is known that some cells invade the zone at late stages of photoreceptor degeneration (Dowling, Sidman 1962; Bok, Hall 1971; Essner et al. 1978; Essner, Gorrin 1979; LaVail 1979). In the present study, we have demonstrated that the debris zone becomes highly cellular in composition as photoreceptors degenerate. The cellular components are expanded and elongated processes of the RPE cells, microvillous Müller cell processes, fragments of photoreceptor cells, macrophage cell processes and cellular processes of undetermined origin, although we assume these are derived from the aforementioned cells. The major cellular constituents are the expanded RPE cell processes and macrophage processes. These are evident at least as early as P32, and by P75 the cytoplasmic elements may make up as much as 35% of the volume of the debris zone (Table 1). It should be stressed that in our stereologic analysis, we avoided fields that contained macrophage cell bodies, as the majority do by P75. Had the macrophage cell bodies been included, the cytoplasmic composition of the debris zone most surely would have exceeded 35%.

It has been noticed by light microscopy that in RCS rats at about P32-P45, the outer segment debris membranes near the apical surface of RPE cells lose their staining properties (Dowling, Sidman 1962), a process referred to as "blanching" (LaVail, Battelle 1975). It was also shown that at older ages, the zone of relative lack of osmiophilia and basophilia expanded to encompass the entire thickness of the debris zone, and that the temporal sequence of "blanching" correlated directly with the loss of the membranes to regenerate rhodopsin during dark adaptation (LaVail, Battelle 1975). It is clear from the electron micrographs in the present study that the abundance of expanded, lightly staining RPE cell processes form the basis, at least in part, of the "blanching" of the apical debris zone in RCS rat retinas.

In the most comprehensive biochemical study of the outer segment debris zone of RCS rats, Organisciak and Noell (1976) found that as much as 30% of the phospholipids of the debris zone (predominantly sphingomyelin) appeared to be non-outer-segment in origin. This finding was made using dark-reared, albino dystrophic rats at the age of P100. Since dark-rearing of non-pigmented rats results in the same pattern and rate of retinal degeneration (and same debris zone thickness and appearance) as pigmented dystrophic rats reared in cyclic light (LaVail, Battelle 1975), our observations on the P96 pigmented rats should be directly comparable to the animals used by Organisciak and Noell. One remarkable feature of the P96 rat retinas was the abundance of RPE cell processes in the debris zone and the distribution of those processes even into the most basal part of the debris zone near the outer limiting membrane. This is particularly relevant to the biochemical findings of Organisciak and Noell, who studied predominantly the basal part of the debris zone. Thus, our morphological findings clearly support their suggestion of an RPE origin of the non-outer segment-derived phospholipids.

In a recent study of RCS rats at P66, a broad band of lipofuscin-like autofluorescence was observed in the debris zone, predominantly in the apical region near the RPE cell surface (Katz et al. 1986). It was assumed that the debris membranes were the source of the autofluorescence. We have demonstrated in the present study that the precise zone that demonstrates lipofuscin-like autofluorescence is rich in expanded RPE cell processes. Thus, while there are arguments concerning autooxidative mechanisms to support the debris membranes as a source of the fluorescence (Katz et al. 1986), it seems equally plausible that the source is the RPE cell processes, since the RPE contains significant quantities of lipofuscin that may have been displaced into the expanded cell processes in the dystrophic animals. Consistent with this view, the distribution of the RPE cell processes in the debris zone at P66 would explain each of the retinal separation experiments carried out by Katz and co-workers (Katz et al. 1986). Moreover, the lipofuscin-like material found in the neural retina and associated with retinal capillaries in 16-month-old dystrophic rats (Katz et al. 1986) could be explained on the same basis, since RPE cells migrate into the neural retina along retinal capillaries at older ages (LaVail 1979).

ACKNOWLEDGEMENTS

We wish to thank Douglas Yasumura, Gregg Gorrin, Nancy Lawson and Gloria Riggs for technical, photographic and secretarial assistance. This investigation was supported in part by USPHS research grants EY01919 and EY06842, core grant EY02162, That Man May See, Inc., Research to Prevent Blindness and the Retinitis Pigmentosa Foundation Fighting Blindness. Dr. LaVail is the recipient of a Research to Prevent Blindness Senior Scientific Investigators Award.

REFERENCES

Bok D, Hall MO (1971). The role of the pigment epithelium in the etiology of inherited retinal dystrophy in the rat. J. Cell Biol. 49:664-682.

Burden EM, Yates CM, Reading HW, Bitensky L, Chayen J (1971). Investigation into the structural integrity of lysosomes in the normal and dystrophic rat retina. Exp. Eye Res. 12:159-165.

Delmelle M, Noell WK, Organisciak DT (1975). Hereditary retinal dystrophy in the rat: rhodopsin, retinol, vitamin A deficiency. Exp. Eye Res. 21:369-380.

Dewar AJ, Reading HW (1975). The role of retinol in, and the action of anti-inflammatory drugs on, hereditary retinal degeneration. In Cristofalo VJ, Holeckova E (eds): "Cell Impairment in Aging and Development," New York: Plenum Press, pp 281-295.

Dowling JE, Sidman RL (1962). Inherited retinal dystrophy in the rat. J. Cell Biol. 14:73-109.

Eckmiller MS, Steinberg RH (1981). Localized depigmentation of the retinal pigment epithelium and macrophage invasion of the retina in the bullfrog. Invest. Ophthalmol. Vis. Sci. 21:369-394.

Essner E, Gorrin G (1979). An electron microscopic study of macrophages in rats with inherited retinal dystrophy. Invest. Ophthalmol. Vis. Sci. 18:11-25.

Essner E, Gorrin GM, Griewski RA (1978). Localization of lysosomal enzymes in retinal pigment epithelium of rats with inherited retinal dystrophy. Invest. Ophthalmol. Vis. Sci. 17:278-288.

Fine BS, Zimmerman LE (1963). Observations on the rod and cone layer of the human retina: a light and electron microscopic study. Invest. Ophthalmol. 2:447-459.

Goldman AI, O'Brien PJ (1978). Phagocytosis in the retinal pigment epithelium of the RCS rat. Science 201:1023-1025.

Herron WL, Riegel BW, Myers OE, Rubin ML (1969). Retinal dystrophy in the rat-a pigment epithelial disease. Invest. Ophthalmol. 8:595-604.

Katz ML, Drea CM, Eldred GE, Hess HH, Robison WG Jr. (1986). Influence of early photoreceptor degeneration of lipofuscin in the retinal pigment epithelium. Exp. Eye Res. 43:561-573.

LaVail MM (1976). Rod outer segment disc shedding in rat retina: relationship to cyclic lighting. Science 194:1071-1074.

LaVail MM (1979). The retinal pigment epithelium in mice and rats with inherited retinal degeneration. In Zinn KM, Marmor MF (eds): "The Retinal Pigment Epithelium," Cambridge: Harvard University Press, pp 357-380.

LaVail MM (1981). Photoreceptor characteristics in congenic strains of RCS rats. Invest. Ophthalmol. Vis. Sci. 20:671 675.

LaVail MM, Battelle BA (1975). Influence of eye pigmentation and light deprivation on inherited retinal dystrophy in the rat. Exp. Eye Res. 21:167-192.

LaVail MM, Pinto LH, Yasumura D (1981). The interphotoreceptor matrix in rats with inherited retinal dystrophy. Invest. Ophthalmol. Vis. Sci. 21:658-668.

LaVail MM, Sidman M, Rauzin R, Sidman RL (1974). Discrimination of light intensity by rats with inherited retinal degeneration: a behavioral and cytological study. Vision Res. 14:693-702.

LaVail MM, Sidman RL, O'Neil DA (1972). Photoreceptor-pigment epithelial cell relationships in rats with inherited retinal degeneration. Radioautographic and electron microscope evidence for a dual source of extra lamellar material. J. Cell Biol. 53:185-209.

LaVail MM, Yasumura D, Gorrin G, Pinto LH (1982). The interphotoreceptor matrix in RCS rats: possible role in photoreceptor cell death. In Clayton RM, Haywood J, Reading HW, Wright A (eds): "Problems of Normal and Genetically Abnormal Retinas," New York: Academic Press, pp 215-222.

Lolley RN, Farber DB (1976). A proposed link between debris accumulation, guanosine 3', 5' cyclic monophosphate changes and photoreceptor cell degeneration in retina of RCS rats. Exp. Eye Res. 22:477-486.

Noell WK (1963). Cellular physiology of the retina. J. Opt. Soc. Amer. 53:36-48.

Organisciak DT, Noell WK (1976). Hereditary retinal dystrophy in the rat: lipid composition of debris. Exp. Eye Res. 22:101-113.

Organisciak DT, Wang H-, Kou AL (1982). Rod outer segment lipid-opsin ratios in the developing normal and retinal dystrophic rat. Exp. Eye Res. 34:401-412.

Porrello K, Yasumura D, LaVail MM (1986). The interphotoreceptor matrix in RCS rats: histochemical analysis and correlation with the rate of retinal degeneration. Exp. Eye Res. 43:413-429.

Reading HW (1980). Biochemistry of retinal degeneration in rats and mice: a short review. Neurochemistry 1:477-485.

Röhlich P (1970). The interphotoreceptor matrix: electron microscopic and histochemical observations on the vertebrate retina. Exp. Eye Res. 10:80-96.

Sidman RL (1958). Histochemical studies on photoreceptor cells. Ann. N.Y. Acad. Sci. 74:182-195.

Sidman RL, Pearlstein R (1965). Pink-eyed dilution (p) gene in rodents: increased pigmentation in tissue culture. Dev. Biol. 12:93-116.

Weibel ER (1979). Point counting method. In Weibel ER (ed): "Stereological Methods. VI. Practical Methods for Biological Morphometry," San Francisco: Academic Press, pp 101-159.

Weibel ER, Kister GS, Scheele WF (1966). Practical stereological methods for morphometric cytology. J. Cell Biol. 30:23-38.

Zimmerman LE, Eastham AB (1959). Acid mucopolysaccharide in the retinal pigment epithelium and visual cell layer of the developing mouse eye. Am. J. Ophthal. 47:488-499.

Inherited and Environmentally Induced
Retinal Degenerations, pages 331–341

EFFECT OF DIETARY RIBOFLAVIN ON RETINAL DENSITY AND FLAVIN CONCENTRATIONS IN NORMAL AND DYSTROPHIC RCS RATS

Curtis D. Eckhert, Mei-Hui Hsu and David W. Batey

Division of Nutritional Sciences
School of Public Health
University of California
Los Angeles, California 90024

INTRODUCTION

The identification of environmental factors that influence the rate of disease can afford the basis for developing health promotion strategies which are both economical and applicable to use in public health programs.

Several years ago my laboratory obtained the RCS rat strain for the purpose of evaluating the ability of nutrients to influence the rate of retinal degeneration. Prior to beginning our studies it was necessary to ensure that the basal diet provided to rats contained all the nutrients in amounts necessary for healthy growth, development and reproductive function. Diets that do not meet basic nutrient requirements result in alterations removed from the primary genetic defect. Meeting this criterion has provided interesting insights into the RCS model. Some of our observations will be presented in this report.

PROCEDURES AND RESULTS

RCS dystrophic pups have a high rate of mortality during the first week of life. Survival was therefore selected as the initial focus of our investigation. The approach to this problem was to provide female rats a purified diet containing all nutrients at recommended levels immediately following a night of successful copulation. Variations on this complete basal diet were simultaneously evaluated to determine if dietary components were a contributing factor in the death of the dystrophic rats. The survival of the F_2 generation, that is the maternal grandchildren, was determined during the first week of life. This design exposed all periods of tissue development to the physiological conditions set up by the different diets. The results of the first experiment are given in Table 1. Variations on the basal diet consisted of doubling a single component. Doubling the vitamin mix raised survival from 73% to the acceptable level of 91.7% (Eckhert, 1982), however doubling the level of protein and the mineral mix was lethal. The next question to be addressed was - which vitamin was limiting at basal vitamin mix concentrations?

TABLE 1. Survival of F_2 RCS dystrophic rats

Diet	F_0 females	F_2 survival
Basal modified AIN-76	8	73.0 ± 11.0
Basal + 1 x vitamin mix	8	91.7 ± 4.9*
Basal + 1 x fat	8	81.2 ± 10.0
Basal + 1 x mineral mix	8	45.8 ± 12.2*
Basal + 1 x protein	8	21.0 ± 9.7*

*Significantly different from basal modified AIN-76 diet, $p<0.05$. (Eckhert, 1982).

To determine the limiting vitamin, the breeding procedure was repeated and each individual vitamin in the vitamin mix doubled one at a time. The results are given in Tables 2 and 3. It can be seen from Table 2 that when the individual fat soluble vitamins were doubled, vitamin E was the one which proved to be required in higher amounts (Eckhert, 1987). Doubling the level of vitamin E raised the level from 50 to 100 IU/kg diet. This increase raised the F_2 survival level to that obtained by doubling the entire vitamin mix.

However, to our surprise when the water soluble vitamins were evaluated, riboflavin proved detrimental. The data in Table 3 show that when riboflavin was raised from 6 to 12 mg/kg diet, the survival rate at the end of the first week of life of F_2 offspring dropped to only 44.6% (Table 2) (Eckhert, 1987). Riboflavin is not generally thought to be

TABLE 2. Survival of F_2 RCS dystrophic rats receiving fat soluble vitamin supplements

Diet	F_0 females	F_2 survival
		%
Basal modified AIN-76	8	72.3 ± 11.0
Basal + 1 x vitamin A	8	79.2 ± 11.9
Basal + 1 x vitamin D	8	78.6 ± 5.9
Basal + 1 x vitamin E	8	92.5 ± 3.8*
Basal + 1 x vitamin K	8	73.4 ± 11.6

*Significantly different than basal diet, $p<0.005$. (Eckhert, 1987).

TABLE 3. Survival of F_2 RCS dystrophic rats receiving water-soluble vitamin supplements

Diet	F_0 females	F_2 survival
		%
Basal modified AIN-76	8	64.0 ± 9.5
Basal + 1 x thiamin	8	66.9 ± 15.0
Basal + 1 x riboflavin	8	44.6 ± 14.0*
Basal + 1 x pyridoxine	8	60.6 ± 14.7
Basal + 1 x nicotinic acid	8	66.2 ± 10.6
Basal + 1 x pantothenate	8	66.6 ± 11.2
Basal + 1 x folic acid	8	54.6 ± 15.8
Basal + 1 x biotin	8	56.5 ± 12.9
Basal + 1 x vitamin B-12	8	68.4 ± 9.1

*Significantly different than basal diet, $p<0.01$. (Eckhert, 1987).

toxic and the results raised the question of whether the rdy mutation involved an error in flavin metabolism. If riboflavin was detrimental to the survival of the F_2 off-spring, was the retina a target tissue?

The effect on the retina of doubling riboflavin, from the recommended level of 6 mg/kg diet (American Institute of Nutrition, 1977) to 12 mg/kg, was evaluated in the F_2 generation. The diets were supplemented with double the recommended level of vitamin E, receiving 100 IU RRR-α-tocopherol/kg diet. The first criterion chosen for evaluation was the physical parameter, specific gravity. A gradient of kerosene and monobromobenzene was prepared and allowed to equilibrate for 24 hrs (Fujiwara et al, 1981). F_2 rats were killed and their eyes immersed immediately in a water treated kerosene bath. Retinas were carefully removed within the kerosene bath and allowed to equilibrate on the gradient. Reference standards of known density were then applied to the column to envelope the retinal tissues.

The results are given in Table 4. Dietary riboflavin altered the specific gravity of normal retinas from congenic RCS rats but not dystrophic retinas. The difference between

TABLE 4. Specific gravity: retina RCS congenic strains

Riboflavin	Normal (rdy^+, p^+)	Dystrophic (rdy/rdy, p^+)
mg/kg		
6	1.0318 ± 0.0032 (24)	1.0318 ± 0.0033 (28)
12	1.0291 ± 0.0026 (15)*	1.0326 ± 0.0021 (19)

Mean ± SD (h)
Values are presented at an age of 46 days.
*Significantly less than normal provided 6 mg riboflavin/kg diet and dystrophic provided 12 mg/kg diet, $p<0.001$.

normal and dystrophic RCS retinas is that whereas in 46 day old normal retinas the rod outer segments are intact, in dystrophic retinas they are disorganized and degenerating. The specific gravities of rat rod outer segments and pigment epithelial cells were then determined on a Percoll gradient against Pharmacia density standards. In order of increasing density, values for normal retina (6 mg riboflavin/kg) were: rod outer segments 1.04; and pigmented epithelium, 1.06-1.08. Therefore a loss of outer segments and attached pigment epithelial cells would be expected to shift the specific gravity of the less dense intact retina to lower values.

A determination was then made to evaluate the effect of riboflavin supplementation on flavin concentrations in the retina. Flavins were determined using reverse phase HPLC. An ammonium phosphate acetonitrile, isocratic gradient was used to elute the flavins. Flavins were detected by fluorescence using an excitation at 450 nm and emission at 530 nm.

TABLE 5. Retinal flavin concentrations

Riboflavin	Riboflavin	FMN	FAD
mg/kg diet	µg/mg protein		
		NORMAL (rdy^+, p^+)	
6	0.003±0.001 (4)	0.013±0.006 (4)	0.034±0.019 (4)
12	0.003±0.001 (3)	0.012±0.005 (3)	0.033±0.005 (3)
		DYSTROPHIC (rdy/rdy, p^+)	
6	0.002±0.002 (4)	0.011±0.004 (4)	0.034±0.015 (4)
12	0.002±0.002 (2)	0.013±0.003 (2)	0.042±0.007 (2)

Mean ± SD (n)
n = number of separate determinations using pooled retinas from 12-34 eyes.

Dietary riboflavin did not alter the concentration of riboflavin, FMN or FAD in the normal retinas of the F_2

generation. The values for dystrophic rats suggest that FAD is elevated by riboflavin supplementation. We are in the process of increasing the number of determinations in order to make a valid comparison. These results however do not explain the observation that riboflavin supplementation decreased the density of normal retinas.

The solution to the density changes became obvious when retinal tissue was evaluated histologically. Retinas from the F_2 generation at 46 days of age were embedded in JB-4. Sections 1-3 microns thick were stained with hematoxylin-PAS. Figures 1(a) and 2(a) show a normal retina from a rat

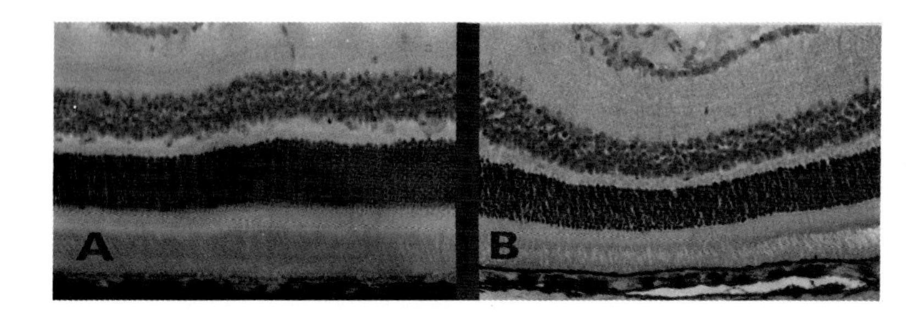

Figure 1. Retinas from normal RCS rats provided: (a) 6 mg riboflavin/kg diet and (b) 12 mg riboflavin/kg diet. The outer segments of rats provided 12 mg/kg appears mottled due to the loss of outer segments Original magnification x250.

provided 6 mg riboflavin/kg. The photoreceptors with their inner and outer segments are clearly defined. Figures 1(b) and 2(b) and (c) are sections through the retina of a normal rat provided 12 mg riboflavin/kg. Note the less defined, mottled appearance of the outer segment structure. At higher power this can be seen to be due to the tearing and subsequent loss of outer segments. Seventy-one percent of the embedded tissue blocks from the 12 mg riboflavin/kg diet group exhibited this phenomenon.

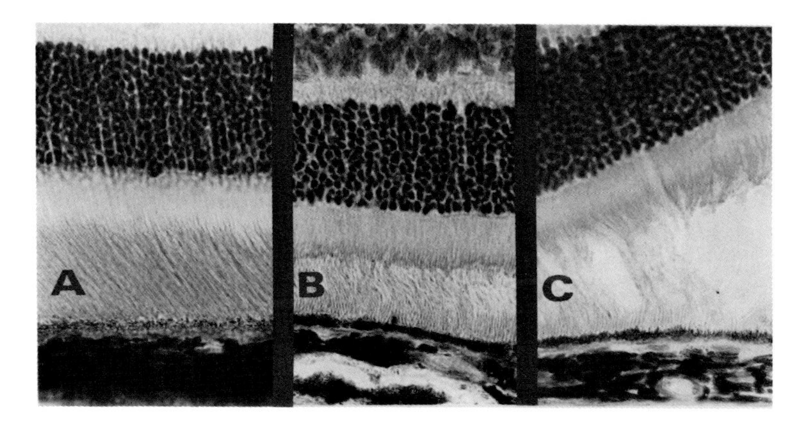

Figure 2. Retinas from normal congenic RCS rats: (a) normal outer segment from rats provided 6 mg riboflavin/kg diet; (b) and (c) 12 mg riboflavin/kg diet showing loss of outer segments and tearing of photoreceptor. Original magnification x630.

Figure 3 shows a comparison of the separation of normal retina from the pigment epithelium in rats provided 6 mg riboflavin/kg diet and the abnormal separation through the outer segment layer in retina from rats provided 12 mg riboflavin/kg diet. At lower levels of riboflavin the retina separates at the junction with the pigment epithelium. However at the higher level it separates near the junction between the inner and outer segments. The reason for this we interpret as increased fragility induced by the riboflavin. The decrease in retinal density was due to this fragility which resulted in the loss of dense outer segments and attached pigment epithelium in the retinas from rats provided 12 mg riboflavin/kg diet.

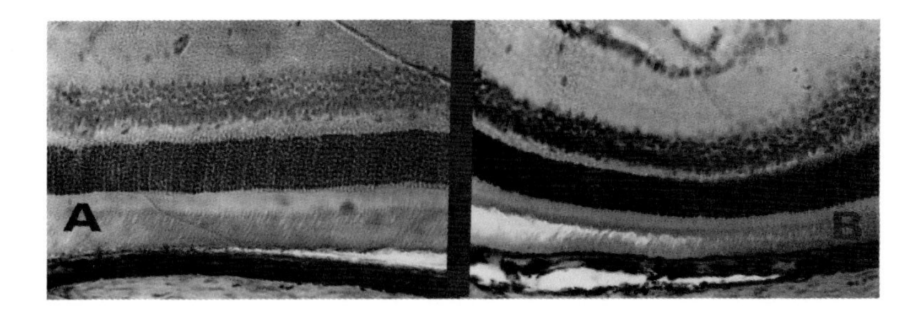

Figure 3. Separation of retina from pigment epithelium: (a) retina from rats provided 6 mg riboflavin/kg diet separate at the junction of the pigment epithelium; (b) retina from rats provided 12 mg riboflavin/kg diet separate near junction of inner and outer segments. Original magnification x250.

Sections of dystrophic retina from rats provided 6 and 12 mg riboflavin/kg are shown in Figures 4 and 5. Dark staining cells were frequently present in the outer segment region in the 12 mg/kg group.

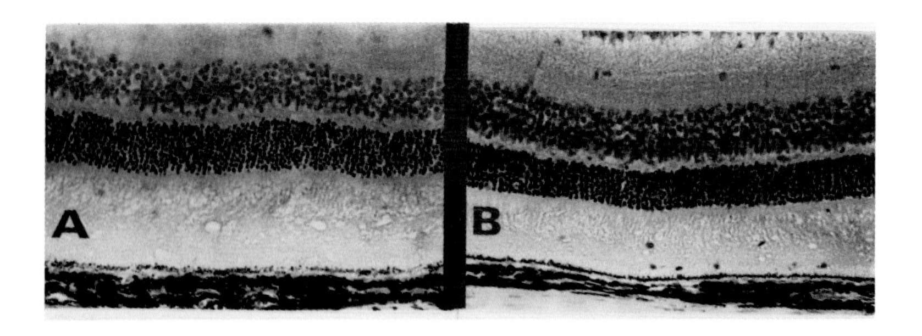

Figure 4. Retinas from RCS dystrophic rats provided (a) 6 mg riboflavin/kg diet and (b) 12 mg riboflavin/kg diet. Many dark staining cells were apparent in the outer segment region of the 12 mg/kg retinas. Original magnification x250.

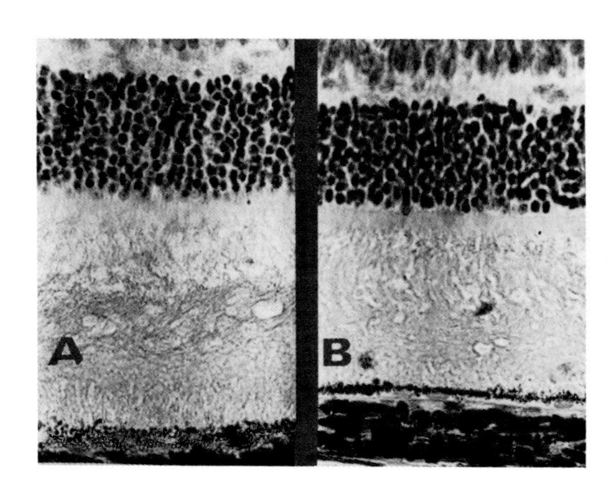

Figure 5. Retinas from RCS dystrophic rats provided (a) 6 mg riboflavin/kg diet and (b) 12 mg riboflavin/kg diet. Original magnification x630.

DISCUSSION

The requirement for riboflavin is about 0.6 mg/1000 kcal for both the rat and human (National Research Council, 1972; 1980). The level of 6 mg/kg established by the American Institute of Nutrition is actually 2.4 times this amount (American Institute of Nutrition, 1977). The level of 12 mg/kg is therefore 5 times the physiological requirement.

It has been known since the 1930s that free riboflavin accumulates in the pigment epithelium of fish (Adler and Euler, 1938). Considering riboflavin's diverse functions, one can only speculate as to why the vitamin resulted in damage to the outer segment. One possible explanation is that exposure to light induced the formation of oxidants from noncovalently bound flavins. When a photon strikes riboflavin, and to a lesser degree FMN and FAD, the ribityl side chain or an adduct serves as an electron donor to cause a photo reduction of the isoalloxine ring system. The primary products are formylmethylflavin, carboxymethylflavin, lumiflavin and lumichrome (Cairns and Metzler, 1971). It is known that photolytic products of riboflavin result in oxidative damage to illuminated gut epithelial cells, breaking down their barrier to absorption

(Ghazy, 1974). This ability of photodegradative products of flavins to oxidize other compounds has been put to good use in pediatrics. When riboflavin is given to infants with jaundice its photodegradative products facilitate the degradation and removal of bilirubin.

In its coenzyme forms, FMN and FAD, riboflavin serves as an intermediary in the transfer of electrons in oxidation-reduction reactions. Many of the flavin dependent dehydrogenase-electron transferases can reduce oxygen to superoxide anion. Flavin oxidase catalyses the formation of H_2O_2 from oxygen. In addition to its importance in reactions involved in potential oxidative damage, three types of fatty acyl-CoA dehydrogenases are flavoproteins and 5' nucleotidase is also a flavin dependent enzyme. It is conceivable that a genetic alteration resulting in the failure of FAD to activate its apoenzymes underlies many of the observations in enzyme activities reported in retinal degeneration.

REFERENCES

Adler E and Euler Hv (1938). Lactoflavin in the eyes of fish. Nature 141:790-791.

American Institute of Nutrition (1977). Report of the American Institute of Nutrition ad hoc committee on standards for nutritional studies. J Nutr 107:1340-1348.

Cairns WL, Metzler DE (1971). Photochemical degradation of flavins. VI. A new photoproduct and its use in studying the photolytic mechanism. J Amer Chem Soc 93:2772-2777.

Eckhert CD (1982). Growth and survival of RCS dystrophic rats fed modifications of the AIN diet. J Nutr 112:2374-2380.

Eckhert CD (1987). Differential effects of riboflavin and RRR-α-tocopherol acetate on the survival of newborn RCS rats with inheritable retinal degeneration. J Nutr 117: 208-211.

Fujiwara K, Nitsch C, Suzuki R, Klatzo I (1981). Factors in the reproducibility of the gravimetric method for evaluation of edematous change in the brain. Neurol Res 3:345-361.

Ghazy FS, Kimura T, Muranishi S, Sezaki H (1974). An anomalous effect of the intermediate products of riboflavine photolysis on the intestinal absorption of poorly absorbed water soluble drugs in rats. J Pharm Pharmac 27:268-272.

National Research Council (1972). Nutrient Requirements of Laboratory Animals No. 10, 2nd revised ed. Washington, D.C.: National Academy of Sciences.

National Research Council (1980). Recommended Dietary Allowances, 9th revised ed. Washington, D.C.: National Academy of Sciences.

Inherited and Environmentally Induced
Retinal Degenerations, pages 343–355

THE EFFECT OF ANTI-INFLAMMATORY DRUG ADMINISTRATION ON THE COURSE OF RETINAL DYSTROPHY IN RCS RATS

E. El-Hifnawi and W. Kühnel

Department of Anatomy, Medical University of Lübeck
Ratzeburger Allee 160, Lübeck, F.R.G.

INTRODUCTION

A recessive gene defect leading to a gradual decline in the number of photoreceptor cells has been implicated in the etiology of retinal dystrophy in the RCS rat. Numerous experiments have indicated that this defect has its primary influence on the retinal pigment epithelium (RPE) cells. The RPE cells of a dystrophically altered rat retina are unable to phagocytose discarded photoreceptor outer segments. Studies on experimental chimeras of normal and dystrophic rats show that RPE from dystrophic animals is unable to normally phagocytose outer segment material of any origin, whether from normal photoreceptor cells or from cells deriving from dystrophic animals (Mullen and LaVail, 1976). This further strengthens the hypothesis that the gene defect has a potent impact on the RPE cells.

Degenerative changes in the photoreceptor layer of the dystrophic rat retina, in the form of accumulation of discarded outer segment discs in the subretinal space, become discernable as early as the fourteenth postnatal day. This accumulation may be responsible for the deterioration of the photoreceptor layer due to inhibition of metabolic exchange through creation of a physical barrier between the photoreceptor cells and the choriocapillaris (Herron et al., 1969). The photoreceptor cell nuclei gradually become pycnotic beginning on about day 15 and then diminish rapidly in number until the outer nuclear layer has been virtually eliminated (LaVail, 1979; El-Hifnawi, 1985). A few photoreceptor nuclei persist through the sixth month or even throughout the lifespan of the animal although it is difficult to identify the isolated survivors as rods or cones (LaVail et al., 1974; LaVail and Battelle, 1975).

Factors responsible for the failure of the RPE cells to phagocytose discarded outer segment material have not been completely determined. It has been suggested that a failure of internalization after attachment could be responsible for the phagocytotic disorder. Experiments done by Chaitin and Hall (1983) on cultivated, normal and rdy-RPE cells indicate that attachment takes place, but that some decisive steps leading to internalization fail to occur. In cytochemical studies done by Cohen and Nir (1984), it was noted that, in the normal rat retina, the surface of the rod outer segments was inaccessible to labeling with a colloidal iron marker for sialic acid because it was so closely enveloped by the RPE microvilli. In contrast, the rod outer segments of dystrophic rats could be labeled. This implies that the immediate engulfment of discarded outer segment tips characteristic of the normal rat RPE fails to occur in RCS rats.

Immunohistochemical studies done by McLaughlin et al. (1984) further show that an irregularity is present in the glycoconjugate components of the RPE cell membranes of RCS rats. This unusual surface composition provides one possible explanation for the disruption in phagocytosis of discarded outer segment discs. An additional factor which may play a significant role in altering the RPE cell membrane glycocalyx and the interphotoreceptor cell matrix leading to the phagocytotic disorder lies in the pronounced instability of the lysosomal membrane. Lysosomal instability was described by Burden et al. (1971) using biochemical techniques and substantiated by Ansell and Marshall (1974) in histochemical studies done on dystrophically changed retinas. The "lysosome hypothesis", proposed by Burden et al. (1971), has long been neglected in the aftermath of numerous follow-up studies which lead to varying, contradictory and, therefore, inconclusive results. Recent studies have indicated that this instability of the lysosomal membrane in RPE cells may indeed play a role in the etiology of hereditary retinal dystrophy in RCS rats (El-Hifnawi and Kühnel, 1987). Acid phosphatase infiltration of the subretinal space was detected in animals as early as one week in age. Based on these results and those of Dewar et al. (1975) which show that acetylsalicylic acid (ASA) has a stabilizing effect on the lysosomal membrane in rat retinal tissue in vitro, the present study was undertaken to determine the morphological effects of the administration of ASA on the progress of retinal dystrophy in RCS rats. Furthermore, variation in the survival and condition of the photoreceptor cells and the RPE in both treated and untreated animals was followed with particular attention.

MATERIAL AND METHODS

Three groups of 7 each, one-day-old RCS rats of both sexes were orally administered doses of ASA suspended in tylose at 10 mg, 20 mg and 30 mg q.i.d. respectively according to body weight in kilograms for a period of up to 34 days. An appropriate number of untreated control animals were taken from identical litters, kept in the same cages according to their respective groups and subjected to identical conditions of illumination; i.e. 12 hours of light at a room illumination of 12 to 16 foot-candles alternating with 12 hours of darkness. The treatment of animals in this study conformed to the ARVO resolution on animal use in research. The animals were sacrificed after intervals of approximately 7, 14, 21 and 30 days using an i.p. overdose of phenobarbital (Nembutal 0.1 ml/100 g body weight) with removal of eyes immediately upon death.

Fixation for electron microscopy was done using a solution of 2 % Glutaraldehyde, 0.6 % Paraformaldehyde and 0.03 % $CaCl_2$ in 0.06 M cacodylate buffer (pH 7.3). After washing in the same buffering solution, the specimens were post-fixed in 2 % osmium tetroxide. Oriented semi-thin sections were stained according to Richardson et al. (1960). Ultrathin sections were double stained with uranyl acetate and lead citrate (Reynolds, 1963). The electron microscopic examinations were performed on a TEM Zeiss 9S.

Several tissue samples taken from treated and control groups were prepared for acid phosphatase histochemical studies after immersion in a fixative solution of 2.5 % glutaraldehyde in cacodylate buffer at a pH of 7.4 for 30 minutes at room temperature. Further preparation was completed as described in our previous study (El-Hifnawi and Kühnel, 1987).

In addition, half eyes were embedded in paraffin and serially sectioned for routine light microscopic studies using HE staining.

RESULTS

The retinas of untreated 14-day-old RCS rats exhibit disorganization of the photoreceptor outer segments and accumulation of discarded outer segment debris in the subretinal space. No phagosomes can be detected although the RPE shows no obvious signs of degenerative change (Fig. 1).

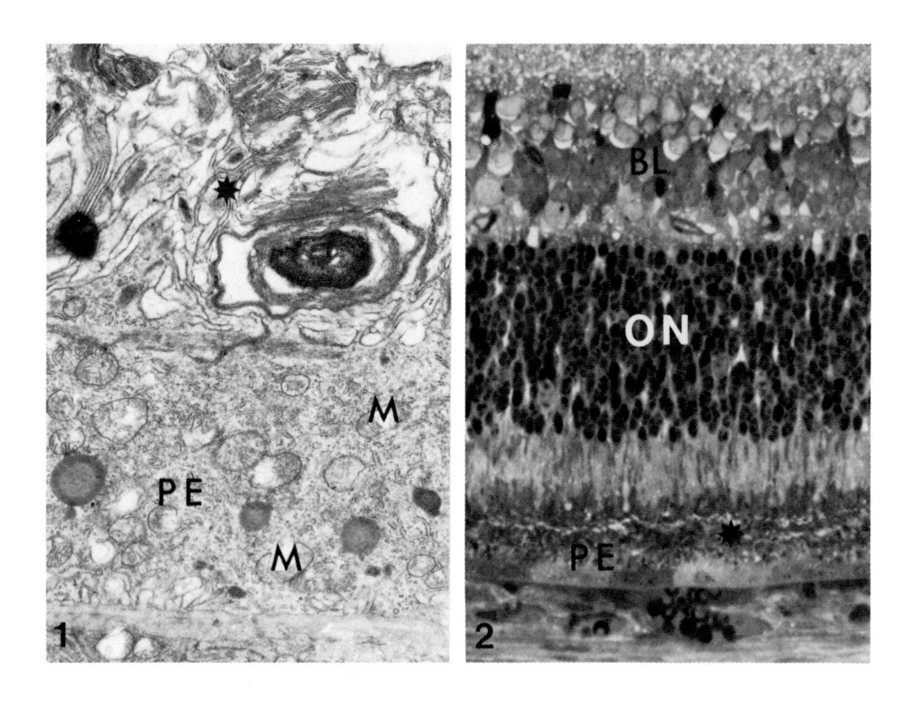

Figures 1 and 2. Micrographs of the retinas of RCS rats of different ages. (Fig. 1 Ultrathin section from a 14-day-old untreated animal; Fig. 2 Light micrograph of an 18-day-old animal treated with 20 mg ASA) Bipolar layer (BL), Discarded outer segment material (*), Outer nuclear layer (ON), Mitochondria (M), Pigment epithelial cell (PE). (Fig. 1, X12,740. Fig.2, X400)

After 18 days, occasional pycnotic nuclei can be found in the photoreceptor nuclear layer of the retinas of untreated animais or animals treated with either 10 or 20 mg ASA q.i.d.. The continuing accumulation of discarded material in the subretinal space, as well as the absence of phagosomes in the RPE cells of these animals, indicates that phagocytosis fails to occur (Figs. 2 and 3).

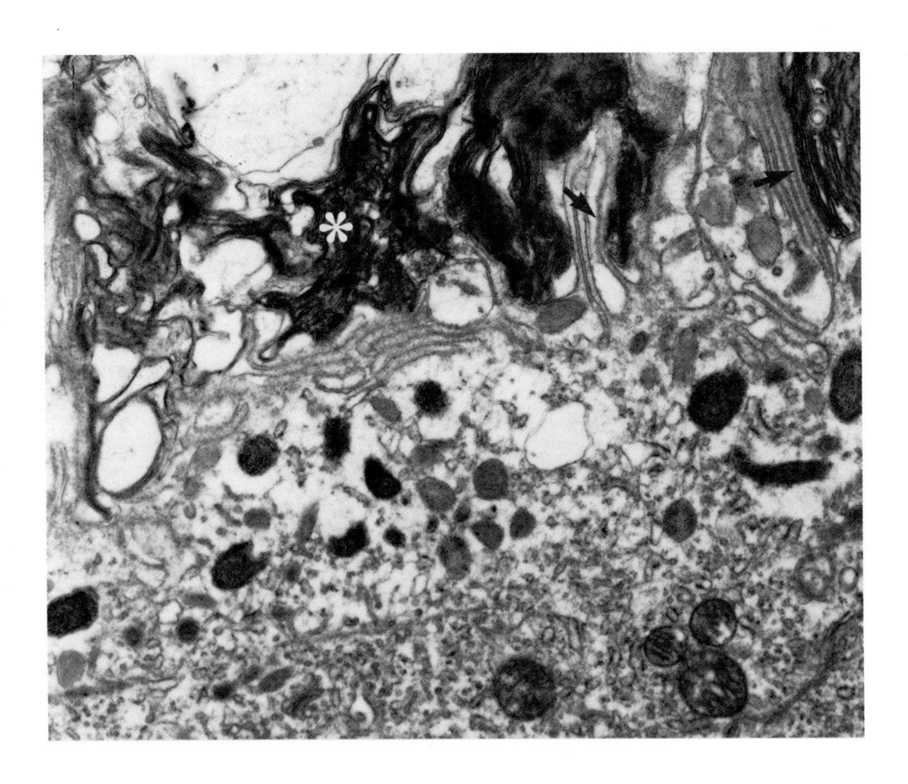

Figure 3. Electron micrograph of the apical RPE of an 18-day-old RCS rat treated with 20 mg ASA. Note the close contact of cytoplasmic processes (arrows) with discarded outer segment material (*). (X15,200)

RCS rat retinas of 31-day-old untreated animals and animals treated with either 10 or 20 mg ASA q.i.d. show no significant differences in the condition of the outer nuclear layer which has been reduced to a minimum; most of

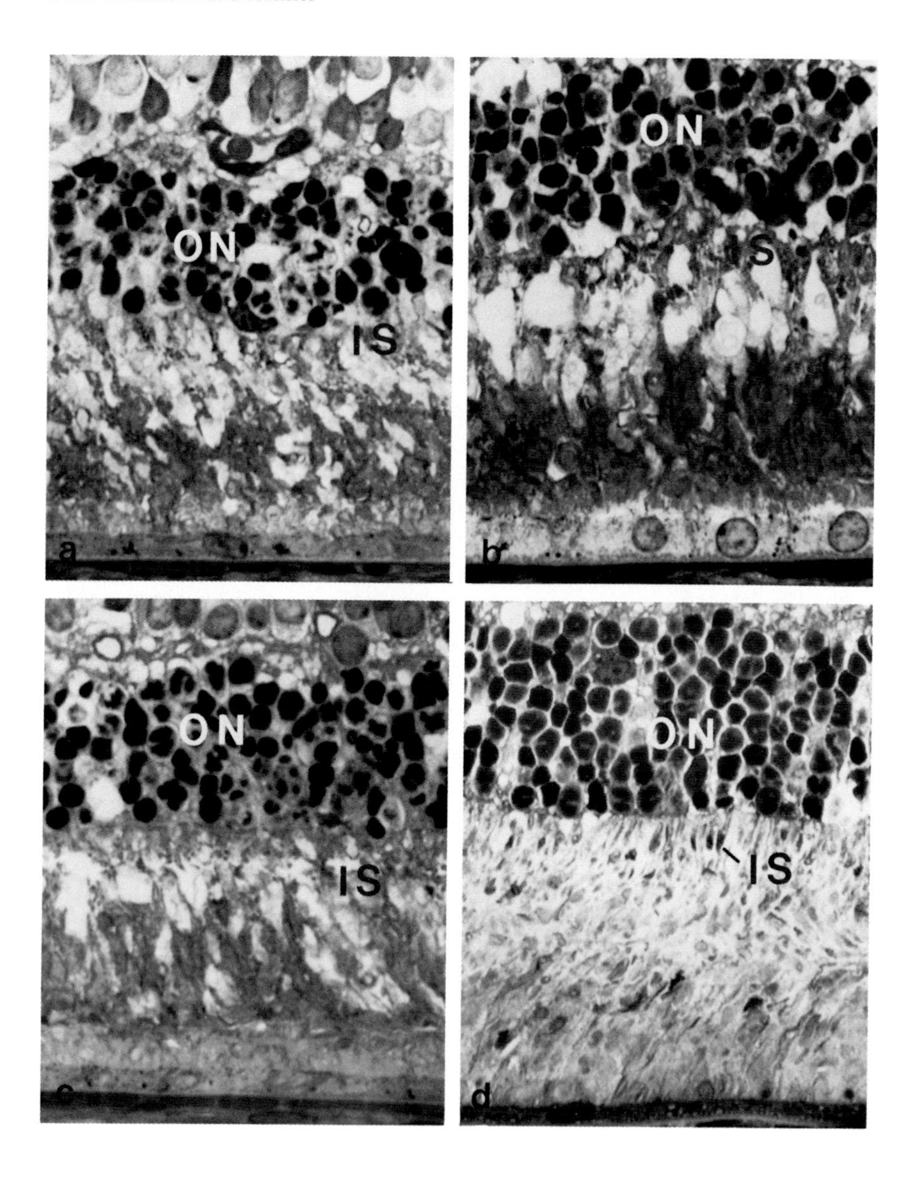

Figures 4 a-d. Light micrographs of retinas from 31-day-old RCS rats at the same magnification. (Fig. a untreated; Fig. b treated with 10 mg; Fig. c treated with 20 mg; Fig. d treated with 30 mg ASA) Note the differences in the condition of the outer nuclear layer (ON) and the inner segments (IS). (X900)

the remaining nuclei are pycnotic (Figs. 4 a-c). In contrast, the photoreceptor nuclear layer in the retinas of 31-day-old animals treated with 30 mg ASA q.i.d. is better conserved, showing infrequent pycnotic nuclei and being noticeably thicker (8-10 rows of nuclei, as opposed to 5-6 rows) (Fig. 4 d). Examination of paraffin serial sections of the entire retina show that these observations are not limited to one particular region, but rather hold true throughout the retina. The inner segments of the photoreceptors in animals treated with 30 mg ASA q.i.d. are clearly in better condition than those of the other groups (Fig. 4 d). The bipolar nuclear layers of animals treated with 20 as well as 30 mg ASA q.i.d. are in better condition than those of either animals treated with 10 mg ASA q.i.d. or control animals.

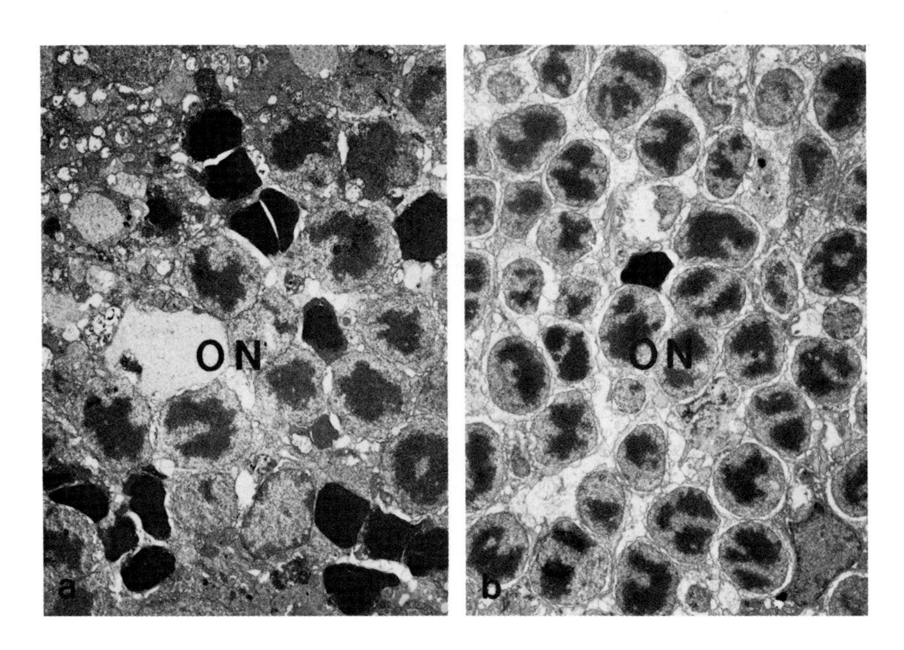

Figures 5 a and b. Electron micrographs of retinas from 31-day-old RCS rats. (Fig. a untreated animal; Fig. b treated with 30 mg ASA) Note the differences in the thickness of the outer nuclear layer (ON) and the frequency of pycnotic nuclei. (Fig. a, X3,400. Fig. b, X3,000)

Electron microscopic examinations of 31-day-old RCS rat retinas of animals, both untreated and treated with 30 mg ASA q.i.d., confirm the previously described light microscopic observations made on the condition of the photoreceptor nuclei (Figs. 5 a and b).

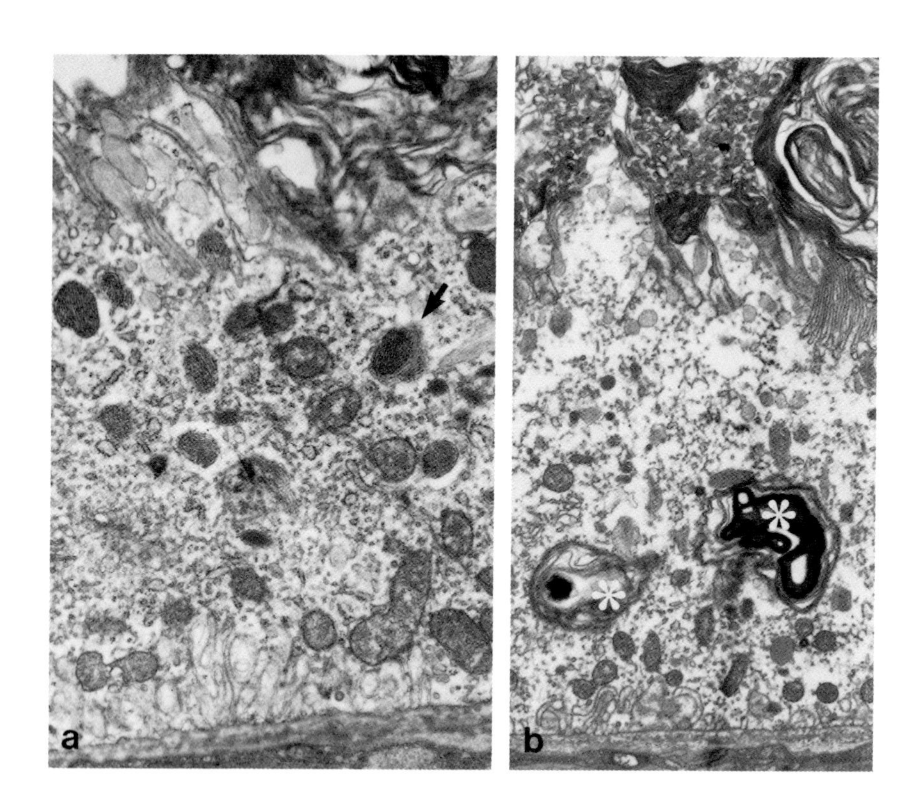

Figures 6 a and b. Electron micrographs of RPE cells from 31-day-old RCS rats treated with 30 mg ASA. Note the presence of phagosomes (arrows). Examination of serial sections confirm the intracellular location of the lamellar whorls (*) in Fig. b. (Fig. a, X11,960. Fig. b, X6,750)

Ultrathin sections of the RPE apical cell surface of retinas from 31-day-old RCS rats, treated with 30 mg ASA q.i.d., reveal an irregular pattern of cytoplasmic processes interspersed with infoldings of varying depths. Discarded outer segment material is found in close proximity to the RPE cells, surrounded by their thin cytoplasmic processes. This material, some of it still retaining lamellar whorl structure, varies in size, form, and electron density. Membrane bound, similarly lamellar and non-lamellar structures of varying electron density can also be detected in the RPE cell cytoplasm, mostly in the apical portion of the cell. In some cases, membranous structures can be found in the central or basal portion of the cells. A review of our serial sections supports the implication that the lamellar material is intracellularly located. Several structures, strongly resembling phagosomes, can also be identified in the perinuclear zone. RPE cell organelles, especially the mitochondria and Golgi apparatus, are unusually well developed for RCS rats and numerous small pinocytotic vesicles can be found primarily in the apical region of the RPE cells (Figs. 6 a and b, Fig. 7).

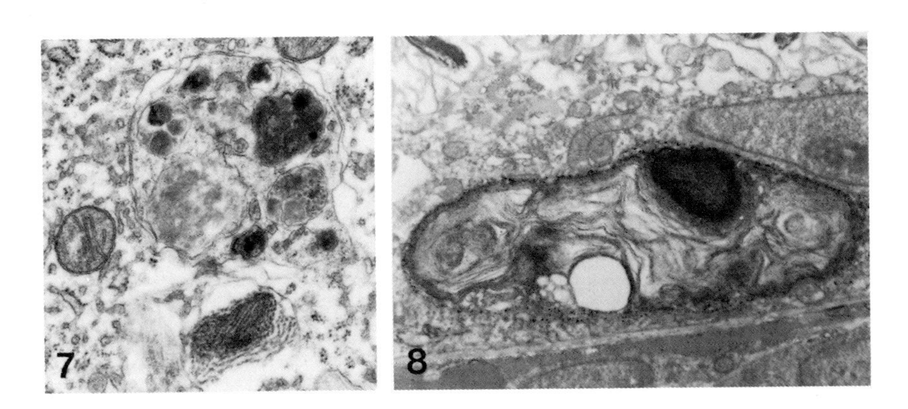

Figures 7 and 8. Electron micrographs of RPE cells from RCS rats treated with 30 mg ASA at different ages. (Fig. 7 31-day-old animal; Fig. 8 unstained section from an 11-day-old animal). Note the phagosome in Fig. 7. The lamellar whorl in Fig. 8 is intracellularly located. Acid phosphatase reaction product is dispersed over the whorl. (Fig. 7, X20,900. Fig. 8, X9,600)

Membrane bound lamellar structures similar to those described above in 31-day-old RCS rats are also found in the basal portion of RPE cells of 11-day-old RCS rats treated with 30 mg ASA q.i.d. As seen in Fig. 8, acid phosphatase activity is observed dispersed throughout the membrane bound material.

DISCUSSION

The present study shows that the photoreceptor nuclear layer after a period of 31 days in RCS rats treated with 30 mg ASA q.i.d. is markedly better conserved than that in untreated animals or animals treated at lower doses. There are fewer pycnotic cell nuclei and the nuclear layer is discernibly thicker in the animals treated with 30 mg ASA q.i.d.. The organization of the outer segment material is not as improved as could have been expected, given the condition of the nuclear layer. The inner segments, however, appear to be in better condition than those of the retinas of animals which are either untreated or have been treated with lower doses of ASA. Most significant, however, is the appearance of acid phosphatase positive membrane bound lamellar structures within the retinal pigment epithelial cells. This material morphologically resembles the discarded outer segment material found in the subretinal space. Phagosomes have never before been reported in the RPE cells of RCS rats (Dowling, 1964; El-Hifnawi, 1985).

Our earlier studies showed that acid phosphatase reaction product is found in abundance in the subretinal space, as well as intercellularly between RPE cells of dystrophic retinas (El-Hifnawi and Kühnel, 1987). In contrast to these previous observations and those made on untreated animals, as well as animals treated with 10 or 20 mg ASA q.i.d., no acid phosphatase reaction product was found outside of the RPE cells of animals treated with 30 mg ASA q.i.d. The increased number of pinocytotic vesicles, and the improvement in the condition of the cell organelles in the RPE cells, as observed in the animals treated with 30 mg ASA q.i.d., indicates an improved cellular activity.

The higher dose of ASA (30 mg q.i.d.) appears to have improved the ability of the RPE cells to phagocytose discarded outer segment material. The absence of such material in the RPE cells of animals treated with lower doses of ASA indicates that the effects of the anti-inflammatory agent are dosage dependent. A dose of 30 mg ASA q.i.d. has been documented to be within the therapeutic range for rats (Strubelt and Zetler, 1980). The continuing disorganization and accumulation of photoreceptor outer segment material in animals treated with 30 mg ASA q.i.d. indicates that the phagocytotic function of the RPE cells has not been sufficiently improved to prevent visual cell degeneration. It may be that the levels of ASA administered are still too low to sufficiently stabilize the lysosomal membrane and increase the phagocytotic activity.

Our results correlate with those of Dewar et al. (1975), who found that in vitro application of ASA improved the stability of rat retinal lysosomes. Based on DNA measurements, they showed that the administration of ASA somewhat slowed the rate of retinal degeneration. The mechanism by which ASA retards the dystrophic process in the retinas of RCS rats remains to be elucidated. It has been well established that ASA acts on the metabolism of arachidonic acid. It is noteworthy that in vitro studies done on cultured RPE cells from RCS rats and a congeneic control group indicate that higher doses of arachidonic acid markedly suppressed the ability of the cells from dystrophic animals to phagocytose carmine particles (Tripathi and Tripathi, 1981). They suggested that the phagocytotic disorder in the RPE cells of RCS rats could be due, at least in part, to an increased sensitivity of the cells to the presence of arachidonic acid.

Further studies are being conducted using immunocytochemical methods in order to determine whether the phagocytosed material within the RPE cells does in fact originate from the outer segments. In addition, morphometric assessments are being carried out in order to quantify the increased survival rate of photoreceptor nuclei found at higher doses of ASA. Other anti-inflammatory agents are also being tested in order to determine whether phagocytotic activity can be increased to a greater extent and whether the photoreceptor cell layer can be maintained in a more intact condition.

REFERENCES

Ansell PL, Marshall J (1974). The distribution of extracellular acid phosphatase in the retinas of retinitis pigmentosa rats. Exp Eye Res 19:273-279.

Burden EM, Yates CM, Reading HW, Bitensky L, Chayen J (1971). Investigation into the structural integrity of lysosomes in the normal and dystrophic rat retina. Exp Eye Res 12:159-165.
Chaitin MH, Hall MO (1983). Defective ingestion of rod outer segments by cultured dystrophic rat pigment epithelial cells. Invest Ophthalmol Vis Sci 24:812-820.
Cohen D, Nir I (1984). Sialic acid on the surface of photoreceptors and pigment epithelium in RCS rats. Invest Ophthalmol Vis Sci 25:1342-1345.
Dewar AJ, Barron G, Reading HW (1975). The effect of retinol and acetylsalicylic acid on the release of lysosomal enzymes from rat retina in vitro. Exp Eye Res 20:63-72.
Dowling JE (1964). Nutritional and inherited blindness in the rat. Exp Eye Res 3:348-356.
El-Hifnawi E (1985). "Pathomorphologische Untersuchungen zum Verlauf der hereditären Netzhaut-Dystrophie bei RCS-Ratten." Stuttgart: Ferdinand Enke Verlag.
El-Hifnawi E, Kühnel W (1987). The role of lysosomes in hereditary retinal dystrophy in RCS rats. In Zrenner E, Krastel H, Goebel H-H (eds): "Research in retinitis pigmentosa," Oxford: Pergamon Journals Ltd., pp 381-395.
Herron WL, Riegel BW, Myers OE, Rubin ML (1969). Retinal dystrophy in the rat - A pigment epithelial disease. Invest Ophthalmol Vis Sci 8:595-604.
LaVail MM (1979). The retinal pigment epithelium in mice and rats with inherited retinal degeneration. In Zinn KM, Marmor MF (eds): "The retinal pigment epithelium," Cambridge, Massachusetts: Harvard University Press, pp 357-380.
LaVail MM, Battelle BA (1975). Influence of eye pigmentation and light deprivation on inherited retinal dystrophy in the rat. Exp Eye Res 21:167-192.
LaVail MM, Sidman M, Rausin R, Sidman RL (1974). Discrimination of light intensity by rats with inherited retinal degeneration: A behavioral and cytological study. Vision Res 14:693-702.
McLaughlin BJ, Boykins LG, Caldwell RB (1984). Lectin-ferritin binding on dystrophic and normal retinal pigment epithelial membranes. J Neurocytol 13:467-480.
Mullen RJ, LaVail MM (1976). Inherited retinal dystrophy: Primary defect in pigment epithelium determined with experimental rat chimeras. Science 192:799-801.

Reynolds ES (1963). The use of lead citrate at high pH as an electron-opaque stain in electron microscopy. J Cell Biol 17:208-212.
Richardson KC, Jarett L, Finke EH (1960). Embedding in epoxy resins for ultrathin sectioning in electron microscopy. Stain Technol 35:313-323.
Strubelt O, Zetler G (1980). Anti-inflammatory effect of ethanol and other alcohols on rat paw edema and pleurisy. Agents Actions 10:279-286.
Tripathi BJ, Tripathi RC (1981). Effect of arachidonic acid on normal and dystrophic retinal pigment epithelium in tissue culture. Invest Ophthalmol Vis Sci 20:553-557.

Inherited and Environmentally Induced
Retinal Degenerations, pages 357–375

ERG OF THE PIGMENTED *RDY* RAT AT ADVANCED STAGES OF HEREDITARY RETINAL DEGENERATION

Werner K. Noell, E. Bradly Pewitt and John R. Cotter.

Department of Ophthalmology, University of Kansas Medical Center, Kansas City, KS 66103

INTRODUCTION

In 1974 LaVail, M. Sidman, Rausin and R.L. Sidman demonstrated that the RCS rat afflicted with hereditary "retinal dystrophy" (*rdy*) could be conditioned to respond to the onset of light up to the age of more than 2 years despite severe photoreceptor cell loss. Histologically, at the age of 254 and 364 days, the degenerative hereditary disease had led to the total disappearance of the rod cell population while scattered surviving cone cells identified by their nucleus lacked the outer segment. The organelles of the inner segment of these remanent cone cells were transposed to the peri-nuclear region. There was a preserved axonal process which made typical synaptic connections with the 2nd order neurons of the retina.

At about the same time electrophysiological work in our laboratory initiated by Dr. Ken Kant (unpublished) revealed that rats afflicted with the same genetic disease of the retina were surprisingly responsive to light flash stimulation when the evoked potentials of the striate area of the cerebral cortex were used to measure retinal function at an age when a substantial fraction of the rod cells had vanished histologically. Stimulated by the report of LaVail et al (1974) our work was greatly extended including not only the responses of the visual cortex but also the study of the receptive field of single retinal ganglion cells and the behavioral discrimination of the orientation of a rectangular bar (Noell et al. 1981, Noell and Salinsky, 1985). In every case retinal function seemed to continue beyond the age of total rod cell loss and despite the changes in the morphology of the surviving cone cells.

The loss of the outer segment membranes of the surviving cone cells in our visually responsive rats was proven by an EM study at the age of 200 days (Cotter and Noell, 1984). We furthermore counted the surviving cone cell nuclei over 40° of the horizontal meridian and up to 20° to either side, from the age of 100 to 500 days using rats studied behaviorally (Noell and Spongr, unpublished). The number of cone cells which in the normal rat comprises 1-2% of all photoreceptors, continuously decreased during this age period, reaching 50% at 250 and 10% at 400 days, the 100 day count set at 100%. This decline seemed to mimic the deterioration in electrophysiological functions and set the limit for the behavioral performance.

The hypothesis emerged that the remnant cone cell lacking the outer segment membranes and destined to degenerate completely, remains a photosensitive cell capable of activating the neuronal retinal network with which it continues to have synaptic connection.

The *rdy*$^{-}$/*rdy*$^{-}$ RCS rats of LaVail are qualitatively identical in the phenomenology of retinal degeneration with the strains of our laboratory (B, BA) denoted here as *rdy* rats. Because of the better optical properties of the darkly pigmented eyes compared to albinos or the "pink-eyed" RCS rats our work on the visual capacities has exclusively used the pigmented *rdy* rat. In order to provide uniformity, all rats were reared in darkness.

METHODS

Animals

The *rdy* rats were hooded, pigmented rats all derived from the same stock of 6 hooded animals, obtained in 1961 through the courtesy of Drs. Katherine Tansley and Clive Graymore from the Institute of Ophthalmology, London. The animals chosen for this study were kept continuously from birth in virtually complete darkness because rearing in the light-deprived condition greatly reduces intra-litter and inter-litter variability. The same breed of animals was used in a variety of other studies. Control animals were of the Long Evans strain, bred for over 10 years in our laboratory under the same conditions as the *rdy* rats.

Anesthesia

The animals were anesthetized by an intraperitoneal injection of 25% (w/v) ethyl urethane, Fisher, 1.2 g/kg body weight. If this dose was insufficient for complete relaxation, 1/3 of it was repeated after 20-30 minutes. Prior to anesthesia, the pupils were maximally dilated by Cyclomydril. Following the induction of general anesthesia and maximal dilatation, the tissue around the eyes was infiltrated with 2% Xylocaine which was also applied to the cornea.

Preparation

Using adhesive paper tape, the upper and lower lids were retracted. Another piece of tape was wrapped around the head so that the eyes protruded slightly through apertures in the tape. The animal was positioned on a heavy metal platform covered with a thick layer of cotton. Fixation of the head was achieved by clamping the upper incisors as in a stereotaxic instrument. The animal was covered with a blanket of cotton. Colonic temperature was continuously monitored and not permitted to fall under 36°C during the experimental procedure. The corneas were inspected under red light at 15 minute intervals and a drop of Xylocaine or Ultratear was applied.

Electrodes

The ERG was recorded from the left eye by a thin cotton strand of a wick electrode touching the center of the cornea. The cotton wick was wrapped by a coil of platinum wire surrounded by an opaque plastic tube filled with saline solution. Only the cotton wick protruded out of the tube. Another electrode was placed on the right cornea but was surrounded by a funnel of black plastic taped to the skin around the right eye, shielding it from light stimulation. Needle electrodes were inserted tangentially over the supraorbital bone piercing its periosteum. The ground electrode was a needle inserted into the midline neck tissue caudal to the occipital bone.

The ERG was routinely recorded from the left cornea in electrical reference to the light-shielded right cornea. In addition, the protocol required recording from both corneas separately in reference to the supraorbital needle electrode over the same eye.

Stimulation

A Grass PS2 Photostimulator was the source of the 10 usec xenon flashes, used at the maximal intensity setting of "16". An 8 mm diameter, one meter long fiber optic was mounted to the front of the parabolic flash unit with an attachment for inserting, at this point, neutral density filters. The other end of the fiber optic was placed directly in front of the left eye at a distance of 1 cm from the cornea. In order to minimize audible clicks, produced simultaneously with the flashes, the flash unit was in a sound attenuating box; and the electrical unit of the photostimulator was on the other side of a wall 8 feet away from the preparation. For the human ear no flash related clicks could be detected.

Stimulation and recording equipment was triggered by a Grass SD5 pulse stimulator. The flash rate was .8/s for routine testing. In order to study approximate steady state conditions, the first 5 flashes or first 5 seconds of high frequency stimulation were not used for averaging.

Amplification, Recording

The electrodes were connected to a Grass P-15 amplifier set at an amplification of 100, bandwidth 0.3 to 1000 Hz. The output of the pre-amplifier was connected to one channel of a Dual Beam, Tektronics #RM-502A oscilloscope at 20 or 10 mv/cm through its DC input. The vertical amplifier output from this oscilloscope was fed directly into a CAT 400C, set routinely at 250 msec sweep time using all 400 points. Recordings from the CAT were made with a Houston Instrument Corporation, model Hr-95 X-Y recorder.

Animal, fiber-optic, pre-amplifier and temperature monitor were enclosed in an electrically shielded, light-proofed box; and the laboratory was dimly illuminated. All animal preparations were made in dim light. A 10 to 20 minute period of no stimulation preceded every phase of the protocol.

ERG Measurements

Amplitude and latency were directly measured from the 25 x 18 cm recorder plot. Total, "absolute" amplitude of a wave complex was measured from the lowest negative to the highest positive point. "Latency" refers to the measurement of the time from stimulus application to the start of a particular deflection.

Quantitation of the State of Retinal Degeneration

This report combines 2 studies, separated by about 3 years, using off-spring of the same breeding stock. During the first study the DNA content of the carefully separated whole retina was used to assess the over-all loss of the rod cells in comparison with simultaneously measured 2-4 adult control retinas of Long Evans rats, 140-250 days old. The average of the control DNA of the same assay run was set at 100%. During the second study, the rat's head was perfused with fixative after an overdose of urethane; black cotton threads were inserted into the limbal sclera as markers for the vertical and temporal meridian; and the eye was dissected and prepared for histology after paraffin or plastic embedding. The eyes were serially sectioned in the sagittal plane. ERG procedures were the same in both studies except that the ERG reference electrode was in the ipsilateral supraorbital tissue and that the routine flash repetition rate was 0.2/s. The ERG measurements of the first study were performed by E.B.P., those of the second one by J.R.C.

DNA Measurement

The DNA measuring technique was the following: At the end of the ERG measurements the animal was killed rapidly by chloroform in a desiccator. The eye was opened, lens and vitreous removed, and the retina very gently moved out of the eye by a fine forceps starting at the posterior pole. The retina was placed in 1 ml of DNA-buffer, consisting of 2 M NaCl, 0.5 M sodium phosphate buffer. The tissue was sonicated 3 times for 10 seconds with intermittent cooling. 20 ul of the homogenate was pipetted to 980 ul of the fluorochromic solution, which contained 1 ug/ml of Hoechst #33258 fluorochrome in DNA buffer. After vortexing, 120-200 ul were placed into round quartz cuvettes. Fluorescence was measured in a Farrand Spectrofluorometer, excitation at 360 nm, emission at 450 nm. Photomultiplier current was recorded on continuously running chart paper with the entrance slit open for 1/2 minute. All measurements were performed in duplicate.

RESULTS

On the basis of previous experiences during studies of the visual capacities of the *rdy* rat after rod cell loss measuring visually evoked responses of the striate area, activities of optic tract axons and behavioral form orientation discrimination (Noell et al. 1981, Noell and Salinsky 1985), it was not surprising to find that small light-flash stimulated electrical responses were recordable at the cornea using conventional ERG techniques. Consistent with the severe changes in retinal histology expressed by the disappearance of the rod cells throughout the retina and also by the loss of the outer segment of the surviving "remnant" cone cells, these ERGs differed not only in amplitude but also in form and temporal characteristics from those of the normal rat. The most characteristic waveform manifesting the retinal responsiveness at such an advanced stage of degeneration is illustrated in Figure 1.

At the age of 286 days (Figure 1, left side) the response is a biphasic negative-positive transient of about 100 msec duration. A slow positive potential lasting longer than 250 msec is present and becomes manifest at different time points of the return to baseline of the positive transient. The total amplitude of the negative-positive transient is 38 uV. The initial negative deflection is small and proved to be more

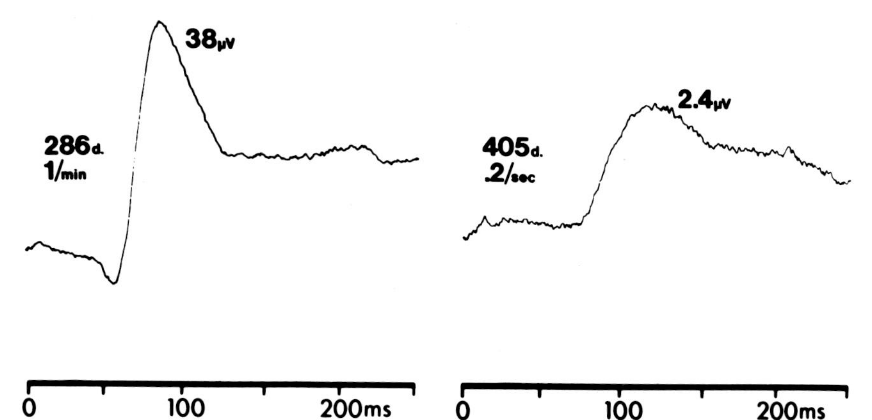

FIGURE 1. The ERG of 2 pigmented *rdy* rats recorded at the age of 286 days (left side) and 405 days (right side). Flash intensity is the maximal one (log I=0). Flash repetition is at 1 per minute (left side) and 0.2 per second (right side). Total number of repetitions is adjusted to give readable records of the write-out. The amplitude of each response is given in microvolts (38, left side, and 2.4, right side). Sweep time in this figure and the following ones is always 250 msec.

variable than the larger positive component. Characteristically, this biphasic transient always appeared with a long latency compared to the normal ERG. In the case illustrated, the latency measured from flash to response was almost 50 msec.

For a 405 day old rat (Figure 1, right side) the latency of the response was about 75 msec. The negative component of the transient is absent and the positive deflection measures only 2.4 uV. It is broader than at the younger age but the slow positive potential is still discernable. The age at which responses from the cornea had become unrecordable was near 450 days.

Figure 2 gives a survey of the *rdy* ERGs recorded from rats ranging in age from 76 to 428 days. The flash intensity is the maximal one with a repetition rate of 0.8/s. From 76 to 105 days the ERG amplitude declined from 234 to 89 uV. The amplitude was 28 uV at 151, 14 uV

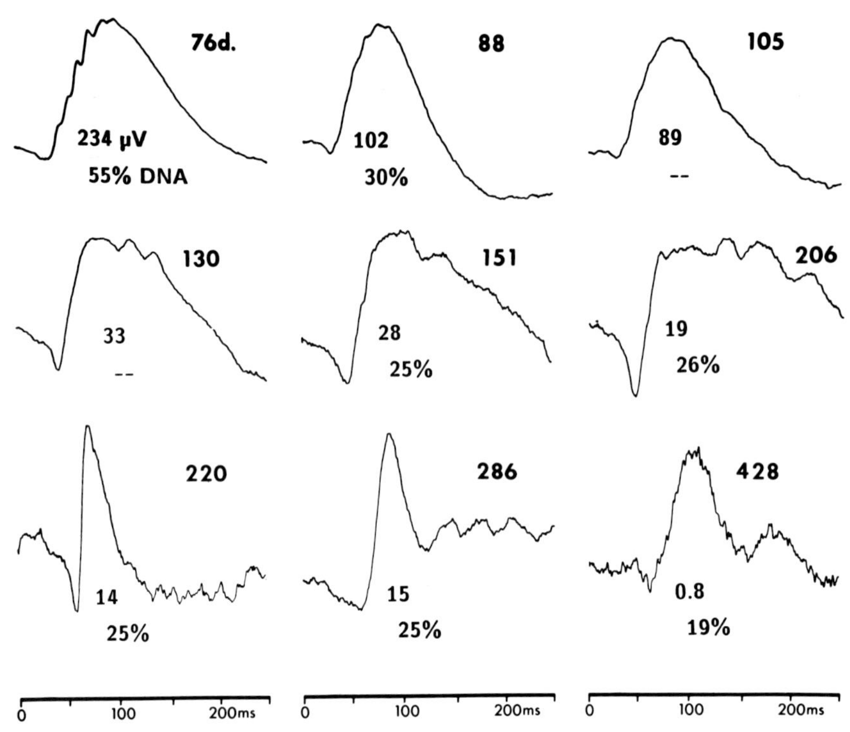

FIGURE 2. The ERG of 9 *rdy* rats recorded at different ages as indicated (76-428 days). Flash intensity is the maximal one. Flash frequency is 0.8/s except for the record from the 428 day rat where it is 0.2/sec. The numbers beneath each tracing give the measurement for the peak amplitude in microvolts and the DNA content of the whole retina of the recorded eyes expressed in % of the retinal DNA of adult control eyes (Long Evans rats) simultaneously measured. The low DNA content at 428 days suggests cell loss within the inner layers.

at 220 and 0.8 at 428 days.

In order to quantitate the disappearance of rod cells, the DNA content of the whole retina was measured in comparison to the DNA of the normal adult Long Evans rat set at 100%. As indicated, retinal DNA measured 55% for the 76 day old *rdy* rat; it was 30% at 88 days, and 25% at 151, 220 and 286 days. Previous measurements after the virtually complete loss of the rod cells by exposure to extensively damaging light (cf. Noell et al. 1987) always gave retinal DNA values between 25 and 28%. We conclude that in the pigmented *rdy* rats of the present study rod cell loss was definitely complete at 200 days, consistent with previous histological studies (Cotter and Noell, 1984).

In Figure 2, the ERGs resemble the normal one up to the age of 105 days. They have a b-wave and wavelets. However, starting with 76 days, a small negative deflection, which is not a-wave like, precedes the cornea positive response. This negative potential increases in relative size with age until the biphasic negative-positive transient of Figure 1 becomes evident at 220 days. Characteristically repetitive positive waves follow the initial deflections.

Quantitative evaluation of the *rdy* ERG is illustrated in Figure 3 for the latency of the response (left) and the threshold (right). The latency measurements give 2 examples for 151 day old rats of different litters. One of these (#14) had much higher ERGs than the other (#21) illustrated in Figure 2. The DNA measurement of the tested eye of #14 gave a value of 38%, while it was 25% for #21. Latency measurement vs log flash intensity gave for #14 a curve close to that of the 76 day old rat while for #21 the curve was close to the one from the 286 day rat, which like #21 had a DNA content of 25%.

Figure 3 suggests that latency is almost twice as long at the age between 200 and 300 days, when the remnant cone cells have lost the outer segment, compared to younger animals with still substantial rod cell survival and presumably better preserved cone cells. Using a criterion response of 3 uV, the threshold of rats 200-300 days old is 1.4 log units higher than at the age of 100 days. Compared to the normal controls this is an increase of 5.2 log units. Following the age of 300 days there is rapid deterioration in all measurements.

Repetitive flash stimulation proved to be a simple procedure for assessing the progress of retinal degeneration in the *rdy* rat. Figure 4 illustrates responses evoked by 8/s stimulation for the age range from 88 to 286 days. The responses are low in amplitude

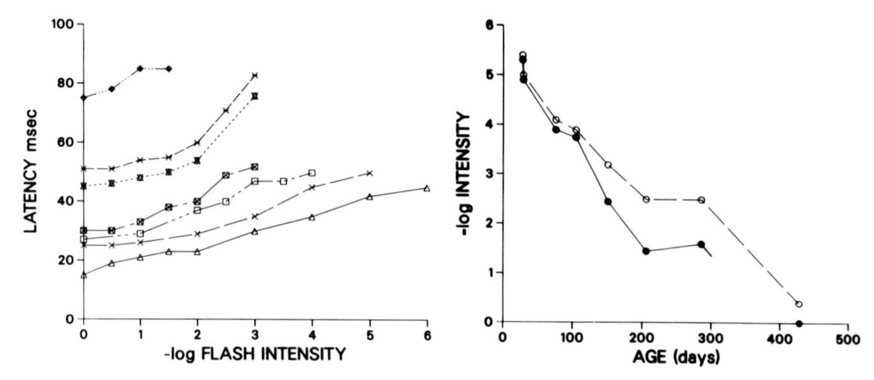

FIGURE 3. <u>Left Side</u>: The latency of the cornea positive transient measured from flash to start of the upswing plotted against -log flash intensity using the records from individual rats. The curves from *rdy* rats are of the following ages in from top downward: 428, 286, 151 (#21), 151 (#14), 76 and 28. The bottom curve is from an adult Long Evans rat. Flash frequency is 0.8/s for all rats except the 428 day old one where it is 0.2/s. <u>Right side</u>: ERG threshold (-log flash intensity) plotted against age for the same rats as on <u>left side</u>. The plot at 151 days is from rat #21. The 2 youngest rats were 28 and 29 days old. The criterion threshold responses were 7.5 (filled circle) and 3 microvolts (open circle).

compared to the normal control, about 15% of normal at the age of 88 days (not shown). They occur after a long time interval from the stimulating flash consistent with the increase in the response latency observed at 0.8/s.

At stimulation frequencies higher than 8/s the waveform of the normal and especially of the *rdy* rat becomes more complex than illustrated in Figure 4. This will not be discussed here. Waves of regularly alternating amplitude occurred at 45/s in the normal and at 22/s in the 100 day old *rdy* rat. For quantitation, we have neglected these variations and always measured the highest response of 2 or 4 successive responses.

In Figure 5 the log amplitude of the individual responses measured from the lowest preceding point to the

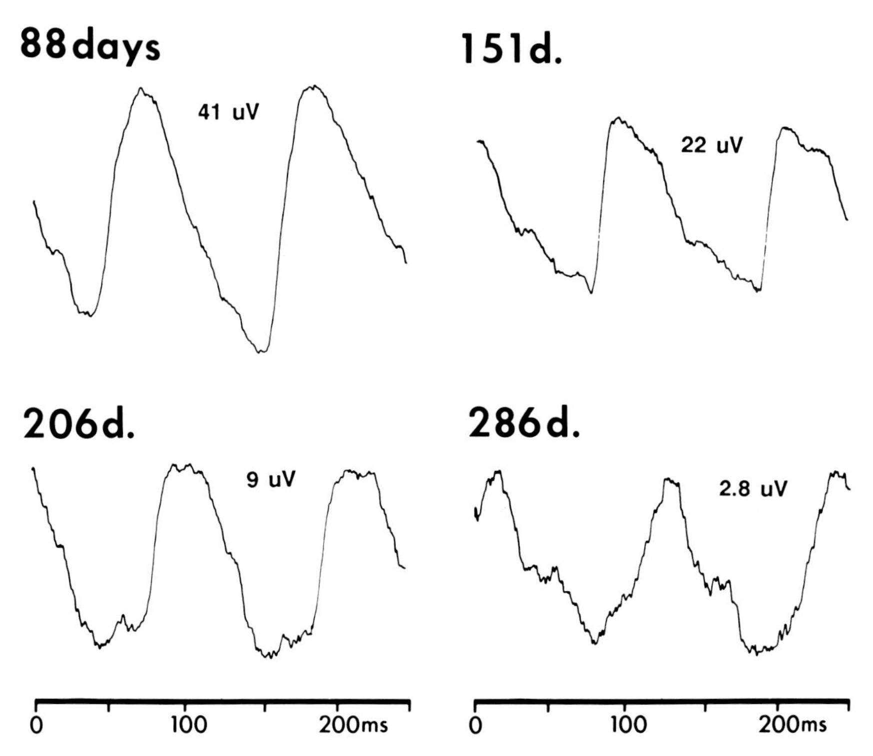

FIGURE 4. ERGs elicited with 8/s flashes of maximal flash intensity at the *rdy* ages as indicated. The average amplitude (total deflection) of each response is given in microvolts. The ERGs of the same rats with a stimulus frequency of 0.8/s are shown in Figure 2.

peak of the positive deflection is plotted against flash frequency. It seems that 3 age groups can be distinguished: One group, up to the age of about 100 days, where responses to repetitive stimulation follow the normal curve; another group around 200 days where the responses rapidly deteriorate at a flash frequency higher than 8/s; and a third group around 300 days and older showing the deterioration at less than 8/s. This is most strikingly seen with the oldest rat of 428 days where every increase in frequency above 1/min is associated with response deterioration so that responses are almost unmeasurable at about 0.8/s.

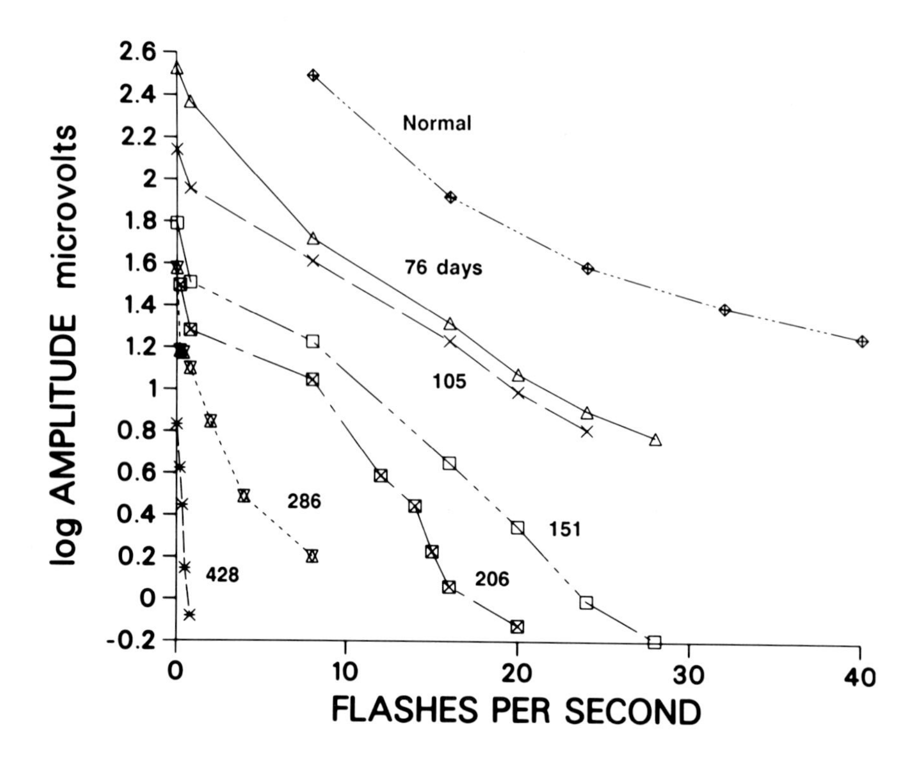

FIGURE 5. Response amplitude (log microvolt) of *rdy* rats plotted against flash frequency at maximal flash intensity. The ages in days are as indicated. The top curve is from a normal Long Evans rat. The amplitude for 1/min. flashes is plotted at the zero vertical.

With about a 3 year interval the study of the pigmented *rdy* rat was repeated using off-spring of the same breeding stock as studied initially. The goal of this 2nd study was to combine the electrophysiological measurements with retinal morphometry of the tested eyes dissected the same day. Procedures and instrumentation for ERG measurements and flash stimulation were the same as in the first study except that the standard flash rate was lowered to 0.2/s from 0.8/s.

Figure 6 gives examples of the ERG recordings in response to a maximal intensity stimulus starting with the ERG of a control Long Evans rat, 136 days old. The *rdy* age ranged from 72 to 362 days. Results were the same as described in the preceding section except that the amplitudes were higher because of the lower flash repetition rate. The positive wave also tended to be broader at around 100 days (see Figure 6B). The negative-positive transient characterizing the *rdy* ERG between 200 and 300 days was expressed during the second study (see Figure 6E) as clearly as during the first.

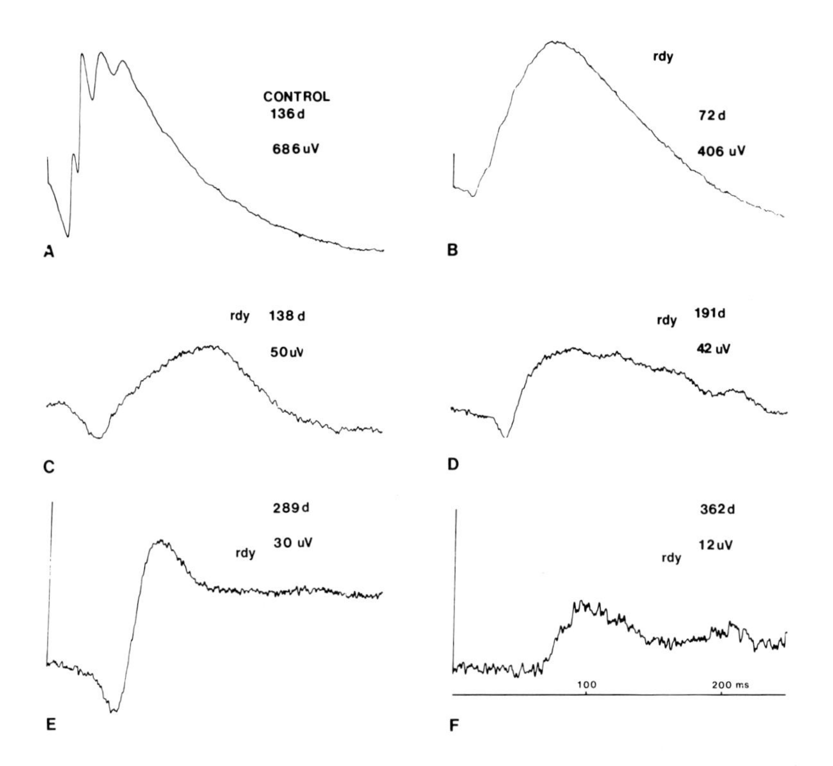

FIGURE 6. ERGs in response to maximal flash intensity recorded during the 2nd study. Flash frequency is 0.2/s

Quantitative ERG data of the second study are illustrated in Figure 7 which shows plots of amplitude (left side) and latency (right side) vs log flash intensity. Because of the higher amplitudes in the second study threshold responses were measurable at lower stimulus intensities.

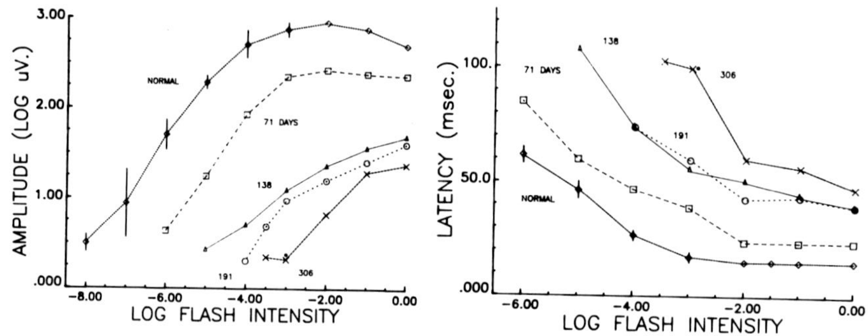

FIGURE 7. Left side: Plots of amplitude vs log flash intensity at the indicated ages of individual *rdy* rats. Control measurements are the means with standard error from 4 Long Evans 75, 99, 136 and 233 days old. Flash rate is 0.2/s except 0.8/s for one point at 306 days marked by a dot. Right side: Latency in milliseconds of the positive wave vs log flash intensity from the same animals as on left side.

The eyes were serially sectioned in the sagittal plane for histology. As expected from the previous study of the retinotopic preservation of the responses of the visual cortex (Noell and Salinsky, 1985), the degeneration process did not proceed with age along the same radial gradient throughout the retina. This is illustrated by the 2 maps of Figure 8 for the age of 72 and 139 days. These maps show the retinal distribution of rod cell disappearance as measured by the width of the outer nuclear layer.

For the retina of the 72 day old *rdy* rat the greatest and most extensive rod cell loss was localized in the periphery of the inferior retina in addition to a small superior area near the zero sagittal plane (Figure 8, left). Rod cell loss was the least pronounced in the nasal periphery and in irregularly shaped areas of the superior retina. It is easy to imagine that with increasing age of this animal the irregular non-uniform

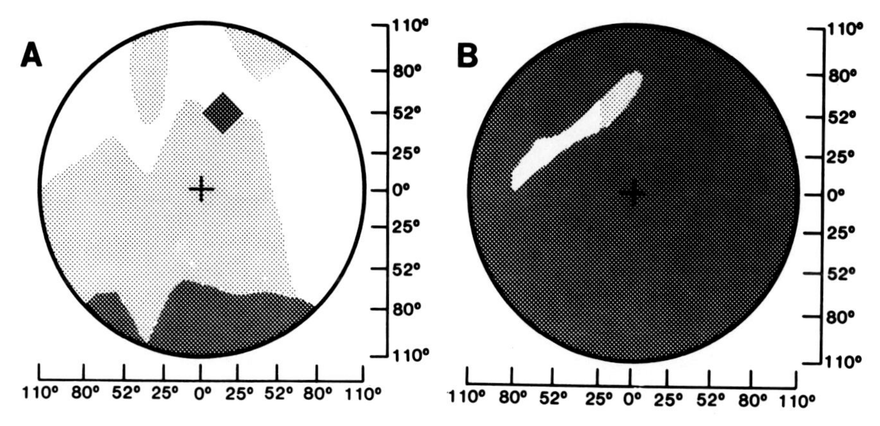

FIGURE 8. Retinal maps reconstructed from sagittal sections of individual eyes of *rdy* rats, 72 (left side) and 138 (right side) days old. The temporal retina is towards the left. The thickness of the outer nuclear layer was measured and is indicated by white, 12-23 um; light stipple, 6-12 um; and dark stipple, less than 6 um. A simultaneously studied retina of a 196 day old *rdy* rat (see Figure 9) showed uniformity in rod cell loss throughout the retina.

distribution of rod cell degeneration would continue to persist to a decreasing degree until the rod cells of every retinal region had disappeared.

The map of the 138 day old *rdy* rat shows the best preservation of the rod cells in a narrow band running through the superior temporal region at about the same distance from the ora serrata while in all other retinal regions the outer nuclear layer has virtually disappeared (Figure 8, right). The same superior temporal region as in the 138 day old rat was consistently better preserved up to the age of about 150 days in several rats.

In Figure 9 photomicrographs of the temporal retina illustrate the histological appearance of the outer retina at the ages of 72, 138 and 196 days comparing the superior temporal region (left panel) with the inferior one in the same sagittal plane at the same distance from the

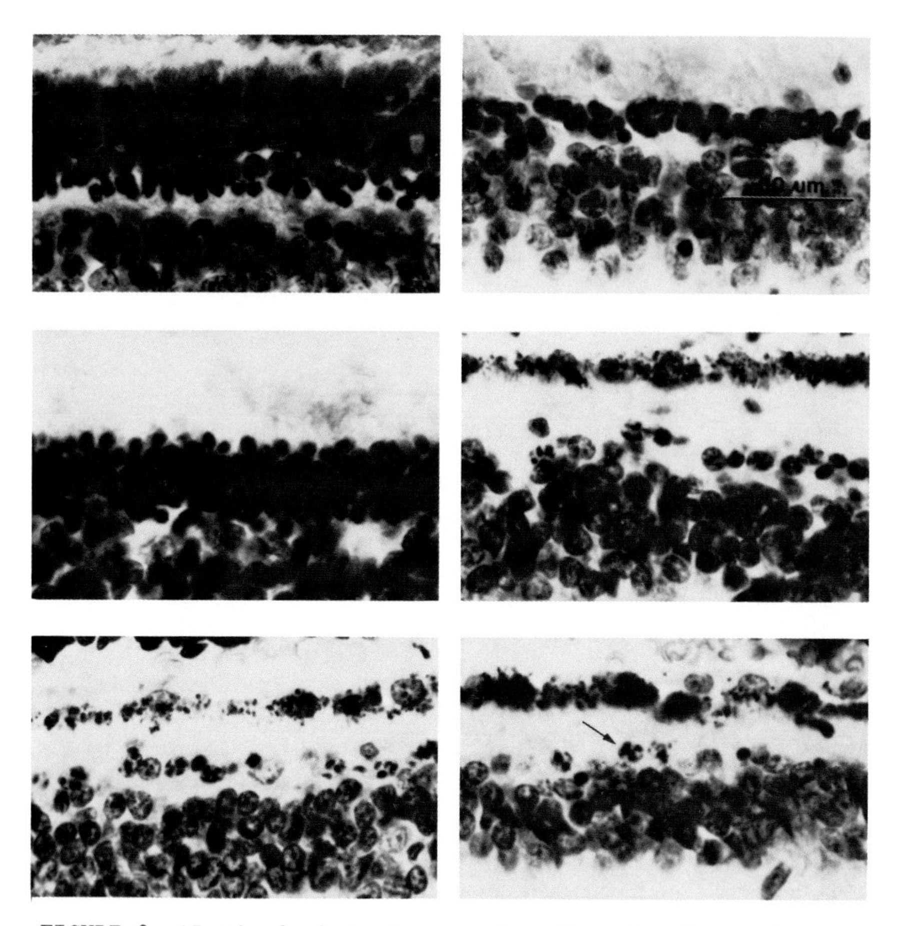

FIGURE 9. Retinal photomicrographs of sagittal sections through the eyes of 3 *rdy* rats 72, 138 and 196 days old, from top to bottom. The photographs are from the superior (left side) and inferior temporal (right side) retina at the same distance from the horizontal meridian. The nucleus of a remnant cone cell is marked by an arrow in the lowest panel on the right.

horizontal meridian (right panel). The differences in outer nuclear layer width are very impressive at 72 and 138 days. However, at the age of 196 days (bottom row) the differences have almost vanished. The same retinal regions then show only a few pyknotic rod nuclei and the nuclei of the remnant cone cells (see bottom right panel - arrow). Only exacting counts of these nuclei would permit a statement on the relative preservation of the remnant cone cells. However one would expect that regional differences in rod cell degeneration are associated with similar differences in cone cell survival.

DISCUSSION

One may first ask whether the small potentials recorded from the cornea in response to the light flash were truly of retinal origin as implied in describing the results. The answer is yes for the following reasons. The pick-up of extraocular potentials was minimized if not eliminated by recording from the cornea of both eyes, the one eye being exposed to the flash emerging from the fiber guide in front of the eye, the other eye being light-shielded. The anatomy of the rat's head is such that both corneas should be on an iso-potential line for current produced within the brain.

The major concern was the electrical spread of responses generated in the visual cortex and optic midbrain. These responses are higher in amplitude than the corneal potentials (see Noell and Salinsky, 1985). Their major source is contralateral to the stimulated eye; but, when measured individually, the light-shielded (contralateral) cornea was electrically silent. Auditory brain responses of *rdy* rats below and above the threshold of human perception were extensively studied in our laboratory, but they had clearly much shorter latencies than simultaneously recorded visually evoked responses of the striate area. Furthermore, the corneal potentials were responsive as the normal ERG to changes in light intensity by inserting a thin filter into the light path. The potentials were very small when the pupil was narrow prior to the application of the mydriatic.

The second question then concerns the site of origin of the potentials within the retina. Two statements can be made. The positive corneal potential up to the age of about 150 days is b-wave like in appearance. This would suggest that any contribution of the input from the surviving cone cells to the generation of this potential is by the same mechanism as with the rod input.

The other statement concerns the positive transient which appears when all rod cells have vanished and when the number of cone nuclei is much reduced. It is proposed that the source of this transient is located within the inner layers. The transient is sharp and brief, and seems to be repetitive during the 250 ms sweep. This would not be expected of a b-wave like process.

We leave open the question regarding the origin of the slow positive wave which follows the transient (see Hardten and Noell, this volume). Occasionally we also observed a slow cornea negative wave (not illustrated). It was most prominent in a rat 130 days old. It required relatively weak flashes for appearance and seemed occluded with 0.8/s stimulation at maximal intensity.

Finally we ask what abnormalities, if any, in the responsiveness of the remnant cone cells did this ERG study reveal? Three factors must be considered prior to any conclusion on the function of the remnant cone cells: (1) the decrease in remnant cone cell number which is a function of age; (2) changes in the anatomy of the layers especially with regard to the Muller cell; (3) "plastic" physiological changes in the neuronal network in compensation for the decrease in photoreceptor input.

Significant and consistent ERG abnormalities in addition to amplitude decline were the following: (a) the increase in latency, (b) the decrease in sensitivity, and (c) the deterioration in responsiveness to repetitive flash stimulation. Of these, (a) and (b) must be significantly dependent on the number of the surviving cone nuclei; but it is reasonable to assume that the increase in latency and threshold is also to some degree a manifestation of a continuously increasing abnormality of the remnant cone cells.

The one abnormality which one would like to assign exclusively to the degenerating remnant cone cells is the progressive failure to respond to repetitive stimulation at the age above 150 days. It may relate to a decreasing efficiency of energy production and/or protein synthesis.

In analogy with denervation pathology in peripheral and central systems "plastic" changes in the inner layers that may be expressed only physiologically would not be surprising in the *rdy* rat at an advanced stage of degeneration. The prominence of the positive transient may possibly be a manifestation of hyperexcitability.

Finally, it is emphasized that a low-voltage ERG <u>is</u> generated by a normally rod dominated retina during its degeneration until surviving cone cells, having lost the outer segment membranes, are reduced in number to a very small fraction of the normal population.

REFERENCES

Cotter JR, Noell WK (1984). Ultrastructure of remnant photoreceptors in advanced hereditary retinal degeneration. Invest. Ophthal. Vis. Sci. 25:1366-1375.

LaVail MM, Sidman M, Rausin R, Sidman RL (1974). Discrimination of light intensity by rats with inherited retinal degeneration: A behavioral and cytological study. Vis Res 14:693-702.

Noell WK, Salinsky MC, Stockton RA, Schnitzer SB, Kan V (1981). Electrophysiological studies of the visual capacities at advanced stages of photoreceptor degeneration in the rat. Doc. Ophthal. Proc. Ser. 27:175-181.

Noell WK, Salinsky MC (1985). The preservation of visual evoked cortical responses at an advanced state of retinal degeneration in the *rdy* rat. In La Vail MM, Hollyfield JG, Anderson RE (eds): "Retinal degeneration", Alan R. Liss, pp 301-320.

Noell WK, Organisciak DT, Ando H, Braniecki MA, Durlin C (1987). Ascorbate and dietary protective mechanisms in retinal light damage of rats: Electrophysiological, histological and DNA measurements. In LaVail, MM, Hollyfield JG, Anderson RE (eds): "Degenerative Retinal Disorders", Alan R. Liss, pp 469-483.

ACKNOWLEDGMENTS

The study was supported by NIH grant R01 EY 04249, Kansas Lions Sight Foundation, and Research for Prevention of Blindness. We are grateful for the invaluable assistance of Dr. R.S. Crockett, especially in the preparation of the manuscript for printing. We thank Mr. Asher Ertel for the photographic work and his pleasant, unfailing association. We thank Dr. Marylee A. Braniecki for expert retinal dissections and DNA determinations.

Inherited and Environmentally Induced
Retinal Degenerations, pages 377–391

ERG OF THE ALBINO *RDY* RAT AND SUSCEPTIBILITY TO LIGHT DAMAGE.

David R. Hardten and Werner K. Noell

Department of Ophthalmology,
University of Kansas Medical Center, Kansas
City, KS. 66103

INTRODUCTION

The first studies of the ERG in the hereditary retinal degeneration of the rat (*rdy*) were reported for the RCS rat by Dowling and Sidman, 1962, and for a pigmented *rdy* strain by Noell, 1963, in conjunction with histology and rhodopsin measurements. In both studies the single flash ERG disappeared in parallel with the disappearance of the photoreceptor nuclei at the age of 60 to 120 days. The assumption that the ERG was extinct at older ages, however, proved to be incorrect when the availability of averaging techniques extended the recordable range of low-voltage responses. As shown in the accompanying paper by Noell, Pewitt and Cotter, this volume, ERGs of characteristic wave form can be elicited when all rod cells have disappeared and when retinal photosensitivity is provided by remnant cone cells which have lost their outer segment membranes (LaVail et al. 1974; Cicerone et al. 1979; Cotter and Noell 1984). The disappearance of these low-voltage ERGs is a function of the decrease in number of the remnant cone cell nuclei and parallels the disappearance of visually evoked responses of the striate area of the visual cortex (Noell and Salinsky, 1985). We will here report that the same characteristic ERGs are generated by the *rdy-/rdy-* albino rat at an advanced stage of the hereditary degeneration.

A separate study of the albino *rdy* rat was dictated by the aim to measure the susceptibility of the remnant cone cells to constant strong light known to destroy the photoreceptors of the normal rat (Noell et al., 1966,

Noell, 1980). Albino rats are preferred to pigmented ones for the study of retinal damage by visible light simply for optical consideration of retinal light irradiation.

Dowling and Sidman (1962) demonstrated that the progress of the hereditary degeneration is greatly slowed when the RCS rats are maintained in continuous darkness instead of light. Noell (1974) extended the study of the effects of light on the progress of the hereditary degeneration by exposing young dark-reared albino *rdy* rats to either cyclic (12 hours per day) or constant light. Weak cyclic light of as little as 2 ft. cdl. for 8 days accelerated step wise and irreversibly the decline of the ERG with age. Constant 150 ft. cdl. light for 10 minutes to 24 hours was effective in destroying acutely a large fraction of the surviving rod cell population to a much greater extent than observed in the normal controls. Massive rod cell loss in a retinal region was associated with the destruction of the pigment epithelium as in the light damage of the controls. Very significantly, however, this destruction of the pigment epithelium did not occur when the *rdy* rats were exposed to constant light at the age of advanced retinal degeneration characterized by the loss of the rod cells.

We asked whether or not the degeneration of the remnant cone cells with age would be accelerated in an irreversible way when the *rdy* rat was exposed to constant strong light at an advanced stage of the hereditary degeneration. We found that no such effect of strong constant light was produced.

METHODS

Animals

The *rdy-/rdy-* rats (*rdy*) were from an albino strain ("BA") bred in our laboratory for over 20 years. The strain was derived from crossings of Sprague-Dawley rats (Charles River Laboratories, Inc.) and the off-spring of six pigmented *rdy* rats obtained in 1961 from the Institute of Ophthalmology, London. The control albino strain originated from the Sprague-Dawley rats used for establishing the albino *rdy* strain. This normal strain is referred to by "CD". All animals were reared in darkness from birth.

ERG Study

All procedures including general and local anesthesia were the same as described in the companion paper on the ERG of the pigmented rat (Noell et al, this volume). The same laboratory set-up and the same instruments were used. The ERG was recorded by cotton wick electrode from the center of the left cornea in reference to a needle electrode in the upper lid. Noell et al (this volume) had established the feasibility of this technique and the absence of interference of non-retinal origin. An initial testing of the ERG preceded the main study in every rat to assure that there were no such interferences. Once the ERG study of the left eye had been completed, the other eye was briefly tested. With very few exceptions both eyes had about the same ERG. Following this the animal was sacrificed by an overdose of I.P. urethane or pentobarbital.

The flash of the Grass P52 photostimulator at maximal setting was presented at a rate of 0.2/s for establishing the ERG responsiveness vs flash intensity. This relationship was obtained over 4-6 log units from the maximal flash intensity (log I=0) at steps of 1.0 and 0.5 log unit by inserting density filters into the light path as in the study of the pigmented *rdy* rats. Three recordings of the responses at every flash intensity were made. Once the flash intensity/response relationship sufficient for the determination of threshold had been obtained, the test continued with the study of the ERG at high flash repetition rates in an ascending frequency series. This included the testing at 4/s, 8/s, 16/s, 24/s, and 32/s. Flash intensity was the maximal one (log I=0).

The number of runs for averaging the ERG of the *rdy* rat was routinely 150. When responses were very low 300 and 600 runs were averaged. Linear additivity of the runs was always ascertained by adding runs of 50. Averaging started after the first 5 flashes at 0.2/s had been given; with flash stimulation at 4/s or higher, averaging began after 5 seconds. Amplifier settings were constant.

Light Exposure

Constant exposure to strong light was performed in the same chambers and under the same conditions as routinely done in our laboratory for many years and as initially developed and evaluated by Noell et al, 1966. Illumination was by three 12 inch circular Norelco "cool-white" fluorescent 32 watt bulbs. A green plastic filter (see Noell et al., 1966) to minimize short-wave light exposure was always located between the bulbs and the wire-mesh of the chamber which housed generally 2 rats during exposure. Chamber illumination varied between 120 and 160 ft cdl in the mid-periphery of the chamber and was known to bleach the rhodopsin of a normal rat by about 90% within 10 minutes. This level of light exposure is referred to in the text as "150 ft. cdl. light". Chamber temperature was monitored and ventilation adjusted for constancy between 23 and 26^{0}C when the animals were in the chamber. All chambers were tested with regard to illumination and temperature control for several hours during the day preceding the use of animals. For the simultaneous use of several chambers, chambers were selected which were about the same in illumination and long term temperature constancy. Moreover, the same chambers were used during the whole study period.

The duration of light exposure was set at 30 hours because preceding studies with the same chambers had demonstrated that 12 hours of exposure of dark-reared rats of the normal albino strain was sufficient to abolish irreversibly the single-flash ERG in the majority of the animals. During exposure the unanesthetized rats were free to move (and sleep) and had ample access to drinking water and food pellets. No difference in behavior during exposure was noted between *rdy* and normal rats. The rats were directly transferred from the dark-environment into the chamber and immediately returned to darkness following exposure.

RESULTS

The following description of the ERG of the albino *rdy* rat is based on a total of 65 rats of the same strain tested between the age of 125 and about 450 days. 63% of these rats were exposed to 30 hours of continuous 150 ft. cdl. light 48 to 120 days prior to the ERG test. The remaining 37% served as the non-exposed controls and were generally litter mates of the light exposed ones. As will be documented no ERG differences between the exposed and non-exposed *rdy* rats could be detected. For describing and illustrating the ERG the 2 groups of rats were combined.

Figures 1 and 2 illustrate the *rdy* ERG at the age of 125 and 233 days, respectively. Figure 3, top row, shows 2 ERGs of a 333 days old *rdy* rat. As estimated from extensive DNA measurements over several years of other rats of the same albino strain (Noell, Saavedra, Braniecki, unpublished), 96% of the rod cells of the whole retina have disappeared at the age of 125 days in the dark-reared albino *rdy* rats while at the age of 200 days all rod cells have vanished histologically. Counts of the nuclei of cone cells over 40° of the horizontal meridian through the posterior pole of the eye were preformed in 3 albino *rdy* rats 196, 270 and 378 days old (Noell and Spongr, unpublished). It is estimated from these measurements that 41% of the cone cells probably have been lost at the age of 233 days, and 76% at 333 days. Only a small number of cone cells is assumed to have completely degenerated at 125 days. We estimated that the cone cells of the pigmented and albino rat normally comprise 1-2% of the total photoreceptor population. LaVail, 1976, 1984 gives a number of 1.3-1.8%. Nevertheless, a small ERG can be recorded in the *rdy* rat even when this number is drastically reduced after rod cell degeneration, as will be demonstrated here for the albino *rdy* rat. Moreover, the surviving cone cells have lost the outer segment membranes (LaVail et al, 1974; Cotter and Noell, 1984). The timing of the outer segment membrane loss is still unknown, but it seems safe to assume that the great majority of surviving cone cells has lost these membranes at the age of 200 days.

The averaged ERGs of the 125 old albino *rdy* rat of Figuro 1 show as the prominent component a b-wave like (cornea-positive) potential with a maximal amplitude of almost 40 uV at a flash repetition rate of 0.2/s. The wave has superimposed wavelets; a small negative deflection precedes its start. The positive wave has a

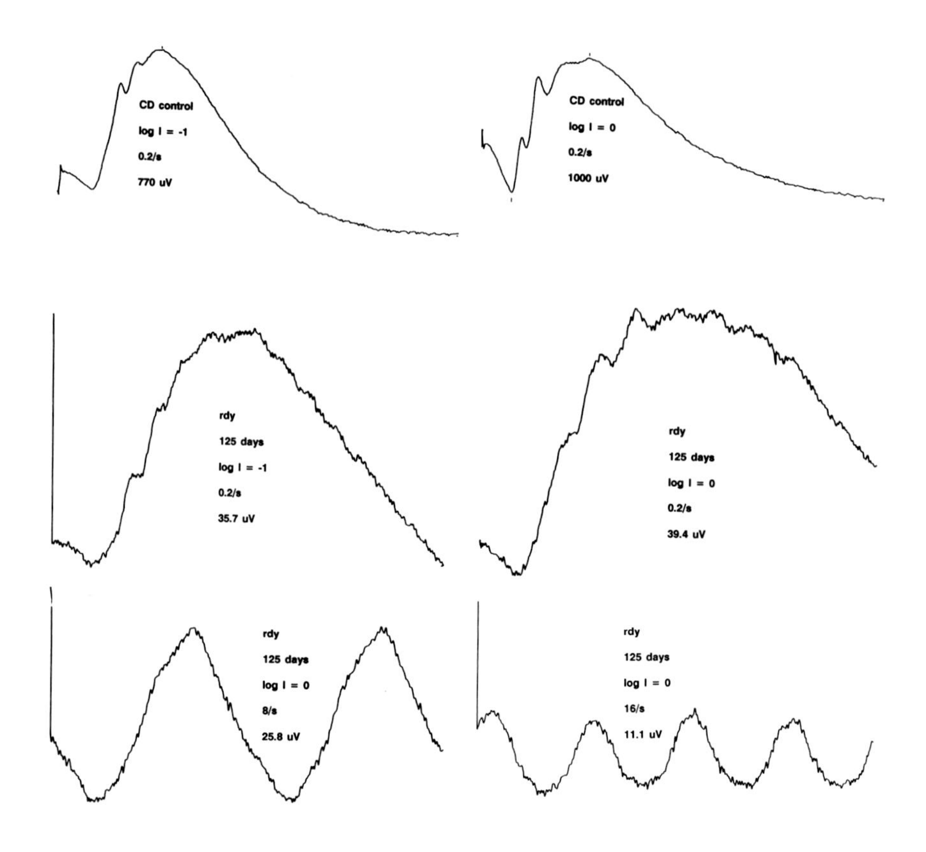

FIGURE 1. Corneal ERGs of a 125 day old albino *rdy* rat in comparison to the ERGs of an adult control albino rat CD. Data given with each record give the rat strain, the flash intensity in arbitrary log units (log maximal intensity is zero), the frequency of flash repetition, and the peak amplitude of the positive wave measured from the bottom of the preceding negative wave if present or the base line. Sweep time is always 250 msec which gives the time base. For the *rdy* rats 150 sweeps were averaged; for the normal CD rat only 4 (left) and 3 (right) runs were needed without changing amplification. The *rdy* ERGs are from a rat of a group of 7 *rdy* rats tested at the age of 123-129 days. This rat had the highest response of the group.

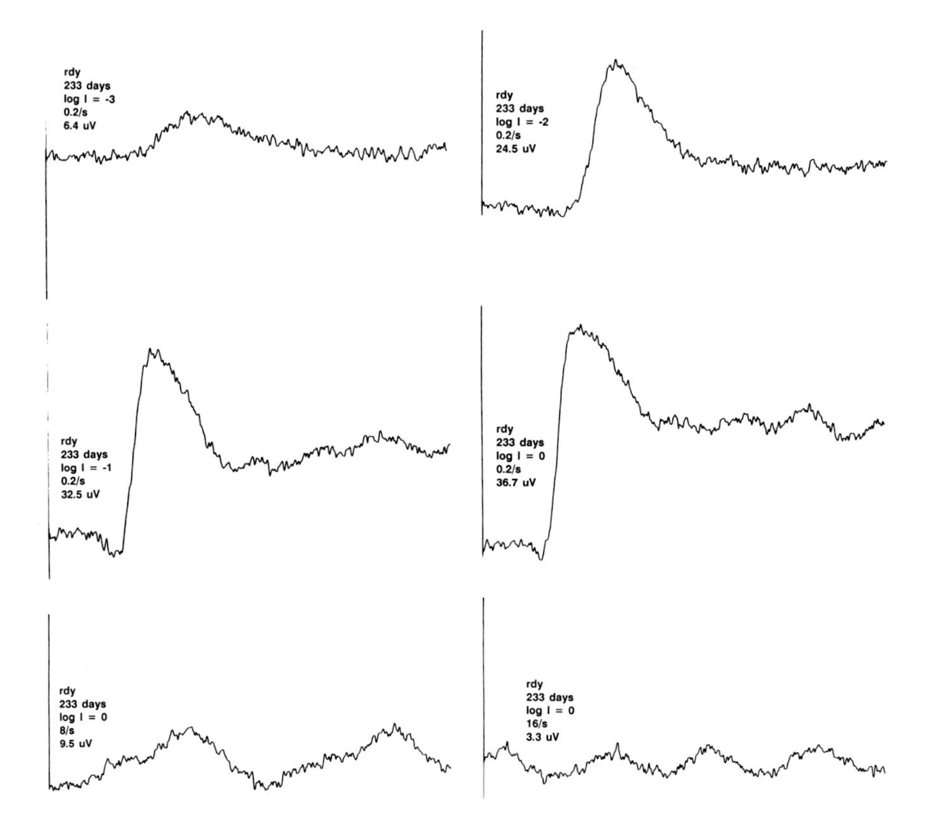

FIGURE 2. ERGs of a 233 days old albino r<u>dy</u> rat at different flash intensities (<u>top and middle row</u>), and different flash frequencies at maximal flash intensity (bottom row). Sweep time is 250 mesc; 150 runs were averaged. The rat is from a group of 9 of about the same age; it had the highest positive transient.

duration of over 250 msec at maximal flash intensity (log I=0). For comparison the ERGs of an adult albino control rat (CD) are illustrated (Figure 1, top row). The b-wave with sharp wavelets superimposed has a maximal amplitude of 1000 uV in this normal rat and is preceded by a short latency a-wave. The duration of the b-wave of the normal is one half the b-wave-like potential of the 125 day old *rdy* rat.

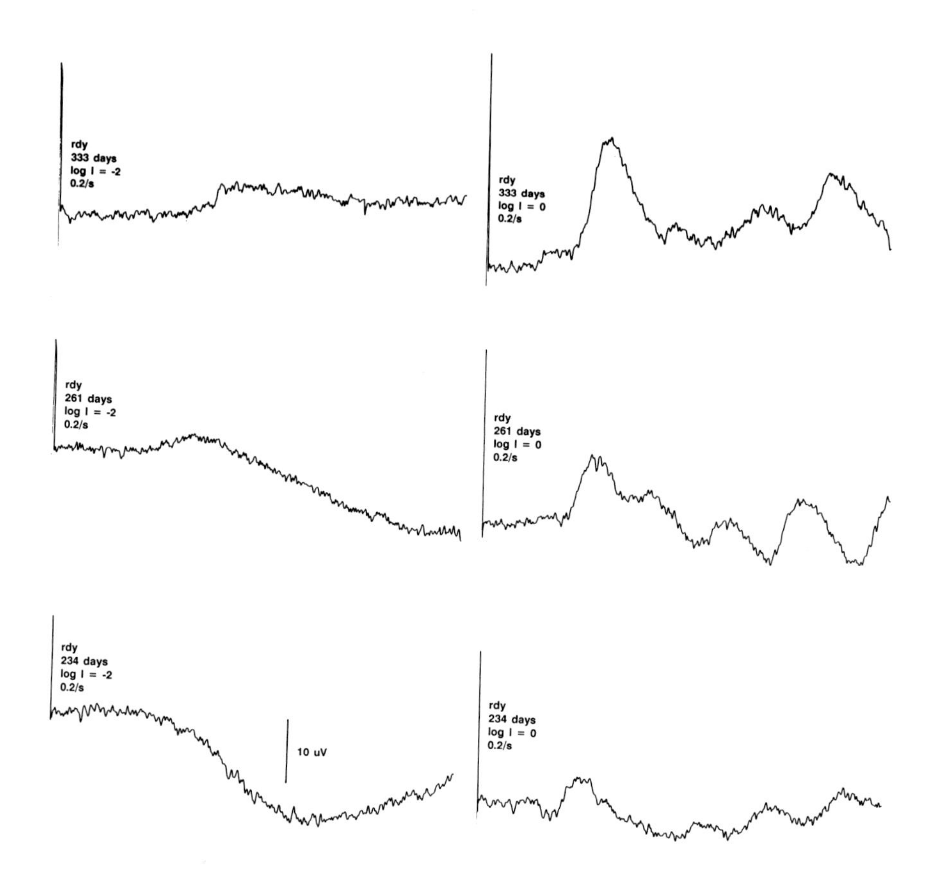

FIGURE 3. TOP ROW: ERGs of a light exposed 333 day albino *rdy* rat. Middle Row: ERGs of a non-exposed 261 day albino *rdy* rat showing the slow negative wave (left) and wave repetition (right). Bottom Row: ERGs of a light exposed, 234 day old albino *rdy* rat showing the slow negative wave at attenuated light intensity (left) and repetitive waves at maximal stimulus intensity (right). Sweep time throughout is 250 msec; 150 runs are averaged; amplitude calibration (10 uV) is given with the ERG of the bottom row at the left.

The *rdy* rat of 125 days is responsive to 8/s and 16/s flash repetition with amplitudes of 26 and 11 uV respectively (Figure 1, bottom). The corresponding amplitudes for the normal rat are 311 uV for 8/s and 153 uV for 16/s (not illustrated). In this rat 24/s flashes elicited a response of 75 uV and one of 26 uV with 32/s. There was no difference in wave form between the adult normal and the 125 old *rdy* rat at 8 and 16/s, but the time interval between flash and the (positive) peak of the corresponding response was increased by a factor of 1.78 from 50 ms in the normal at 8/s and by 1.68 from 44 ms in the normal at 16/s.

The ERGs of Figure 2 recorded at the age of 233 days are quite different in form from those illustrated in Figure 1. The most prominent wave is a sharply peaked positive potential of 65 msec duration preceded by a brief negative deflection and subsiding into a long-duration positive wave upon which repetitive wave activity is superimposed. The maximal amplitude of the phasic response is 37 uV and equals the maximal amplitude of the b-wave like potential of Figure 1. Near threshold (log I= -3) the response is simply a long-latency, rather smooth positive wave, of almost 100 msec duration. Compared with the 125 days old *rdy* rat (Figure 1), flash rates of 8/s and 16/s elicited weaker responses; they measure 9.8 uV and 3.3 uV, respectively.

The phasic initial component of the ERG illustrated in Figure 2 in response to near-maximal stimuli was termed the "negative-positive transient" or "positive transient" in the accompanying paper on the ERG of the pigmented *rdy* rat (Noell et al, this volume). The same expression will be used here in referring to the initial ERG potential of the albino *rdy* rat at an advanced stage of degeneration.

Figure 3, top row, shows ERGs of a 333 days old albino *rdy* rat. Amplitudes are low, 18.9 uV for the maximal initial transient (right) which has a duration of 60 msec. This positive transient is clearly repetitive as was generally observed to varying degree whenever the sharp positive transient was apparent (see also Noell et al; this volume). The positive wave elicited with 1/100 the maximal flash intensity had an amplitude of 4.5 uV (left).

The occasional occurrence of a slow negative wave was briefly mentioned by Noell et al, this volume. The greater number of studied albino *rdy* rats provided more data on this potential as illustrated in Figure 3, middle and bottom rows for 2 albino *rdy* rats, 261 (middle) and 234 (bottom) days old. Characteristically sub-maximal and near-threshold flash stimuli were best in demonstrating this slow negative wave. Stimuli which elicited the sharp positive transient, showed a suppression or masking of this wave. Generally also the amplitude of the positive transient was lower than in the average whenever the negative wave was clearly manifest at a low-intensity stimulus. In one rat out of ten of the 200 - 250 day old group, the slow negative wave was the apparent sole component of the response at 0.2/s stimulation, when the flash was 10 or 100 times weaker than the maximal one which elicited the positive transient, (see bottom row of Figure 3). The maximal amplitude of the negative wave was almost 40 uV.

As in the pigmented *rdy* rat, the latency of the positive transient increased with age. For 55 albino *rdy* rats of the age group of 175 to 425 days, the average shortest latency (maximal stimulus) was an increasing linear function of age with a slope of 8.4 ms per 50 days starting with 25 ms at 175 days. There was little scatter of the latency data in contrast to the amplitude measurement which varied greatly.

For quantitative documentation of ERG effects of 30 hours of continuous 150 ft. cdl. exposure, four measurements were chosen for inclusion into Table 1 and 2. These are (a) the peak amplitude of the positive transient of the largest response at or near maximal intensity of 0.2/s flashes; if a negative deflection preceded the positive transient the measurement was from the lowest negative point at the start of the positive upswing; (b) the shortest latency of the positive transient measured from the start of the 250 msec. sweep which coincided with the delivery of the flash; (c) the threshold of a response whatever its form by extrapolation between the minimally effective flash intensity and the closest flash intensity proven to be ineffective; flash intensity was expressed in log unit attenuation of the maximum intensity set at 0; the criterion amplitude level was 1% of the maximum amplitude at 0.2/s; (d) the highest effective flash

TABLE 1. ERG after Light Exposure of Albino *rdy* Rats at the Age of 9 and 11 Months.

Strain	n	Age at exp. days (S.D.)	Age at test days (S.D.)	Ampl. of posit. trans. uV(S.D.)	Latency of posit. trans. ms(S.D.)	Threshold log I (S.D.)	Highest flash rate (Hz) (S.D.)
rdy	7	-	339	6.8	51.5	-3.0	15.7
			(±5)	(±1.5)	(±4.1)	(±0.6)	(±4.5)
rdy	16	269	340	7.6	52.1	-2.7	16.2
		(±18)	(±12)	(±3.5)	(±8.4)	(±0.5)	(±2.0)
rdy	5	-	387	8.8	60.6	-2.0	12.0
			(±34)	(±3.1)	(±6.0)	(±0.8)	(±3.8)
rdy	7	334	382	10.6	58.4	-2.2	13.1
		(±34)	(±21)	(±7.3)	(±11.9)	(±0.5)	(±4.6)

repetition rate by extrapolating to the repetition rate at which the response would be 1% of the maximum response at 0.2/s.

Using these criteria, and others not listed, a ERG difference between exposed and non-exposed rats was not apparent when 30 hours of continuous 150 ft. cdl. light were applied at the average age of 269 days. The ERG study for determining the effect of light exposure was performed at the age of 340 days in comparison with the unexposed litter mates (generally) of the same age (Table 1, top). The long delay of 71 days between light exposure and ERG study was designed to satisfy our goal to evaluate the irreversible effect light exposure might have on the degenerating cone cells of the *rdy* retina. It also was stipulated to remove the rats from the continuous dark-environment only once for the purpose of light exposure, and to restrict the use of anesthesia and any ERG test related procedure to one study of the ERG. The use of rats older than 250 days was set by the a priori assumption that surviving cone cells could be most vulnerable near the age of their death.

TABLE 2. ERG after Light Exposure of Albino *rdy* Rats at the Age of 4 Months.

Strain	n	Age at exp. days	Age at test days	Ampl. of posit. trans. uV	Latency of posit. trans. ms	Threshold log I	Highest flash rate (Hz)
rdy	1	-	211	18.0	31.1	-4.4	24.0
rdy	1	-	213	7.4	36.1	-3.0	24.0
rdy	1	-	242	11.5	31.6	-4.0	24.0
rdy	1	119	233	33.0	30.0	-4.0	23.0
rdy	1	119	233	29.3	30.6	-4.0	24.0
rdy	1	121	213	20.0	43.9	-4.4	24.0
rdy	1	121	218	18.7	23.1	-4.4	24.0
rdy	1	122	217	6.9	42.3	-3.0	16.0
rdy	1	122	242	16.2	37.5	-4.0	16.0

The second group of *rdy* rats was on the average 334 days old when light exposed. Their ERG was studied about 48 days later. Results are listed in the bottom half of Table 1. Again there was no difference in the ERG of non-exposed and light exposed rats; all four ERG measurements indicated that the disease in the exposed rats had not progressed to a more advanced stage than in the non-exposed ones. For a small number of rats of the same age groups as in Table 1 the duration of exposure was extended to 48, 72 and 96 hours (not listed). Results were the same as with the 30 hours exposure.

Table 2 lists the results of 30 hours, 150 ft. cdl. exposure at the age of 119-122 days. The ERG study was performed about 100 days later. Considering the variability of the ERG measurements none of the listed 6 *rdy* rats seems to have been detrimentally affected by the light exposure.

DISCUSSION

The study of the ERG in the albino *rdy* rat greatly reenforces the finding with the pigmented *rdy* rat by showing that the same qualitative and quantitative changes occur in both strains. Any small quantative difference which was discarded in describing the results can easily be explained by the greater retinal irradiance in the albino rat. It thus seems justified to generalize the previously made conclusion that the ERG is changed in form, amplitude, latency, threshold and "flicker" responsiveness, in relation to the decline in number of the remnant cone cells once the rod cells have disappeard.

The findings of identical changes in both strains also supports the previous conclusion that the site of origin of the low-voltage ERG potentials shifts with the decline in the number of the remnant cone cells from the distal region of the surviving retina near the sites of normal b-wave generation to proximal sites within the inner layers. With this shift the ERG becomes more phasic showing a prominent brief positive transient subsiding into a slow positive wave. The initial transient is repetitive, often in a very striking manner. One is tempted to assume that the positive transient and its apparent repetition is associated with bursts of ganglion cell firing and that the slowly diminishing late positivity is related to a tonic phase of ganglion cell firing. There is no question that ganglion cell firing in response to receptive field stimulation continues when the rod cells of the *rdy* rat have disappeared (see Noell et al. 1981). The initial ganglion cell firing to the ON of receptive field center stimulation in the *rdy* rat also is very strong for a brief period of time (Stockton and Noell, unpublished).

There are several possibilities for discussing the origin of the slow negative wave of the *rdy* ERG especially with regard to the cone driven M-wave of the cat's ERG and the cornea-negative wave near rod threshold (Sieving et al. 1986; Sieving and Nino, 1988). Both waves are proved to be generated within the proximal retina, but a discussion must await further analysis of the *rdy* ERG which should be a worthwhile endeaver.

Our results on the effects of strong constant light are unambiguous in showing that the remnant cone cells of the *rdy* retina are not affected irreversibly by strong light which destroys very effectively the rod cells of the normal retina and even more effectively the rod cells of the young *rdy* rat.

It will be necessary, however, to extend our experiments to a study of the blue-light damage of the renmant cone cells before a general statement on the susceptibility to damage by visible light can be made for the *rdy* retina at an advanced stage of hereditary degeneration (cf. Noell, 1980; Lawwill, 1982). Finally, attention is directed to a study by LaVail (1976) which showed that the cone cells surviving constant exposure to strong light in cyclic light reared normal rats have morphologically the same features as the renmant cone cells of the *rdy* retina, after the rod cells have been lost to a great degree (for the type of light damage in cyclic light reared normal rats, see Noell, 1980).

REFERENCES

Cicerone C.M., Green D.G., Fisher L.J. (1979). Cone inputs to ganglion cells in hereditary retinal degeneration. Science 203:1113-1115.

Cotter J.R. and Noell W.K. (1984). Ultrastructure of remnant photoreceptors in advanced hereditary retinal degeneration. Invest. Ophthal. Vis. Sci. 25:1366-1375.

Dowling J.E., Sidman R.L. (1962). Inherited retinal dystrophy in the rat. J. Cell Biol. 14:73-109.

LaVail M.M. (1976). Survival of some photoreceptor cells in albino rats following long-term exposure to continuous light. Invest. Ophthalm. 15:64-70.

LaVail M.M. (1981). Photoreceptor characteristics in congenic strains of RCS rats. Invest. Ophthalmol. Vis. Sci. 20:671-675.

LaVail M.M., Sidman M., Rausin R., Sidman R.L. (1974). Discrimination of light intensity by rats with inherited retinal degeneration: A behavioral and cytological study. Vis. Res. 14:693-702.

Lawwill T. (1982). Three major pathologic processes caused by light in the primate retina: A search for mechanisms. Tr. Am. Ophth. Soc. 80:517-579.

Noell W.K. (1963). Cellular physiology of the retina. J. Opt. Soc. Amer. 53:36-48.

Noell W.K. (1974). Hereditary retinal degeneration and damage by light. Estratto dagli Atti del Simposio di Optalmologia Pediatrica, Casa Editrice Maccari, Parma, pp. 322-329.

Noell W.K. (1980). Possible mechanism of photoreceptor light damage in mammals. Vis. Res. 20:1163-1172.

Noell W.K., Walker V.S., Kang B.S. and Berman S. (1966). Retinal damage by light in rats. Invest. Ophthalmol. 5:450-483.

Noell W.K., Salinsky M.C., Stockton R.A., Schnitzer S.B., Kan V. (1981). Electrophysiological studies of the visual capacities at advanced stages of photoreceptor degeneration in the rat. Doc. Ophthalmologica. Proc. Ser. 27:175-181.

Noell W.K., Salinsky M.C. (1985). The preservation or visual evoked cortical responses at an advanced stage of retinal degeneration in the rdy rat. In LaVail M.M., Hollyfield J.G., Anderson R.E. (eds): "Retinal degeneration", Alan R. Liss, pp 301-320.

Sieving P.A., Frishman L.J. and Steinberg R.H. (1986). The M-wave of the proximal retina in cat. J. Neurophysiol. 56:1039-48.

Sieving P.A. and Nino C. (1988). Scotopic threshold response (STR) of the human electroretinogram. Invest. Ophthalm. Vis. Sci. 29:1608-14.

ACKNOWLEDGMENTS

The study was supported by the Kansas Lions Sight Foundation, Research for Prevention of Blindness, and a Summer Student stipend by the Medical School to D.R.H. We are very grateful to Dr. R.S. Crockett for generous assistance in the preparation of the manuscript for printing, Mr. Asher Ertel for photography and Ms. Annette Staatz for the typing of the manuscript.

Inherited and Environmentally Induced
Retinal Degenerations, pages 393–407

RPE-ASSOCIATED EXTRACELLULAR MATRIX CHANGES ACCOMPANY RETINAL VASCULAR PROLIFERATION AND RETINO-VITREAL MEMBRANES IN A NEW MODEL FOR PROLIFERATIVE RETINOPATHY: THE DYSTROPHIC RAT

Ruth B. Caldwell, Susan M. Slapnick and Rouel S. Roque

Anatomy Department, Medical College of Georgia, Augusta, GA

INTRODUCTION

Retino-vitreal neovascularization occurs in both spontaneously hypertensive and RCS rats with severe retinal dystrophy (Frank and Mancini, 1986; Frank and Das, 1988). In both models, the new vessels are surrounded by abnormal, proliferating retinal pigment epithelial (RPE) cells. The RPE-associated vessels in RCS retinas represent true neovascularization, because endothelial cell and pericyte nuclei incorporate substantial amounts of [^{3}H]thymidine (Frank and Das, 1988). RPE cell-associated basal lamina changes also occur in the RCS retina. First, the basal laminae of new blood vessels surrounded by abnormal, proliferating RPE cells contain anionic sites not present normally (Caldwell, unpublished). Another change is that more anionic sites are present along the RPE basal membrane than in normal rats (Caldwell, 1987). Because basal lamina changes are among the first structural alterations to occur during neovascular growth (Ausprunk and Folkman, 1977), these data suggested that RPE cell alterations may stimulate neovascularization in the dystrophic retina. We tested this hypothesis using quantitative light and electron microscopic techniques to study the progression of these changes in the RCS retina.

METHODS

Dystrophic (RCS) and control (RCS-*rdy*$^{+}$) rats were injected with the cationic tracer polyethyleneimine (PEI), and their retinas were prepared for light and electron microscopic analysis (see Caldwell *et al.*, 1986). PEI sites/µm of RPE basal membrane and basal lamina thickness were measured in randomly generated electron micrographs using a computer-assisted digitizing tablet. Density of vascular profiles was analyzed in tissue sections extending from the ora serrata to the optic disc in the lower temporal and nasal retina. Vessels were counted under the light microscope and drawings were made using a drawing tube. Retinal area was measured using a digitizer. Differences were tested by ANOVA.

RESULTS

In adult control rats the RPE was a uniform monolayer of cells with highly infolded basal membranes and long apical microvillous processes (Fig. 1A). In rats injected with PEI, electron dense particles were seen in Bruch's membrane along either side of the RPE and choriocapillaris basal laminae as well as around collagen and elastic fibers in the collagenous and elastic layers (see inset, Fig. 1A). The retinal capillaries in adult control rats were lined by a continuous, non-fenestrated endothelium. The endothelial basal lamina was usually split to enclose a pericyte and was never observed to contain PEI binding sites in control animals injected with the tracer (Fig. 1B). The normal retinal vessels were completely surrounded by perivascular glia.

At two weeks in both normal and dystrophic retinas, RPE and vascular morphology and PEI binding patterns were qualitatively similar to those in the adult control retinas. Quantitative analysis showed no differences between RCS and control rats at two weeks (see Table I).

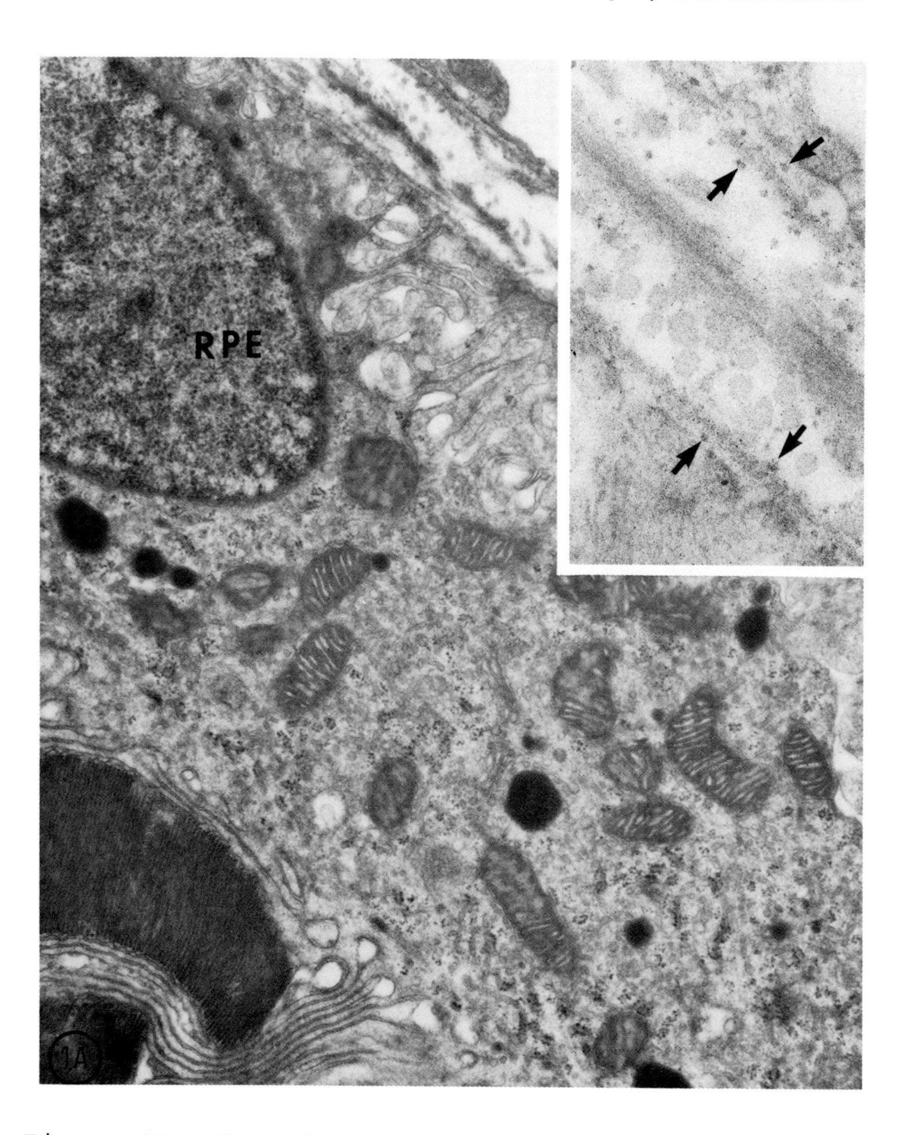

Figure 1A. Normal RPE cell (RCS-*rdy+*, 10 mos) has infolded basal membrane (upper right) and prominent microvilli (lower left). 22,000X. Inset shows Bruch's membrane with PEI-positive sites (arrows) along the RPE and choriocapillaris basal laminae. 62,500X.

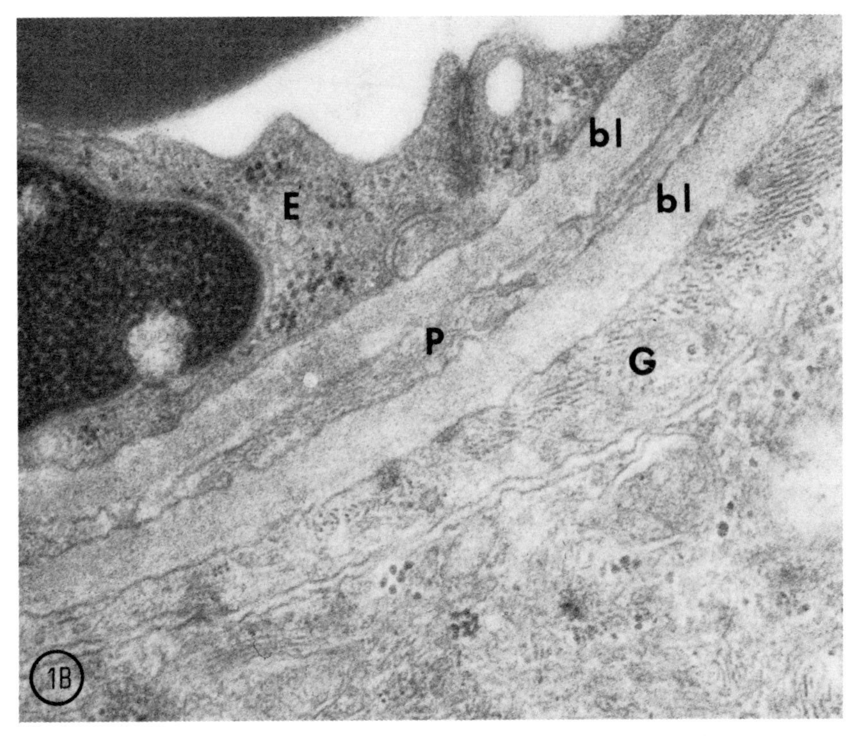

Figure 1B. Normal retinal vessel (RCS-*rdy*$^{+}$,10 mos) with endothelium (E), pericyte (P), and perivascular glia (G). Basal lamina (bl) is free of PEI sites. 55,000X.

By two months in the dystrophic retina, photoreceptor degeneration was virtually complete and only small amounts of debris remained. The RPE was hypertrophic in some areas and hyperplastic in others with abnormally flattened or convoluted surfaces (Fig. 2A). These abnormal RPE cells contained extensive rough endoplasmic reticulum filled with electron dense material. Abnormal deposits of PEI-positive basal lamina material were often present in the basolateral spaces and between superimposed RPE cells (Fig. 2B). Quantitative analysis showed substantial increases in anionic sites along the RPE basal lamina and in basal lamina thickness (see Table I). Thickness of the choriocapillaris basal lamina was similar in RCS and control rats ($p > 0.05$).

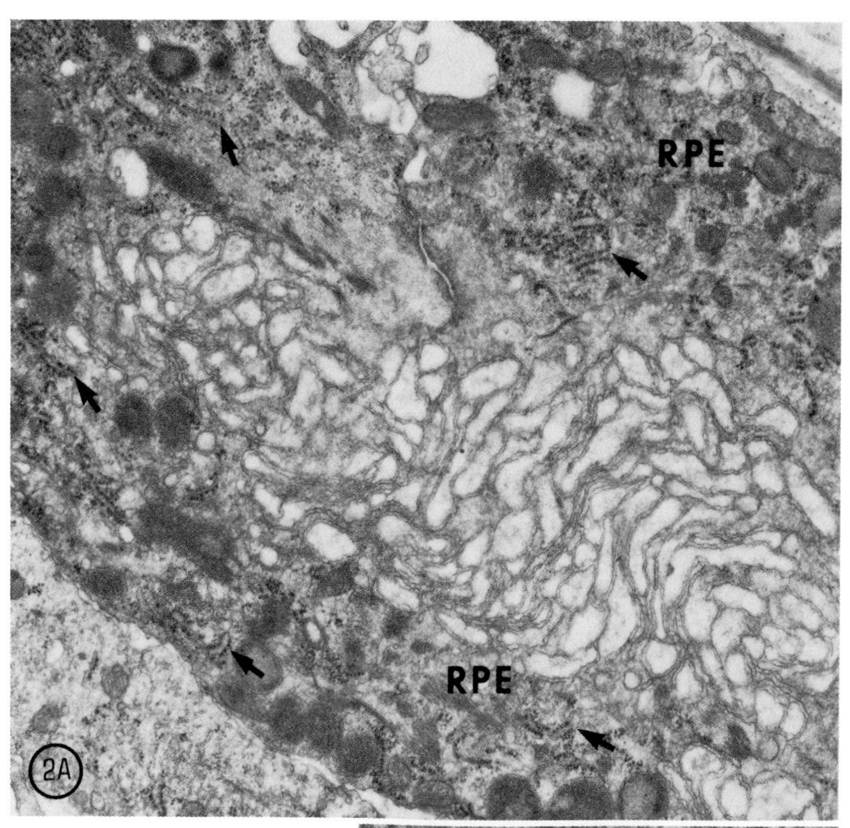

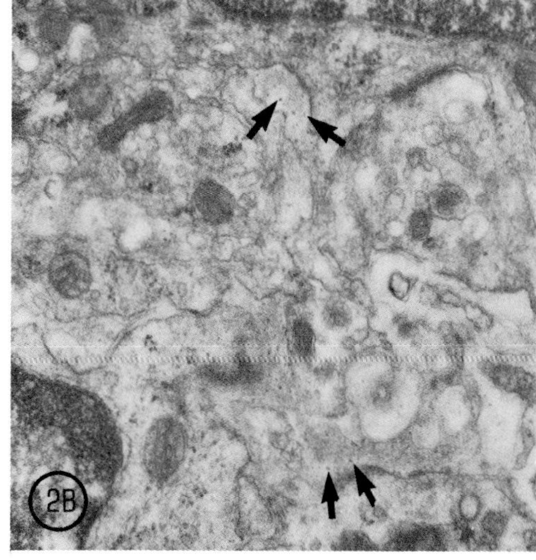

Figure 2. RCS RPE is abnormal at 2 mos. A. Hyperplastic RPE cells contain prominent rough endoplasmic reticulum (arrows). 22,000X. B. PEI-positive (arrows) basal lamina material is present between superimposed RPE cells. 21,300X.

Subtle vascular changes were evident at two months in the dystrophic rats. Vascular "ghosts" consisting of basal lamina remnants surrounding amorphous deposits of electron dense material were sometimes present in the inner retina (Fig. 3A). In addition, perivascular glial processes surrounding endothelial cells, and pericytes appeared to degenerate (Fig. 3B). Quantitative analysis of retinal vascular density showed no change in the number of vessels/mm^2 of retina, however (Table I).

By four months in the dystrophic retina, retinal vessels had begun to appear in the RPE layer. These vessels were often completely surrounded by RPE cell processes. However, even though these vessels were often seen adjacent to Bruch's membrane, they never crossed the RPE basal lamina to reach the choroidal vasculature. The RPE vessels progressively formed elaborate vascular networks, and later extended across the retina to the inner limiting membrane (Fig. 4A). The RPE-associated vessels differed from normal inner retinal vessels in that they were not continuous and often contained diaphragmed fenestrae and channels. Furthermore, unlike the normal inner retinal vessels, their basal laminae contained numerous PEI-positive anionic sites (Fig. 4B).

Vascularization of the RPE increased with age. RPE-associated vessels were occasionally observed at two months, commonly seen at four months, and consistently present at seven months and later (see Table I). This age-related increase in RPE vascularization was correlated with an increase in the number of vessel profiles/mm^2 of retinal tissue. By 7 months vascular density in the dystrophic RPE was 160% of that in the control rats. Because retinal thickness in RCS rats is about 75% of that in the age-matched controls, we examined the relationship between retinal thinning and retinal vascularity by calculating number of vessels/mm of retina without regard to retinal thickness. This analysis showed 38 vessels/retina in dystrophic rats compared with 33 in the controls.

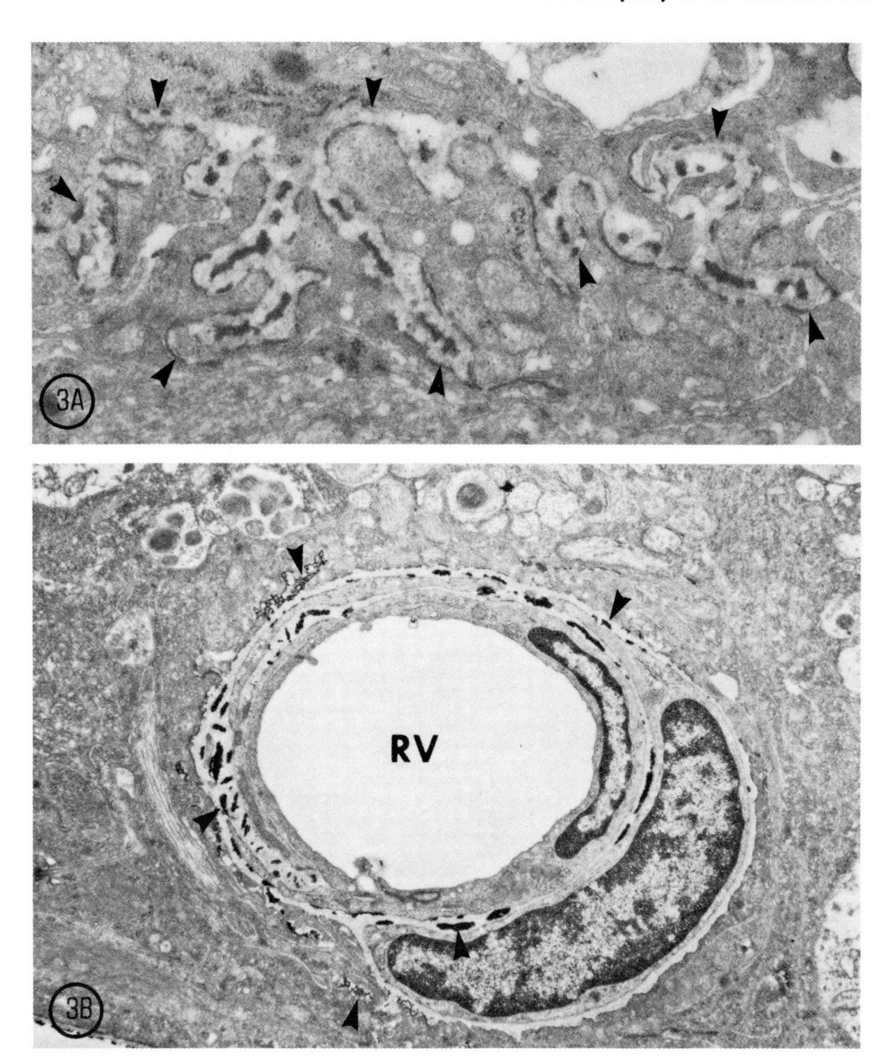

Figure 3. Vascular degeneration in RCS retina (2 mos). A. Vascular "ghost" (arrowheads) consists of basal lamina remnants and electron dense debris. 30,000X. B. Degenerating perivascular processes (arrowheads) surround retinal vessel. 8,550X.

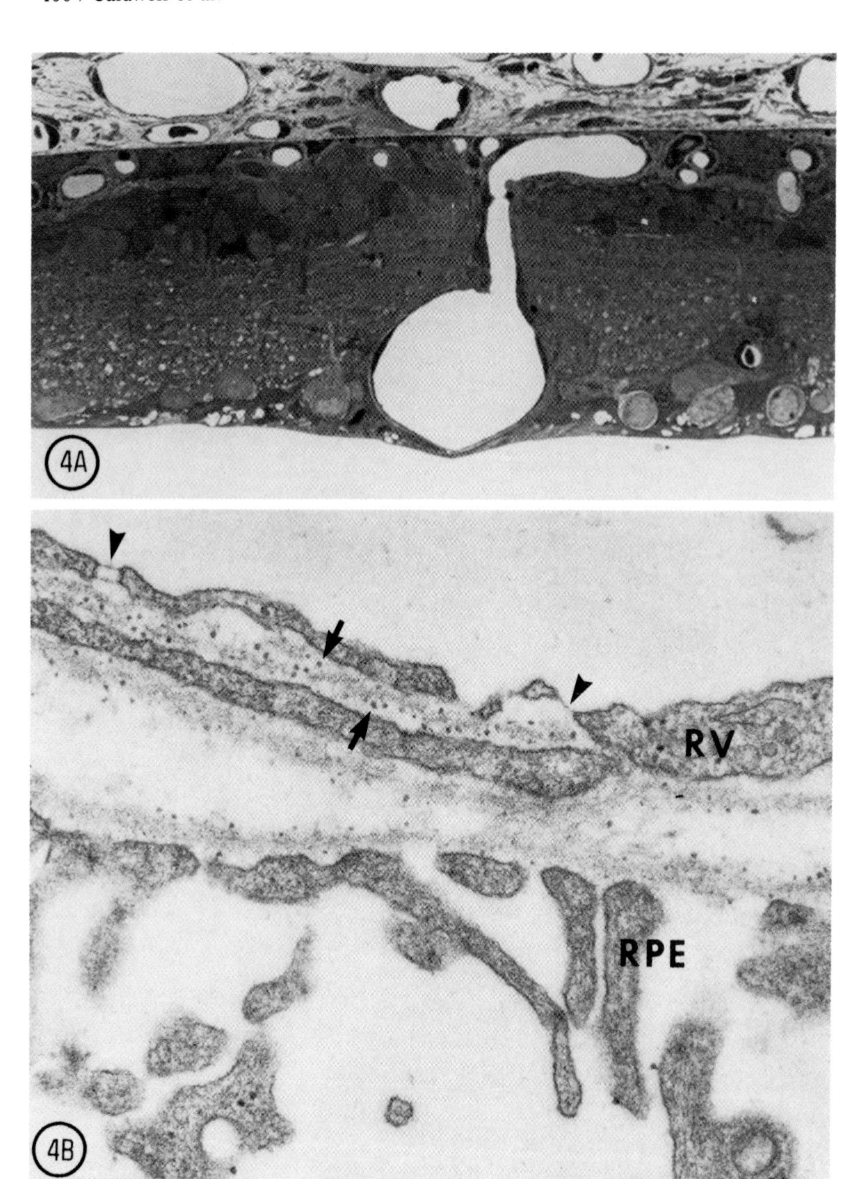

Figure 4. RPE-associated vessels in RCS retina. A. Vessel extends from the RPE to the nerve fiber layer (11 mos). B. Fenestrated (arrowheads) retinal vessel (RV) in RPE has many PEI-sites (arrows) along its basal lamina (2 mos). 60,000X.

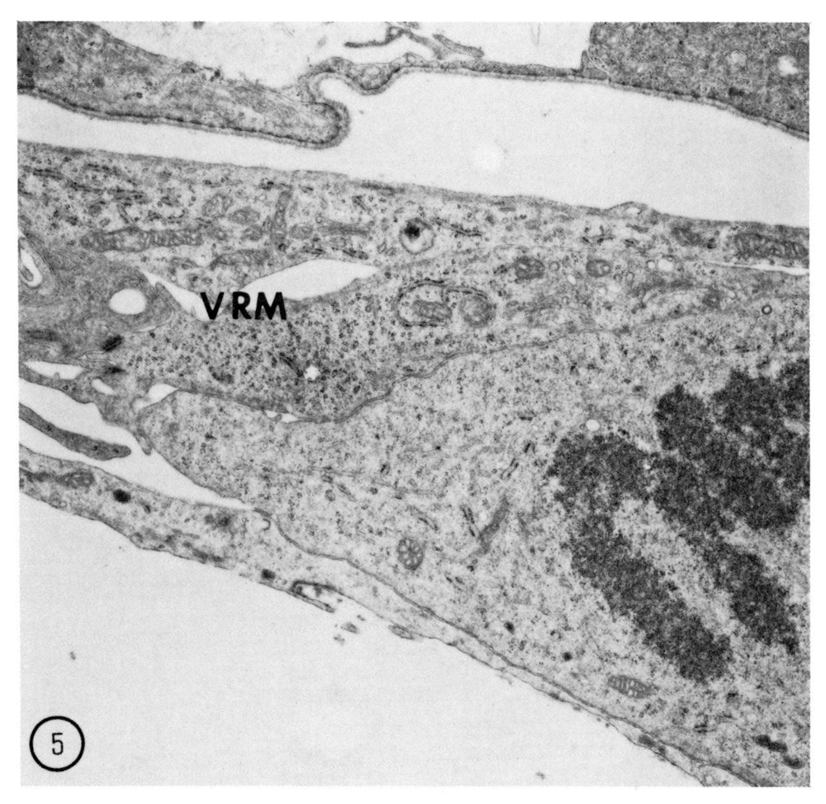

Figure 5. Cell in mitosis in VRM from RCS retina (11 mos). 11,400X.

An additional alteration in dystrophic rats was that proliferating vessels and/or glial cells sometimes grew into the vitreous cavity, forming vitreo-retinal membranes (VRMs, Figs. 5,6). The presence of mitotic figures (Fig. 5) indicated cellular proliferation in these membranes. VRMs were only seen in rats with RPE vascularization. In the dystrophic rats aged two months or older, 5 of 13 rats with RPE vessels also had VRMs (see Table I). Based on size and cellular morphology, two different types of VRMs were identified: one was large, highly vascularized, and projected for some distance into the vitreous cavity (see Fitzgerald *et al.*, this volume); the other was smaller, avascular, and adhered closely to the retinal surface (Fig. 6).

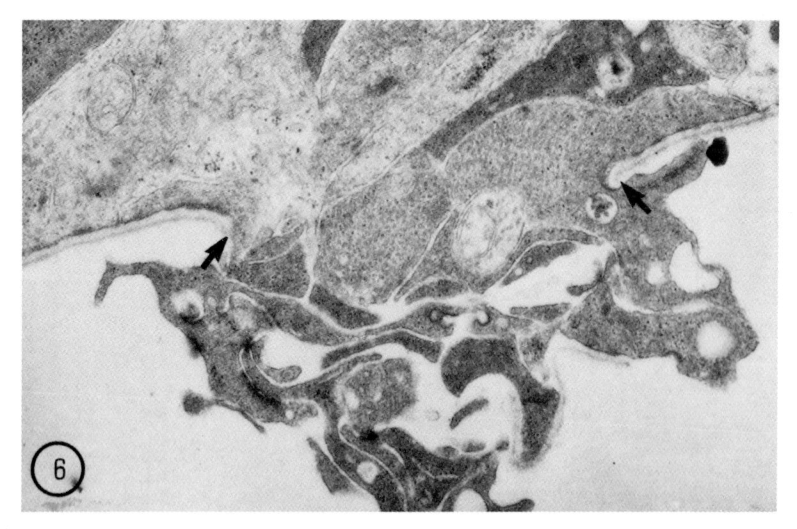

Figure 6. Glial processes in RCS retina (4 mos) penetrate the inner limiting membrane (arrows) to enter the vitreal cavity. 19,000X.

Table I. Summary of quantitative results.

Group	Age (mos)	PEI sites /μm RPE	Basal Lamina (nm)	#Vessels /mm^2	RPE-vessels[a]	VRMs[a]
RCS-rdy$^+$	1/2	4	46	212	0/6	0/6
	2	3	41	275	0/6	0/6
	>7	-	-	228	0/8	0/6
RCS	1/2	3	47	196	0/6	0/6
	2	10*	79*	301	1/6	1/6
	4	-	-	313	4/7	2/7
	>7	-	-	364*	8/8	2/8

Notes: -: not analyzed, a: number of rats/group, *: $p < .001$.

DISCUSSION

Our results show that hyperplasia of the dystrophic RPE is accompanied by increased numbers of anionic sites along the RPE basal membrane, by thickening of the RPE basal lamina, and by abnormal deposits of PEI-positive basal lamina material among displaced RPE cells. These changes are followed by migration of retinal blood vessels into the RPE layer where the RPE-associated retinal vessels develop fenestrations and display PEI-positive anionic sites along their basal lamina. Upon vascularization of the RPE, proliferating RPE cells migrate into the inner retina and VRMs are often observed. A similar association between retino-vitreal neovascularization and RPE cell proliferation in spontaneously hypertensive rats led Frank and Mancini (1986) to suggest that the RPE may stimulate proliferation of vascular endothelial cells. The temporal sequence of the changes we found in the RCS dystrophic retina supports this hypothesis.

Considerable evidence is accumulating to suggest a role for the RPE in regulating vascular growth. Because the choriocapillaris atrophies in regions under abnormal RPE cells, the RPE has been suggested to produce a vascular modulating factor that exerts a trophic effect upon the choriocapillaris (Korte et al., 1984). Proliferation of cultured choroidal endothelial cells treated with RPE-conditioned medium supports this hypothesis (Morse *et al.*, 1986). In retinal dystrophy, RPE cell proliferation may lead to increases in this vascular trophic factor. In addition, breakdown of RPE tight junctions in the RCS retina (see Caldwell, 1987 for review) may allow factors normally restricted to the RPE basal surface and Bruch's membrane to reach the inner retinal vessels.

On the other hand, RPE proliferation has also been suggested to inhibit neovascular growth, since treatment of proliferative diabetic retinopathy by pan retinal photocoagulation results in RPE cell proliferation and regression of neovascularization (see Glaser, 1988). Furthermore, RPE-conditioned medium inhibits vascular growth in the chick chorioallantoic membrane (Glaser, 1986). Cultured RPE cells may also inhibit neovascular growth by inhibition of the protease urokinase. Urokinase plays a central role in degradation of the extracellular matrix which is the first step of endothelial cell migration. Cultured RPE cells have been found to release a specific urokinase inhibitor and this protein has been localized in the RPE and Bruch's membrane of normal human eyes (see Glaser, 1988). Thus, RPE cells can stimulate or inhibit vascular growth under appropriate experimental conditions. Indeed, recent studies have found that human RPE cells in culture release both an inhibitor and a stimulator of endothelial cell proliferation (Conner and Glaser, 1987). Additional *in vivo* studies in the dystrophic rat retina may help clarify the role of RPE cells in the modulation of vascular growth.

One way the RPE could modulate vascular growth is by affecting extracellular matrix molecules. The increased PEI binding in the dystrophic retina suggests that RPE-associated increases in anionic extracellular matrix molecules occur in retinopathy. Heparin sulfate proteoglycan is one RPE-associated anionic molecule (Pino *et al*, 1982; Lin and Essner, 1988) whose alteration could contribute to endothelial cell proliferation, because basal lamina heparin sulfates have been demonstrated to bind growth factors (Jeanny et al., 1987). Thus, heparin sulfate at the RPE basal membrane and within the displaced basal lamina material produced by the abnormal RPE cells could bind increased amounts of retinal growth factors, stimulating retinal endothelial cell migration and proliferation in the RPE cell layer.

In addition to inducing changes in vascular growth, RPE alterations may also affect vascular function. In phototoxic and urethane retinopathies retinal vessels associated with RPE cells are transformed from a continuous, non-permeable endothelium to a fenestrated permeable endothelium (see Burns et al., 1986). Observation of PEI sites along the basal laminae of the fenestrated RPE-associated vessels in RCS retinas suggests increased permeability in these vessels, since PEI (mol wt 70,000) does not penetrate the normal retinal vessels and PEI sites are not present abluminally. In addition, our horseradish peroxidase and lectin studies of RPE-associated vessels show increased permeability and decreased concanavalin-A binding. This suggests that development of fenestrations and increased permeability in these vessels are also associated with an alteration in plasma membrane composition (see Fitzgerald *et al.*, this volume).

The dramatic increases in vascular density in the dystrophic retina were unexpected because retinal thickness and retinal vascularity are normally positively correlated. Vascularization of the pigment epithelium has been reported late in dystrophic retinas (Dowling and Sidman, 1962; Bok and Hall, 1971, Bok and Heller, 1980, Matthes and Bok, 1985), but previous studies using trypsin digest and whole-mount techniques found overall decreases in retinal vascularization (Gerstein and Dantzker, 1969; Matthes and Bok, 1985). The differences between our results and these previous studies may be due to methodology. The neural retina must be detached from the pigment epithelium with either technique, and since most new blood vessels are in the RPE layer or associated with displaced RPE cells, they could be lost or damaged during detachment. Furthermore, vascular density may have been underestimated in the previous studies because measurement of retinal volume is not possible with whole-mount techniques. Additional work is needed to resolve this conflict.

ACKNOWLEDGEMENTS

We wish to thank Libby Perry and Daniel Lum for their help in preparing the illustrations. Supported in part by NIH Grant EY-04618 and a research grant from the Juvenile Diabetes Foundation, International.

REFERENCES

Ausprunk DH, Folkman J (1977). Migration and proliferation of endothelial cells in preformed and newly formed blood vessels during tumor angiogenesis. Microvasc Res 14:53-65.

Bok D, Hall MO (1971). The role of the pigment epithelium in the etiology of inherited retinal dystrophy in the rat. J Cell Biol 49:664-682.

Bok D, Heller J (1980). Autoradiographic localization of serum retinol-binding protein receptors of the pigment epithelium in dystrophic rat retinas. Invest Ophthal Mis Sci 19:1405-1414.

Burns MS, Bellhorn RP, Korte GE, Heriot WJ (1986). Plasticity of the retinal vasculature. In Osborne N, Chader G (eds): "Progress in Retinal Research," New York: Pergamon Press, pp 253-307.

Caldwell, R.B. (1987) Blood-retinal barrier changes in the retinal pigment epithelium of RCS rats with inherited retinal degeneration. In JG Hollyfield, RE Anderson, MM LaVail (eds): "Degenerative Retinal Disorders: Clinical and Laboratory Investigations," New York: Alan R Liss, pp 333-347.

Caldwell RB, Slapnick SM, McLaughlin BJ (1986). Decreased anionic sites in Bruch's membrane of spontaneous and drug-induced diabetes. Invest Ophthalmol Vis Sci 12:1691-1697.

Conner T, Glaser (1987). RPE cells can simultaneously release inhibitors and stimulators of endothelial cell proliferation. Invest Ophthalmol Vis Sci 28(Suppl):203.

Dowling JE, Sidman RL (1962). Inherited retinal dystrophy in the rat. J Cell Biol 14:73-109.

Fitzgerald MEC, Caldwell RB (1988). Alterations in lectin binding accompany increased permeability in the dystrophic rat model for proliferative retinopathy. In LaVail MM, Anderson GE, Hollyfield JG (eds): "Inherited and Environmentally Induced Retinal Degenerations," New York: Alan R. Liss, this volume.

Frank RN, Mancini MA (1986). Presumed retinovitreal neovascularization in dystrophic retinas of spontaneously hypertensive rats. Invest Ophthalmol Vis Sci 27:346-355.

Frank RN, Das A (1988). Models of retinal neovascularization in the rat. Invest Ophthalmol Vis Sci 29(Suppl): 245.

Gerstein DD, Dantzker DR (1969). Retinal vascular changes in hereditary visual cell degeneration. Arch Ophthal 81:99-105, 1969.

Glaser BM (1986). Cell biology and biochemistry of endothelial cells and the phenomenon of intraocular neovascularization. In Adler R, Farber D (eds): "The Retina, Part II," London: Academic Press, pp 215-243.

Glaser BM (1988). Extracellular modulating factors and the control of intraocular neovacularization. Arch Ophthalmol 106:603-607.

Jeanny J-C, Fayein N, Moenner M, Chevallier DB, Courtois Y (1987). Specific fixation of bovine brain and retinal acidic and basic fibroblast growth factors to mouse embryonic eye basement membranes. Exp Cell Res 171:63-75.

Korte GE, Reppucci V, Henkind P (1984). RPE destruction causes choriocapillary atrophy. Invest Ophthalmol Vis Sci 25:1135-1145.

Lin W-L, Essner E (1988). Immunogold labeling of basal lamina components in rat retinal blood vessels and Bruch's membrane. Proceedings of the International Society for Eye Research, Vol V: 50.

Matthes MT, Bok D (1985). Blood vascular abnormalities in animals with inherited retinal degeneration. In LaVail MM, Hollyfield JG, Anderson RE (eds): "Retinal Degeneration: Experimental and Clinical Studies," New York: Alan R Liss, pp 209-237.

Morse LS, Sidikaro Y, Terrell J (1986). Retinal pigment epithelium promotes proliferation of choroidal microvessel endothelium in vitro. Invest Ophthalmol Vis Sci 27(Suppl):327.

Pino RM, Essner E, Pino LC (1982). Location and chemical composition of anionic sites in Bruch's membrane of the rat. J Histochem Cytochem 30:245-252.

Inherited and Environmentally Induced
Retinal Degenerations, pages 409–425

ALTERATIONS IN LECTIN BINDING ACCOMPANY INCREASED PERMEABILITY IN THE DYSTROPHIC RAT MODEL FOR PROLIFERATIVE RETINOPATHY

Malinda E.C. Fitzgerald, Susan M. Slapnick,and Ruth B. Caldwell

Dept. of Anatomy and Neurobiology, The Univ. TN, Memphis, TN, (MECF) Dept. of Neurosurgery, Univ of WI, Madison, WI,(SMS) and Dept. of Anatomy, The Medical College of GA, Augusta,GA, (RBC)

In the dystrophic Royal College of Surgeons (RCS) rat, migration of vessels from the inner retina into the retinal pigment epithelium (RPE) is associated with neovascular proliferation and formation of vitreo-retinal membranes (VRMs), (Caldwell et al., 1988; Frank and Das, 1988). We studied permeability and luminal membrane glycoconjugates in these vessels using horseradish peroxidase (HRP) and lectin-ferritin (Fe) techniques. RCS and genetic control rats were injected with HRP, their retinas were fixed, incubated in Fe conjugates of wheat germ agglutinin (WGA-Fe) or concanavalin-A (ConA-Fe), reacted to demonstrate HRP, and prepared for electron microscopy. The RPE and VRM vessels in RCS retinas were compared with the normal inner retina and choriocapillaris vessels in RCS and genetic control rats. In both groups inner retinal vessels formed a barrier to HRP, while fenestrated choriocapillaris (CE) vessels were permeable to the tracer. In both of these vascular beds plasma membrane WGA-Fe binding was dense and uniform, while ConA-Fe binding was sparse and patchy. Studies with competitive sugars showed that WGA-Fe binding was primarily to N-acetylglucosamine (NAG) and that ConA-Fe was to mannose. In both RPE and VRM vessels tight junctions appeared intact, but both vessel types were permeable to HRP with the RPE vessels often containing fenestrae and channels. As compared with binding in the inner retina and CE vessels, WGA-Fe binding was lower in VRM vessels and normal in RPE vessels, while ConA-Fe binding was higher in both RPE and VRM vessels. Thus, increased permeability is accompanied by alterations in both NAG and mannose residues in the VRM vessels and with alterations in mannose residues and the presence of fenestrations and channels in the RPE vessels.

INTRODUCTION

Neovascularization during proliferative diabetic retinopathy results in blindness in some patients. An increase in blood-retinal barrier permeability is thought to contribute to the development of this pathology (Cunha-Vaz, 1976; 1979; 1980; Goldberg, 1980), but relatively little is known about the cellular events that occur during these changes. Experiments in diabetic rats suggest that the barrier function of the outer retina may be abnormal (Blair et al., 1980; Caldwell et al., 1985; 1986; 1987; Kirber et al., 1980; Vinores et al., 1988), and that, in the retinal endothelium, tight junction breakdown and/or increased pinocytosis may occur (Ishibashi et al., 1980; 1983; Wallow and Engerman, 1977). We have reported increased endothelial cell uptake of HRP associated with decreases in plasma membrane binding of cationized ferritin (Fitzgerald and Caldwell, 1987) and WGA-Fe binding in some retinal vessels of spontaneously diabetic Bio-Breeding (BB) rats (Fitzgerald and Caldwell, 1988). This suggests that increased pinocytosis may be associated with luminal surface alterations in the diabetic retinal microvasculature. However, while the spontaneously diabetic BB rats exhibit background retinopathy (Sima et al., 1985) with regional variation in permeability, luminal surface charge, and composition (Fitzgerald and Caldwell, 1987; 1988), proliferative retinopathy does not occur.

RCS rats with severe retinal dystrophy develop proliferative retinopathy (Dowling and Sidman, 1962; Frank and Das, 1988). In addition to retinal neovascularization, the RCS rats develop VRMs, (Caldwell et al., 1988; Frank and Das, 1988) that are often vascularized. We sought to test our hypothesis that increased permeability correlates with changes in membrane composition in this animal model which exhibits greater vascular plasticity than the diabetic BB rats. We investigated the relationship between vascular permeability and endothelial cell luminal membrane microdomains in the RCS rat using HRP and lectin-Fe techniques.

METHODS

RCS and age-matched genetic control (RCS-rdy$^+$) rats between the ages of 12 and 18 months were used in this study. Animals were maintained on a 12 hr light (illumination level of 5 ft candles)/12 hr dark cycle, and experiments were conducted during the light period.

Animals were anesthetized with 30% chloral hydrate (0.1 ml/100 g) with a metofane supplement, and HRP was injected into the tongue vein. After a circulation time of 30 min., the animals were transcardially perfused with fixative and eye cups removed (Fitzgerald and Caldwell, 1988). The eyes were quartered, the sclera was removed, and the posterior retina was tissue chopped (40 µm). Prior to lectin-Fe incubations the tissue sections were rinsed in buffer, placed in 0.1 M glycine for 30 min to quench free aldehydes, rinsed, incubated in 1% bovine serum albumin for 1 hour to block non-specific lectin binding, and rinsed. The buffer was then changed to one compatible with the lectin. Lectin-Fe conjugates of either WGA (specific for NAG, sialic acid, 400 µg/ml, Polysciences) or ConA (specific for mannose, 1000 µg/ml, Polysciences) were used. WGA binding was conducted in a 0.1 M phosphate buffered saline solution, pH 7.45, and ConA binding was conducted in a 0.005 M Tris with 0.015 M NaCl, 0.001 M

$CaCl_2$, and 0.001 M $MnCl_2$, pH 7.0. In lectin control studies, WGA was blocked with 0.03 M N, N' N"-triacetylchitotriose (CIT, for NAG) and ConA was blocked with 0.3 M alpha-D-mannose. All lectin incubations were carried out at 37°C for 2 hours. After rinsing, the tissue was processed for peroxidase localization (Graham and Karnovsky, 1966) followed by post-fixation in 1% OsO_4, dehydration, and plastic embedment for electron microscopy.

RESULTS

Vascular beds

Permeability and the endothelial luminal surface lectin binding characteristics were investigated in the capillaries of the CE and the inner retina. In normal rats, the inner retinal vessels were concentrated in three zones between the outer plexiform layer (OPL) and the nerve fiber layer (NFL). In the RCS rat following degeneration of the photoreceptors, the laminar distribution of the inner retinal vessels was altered and the OPL vessels were no longer distinguishable. Numerous vessels were observed at the level of the RPE. These vessels were surrounded by proliferating RPE cells and were derived from the inner retinal vasculature (Caldwell et al., 1988). Another alteration observed in the retinal vessels of the RCS rat was vascular proliferation in the VRMs (Fig. 1a). The VRMs contained a variety of cells including endothelium, pericytes, neuronal processes, and cells containing membrane-bound, striated dense bodies (Fig. 1b). Thus, four types of vessels in the RCS rats (CE, RPE-associated, inner retina, and VRM) were compared with two types in the control animals (CE and inner retina).

Vessel morphology

The CE and inner retinal vessels in the RCS rats were structurally identical to those in normal rats. The CE consisted of a fenestrated endothelium; and the endothelium of the inner retinal vessels was continuous and non-fenestrated. The RPE-associated vessels, while apparently originating from the inner retinal vessels, were structurally similar to the CE vessels. They contained diaphragmed fenestrae and channels (see Fig. 2c and 3c) that were not normally observed in the inner retinal vessels. The endothelium of VRM vessels was similar to the continuous endothelium of the inner retina, and fenestrae were not observed. The VRM vessels were associated with multilayered basal lamina and large amounts of extracellular matrix (Fig. 1b).

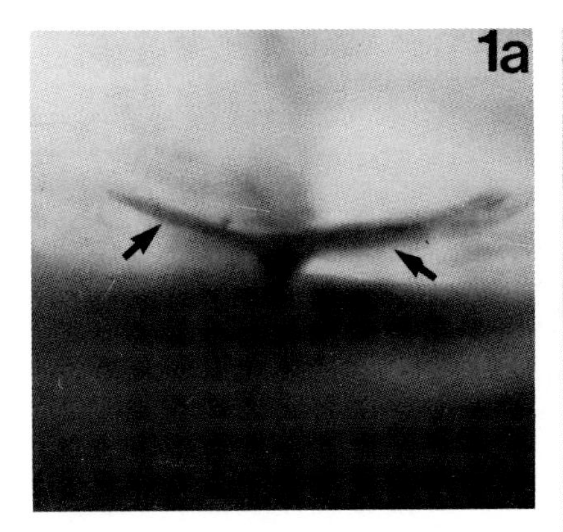

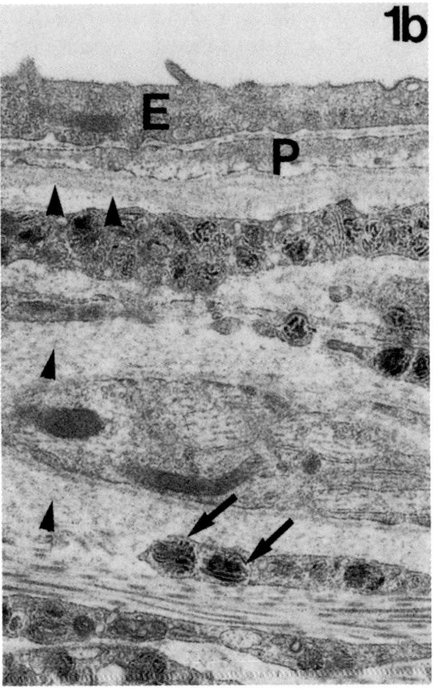

Figure 1. Light micrograph (1a) of a vascularized VRM from an RCS rat. (Mag x 120) Electron micrograph (1b) of the same VRM shows a variety of cells including: endothelium (E), pericyte (P), and cells containing membrane bound striated dense bodies (arrows). Note large amount of extracellular matrix and multilayed basal lamina (arrowheads) (Mag x 81,000)

Membrane microdomains

The luminal membrane microdomains of the retinal capillary endothelium were: coated vesicles or pits (not shown) and uncoated vesicles, diaphragms of uncoated vesicles, and the plasma membrane (see Figs. 2 and 3). Vessels of the CE and RPE also contained diaphragmed fenestrae and channels (see Figs. 2 and 3).

ConA-Fe binding

In the control vessels (not shown) and in the dystrophic CE, and "normal" inner retinal vessels, ConA binding was similar. Binding on the plasma membrane of these vessels was sparse. Small patches of ConA-Fe particles were separated by irregular areas of particle-free membrane (Figs. 2a and 2b). The ConA-Fe binding in the RPE (Fig. 2c) and VRM (Fig. 2d) vessels was higher than in either the control or "normal" vessels of the RCS retina. The binding appeared more uniform with fewer areas of particle-free membrane. All endothelial luminal membrane microdomains were labeled with ConA-Fe, except for some uncoated vesicles and fenestral diaphragms.

WGA-Fe binding

In the control vessels (not shown) and in the dystrophic CE, "normal" inner retinal vessels, and most of the RPE-associated vessels, WGA-Fe binding on the plasma membrane was dense and uniform (Figs. 3a, b, and c). WGA-Fe binding was greatly reduced in many VRM vessels of the RCS retina and was often accompanied by increased HRP permeability (Fig. 3d). All membrane microdomains contained label, except for some uncoated luminal vesicles (see Fig. 3b).

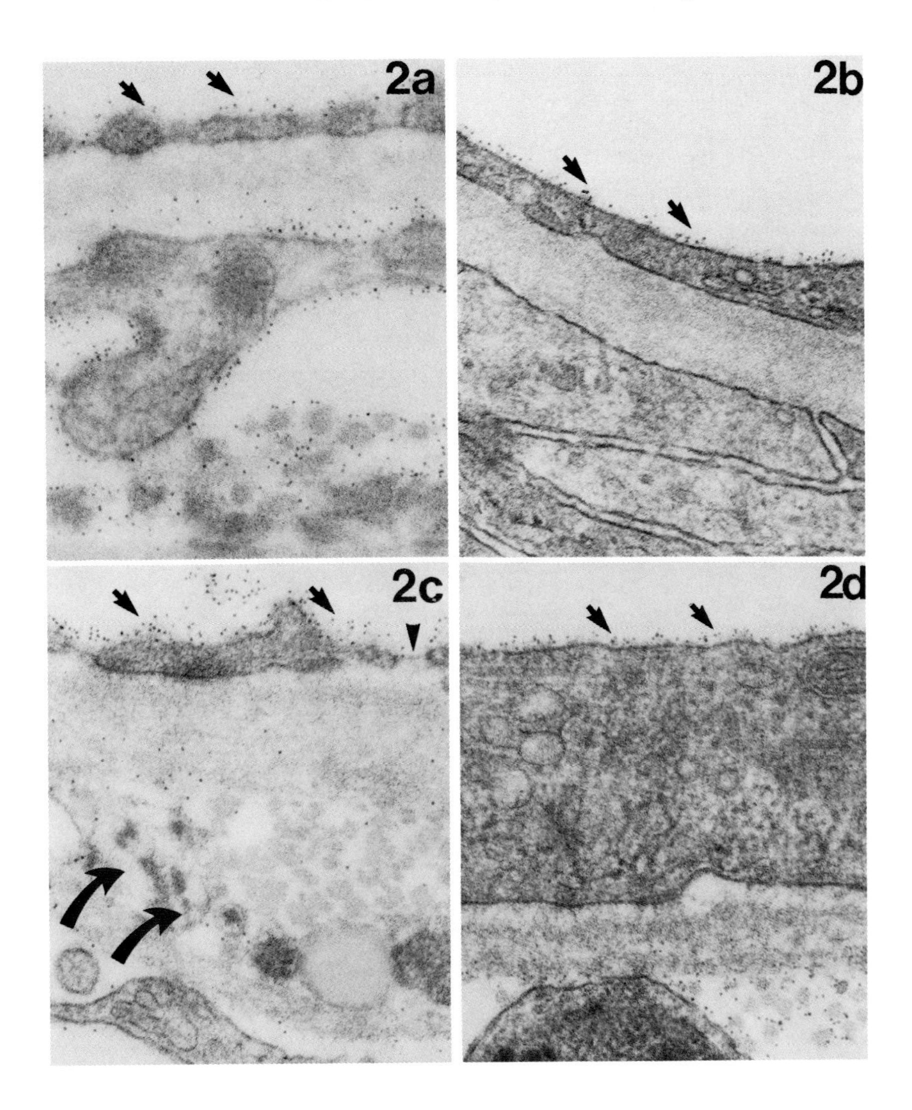

Figure 2. ConA-Fe binding (arrows) is sparse on the luminal plasma membrane of dystrophic CE (2a) and "normal" inner retinal vessels (2b). ConA-Fe binding in the RPE (2c) and VRM (2d) vessels is higher. The RPE vessel contains fenestrae (arrowheads) and is permeable to HRP since reaction produst is present abluminally (curved arrows). (Mag x 81,000)

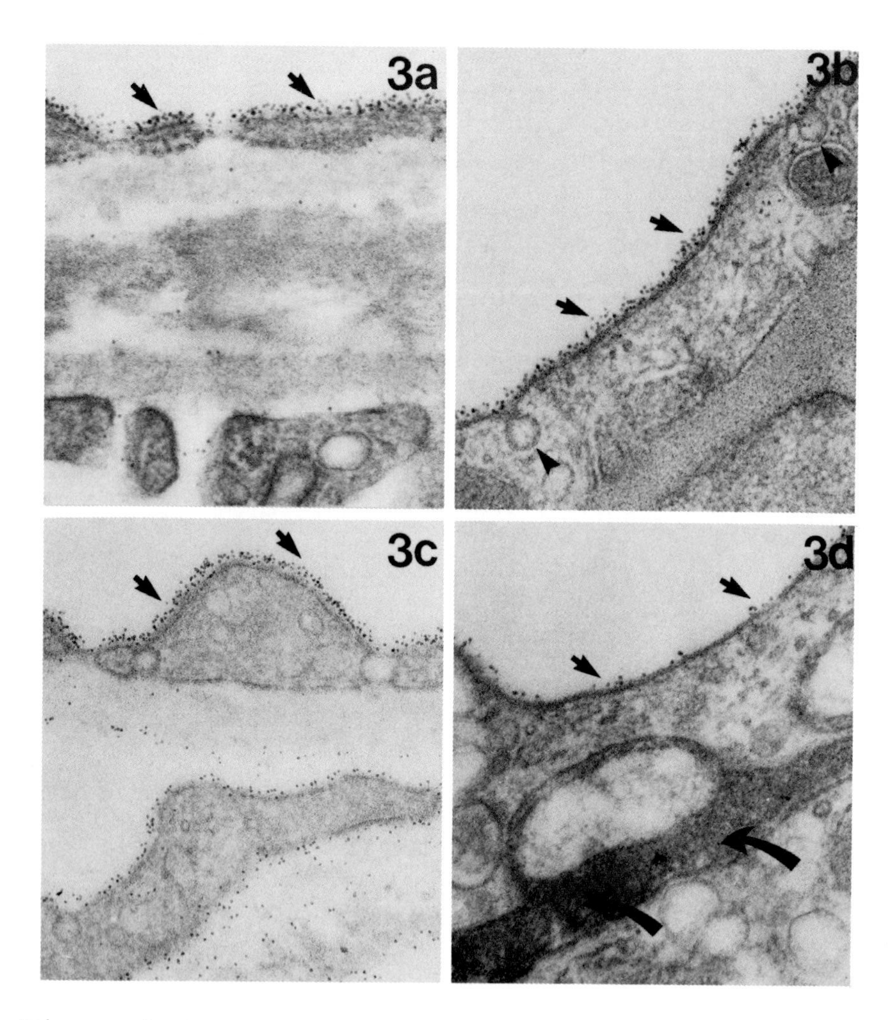

Figure 3. WGA-Fe binding on the luminal plasma membrane (arrows) is uniform in the dystrophic CE (3a), "normal" inner retinal vessels (3b) and the RPE-associated vessels (3c). WGA-Fe binding is greatly reduced in most VRM vessels of the RCS retina (3d). This change is often accompanied by increased HRP permeability in the VRM vessels (curved arrows). Note labeled and unlabeled uncoated luminal vesicles (arrowheads, 3b) (Mag x 81,000)

Competitive sugars

Addition of the competitive sugar mannose greatly reduced binding in both the control and dystrophic vessels. Only an occasional particle was observed on the plasma membrane (Fig. 4a). The competitive sugar CIT reduced, but did not abolish, WGA-Fe binding on the plasma membrane, the coated vesicles or pits, or the uncoated vesicles. A few particles remained on the plasma membrane, at the edge of vesicles, and on some of the vesicle diaphragms (Fig. 4b).

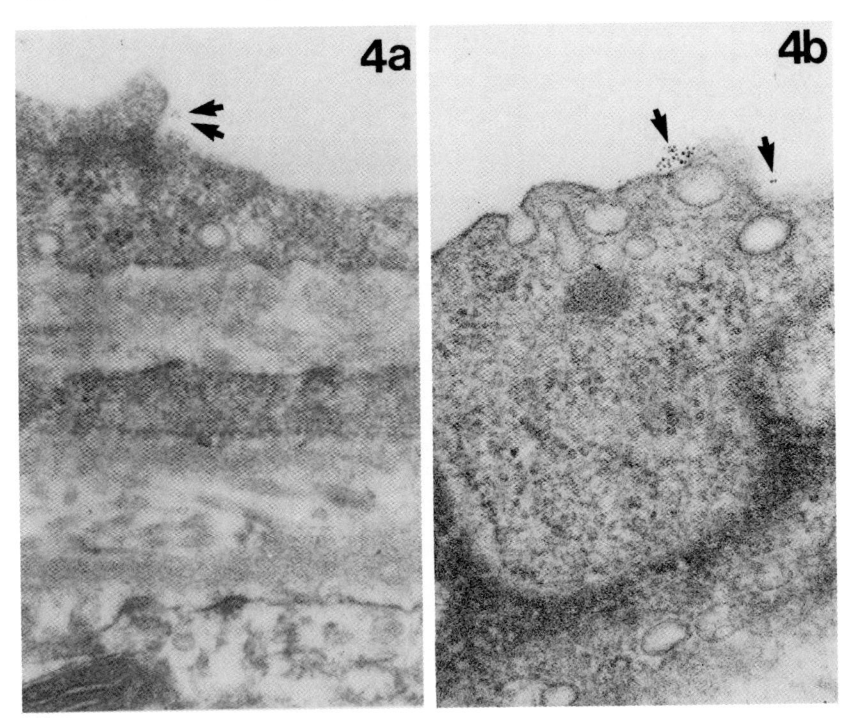

Figure 4. The competitive sugar, mannose, greatly reduces Con A-Fe binding (arrows, 4a). The competitive sugar CIT reduces, but does not abolish WGA-Fe binding (4b) with some particles remaining on vesicle diaphragms and on the plasma membrane (arrows). (Mag x 81,000)

HRP permeability

The control and "normal" RCS inner retinal vessels remained an intact barrier to HRP. However, the RPE and VRM vessels in the RCS rat were permeable to HRP, and the tracer was often observed in the basal lamina (Figs. 2c and 3d). The tight junctions in these permeable vessels appeared intact. The CE in both the control and dystrophic animals was also permeable to HRP (not shown). The WGA-Fe binding was reduced in the permeable VRM vessels (Fig. 3d).

DISCUSSION

In the RCS rat with inherited retinal dystrophy (Bourne et al., 1938), the RPE is defective and fails to phagocytize the photoreceptor outer segments, leading to debris build-up in the subretinal space, and eventual photoreceptor degeneration (Bok and Hall, 1971; Dowling and Sidman, 1962; Edwards and Szamier, 1977; Hall, 1977; Mullen and Lavail, 1976; Lavail, 1979; 1981). A breakdown of the blood-retinal barrier begins in the RCS rats at the level of the RPE after three weeks of age (Caldwell and McLaughlin, 1983). This permeability change is thought to be caused by a breakdown of the RPE tight junctions (Caldwell et al., 1982; 1984). An increase in vascular permeability is observed in some vessels of the outer retina after 10 weeks (Essner et al., 1979). This number of permeable vessels increases with age until at 28 weeks almost all of the outer retinal capillaries are permeable. The inner retinal vessels remain impermeable to HRP regardless of age or stage of retinal degeneration (Essner et al., 1979; 1980). During the later stages of the dystrophy, the inner retinal blood vessels degenerate (Gerstein and Dantzker, 1969; Matthes and Bok, 1984), followed by vascular proliferation in the RPE (Bok and Hall, 1971; Bok and Heller, 1980; Caldwell et al., 1988; Dowling and Sidman, 1962) and sometimes in the vitreous (Caldwell et al., 1988; Frank and Das, 1988).

The luminal microdomains of the inner retinal vessel non-fenestrated endothelium were: coated vesicles or pits and uncoated vesicles, diaphragms of uncoated vesicles, and the plasma membrane. Similar microdomains were observed in the CE as seen in the inner retinal vessels; in addition diaphragmed fenestrae and channels were observed. Membrane heterogenity has been suggested to regulate the uptake of a variety of molecules by endothelial cells (Simionescu et al., 1982; Vorbrodt, 1986). An alteration in either the distribution or number of anionic binding sites (Fitzgerald and Caldwell, 1987; Nag, 1984) or lectin binding sites (Fitzgerald and Caldwell, 1988; Vorbordt, 1986) on the endothelial cell luminal surface has been observed to be associated with increased uptake of HRP and/or permeability increases. These vascular alterations occurred during acute hypertension, cold-induced injury, or diabetes. This reorganization of the luminal surface microdomains during increased permeability suggests that normal organization of luminal surface microdomains contribute to the integrity of the blood-retinal and blood-brain barriers.

Our results showed that the RPE-associated vessels contained diaphragmed fenestrae and channels, these membrane microdomains were not present in normal inner retinal vessels. This alteration has been observed previously in proliferating blood vessels associated with abnormal RPE in the RCS rat (El-Hifnawi, 1987) as well as other retinal degenerative models (Bellhorn et al., 1980; Bellhorn and Korte 1983; Korte et al., 1983; Mancini et al., 1986).

ConA-Fe binding was increased on the plasma membrane of both the RPE and VRM vessels. WGA-Fe plasma membrane binding was reduced in most of the VRM vessels. WGA-Fe binding on most of the RPE-associated vessels appeared similar to the control and the RCS inner retinal vessels and CE. The ConA-Fe and WGA-Fe binding on the plasma membrane in both control and RCS inner retinal vessels and CE appeared similar to that seen by previous investigators (Fitzgerald and Caldwell, 1988; Pino, 1986; Pino and Thourton, 1986).

Control binding of the lectin-Fe conjugates in the presence of their hapten sugars demonstrated that ConA-Fe bound to mannose and that virtually all of the WGA-Fe binding was to NAG. While the CE and RPE-associated vessels were structurally similar, greater amounts of ConA-Fe binding in the RPE-associated vessels suggested an increase in the amount or accessibility of mannose residues. The VRM vessels also exhibited increased ConA-Fe binding. Both RPE-associated and VRM endothelium demonstrate characteristics of cell migration and proliferation such as endothelial microvilli and a multilayered basal laminia (Williams et al., 1988). These changes may be related to membrane differences between migrating and proliferating endothelial cells and the stable, mature endothelium of the inner retinal and CE vessels. Lectin binding studies on cultured neurons showed increased binding on the proliferating growth cones when compared with the cell body (Pfenninger and Maylie-Pfenninger, 1981). These results suggest that migration and growth influences the membrane sugar residues on the cell membrane. In the RPE and VRM vessels, either the number of mannose residues present, or the conformation of a glycoconjugate may be altered, allowing greater binding with the lectin probe. These changes could be due to a change in a basic metabolic pathway or extrinsic factors modifing or masking luminal surface sugar residues.

Reduction of the WGA-Fe binding on the VRM vessels is reminiscent of our previous finding in the BB rats, in which we observed decreased WGA-Fe binding associated with increases in HRP uptake (Fitzgerald and Caldwell, 1988). This reduction of WGA-Fe binding may reflect increased membrane turnover, masking of the surface sugar residues, or degradation of terminal sugar residues. WGA-Fe binding was not altered in the RPE-associated vessels. The increase in HRP permeability of the RPE-associated vessels may relate to the presence of fenestrae, since the fenestrated CE is also permeable. We did not observe fenestrae in the VRM vessels. This is probably due to our sampling; because, they have been observed by other investigators (Frank and Mancini, 1986; Wallow and Geldner, 1980; Williams et al., 1988). These data support the hypothesis that increases in endothelial cell permeability are associated with alterations in plasma membrane composition as well as in the distribution of luminal membrane microdomains.

REFERENCES

Bellhorn RW, Burns MS, Benjamin JV (1980). Retinal vessel abnormalities of phototoxic retinopathy in rats. Invest Ophthalmol Vis Sci 19:584-585.

Bellhorn RW, Korte GE (1983). Permeability of abnormal blood-retinal barrier in rat photoxic retinopathy: A clinicopathologic correlation study using fluorescent markers. Invest Ophthalmol Vis Sci 24:972-975.

Blair NP, Tso MOM, Dodge JT (1984). Pathological studies of the blood-retinal barrier in the spontaneously diabetic BB rat. Invest Ophthalmol Vis Sci 25:302-311.

Bok D, Hall, MO (1971). The role of the pigment epithelium in the etiology of inherited retinal dystrophy in the rat. J Cell Biol 49:664-682.

Bok D, Heller J (1976). Autoradiographic localization of serum retinol-binding protein receptors on the pigment epithelium of dystrophic rat retinas. Invest Ophthalmol Vis Sci 19:1405-1414.

Bourne MC, Campbell DA, Tansley K (1938). Hereditary degeneration of the rat retina. Brit J Ophthalmol 22:613-623.

Caldwell RB, McLaughlin BJ (1983). Permeability of retinal pigment epithelial cell junctions in the dystrophic rat retina. Exp Eye Res 36:415-427.

Caldwell RB, McLaughlin BJ, Boykins LG (1982). Intramembrane changes in retinal pigment epithelial cell junctions of the dystrophic rat retina. Invest Ophthalmol Vis Sci 23:305-318.

Caldwell RB, Slapnick SM, McLaughlin BJ (1985). Lanthanum and freeze-fracture studies of retinal pigment epithelial cell junctions in the streptozotocin diabetic rat. Curr Eye Res 4:215-227.

Caldwell RB, Slapnick SM, McLaughlin BJ (1986). Decreased anionic sites in Bruch's membrane of spontaneous and drug-induced diabetes. Invest Ophthalmol Vis Sci 27:1691-1697.

Caldwell RB, Slapnick SM, McLaughlin BJ (1987). Freeze-fracture quantitative study of retinal pigment epithelial cell basal membranes in diabetic rats. Exp Eye Res 44:245-259.

Caldwell RB, Wade LA, and McLaughlin BJ (1984). A quantitative study of intramembrane changes during cell junctional breakdown in the dystrophic rat retinal pigment epithelium. Exp Cell Res 150:104-117.

Caldwell RB, Slapnick SM, Roque, RS (1988). RPE-associated extracellular matrix changes accompany retinal vascular proliferation and retino-vitreal membranes in a new model for proliferative retinopathy: the dystrophic rat. Contained in this volume.

Cunha-Vaz JG (1976). The blood-retinal barriers. Doc Opthalmol, 41:287-327.

Cunha-Vaz JG (1979). The blood-ocular barriers. Survey Opthalmol, 23:279-296.

Cunha-Vaz JG (1980). Sites and functions of the blood-retinal barriers. In Cunha-Vaz JG (ed): "The blood retinal barriers," New York:Plenum Press, pp.101-117.

Dowling JE, Sidman (1962). Inherited retinal dystrophy in the rat. J Cell Biol 14:73-109.

Edwards RB, Szamier RB (1977). Defective phagocytosis of isolated rod outer segments by RCS rat retinal pigment epithelium in culture. Science 197:1001-1003.

El-Hifnawi E (1987). Pathomorphology of the retina and its vasculature in hereditary retinal dystrophy in RCS rats. In Zenner E, Goebel H, Krastel H (eds): "Research in Retinitis Pigmentosa" Oxford:Pergamon Press, pp 417-434.

Essner E, Pino RM, and Griewski, RA (1979). Permeability of retinal capillaries in rats with inherited retinal degeneration. Invest Ophthalmol Vis Sci 18:859-863

Essner E, Pino RM, and Griewski, RA (1980). Breakdown of blood retinal barrier in RCS rats with inherited retinal degeneration. Laboratory Invest 43:418-426.

Fitzgerald MEC, Caldwell RB (1987). Permeability increases correlate with changes in luminal surface anionic binding sites in the retinal microvasculature of diabetic rats. Soc Neurosci Abst 13:773.

Fitzgerald MEC, Caldwell RB (1988). Lectin-ferritin binding on spontaneously diabetic and control rat retinal microvasculature. Submitted Curr Eye Res.

Frank RN, Das A (1988). Models of retinal neovascularization in the rat. Invest Ophthalmol Vis Sci 29:245.

Frank RN, Mancini MA (1986). Presumed retinovitreal neovascularization in dystrophic retinas of spontaneously hypertensive rats. Invest Ophthalmol Vis Sci 27:346-355.

Gerstein DD, Dantzker DR (1969). Retinal vascular changes in hereditary visual cell degeneration. Arch Ophthalmol 81:99-105.

Goldberg F (1980). Diseases affecting the inner blood-retinal barrier. In Cunha-Vaz JG (ed.): "The blood retinal barriers", New York:Plenum Press, pp.309-364.

Graham RC, Karnovsky MJ (1966). The early stages of absorption of injected horseradish peroxidase in the proximal tubules of mouse kidney: ultrastructural cytochemistry by a new technique. J Histochem Cytochem 14:291-302.

Hall MO (1977). Phagocytosis by rat pigment epithelial cells in tissue culture. Invest Ophthalmol Vis Sci (Suppl) 16:111.

Ishibashi, T., Tanaka, K. and Taniguchi, Y. (1980). Disruption of blood-retinal barrier in experimental diabetic rats: an electron microscopic study. Exp Eye Res 30:401-410.

Ishibashi, T., Tanaka, K. and Taniguchi, Y. (1983). Electron Microscopic examination of retinal vascular changes in diabetic rats. In Abe H, Hoshi M (eds.) : "Diabetic Microangiopathy" Tokyo: Japan Medical Res Foundation, 20:445-455.

Kirber WM, Nichols CW, Grimes PA, Wingrad AI, Laties AM (1980). A permeability defect of the retinal pigment epithelium. Occurrence in early streptozotocin diabetes. Arch Ophthalmol 98:725-728.

Korte GE, Bellhorn RW, Burns MS (1983). Ultrastructure of blood-retinal barrier permeability in rat phototoxic retinopathy. Invest Ophthalmol Vis Sci 24:962-971.

Lavail MM (1979). The retinal pigment epithelium in mice and rats with inherited retinal degeneration. In Zinn KM and Marmor MF (eds): "The Retinal Pigment Epithelium," Cambridge: Harvard University Press, p. 372.

Lavail MM (1981). Analysis of neurological mutants with inherited retinal degeneration. Invest Ophthalmol Vis Sci 21:638-657.

Manncini MA, Frank RN, Keirn RJ, Kennedy A, Khoury JK (1986). Does the retinal pigment epithelium polarize the choriocapillaris. Invest Ophthalmol Vis Sci 27:336-345.

Matthes MT, Bok D (1985). Blood vascular abnormalities in animals with inherited retinal degeneration. In LaVail MM, Hollyfield JG, Anderson RE (eds): "Retinal Degeneration Experimental and Clinical Studies," New York: Alan R. Liss, Inc., pp.209-237.

Mullen RJ, Lavail MM (1976). Inherited retinal dystrophy: Primary defect in pigment epithelium determined with experimental rat chimeras. Science 192:799-801.

Pfenninger KH, Maylie-Pfenninger MF (1981). Lectin labeling of sprouting neurons. I. Regional distribution of surface glycoconjugates. J Cell Biol 89:536-546.

Pino RM, Thouron CL (1986). Identification of lectin-receptor monosaccharides on the

endothelium of retinal capillaries. Curr Eye Res 5:625-628.
Pino RM (1986). The cell surface of a restrictive endothelium. I. Distribution of lectin-receptor monosaccharides on the choriocapillaris. Cell Tissue Res 243:455-155.
Sima AAF, Chakrabarti S, Garcia-Salinas R, Basu PK (1985). The BB rat an authentic model of human diabetic retinopathy. Curr Eye Res 4:1087-1092.
Simionescu, M., Simionescu, N. and Palade, G.E. (1982) Differentiated microdomains on the luminal surface of capillary endothelium: Distribution of lectin receptors. J. Cell Biol. 94:406-413.
Vinores SA, Campochiaro PA, May EE, Blaydes SH (1988). Progressive ultrastructural damage and thickening of the basement membrane of the retinal pigment epithelium in spontaneously diabetic BB rats. Exp Eye Res 46:545-558.
Vorbrodt, A.W. (1986) Changes in the distribution of endothelial surface glycoconjugates associated with altered permeability of brain micro-blood vessels. Acta Neuropathol. 70:103-111.
Wallow IHL, Geldner PS (1980). Endothelial fenestrae in proliferative diabetic retinopathy. Invest Ophthalmol Vis Sci 19:1176-1183.
Wallow IHL, Engerman RL (1977). Permeability and patency of retinal blood vessels in experimental diabetes. Invest Ophthalmol Vis Sci 16:447-461.
Williams JM, deJuan E, Machemer R (1988) Ultrastructural characteristics of new vessels in proliferative diabetic retinopathy. Am J Ophthalmol 105:491-499.

Acknowledgements:

Supported by NIH-EY04618 to RBC and TN Neuroscience Center of Excellence Fellowship to MECF. We express appreciation for the excellent technical assistance of S. Frase, D. Lum, and L. Perry.

Inherited and Environmentally Induced
Retinal Degenerations, pages 427–439

FATTY ACID METABOLISM IN NORMAL MINIATURE POODLES AND THOSE AFFECTED WITH PROGRESSIVE ROD-CONE DEGENERATION (prcd)

M.G.Wetzel*[1], C.Fahlman[2], M.B.Maude*[3], R.A.Alvarez*[3], P.J.O'Brien[1], G.M.Acland[2], G.D.Aguirre[2], and R.E.Anderson[3]

[1]National Eye Institute, NIH, Bethesda, MD, [2]School of Veterinary Medicine, University of Pennsylvania, Philadelphia, PA and [3]Baylor College of Medicine, Houston, TX
*Supported by the RP Foundation Fighting Blindness

INTRODUCTION

The current study describes recent research on fatty acids in blood plasma and retinas of miniature poodles with the genetic mutation known as progressive rod-cone degeneration (prcd). This animal model of retinitis pigmentosa has many similarities to the human disease, particularly the slow progressive loss of retinal function and disease which initially affects the photoreceptor outer segments. Previous studies (Aguirre et al., 1982, Aguirre and O'Brien, 1986) have shown that rod outer segments in the affected dogs first appear morphologically abnormal in the posterior pole and equator by about three months after birth. Intact rod and cone outer segments are seen throughout the retina in electron micrographs until this time. In stage 1 of the disease, disorientation of lamellar disc membranes and some membrane vesiculation is evident in rod outer segments. Stage 2 is characterized by the degeneration of rods and the reduction in width of the outer nuclear layer. In the affected poodles, rods degenerate more rapidly than cones. Eventually there is complete loss of the electroretinogram between two and three years of age and end stage retinal degeneration subsequently develops.

Prior studies of rod outer segment renewal in affected poodles using [^{3}H] leucine-(Aguirre et al., 1982) or fucose-(Aguirre and O'Brien, 1986) labeled opsin showed a slower renewal rate, initially 35-45% of normal, even before morphologic or electroretinographic abnormalities were detectable. It would thus appear that a defect in the elaboration of some constituent uniquely necessary for outer segment renewal might be involved in this type of retinal degeneration. In the previous studies, no abnormalities in either opsin synthesis or glycosylation could be detected 24 hours after intravitreal injection of the leucine or fucose. Consequently, the present investigation has begun to assess the synthesis of lipids, the other major component of outer segment membranes, following intravitreal injection of labeled precursors. In addition, we have adapted the technique of in vitro incubation of trephined retinal punches for the short term study of both protein and lipid synthesis. Plasma samples from normal and affected dogs have also been analyzed for fatty acid and vitamins A and E (Anderson et al, 1988). The results suggest that a defect or deficiency in lipid metabolism could be involved in the prcd mutation in poodles.

The composition of rod outer segment membranes is known to be unique in several respects including the presence of phospholipids containing exceptionally large amounts of polyunsaturated fatty acids. In particular, the essential fatty acid docosahexaenoic acid, a 22 carbon chain with six double bonds (22:6w3), makes up about half of the esterified fatty acids of outer segment phospholipids (Anderson and Maude, 1970, 1972). The functional significance of 22:6w3 in the outer segment is presently unknown although dietary deficiency of linolenic acid (18:3w3), the precursor of 22:6w3, has been shown (Wheeler et al., 1975) to cause altered electrical responses in rat retinas.

Little is known about the biological pathways involved in 22:6w3 synthesis and transport. Although earlier experiments demonstrated some chain elongation and desaturation of 20:5w3 to 22:6w3 in rat retina in vivo (Bazan et al., 1982), recent evidence (Cai et al., 1988) indicates that 18:3w3, the major dietary precursor of 22:6w3, is converted to 22:6w3 in the liver. The 22:6w3 is then distributed to other tissues via the blood. The present investigation focused on the distribution of fatty acids including 22:6w3 in the blood plasma and the incorporation of radiolabeled 18:3w3 and 22:6w3 in the photoreceptor outer segments of normal and affected poodles.

RESULTS

In the current study, fasting plasma was obtained from normal and prcd affected miniature poodles. Samples were extracted and phospholipids separated from neutral lipids by preparative thin-layer chromatography. A portion of the phospholipid was used for phosphorus analysis and an equal amount taken for methyl ester analysis (Wiegand and Anderson, 1982). Initially a total of twenty affected and twenty controls were examined (Table 1)(Anderson et al., 1989). The amount of 22:6w3 was significantly lower in the affected dogs compared to controls while other fatty acids did not differ from controls. This difference was consistently observed whether the concentration of 22:6w3 was expressed per ml of plasma or per ug of lipid phosphorus. These analyses were repeated on sixteen affected and sixteen controls 4-5 months later with similar decreases noted in the level of 22:6w3 (Anderson et al., 1989). A third group of samples has also shown a significant decrease in plasma 22:6w3 in affected poodles. No differences were seen in the levels of 20:4w6, 22:5w3, or vitamins E and A (Anderson et al, 1988, 1989). In addition, the ratios of 22:6w3/22:5w3 were significantly different between affected and controls. Since 22:5w3 is the immediate precursor of 22:6w3 through a delta-4 desaturating enzyme, these results are consistent with a decrease in delta-4 desaturase activity. If this is true, then this would be the first instance of a specific enzyme defect in any form of retinitis pigmentosa or in an animal model of the human disease.

The second part of the current study of lipids in prcd poodles focused on the biochemical characterization of canine outer segment lipids of both normal and affected poodles. The distribution of individual lipids was determined by extraction of crude or gradient purified outer segments and separation of neutral lipids by 1-dimensional thin layer chromatography and phospholipid

TABLE 1

PLASMA LIPIDS IN MINIATURE POODLES (n=20)

PARAMETER	AFFECTED	CONTROLS	P-VALUE
ugm lipid-P/ ml plasma	83.9 ± 12.8	90.5 ± 15.1	NS
nmol 22:6w3/ ml plasma	32.9 ± 15.3	44.9 ± 12.6	<0.005
nmol 22.5w3/ ml plasma	52.4 ± 34.0	49.3 ± 15.1	NS
nmol 20:4w6/ ml plasma	867 ± 258	903 ± 209	NS
nmol 22:6w3/ ugm P	0.39 ± 0.13	0.50 ± 0.12	<0.005
nmol 20:4w6/ ugm P	10.2 ± 1.9	10.0 ± 2.0	NS
22:6w3/22:5w3	0.57 ± 0.17	0.95 ± 0.33	<0.005
nmol vitA/ ml plasma	2.76 ± 0.91	2.98 ± 0.69	NS
nmol vitE/ ml plasma	21.1 ± 4.0	22.6 ± 8.3	NS

classes by 2-dimensional thin layer chromatography (Wetzel et al., 1988). Table 2 summarizes the phospholipid composition of outer segments from normal dog, rat, cow and frog. It may be seen that phosphatidylethanolamine (PE) and phosphatidylcholine (PC) are the two major constituents. There is slightly more PC and less PE in dog outer segments than in rat or cow. The overall composition of dog outer segments more closely resembles that of frogs.

Retinal lipids were labeled by intravitreal injection with [^{3}H] palmitic acid and [^{14}C] 22:6w3 or 18:3w3 for 24 hours prior to sacrifice. Table 3 presents data on the distribution of radiolabeled palmitic acid, 22:6w3 and 18:3w3 in the outer segments of normal and affected animals (Wetzel et al., 1988, 1989). No significant difference in uptake was noted between the normal and affected retinas. Most of the labeled fatty acids were esterified to phospholipids or remained in a pool of unesterified fatty acids. About 6 to 9% of each fatty acid was esterified to neutral lipid (either mono-, di- or

TABLE 2

PHOSPHOLIPID CLASS COMPOSITION OF VERTEBRATE ROS

PHOSPHOLIPID CLASS	PERCENTAGES DOG[a]	RAT[b]	COW[c]	FROG[d]
PI	2.6	2.3	2.5	2.2
PS	15.4	14.2	12.9	12.8
PC	46.7	29.4	41.2	45.3
PE	34.1	46.8	39.0	34.6
SPH	1.1	1.9	2.9	1.9

a Wetzel et al., 1988 c Anderson and Maude, 1970
b Wetzel and O'Brien, 1986 d Anderson and Risk, 1974

TABLE 3

PERCENT DISTRIBUTION OF FATTY ACIDS IN ROS LIPIDS LABELED INTRAVITREALLY FOR 24 HOURS

FATTY ACID	UNESTERIFIED FATTY ACID	NEUTRAL LIPID	PHOSPHOLIPID
[^{3}H] PALMITIC			
NORMAL (4)	10.7±10.0	8.1±1.6	80.1±11.3
AFFECTED (4)	10.6±9.3	8.1±2.8	80.4±12.3
[^{14}C] 22:6w3			
NORMAL (4)	8.5±5.1	5.7±0.9	84.9±5.5
AFFECTED (4)	9.9±3.8	6.1±1.4	83.2±4.7
[^{14}C] LINOLENIC			
NORMAL (3)	9.7±3.4	9.3±0.8	78.9±3.8
AFFECTED (4)	17.3±19.7	8.0±1.2	72.5±20.4

(n) = Number of eyes; Values are means ± standard deviation.

TABLE 4

PERCENT DISTRIBUTION OF FATTY ACIDS IN ROS PHOSPHOLIPIDS LABELED INTRAVITREALLY FOR 24 HOURS

FATTY ACID	PHOSPHOLIPID CLASS			
	PI	PS	PC	PE
[^{3}H] PALMITIC				
NORMAL (7)	3.4±0.4	3.1±0.7	64.6±3.1	22.3±4.1
AFFECTED (8)	3.2±0.6	3.0±0.5	67.9±3.8	17.3±2.0
[^{14}C] 22:6w3				
NORMAL (2)	2.5±0.9	11.6±2.8	26.6±2.5	49.2±1.0
AFFECTED (2)	3.8±1.0	12.4±0.2	24.5±2.3	51.1±0.4
[^{14}C] LINOLENIC				
NORMAL (3)	6.0±0.4	9.6±0.7	48.9±1.7	33.2±1.8
AFFECTED (4)	4.3±0.9	8.1±1.2	53.5±3.4	29.4±3.3

(n) = Number of eyes sampled
Values are means ± standard deviation

triglyceride) in the outer segment. Table 4 summarizes the incorporation of palmitic acid, 22:6w3 and 18:3w3 into individual phospholipid classes in outer segments from these same intravitreally injected retinas (Wetzel et al., 1988, 1989). Palmitic acid was initially esterified primarily to PC while 22:6w3 was enriched in PE and PC and to a lesser extent in phosphatidylserine (PS). Different quadrants of the individual eyes varied in the total amount of isotope incorporated, while the percent distribution remained remarkably constant. This variation in uptake was presumably due to differences in the intravitreal injection, as in vitro incubations did not exhibit this variability in incorporation. No significant difference in uptake was noted between the normal and affected retinas. What was striking, however, was the fact that intravitreally injected 18:3w3 was not distributed in the same pattern as 22:6w3. The characteristic labeling pattern for 22:6w3 was a very high concentration in PE while 18:3w3 was esterified predominantly to PC.

Lipids were extracted from pieces of retina from control and affected animals and subjected to acetolysis (Renkonen, 1965; Choe, et al., 1989). In this reaction, the phosphate-base groups of phospholipids are replaced by acetate, yielding a diacylglycerol acetate (DGAC), which can be separated by high performance liquid chromatography (HPLC) into various molecular species. An HPLC tracing of DGAC from a control retina is shown in Figure 1. The solid line depicts absorbance at 210, characteristic of double bonds, and the dashed lines represent label from [^{3}H] palmitate (short dashes) and [^{14}C] linolenate (long dashes). Several points are noteworthy. There was no elongation of 18:3w3 to 22:6w3, evidenced by the absence of [^{14}C]-radioactivity in the 22:6-22:6 and 18:0-22:6 molecular species. Rather, most of the 18:3w3 appears to be incorporated unchanged into phospholipids. The largest amount of [^{14}C]-radioactivity (see arrow) eluted in an area that contains no identified molecular species of retinal phospholipids. The absence of elongation and desaturation was confirmed by silver nitrate thin layer chromatography (Dudley and Anderson, 1975) of methyl esters of the retinal phospholipids, which showed that most of the radioactivity still remained in the trienoic fraction, with only small amounts in the more polyunsaturated fatty acids. An interesting feature of palmitate incorporation was that most was incorporated into molecular species containing none or one double bond. Also, there was no elongation of palmitate to stearate, evidenced by the absence of [^{3}H]-radioactivity in 18:0-22:6. Lipids from affected dog retinas had a pattern of distribution of radioactivity in individual molecular species similar to that of the control. We therefore conclude that the retinas from both control and affected animals have limited capacity for elongation and desaturation of 18:3w3 to 22:6w3. The retina may therefore be dependent upon blood lipids to supply the essential 22:6w3 necessary for normal outer segment phospholipid synthesis. Whether this is a characteristic of dogs or is a general finding among vertebrates needs to be investigated in greater detail.

In addition to in vivo labeling, we have also incubated trephined punches of retina in vitro for 1 to 4 hours with [^{3}H] palmitate and [^{14}C] 22:6w3 or 18:3w3. Figure 2 summarizes data from a pulse-chase experiment in which [^{3}H] palmitic acid was added for one hour, followed by a buffer chase of one or three hours. In vitro, much more of the fatty acid remains as free fatty acid compared with retinas extracted 24 hours after intravitreal injection. Here it may be seen that palmitate label increases in PC during the three hour chase while the

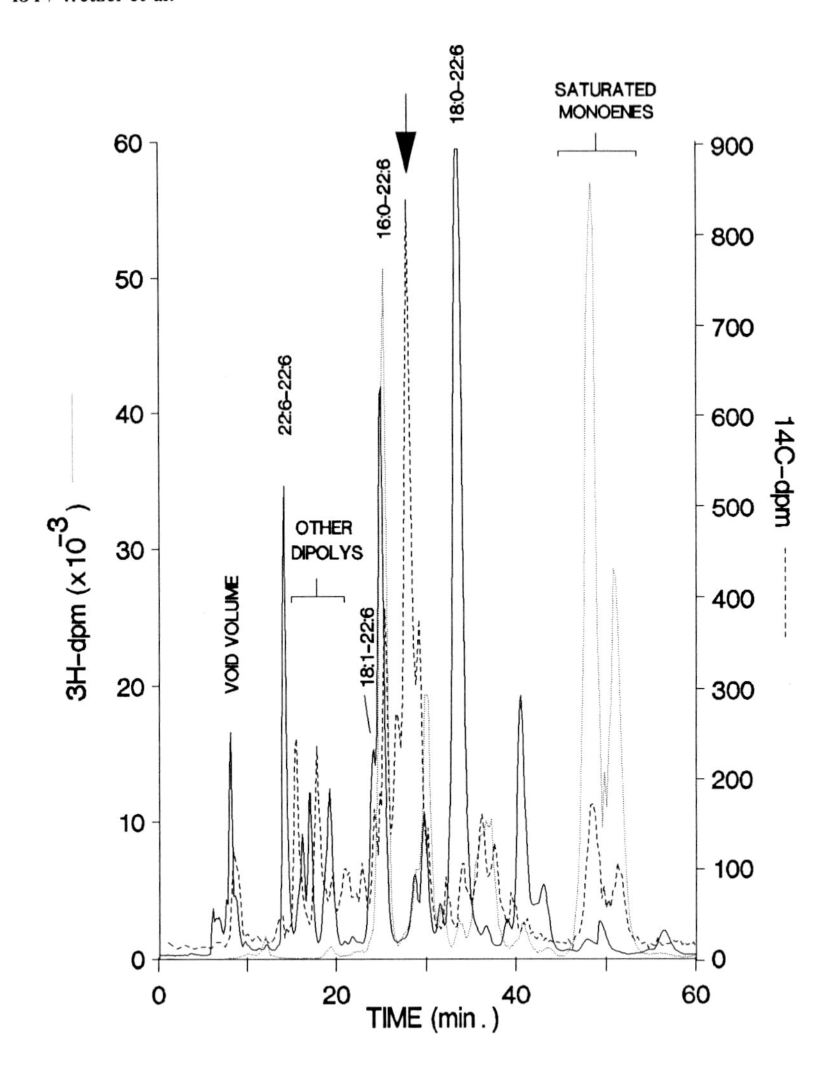

FIGURE 1: HPLC tracing of DGAC molecular species of lipids extracted from retinas of a control dog. Absorbance at 210 nm (fatty acid double bond) is depicted by the solid line. Radioactivity derived from [^{3}H] palmitate is depicted by short dashed lines and that from [^{14}C] linolenate is depicted by long dashed lines.

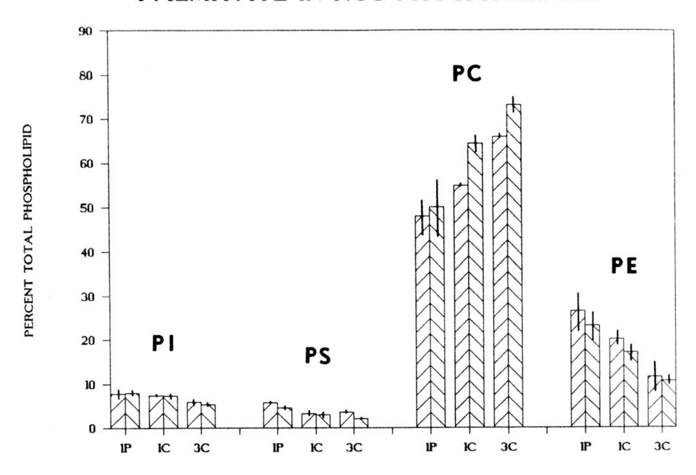

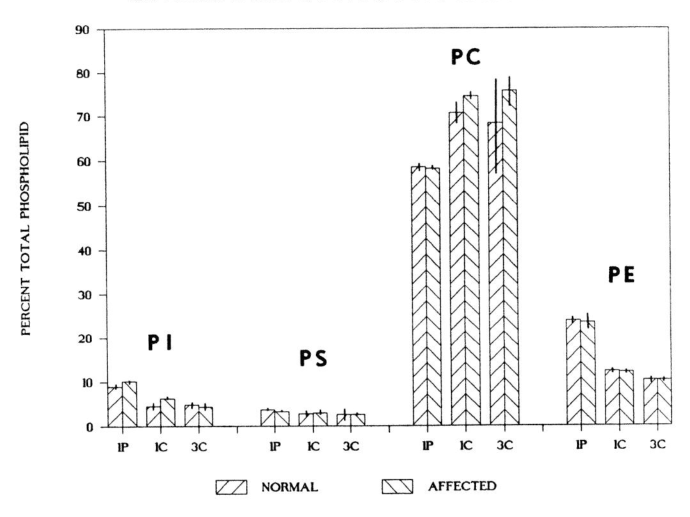

FIGURES 2 and 3: Four 3mm trephine punches of normal or affected poodle retinas per vial were incubated for a 1 hour pulse (1P) in [^{3}H] palmitic acid (Fig. 2) or [^{14}C] linolenic acid (Fig. 3) at 37°C and then in unlabeled medium for a 1 hour (1C) or 3 hour (3C) chase period. Values represent the average of two samples. There was an increase in labeling of PC and decrease in labeling of PE during the three hour chase period for both fatty acids.

TABLE 5

INCORPORATION OF RADIOLABELED FATTY ACIDS AND GLYCEROL INTO ROS PHOSPHOLIPIDS FOLLOWING 2 HOUR INCUBATION

ISOTOPE	PHOSPHOLIPID			
	PI	PS	PC	PE
[^{3}H] PALMITIC				
NORMAL (8)	4.8±0.5	0.8±0.2	71.7±2.8	9.9±0.6
AFFECTED (4)	4.5±0.4	0.6±0.1	76.0±1.5	8.0±0.5
[^{14}C] 22:6w3				
NORMAL (3)	12.1±13.0	7.0±1.4	48.6±5.9	24.4±3.7
AFFECTED (3)	14.8±8.9	4.5±0.8	45.7±3.3	28.9±3.2
[^{3}H] GLYCEROL				
NORMAL (6)	20.5±10.6	0.9±0.5	37.7±7.0	19.8±4.7
AFFECTED (3)	15.7±2.7	0.9±0.3	42.6±4.6	18.0±1.0

(n) = Number of samples, Four 3mm punches per sample.
Values are means ± standard error.

label decreases relatively in phosphatidylinositol (PI), PS and most dramatically in PE during this period. The percent distribution in phospholipids at the three hour chase timepoint is approaching levels just described for retinas labeled for 24 hours in vivo. Figure 3 shows similar data for trephined punches incubated with [^{14}C] 18:3w3 and shows very similar labeling patterns to those just described for palmitate. Again we see the increase in PC and relative decrease in PI and PE labeling with time. No significant difference in distribution of fatty acids was noted in the affected compared with normal retinas for either palmitic or linolenic acid.

Table 5 summarizes data on trephined punches incubated for two hours with palmitic acid, 22:6w3 or glycerol. The resulting phospholipid labeling patterns resemble those seen in vivo except that 22:6w3 now labels PC predominantly just like palmitate, then PE, then PI and lastly PS. Glycerol labels PI with the highest specific activity, PI constituting only 2-3% of the total ROS phospholipid. Glycerol labels PC and PE with about equal specific activity indicating de novo synthesis of these two phospholipids is similar while PS turnover is very slow. Much of the [^{3}H] palmitate incorporation represents acyl exchange to preexisting phospholipids rather than de novo synthesis (Zimmerman and Keys, 1988).

Again, no significant difference in incorporation of palmitate, 22:6w3 or glycerol was noted between affected and normal dogs. These experiments do not rule out the possibility that de novo synthesis of di-polyenoic species of phosphatidic acid (Bazan et al., 1982) may be abnormal in affected animals due to low serum levels of 22:6w3. The species of phosphatidic acid were not measured in the current study.

SUMMARY

It is possible that a genetic defect observed in the prcd poodle involves the abnormality of an enzyme which functions in phospholipid or lipoprotein metabolism. In our studies thus far, we have been unable to detect any defect in retinal phospholipid biosynthesis, but we have noted a decrease in plasma levels of 22:6w3 which may be a result of an enzyme defect in liver biosynthesis of 22:6w3 from its dietary precursor, linolenic acid, or some defect in the blood lipoprotein transport of this essential fatty acid. If 22:6w3 is essential to the normal elaboration and functioning of the photoreceptor outer segment, it is possible that decreased access to this fatty acid due to lower blood levels of 22:6w3 could cause photoreceptor abnormalities. Further studies are needed to confirm the possible defect in delta-4 desaturase activity and possible dietary modification of the course of this prcd retinal degeneration.

ACKNOWLEDGEMENTS

Supported in part by grants from the National Eye Institute (G.D.A. and R.E.A.) and the RP Foundation Fighting Blindness.

REFERENCES

Aguirre G, Alligood J, O'Brien P, Buyukmihci N, (1982). Pathogenesis of progressive rod-cone degeneration in miniature poodles. Invest Ophthal Vis Sci 23:610-630.

Aguirre G, O'Brien P, (1986). Morphological and biochemical studies of canine progressive rod-cone degeneration. Invest Ophthal Vis Sci 27:635-655.

Anderson RE, Maude MB, (1970). The phospholipids of bovine rod outer segments. Biochem 9:3624-3628.
Anderson RE, Maude MB (1972). Lipids of ocular tissues VIII. The effects of essential fatty acid deficiency on the phospholipids of the photoreceptor membranes of rat retina. Arch Biochem Biophys 151:270-276.
Anderson RE, Maude MB, Alvarez RA, Acland GM, Aguirre GD, (1988) Plasma levels of docosahexaenoic acid in miniature poodles with an inherited retinal degeneration. Invest Ophthal Vis Sci 29:169.
Anderson RE, Maude MB, Alvarez RA, Acland GM, Aguirre GD, (1989). Plasma levels of polyunsaturated fatty acids in dogs affected with inherited retinal degenerations. Biochem Biophys Res Communs (submitted for publication).
Anderson RE, Risk M, (1974). The phospholipids of frog photoreceptor membranes. Vision Res 14:129-131.
Bazan HEP, Careaga, MM, Sprecher H, Bazan NG, (1982). Chain elongation and desaturation of eicosapentaenoate to docosahexaenoate and phospholipid labeling in the rat retina in vivo. Biochim Biophys Acta 712:123-128.
Cai F, Scott BL, Bazan NG, (1988). Delivery of omega-3 fatty acids to developing photoreceptor cells. Invest Ophthal Vis Sci 29:245.
Choe H-G, Wiegand RD, Anderson RE (1989). Quantitative analysis of retinal glycerolipid molecular species acetylated by acetolysis. J Lipid Res (in press).
Dudley PA, Anderson RE, (1975). Separation of polyunsaturated fatty acids by argentation thin layer chromatography. Lipids 10:113-114.
Renkonen O, (1965). Individual molecular species of different phospholipid classes. J Am Oil Chem Soc 42:298-304.
Wetzel MG, Fahlman C, Alligood JP, O'Brien PJ, Aguirre GD, (1988). Metabolic labeling of normal canine rod outer segment phospholipids in vivo and in vitro. Exp Eye Res (in press).
Wetzel MG, O'Brien PJ, (1986). Turnover of palmitate, arachidonate and glycerol in phospholipids of rat rod outer segments. Exp Eye Res 43:941-954.
Wetzel MG, Fahlman C, O'Brien PJ, Aguirre GD, (1989). Metabolic labeling of rod outer segment phospholipids in miniature poodles with progressive rod-cone degeneration (prcd). Exp Eye Res (submitted for publication).

Wheeler, TG, Benolken, RM, Anderson, RE (1975). Visual membranes: Specificity of fatty acid precursors for the electrical response to illumination. Science 188:1312-1314.

Wiegand RD, Anderson, RE, (1982). Determination of molecular species of rod outer segment phospholipids. Methods in Enzymology 81:297-304.

Zimmerman WF, Keys S, (1988). Acylation and deacylation of phospholipids in isolated bovine rod outer segments. Exp Eye Res 47:247-260.

Inherited and Environmentally Induced
Retinal Degenerations, pages 441–454

ANALYSIS OF NORMAL AND rcdl IRISH SETTER RETINAL PROTEINS

Jess Cunnick and Dolores Takemoto

Department of Biochemistry, Kansas State University, Manhattan, Kansas 66506

INTRODUCTION

The retina, an integral part of the central nervous system, is a primary visual transduction system for vertebrates. The photoreceptor layer of the retina is made up of rods and cones which each serve a specific function. The cones are primarily distributed in the fovea of the eye, and, as such, receive intense radiation. The rods are found in the peripheral regions of the eye, receive less radiation, and function in dark-light vision.

There are a number of inherited diseases of these photoreceptor cells which often lead to retinal degeneration and blindness in the human population. The causes of these diseases are not known, but the specialized nature of the cells involved and the constant exposure to radiation damage make them especially susceptible to damage from environmental causes or from inherited defects.

Animal mutants with inherited retinal degenerations or dysplasias have been studied for a number of years as possible models for human retinal diseases. These animal models are varied and can be broadly grouped into two classes. The first is the early onset model which includes the <u>rd</u> (LaVail and Sidman, 1974) mouse and the rcdl (Aguirre, et al., 1982a) Irish Setter. The early onset diseases occur at the time of photoreceptor maturation and are referred to as dysplasias since defects occur in tissue development and growth. The second broad class includes the late onset animal models which exhibit degeneration of retinal cells after photoreceptor maturation. This class includes the prcd miniature poodle, among others (Aguirre, et al., 1982). It should be stressed that, although the time course of the disease may be similar within a class, we must assume at this time that this is fortuitous. Having no absolute knowledge of the biochemical cause of any of these specific diseases, we must assume that each has its own cause, until proven otherwise.

The defects which cause the degeneration or dysplasia could be in the phototransduction system, itself, or in something which is distal to this system. For example, it has been reported that elevated cyclic GMP levels lead to degeneration of human retinal tissue in culture (Ulshafer, et al., 1984). In the early onset models which exhibit this defect, the elevation in this cyclic nucleotide could be caused by alterations in the protein structure of a visual transduction component, including rhodopsin, transducin or phosphodiesterase (PDE).

Alternatively, defective membrane assembly of these proteins, which are otherwise normal, could lead to a similar defect. For example, in the prcd poodle the rod outer segments are shorter than normal but the disc morphology appears normal (Aguirre, et al., 1982; Aguirre and O'Brien, 1986). Defective membrane assembly is noted to occur in both of the above mentioned early onset animal models. In the Irish Setter and the rd mouse, the outer segments fail to develop the normal disc structures. This, once again, could be due to a specific defect in one protein or to a general defect in membrane assembly or maturation processes.

Some of these maturation processes would include protein sorting and posttranslational processing of specific proteins. There is much information concerning the necessity of posttranslational processing of proteins which are targeted for specific compartments within a cell. Glycosylation, for example, is a known covalent modification of cell receptors (Bretscher, 1985). Palmitlyation and inositol-phosphate anchoring are also employed by cells in order to assemble proteins properly onto membrane structures (Ferguson and Williams, 1988; Unwin and Henderson, 1984). Defects in any of these systems would cause a failure to develop normal discs in the rod outer segment or could cause later degenerations.

Likewise, protein sorting and assembly at the base of the outer segment certainly utilizes systems which are similar to other cells, including glycosylation, phosphorylation, ubiquitination, palmitlyation and numerous other covalent protein modifications (for a review on this, see Wickner and Lodish, H. 1985; Schekman, 1985). One would assume that misdirected or improperly modified proteins would be rapidly degraded in these systems, accounting for some of the reported lower levels of specific proteins.

At this time, one cannot generally rule out changes in lipid composition in some of these mutants, either. Certainly, changes in membrane fluidity are well known to alter enzymatic processes.

In the present report, we wish to examine one of the early onset animal models for rod dysplasia, the Irish Setter dog.

The rcdl Irish Setter

In 1975, Aguirre and Rubin identified an autosomal recessive mutation expressed in the retina of the Irish Setter (Aguirre and Rubin, 1975). This was an early onset disease, classified as a dysplasia, in which postnatal maturation of rod visual cells was arrested at the stage of development in which the outer segment of the rods begin to develop. The resulting outer segments of the affected dogs are short and are disorganized with few real disc structures. Although the normal and rcdl affected retina are histologically and functionally indistinguishable until approximately 16 days of age, between 16 and 23 days of age the development of the photoreceptor layer ceases in the rcdl retina. Beginning at 16 days, the rod nuclei appear pyknotic and by 20 days it is apparent that the rod outer segments (ROS) are reduced in number and are disorganized. It is also at this time that a defect in ROS disc assembly is observed. The retina of the rcdl Irish Setter never develops detectable rod mediated responses like those found in normal dogs (Buyukmihci, et al. 1980).

Defects in cyclic GMP levels have been observed just prior to the arrest of development and the onset of degeneration follows shortly thereafter (Aguirre, et al., 1982a). Differences have been observed in the localization of cyclic GMP within the rcdl retina beginning at 13 days of age, and a 5 to 10 fold overall increase in cyclic GMP has been shown to peak at approximately 3 weeks of age in the rcdl affected retina (Barbehenn, et al., 1988). While PDE is not light-activated in the rcdl dogs, it is present and is activated by histone (Lee, et al., 1985). Guanylate cyclase activity appears to be normal (Barbehenn, et al., 1988), as are the transducin (Navon, et al., 1987) and S-antigen proteins (Long, et al., 1986). While it is tempting to speculate that the elevated levels of cyclic GMP are due to an altered phosphodiesterase, evidence to confirm this theory is not yet available. Since a defect in any of the components of the visual cascade could cause an altered cyclic GMP level, at present, nothing can be ruled out.

Also, associated with this developmental and degeneration defect is a reduction of opsin phosphorylation (Schmidt, et al., 1986). This defect may be due to increased phosphatase activity during degeneration (Takemoto, et al., 1986), or, to a specific component of the disease etiology. We have detected other alterations in opsin using antisera raised against synthetic peptides which correspond to various regions of bovine rhodopsin (Takemoto, et al., 1985, and Cunnick, et al., 1988). Using antisera raised against a synthetic peptide which corresponds to the carboxyl terminal region of bovine rhodopsin, we have observed a decreased reactivity of this antisera with the undenatured rhodopsin of the rcdl affected retina. In this report, we have reviewed this information and additional data regarding alterations in rod outer segment proteins in the rcdl Irish Setter is presented.

MATERIALS AND METHODS

All Irish Setter eyes were made available from the colony, "Canine Models of Hereditary Retinal Degenerations", sponsored by NEI/NIH grant EY06855. Animals varied from postnatal day 10 through 74, and were designated as N (homozygous, normal), or rcdl (homozygous, affected).

Specific methods concerning the production of antipeptide antisera, the synthesis of peptides, and the solid-phase RIA have been published previously (Cunnick, et al., 1988).

For purification of tryptic peptides by hplc and for sequencing, rod outer segment discs were prepared from 31 day old rcdl and 32 day old normal Irish Setter retinas as previously described (Cunnick, et al., 1988) Isolated discs were depleted of associated proteins by repeated washing with 10 mM Tris (pH 7.4) plus 100 μM GTP. Tryptic peptides were generated by incubating the discs at 37°C for 4 hours in 50 mM ammonium bicarbonate (pH 7.8) with TPCK-treated trypsin at a 1:50 trypsin:disc protein ratio. After incubation, the reaction was terminated by addition of a two-fold excess of soybean trypsin inhibitor. The membranes were pelleted and the peptides from the supernatant were lyophilized.

The trypsin-generated peptides were then separated by reverse phase hplc on a C18 column using an acetonitrile gradient of 0-70% containing 0.1% trifluoroacetic acid. One ml fractions were collected and identified by a radioimmunoassay (Takemoto, et al., 1985). Positive controls included the synthetic peptide and a bovine rod outer segment preparation.

Following identification, each peptide was sequenced on a gas phase sequencer (Applied Biosystems).

RESULTS AND DISCUSSION

The affected photoreceptor cells of the rcdl dogs appear to possess the components of the visual cascade. At 16 days of age, rhodopsin is assembled into the outer segment membrane and it is synthesized at a rate near that of the normal dogs (Aguirre, et al., 1982a). However, by 20 days of age, rods show abnormal outer segment formation.

The transducin levels have been measured and found to be present in 34 day-old affected dogs, but at reduced levels (Navon, et al., 1987). In addition, S-antigen has also been measured and is present in these animals (Long, et al., 1986). Although these proteins are present, the ability of the transducin to interact with phosphodiesterase, or the S-antigen to interact with rhodopsin has not been investigated.

The phosphodiesterase has been shown to be present in its inhibited form by its interaction with a specific monoclonal antibody (Lee, et al., 1985). In addition, this protein is histone-activated as is the phosphodiesterase from the normal Irish Setter retinal photoreceptors. We have used an antipeptide antisera directed against a synthetic peptide which corresponds to the "conserved" region of the phosphodiesterase alpha subunit. This region has been hypothesized to contain the catalytic region since it shows extensive homology with all other known phosphodiesterases from all sources thus far sequenced (Charbonneau, et al., 1986). Table 1 lists the results of densitometric scans of Western blots which were reacted with this antisera (anti-740: corresponds to antisera directed against amino acid residues 740-755 of the bovine phosphodiesterase alpha sequence [Ovchinnikov, et al., 1987) and with an antisera directed against the last 15 amino acid residues of the bovine transducin alpha protein (Medynski, et al., 1985). Less reaction was observed with both antisera for the rcdl retinal samples, even at 10 days of age. If an individual retinal protein were being lost faster than another in the rcdl samples, then comparison of the ratios of these scans should indicate differences. For example, if all of the proteins were less, but no differences in individual losses were observed in the rcdl samples, then the ratios of the scans should be the same as those of the normal proteins. This was not the case. The ratio of the normal sample at ten days of age, when expressed as 740/Tα (reaction of the phosphodiesterase antisera/transducin alpha antisera) was about 19, while that of the rcdl sample was around 3. At 26 days of age, these ratios are around 7 for the normal dogs and 1 for the rcdl dogs. For both ages, the difference between the normal and rcdl ratios is approximately six fold. This indicates that PDE is present in smaller quantities than Tα. Comparative ratios of PDE to rhodopsin, Tβ & S-antigen were also indicative of reduced PDE levels observed in rcdl retinas (data not shown). Low remaining immunoreactive PDE is detectable even to 68 days, presumably because synthesis still takes place at this time. These low levels, observed as early as 10 days, may reflect steady state levels reached between the synthesis and, most likely, extensive degradation occurring in the rcdl dog retinas. We are uncertain as to what is the cause of the more rapid degradation of the phosphodiesterase. Similar enhanced loss of this protein has been reported to occur in the rd mouse (Farber, et al., 1988). In this case, a similar rate of synthesis was observed. If the phosphodiesterase does not form a proper complex, as in the rd mouse (Lee, et al., 1988), the protein may be targetted for destruction by

Table 1

Densitometric Scans of Autoradiograms of Western Blots

Sample	α740	αTα	α740/αTα	Fold Diff.
N10	823645	43686	18.85	
R10	63677	21156	3.00	6.3
N15	666262	72753	9.16	
R14	ND	13421	ND	
N26	192942	28212	6.84	
R27	15007	14148	1.06	6.5
N27	351759	41620	8.45	
R27	25330	16812	1.51	5.6
N29	189572	26066	7.27	
R28	49790	30365	1.64	4.4
N32	588875	38178	15.42	
R31	50332	36506	1.38	11.1
R31	48056	40016	1.20	12.9
R35	23861	19943	1.20	
R35	48084	17678	2.72	
N49	506232	32049	15.80	
R49	38680	ND	ND	
R55	31415	11426	2.75	
R68	33598	17564	1.91	
R74	18446	11516	1.60	

N = Normal Irish Setter
R = rcdl Irish Setter
Numbers after N and R indicate age in days
ND = Not done
α = anti; 740 = antipeptide antisera directed against bovine PDEα residues # 740–755; Tα = antipeptide antisera directed against bovine Tα residues # 336–350; T = transducin; Scans were accomplished using a Shimadzu system; Fold diff. = difference of ratios for normal vs. affected.

specific proteases, as occurs in many cases as a common cellular repair mechanism. Alternatively, one can envision a mechanism by which the individual proteins might be degraded at different rates if the membrane assembly process were not functioning properly. Then, rates of degradation would reflect the degree to which they were abnormally assembled and/or the individual accessibility of each protein to cellular proteases. In the first case, an altered specific component of the visual cascade is suggested. While, in the latter case, the specific components could all be normal, but the membrane assembly would be defective. We have no direct proof for either hypothesis at present.

Rhodopsin has also been found to be present in the rcdl retinas. However, two reports suggest that the phosphorylation of this protein is altered, perhaps due to increased phosphatase activity (Schmidt, et al., 1986; Takemoto, et al., 1986).

We have reported that the reaction of this protein is altered when using antisera directed to the C-terminal amino acid residues of rhodopsin (Takemoto, et al., 1985; Cunnick, et al., 1988). Table 2 reviews these results, as previously reported (Takemoto, et al., 1985). When the reaction with several antipeptide antisera are equalized the reaction with the rhod-4 antisera is decreased for the rcdl samples. This has been observed as early as ten days postnatally. This decreased reaction is apparent in the undenatured samples, and is probably not due entirely to an altered degree of phosphorylation in this region. We have previously reported that treatment of the rcdl samples with alkaline phosphatase does not increase the reaction of rcdl with rhod-4 antisera to the level of the normal retinal samples (Cunnick, et al., 1988). Other possibilities for the altered antisera reactivity could be, clipping of the C-terminus, alteration of the primary amino acid sequence, or altered orientation of the rhodopsin in the membrane (masking of the epitope).

In order to determine which of these possibilities occurs, we have isolated the C-terminal peptide from the rhodopsin proteins of the rcdl and normal dog samples. In Figure 1 are shown the hplc profiles of the tryptic peptides generated from digestion of the depleted disc membranes, as described above. The profiles for the bovine

Table 2

Binding of Anti-Rhodopsin Antisera to Retinal Tissue[a]

Sample	cpm Bound		N/rcdl Ratio	
	Exp. 1	Exp. 2	Exp. 1	Exp. 2
		Anti-Rhod-3		
rcdl	859±28(4)	1250±114(3)	1.11	1.49
Normal	958±66(4)	1864±34(3)		
		Anti-Rhod-4		
rcdl	2374±162(4)	2347±77(3)	2.06	3.22
Normal	4904±194(4)	7566±303(3)		
		Anti-Rhod-5		
rcdl	760±47(4)	1092±123(3)	1.10	1.33
Normal	836±84(4)	1456±33(3)		
		Anti-Rhod-7		
rcdl	785±57(4)	1403±102(3)	0.99	0.83
Normal	782±50(4)	1171±84(3)		
		Anti-Rhod-10		
rcdl	1880±103(4)	1504±174(3)	1.45	1.86
Normal	2730±162(4)	2798±111(3)		

a) From Takemoto, *et*. *al*. (1985).

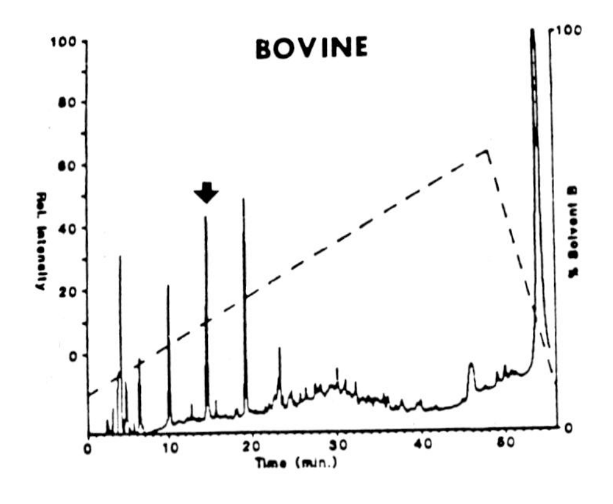
BOVINE
Rel. Intensity
% Solvent B
Time (min.)

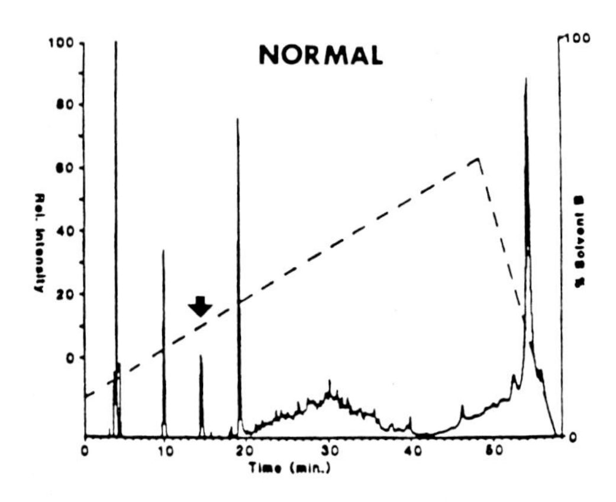
NORMAL
Rel. Intensity
% Solvent B
Time (min.)

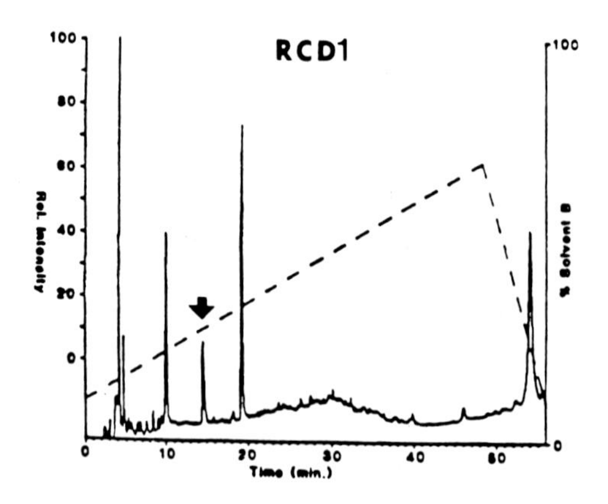
RCD1
Rel. Intensity
% Solvent B
Time (min.)

retinal digest and the normal and rcdl dogs were remarkably similar, giving four major peaks. The identity of the rhod-4 region was determined by reaction with the specific antisera in an RIA (see above). Peak two was identified as containing the rhod-4 region and was subsequently sequenced on a gas-phase sequencer. The sequence for this region of rhodopsin was identical to the bovine sequence and was not altered in the rcdl rhodopsin molecule (THR-GLU-THR-SER-GLN-VAL-ALA-PRO-ALA). Likewise, when adjusted for protein, a similar amount of this peak was isolated from the normal and the rcdl retinal homogenates suggesting that this region was not clipped from the affected rhodopsin protein. It is possible that another region near the rhod-4 region is altered in the rcdl protein, causing epitope masking of the former. Likewise, defective membrane assembly, in general, could cause improper insertion and resultant masking of the reactive rhod-4 epitope.

SUMMARY

In both of the early onset systems, the _rd_ mouse and the rcdl Irish Setter, early elevation of cyclic GMP may be the ultimate cause of accelerated photoreceptor degeneration. This would be consistant with the data utilizing _in vitro_ systems in which retinal samples, in culture, undergo degeneration in response to constant exposure to high levels of this nucleotide. However, the ultimate cause of the elevated cyclic GMP in the _rd_ mouse or in the rcdl Irish Setter still remains a mystery. It appears that all of the necessary proteins of the visual cascade are produced, although they are lost at different rates. The phosphodiesterase appears to be reduced faster than other proteins. This may, in turn, account for the elevation in cyclic GMP levels. The cause of this enhanced

Figure 1. HPLC profile of tryptic peptides isolated from depleted retinal membranes. Profile is from a C-18 reverse-phase column (Phenomenex). Peaks are scanned at 215 nm. The total protein content of the digests were: bovine, 0.15 mg; normal Irish Setter, 0.15 mg; rcdl Irish Setter, 0.3 mg. Samples were from 4-week-old dogs. Arrows indicate peaks which were positive for antisera to the Rhod-4 region.

disappearance could reside in the phosphodiesterase protein itself, or in other more distal components. The alteration in rhodopsin reaction to the specific rhod-4 antisera suggests that this protein is not properly oriented in the disc membrane. Although this may or may not alter the visual cascade, it does suggest that these membranes are not identical to those of the normal dog retina.

Future studies should focus on the individual functional activities of each component, on their structures, and on their proper assembly within the disc.

REFERENCES

Aguirre, G. D., Alligood, J., O'Brien, P., Buyukmihci, N. (1982) Pathogenesis of progressive rod-cone degeneration in miniature poodles. Invest. Ophthalmol. Vis. Sci. 23:610-630.

Aguirre, G. D., Farber, D., Lolley, R., O'Brien, P., Alligood, J.(1982) Retinal Degenerations in the Dog:III. Abnormal cyclic nucleotide metabolism in rod-cone dysplasia. Exp. Eye Res. 35:625-642.

Aguirre, G., and O'Brien, P. (1986) Morphological and biochemical studies of canine progressive rod-cone degeneration. Invest. Ophthalmol. Vis. Sci. 27:635-655.

Aguirre, G., Rubin, L. (1975) Rod-cone dysplasia (progressive retinal atrophy) in Irish Setters. J. Amer. Vet. Med. Assoc. 166:157-164.

Barbehenn, E., Gagnon, C., Noelken, D., Aguirre, G., and Chader, G. (1988) Inherited rod-cone dysplasia:abnormal distribution of cyclic GMP in visual cells of affected Irish Setters. Exp. Eye Res. 46:149-159.

Bretscher, M. S. (1985) The molecules of the cell membrane Sci. Amer. 253:100-108.

Buyukmihci, N., Aguirre, G., and Marshall, J. (1980) Retinal degeneration in the dog. II Development of the retina in rod-cone dysplasia. Exp. Eye Res. 30:575-591.

Charbonneau, H., Beier, N., Walsh, K., and Beavo, J. (1986) Identification of a conserved domain among cyclic nucleotide phosphodiesterases from diverse species. Proc. Natl. Acad. Sci. 83:9308-9312.

Cunnick, J., Rider, M., Takemoto, L., and Takemoto, D. (1988) Rod/cone dysplasia in Irish Setters:presence of an altered rhodopsin. Biochem. Journal 250:335-341.

Farber, D., Park, S., and Yamashita, C. (1988) Cyclic GMP-phosphodiesterase of rd retina. Biosynthesis and Content. Exp. Eye Res. 46:363-374.

Ferguson, M. and Williams, A. (1988) Cell-surface anchoring of proteins via glycosylphosphatidylinositol structures. Ann. Rev. Biochem. 57:285-320.

LaVail, M. M., Sidman, R. L. (1974) C57BL/6J mice with inherited retinal degeneration. Arch Opthal. 91:394-400.

Lee, R., Lieberman, B., Hurwitz, R., and Lolley, R. (1985) Phosphodiesterase-probes show distinct defects in rd mice and Irish Setter dog disorders. Invest. Ophthalmol. Vis. Sci. 26:1569-1579.

Lee, R., Navon, S., Brown, B., Fung, B., and Lolley, R. (1988) Characterization of a phosphodiesterase immunoreactive polypeptide from rod photoreceptors of developing rd mouse retinas. Invest. Ophthalmol. Vis. Sci. 29:1021-1027.

Long, K., Philp, N., Gery, I., and Aguirre, G. (1986) Immunocytochemistry of retinal S-antigen (48K protein) in canine progressive retinal atrophy. Invest. Ophthalmol. Vis. Sci. Suppl. 27:56.

Medynski, D., Sullivan, K., Smith, D., Van Dop, C., Chang, F., Fung, B., Seeburg, P., and Bourne, H. (1985) Amino acid sequence of the α subunit of transducin deduced from the cDNA sequence. Proc. Natl. Acad. Sci. U.S.A. 82:4311-4315.

Navon, S., Lee, R., Lolley, R., and Fung, B. (1987) Immunological determination of transducin content in retinas exhibiting inherited degeneration. Exp. Eye Res. 44:115-125.

Ovchinnikov, Y., Gubanov, V., Khramtsov, N., Ischenko, K., Zagranichny, V., Muradov, K., Shuvaeva, T., and Lipkin, V. (1987) Febs. Lett. 223:169-173.

Schekman, R. (1985) Protein localization and membrane traffic. Ann. Rev. Cell Biol. 1:115-143.

Schmidt, S., Andley, U., Heth, C., and Miller, J. (1986) Deficiency in light-dependent opsin phosphorylation in Irish Setters with rod-cone dyplasia. Invest. Ophthalmol. Vis. Sci. 27:1551-1559.

Takemoto, D., Cunnick, J., and Takemoto, L. (1986) Reduced rhodopsin phosphorylation during retinal dystrophy. Biochem. Biophys. Res. Comm. 135:1022-1028.

Takemoto, D., Spooner, B., and Takemoto, L. J. (1985) Antisera to synthetic peptides of bovine rhodopsin:use as site-specific probes of disc membrane changes in retinal dystrophic dogs. Biochem. Biophys. Res. Comm. 132:438-444.

Ulshafer, R. J., Fliesler, S. J., and Hollyfield, J. G. (1984) Differential sensitivity of protein synthesis in human retina to a phosphodiesterase inhibitor and cyclic nucleotides. Current Eye Res. 3:383-392.

Unwin, N. and Henderson, C. (1984) The structure of proteins in biological membranes. Sci. Amer. 250:78-94.

Wickner, W. and Lodish, H. (1985) Multiple mechanisms of protein insertion into and across membranes. Science 230:400-407.

Inherited and Environmentally Induced
Retinal Degenerations, pages 455–465

TWO-DIMENSIONAL GEL ELECTROPHORETIC ANALYSIS OF PROTEINS IN THE *rd* CHICK RETINA

Susan L. Semple-Rowland

Department of Ophthalmology, University of Florida, College of Medicine, Gainesville, Florida 32610-0284

INTRODUCTION

The *rd* (retinal degeneration) chicken possesses an autosomal recessive mutation which causes blindness at hatch (Ulshafer et al., 1984). This defect is unusual because ERG signals are absent at hatch despite the fact that the ultrastructure of the retina appears normal (Ulshafer and Allen, 1985a; Ulshafer et al., 1984), a situation which suggests that the *rd* mutation affects some aspect of photoreceptor function. Approximately 7-10 days after hatch, the first signs of photoreceptor degeneration appear in the central retina. Ultrastructural changes in the pigment epithelium are observed after the photoreceptor cells begin degenerating (Ulshafer and Allen, 1985b). By 8 months of age, virtually all of the photoreceptors in the retina have degenerated.

In the present study, the technique of two-dimensional gel electrophoresis was used to study and compare retina/pigment epithelium/choroid protein patterns of homozygous blind [rd/rd], heterozygous sighted [+/rd] and normal sighted [+/+] Rhode Island Red x Barred Rock hatchling chicks in order to find proteins affected by the *rd* mutation. Analyses of several protein patterns revealed a group of four proteins which consistently distinguished [+/+] retina from [+/rd] and [rd/rd] retina. Data concerning these proteins is presented along with a description of the two-dimensional gel electrophoresis procedure used in these experiments.

MATERIALS AND METHODS

Animals

Colonies of homozygous blind [rd/rd], heterozygous sighted [±/rd] and Rhode Island Red x Barred Rock normal sighted [±/±] breeder chickens are maintained at the University of Florida Animal Care Facility under NIH guidelines. Controlled matings occurred which produced known [rd/rd], [±/rd] or [±/±] chicks. Eggs were incubated in an automatic rotating, forced draft incubator until 1-2 days before hatching and then transferred to a brooder until day of hatch when they were used in this study.

Sample Preparation

Eyes obtained from embryonic day 21 (E21) (hatchling) chicks were opened along the ora serrata and the vitreous and pectin were removed. The neural retina, pigment epithelium and choroid were then gently scraped away from the sclera and either processed immediately for two-dimensional gel electrophoresis or frozen in liquid nitrogen and stored at -70°C for later use. The eyes from 7 [±/±], 4 [±/rd] and 7 [rd/rd] chicks were examined.

Two-dimensional Gel Electrophoresis

Tissue samples were sonicated for 20 sec in 0.5 ml cold Tris buffer (0.062 M Tris-HCl, pH 6.8). A small portion of each sample was removed for determination of protein concentration using the Pierce BCA Protein Assay Reagent (Pierce Chemical Co., Rockford, IL) and bovine serum albumin as a standard. The remaining sample was immediately solubilized in 8.7 M urea which contained 2% NP-40, 0.2% sodium dodecyl sulfate (SDS) and 2 mM dithiothreitol (DTT).

Two-dimensional gel electrophoresis was performed according to the method of O'Farrell (1975) with the following modifications. For the first dimension, isoelectric focusing was done in 11-cm tube gels (inner diameter= 3.0-mm) containing 9 M urea, 4% acrylamide, 2% NP-40, 0.07% tetramethyl-ethylenediamine (TEMED) and 2%

ampholytes (4:1 ratio of pH 5-8 and pH 3-10, Pharmacia, Inc., Piscataway, NJ). This particular ratio of ampholytes was chosen since it gave the most uniform distribution of proteins across the two-dimensional gels.

Prior to gel polymerization, an aliquot of the sample to be analyzed was added to the gel mixture to yield a final protein concentration of 65 µg protein/gel. The technique of polymerizing protein samples into isoelectric focusing gels has been described previously (Wrigley, 1968; Adamus and Romanowska, 1980). To initiate polymerization, 1 mg ammonium persulfate was added to each 10 ml volume of degassed gel solution. Gels were overlayed with distilled water and allowed to polymerize for 3 hours before electrophoresis. The anode (0.01 M phosphoric acid) and cathode (0.02 M sodium hydroxide) solutions were thoroughly degassed prior to electrofocusing. Electrofocusing was carried out at room temperature, 350 V for 18 hours followed by 800 V for 1 hour. When electrofocusing was completed, the gels were extruded from the tubes and gently agitated in equilibration buffer (2.3% SDS, 5% ß-mercaptoethanol and 10% glycerol in 0.062 M Tris-HCl, pH 6.8) for 10 min or were quickly frozen in buffer and stored at -70°C until use. Approximate isoelectric points along one gel from each run were determined using a surface electrode (Bio-Rad).

For the second dimension, samples were separated by slab polyacrylamide gel electrophoresis in the presence of SDS on separating gels (1.5-mm thickness, 14.5-cm length) containing 10% acrylamide with a 4% acrylamide stacking gel according to the method of Laemmli (1970). The first dimension gels and molecular weight standards, which were embedded in 1% agarose in equilibration buffer, were attached to the stacking gel using 1% agarose in 0.125M Tris-HCl, pH 6.8. Electrophoresis was carried out at a constant current of 25 mA/gel for 4-5 hours.

The separated proteins within the slab gels were fixed overnight in a 40% methanol, 10% acetic acid solution. Fixed gels were silver stained according to the method of Wray et al. (1981) with the following modifications. Fixed gels were washed for a minimum of 1 hour in three changes of 20% ethanol prior to staining. The staining solution was prepared as described by Wray et al. (1981) except that we added 2.0 ml ammonium hydroxide to solution B instead of

1.4 ml. Gels were allowed to stain for 30-45 minutes with constant gentle agitation. The stained gels were then washed for 80 minutes in four changes of 20% ethanol prior to developing the stain. Stain development was stopped by placing the gels in 20% ethanol, 10% acetic acid. Stained gels were dried at room temperature between BioGelWrap (BioDesign Inc., Carmel, NY), and photographed using Kodak Electrophoresis Duplicating Film (EDF).

Analysis of Two-dimensional Gels

Kodak EDF film produces a black image of silver stained proteins on a clear acetate sheet. This film provides a permanent record of the proteins found in a gel and permits visual matching of protein patterns across several gels with the aid of a light box. In the present series of experiments, Kodak EDF films of [+/+], [+/rd] and [rd/rd] retinal protein patterns were visually matched. In a few cases, gels were photographed using Kodak Ekatapan 4x5 inch sheet film and printed onto 16x20 inch Orthofilm which provided large black protein spots on a clear film base. This larger format was easier to examine visually and allowed accurate records of protein pattern differences to be noted on the Orthofilm print.

RESULTS

The two-dimensional gel procedure outlined above was used to separate proteins extracted from [+/+], [+/rd] and [rd/rd] retina/pigment epithelium/choroid (RET/PE/CH). In each of the RET/PE/CH samples examined, over 500 proteins were resolved with a high degree of reproducibility. Very few differences were observed in the protein patterns obtained from the RET/PE/CH of the three groups of animals examined. All proteins were well-focused and very few of the streaking artifacts commonly associated with traditional methods of two-dimensional gel electrophoresis were present. The pH gradient obtained using a 4:1 ratio of pH 5-8 and pH 3-10 ampholytes was very reproducible between pH 6.0 and pH 8.0 and produced an even distribution of proteins across the entire second-dimension gel. Extension of the pH gradient in either the acid or basic direction resulted in a significant loss of protein spot resolution with few new spots appearing on the two-dimensional gel.

The protein pattern shown in Figure 1 for proteins extracted from E21 [+/rd] RET/PE/CH is representative of the protein patterns obtained for all [+/rd] tissue samples examined. Analysis of two-dimensional gels containing proteins extracted from E21 [rd/rd], [+/rd] and [+/+] RET/PE/CH revealed a set of four proteins which consistently distinguished [+/+], [+/rd] and [rd/rd] tissue. The location of these four proteins is shown in Figure 1.

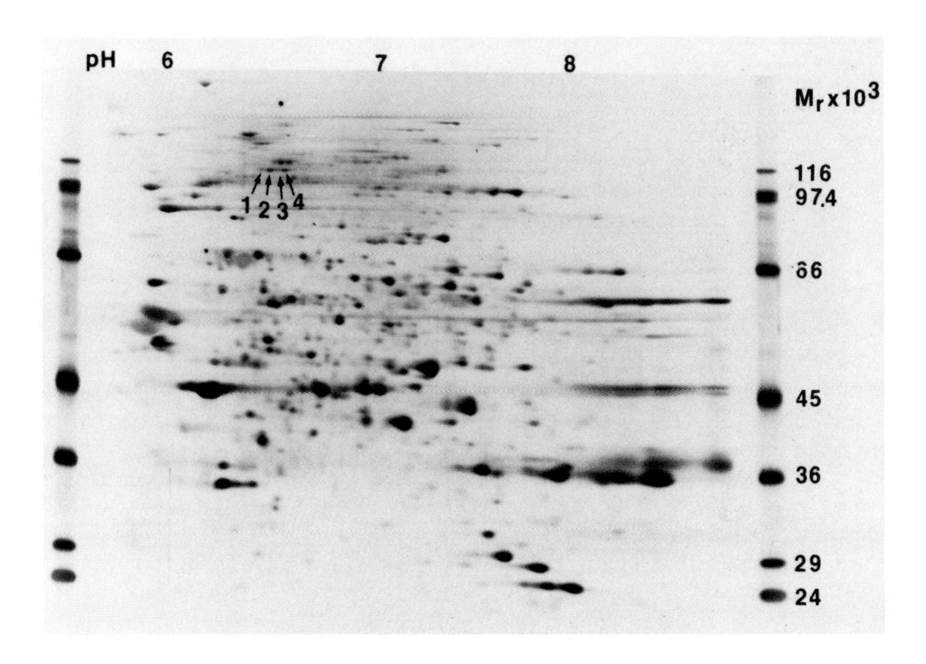

Figure 1. Photograph of two-dimensional silver stained gel containing proteins extracted from E21 [+/rd] retina/pigment epthelium/choroid. Estimated pI values are shown at the top of the gel (first dimension) and MW standards are shown on both sides (second dimension). All four 98 kDa proteins were found in E21 [+/rd] RET/PE/CH and are numbered: 1:P98-6.53 (MW-pI), 2:P98-6.57, 3:P98-6.62, 4:P98-6.66.

All four proteins have an apparent MW of 98 kDa with isoelectric points ranging from 6.53 to 6.66 (Table 1). While there was some variability in the estimated pI values for the proteins between gels, the pI values for the individual proteins within gels alwayss differed by approximately 0.04 pH units.

TABLE 1. Apparent Molecular Mass and Isoelectric Points of Proteins Affected by the *rd* Mutation

	Protein	#Gels Examined	Mass	pI*
1	P98-6.53	10	98 kDa	6.53 ± .10
2	P98-6.57	11	98	6.57 ± .09
3	P98-6.62	9	98	6.62 ± .09
4	P98-6.66	8	98	6.66 ± .11

* Mean ± S.D.

Enlargements of this region, taken from representative gels containing proteins extracted from E21 [+/+], [+/rd] and [rd/rd] RET/PE/CH are shown in Figure 2. The RET/PE/CH tissues of all of the [rd/rd] chicks examined (n=7) contained only proteins P98-6.53 and P98-6.57. Two of the proteins, P98-6.62 and P98-6.66, were never found in [rd/rd] RET/PE/CH. The majority (3 out of 4) of the E21 [+/rd] RET/PE/CH examined contained all four of the proteins. The remaining [+/rd] chick RET/PE/CH contained P98-6.53, P98-6.57 and P98-6.62. Finally, the majority (5 out of 7) of the E21 [+/+] chick RET/PE/CH contained only P98-6.62 and P98-6.66. One of the remaining two [+/+] chick RET/PE/CH contained all four of the proteins, the other contained P98-6.57, P98-6.62 and P98-6.66.

In addition to the electrophoretical characterization of the four 98 kDa proteins, prelimary experiments were conducted to determine the chemical nature of these proteins. To determine the solubility of the proteins thefollowing experiment was done. The proteins contained in[+/+] RET/PE/CH tissues were gently homogenized in 20 mM HEPES, pH 7.6, containing 0.25 M sucrose and centrifuged at

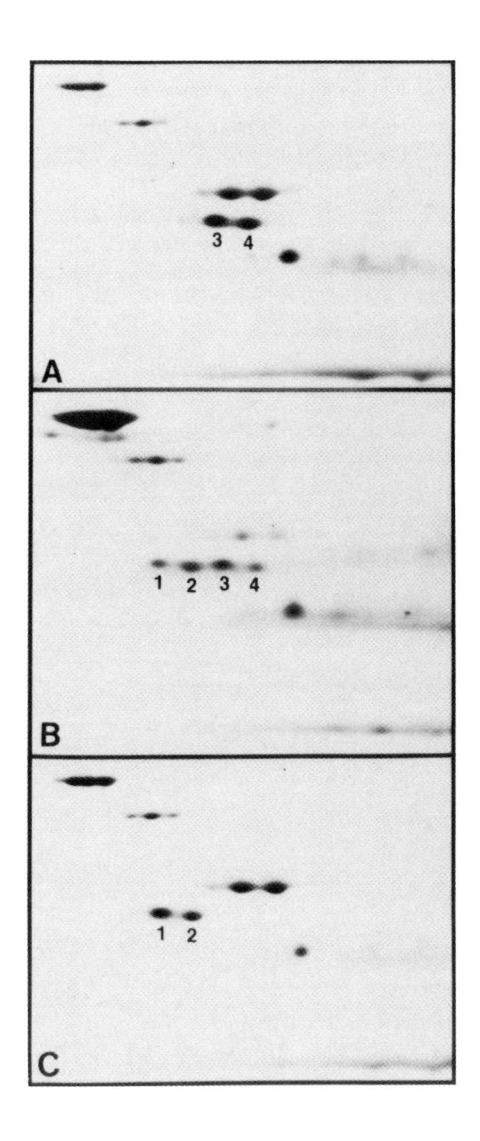

Figure 2. Representative photographs of regions of gels containing the four 98 kDa proteins extracted from E21 [+/+], [+/rd] and [rd/rd] RET/PE/CH. A. [+/+] gel containing proteins P98-6.62 (3) and P98-6.66 (4). B. [+/rd] gel containing proteins (1) P98-6.53, (2) P98-6.57, (3), and (4). C. [rd/rd] gel containing proteins (1) and (2).

35K rpm for 20 min. The resulting pellet was resuspended in buffer and centrifuged at 35K rpm for an additional 60 min. The proteins contained in the combined supernatants and pellet were analyzed separately on two-dimensional gels. Proteins P98-6.62 and P98-6.66 were only present in the supernatant protein pattern (data not shown). In addition, [+/+] RET/PE/CH proteins were electrotransferred to nitrocellulose. Binding with lectins specific for Man, Glu, NAcGlu and NAcGal was performed to determine whether the proteins might be glycoproteins. The 98 kDa proteins did not stain with the lectins tested in spite of the fact that several other proteins showed a positive reaction (data not shown).

DISCUSSION

The two-dimensional gel electrophoresis procedure outlined in this chapter is a powerful tool for the analysis and detection of proteins affected by genetic mutations. Using this procedure, a group of four proteins (P98-6.53, P98-6.57, P98-6.62 and P98-6.66) has been identified which distinguishes [+/+] , [+/rd] and [rd/rd] RET/PE/CH. Proteins P98-6.62 and P98-6.66, which were present in [+/+] and [+/rd] RET/PE/CH at hatch, were never found in [rd/rd] RET/PE/CH. The results of preliminary experiments suggest that these proteins are soluble and that they are not glycoproteins.

The primary biochemical site(s) for the *rd* mutation is not known. Quantities of rhodopsin, 11-cis retinyl esters, 11-cis retinol and interphotoreceptor retinoid-binding protein (IRBP) are comparable in 3 day old [rd/rd] and [+/rd] chick retinas (Bridges et al., 1987), suggesting that blindness in the *rd* chick is not due to an inability to synthesize and secrete IRBP, or to form 11-cis isomer and rhodopsin.

One aspect of photoreceptor cell function which appears to be affected by the mutation is transduction. Existing electrophysiological data suggest that both rod and cone transduction processes are abnormal in *rd* retina (Lee et al., 1988; Ulshafer and Allen, 1985a; Ulshafer et al., 1984). Recent immunocytochemical data (Ulshafer, personal communication) suggest that the absence of the early receptor potential in *rd* (Lee et al., 1988), which is generated by the light-induced conformational changes in cone pigment molecules (Pak, 1968; Goldstein, 1969; Sieving and Fishman, 1982), may be due to a change in the structure and/or insertion of the cone pigments into the outer segment membranes.

Biochemical assays in *rd* retina also suggest that photoreceptor function is abnormal in this mutant. Lee et al. (1987) have obtained experimental evidence which shows that cGMP levels in 2 day old, dark-adapted [rd/rd] chick retina-RPE are significantly lower than those found in dark-adapted [+/rd] and [+/+] chick retina-RPE. Glucose metabolism also appears to be abnormal in *rd* retina. Dark-adapted *rd* chick retinas have been found to utilize significantly less glucose than dark-adapted retinas of sighted chicks; their glucose levels being comparable to those found in the light-adapted retinas of sighted chicks (Ruth et al., 1985).

The function of proteins P98-6.53, P98-6.57, P98-6.62 and P98-6.66 in retina is not yet known. The absence of P98-6.62 and P98-6.66 from *rd* retina at hatch, prior to the onset of photoreceptor pathology, suggests that these proteins may be important for maintenance of retinal function and viability. Preliminary developmental studies suggest that P98-6.62 and P98-6.66 are also absent from embryonic *rd* retina. Studies are currently underway to further define the biochemical characteristics of the 98kDa proteins, identify their location within the retina and understand the genetic mechanisms which control their expression in normal and mutant tissue.

Acknowledgements

I would like to thank Drs. Robert J. Ulshafer, Robert J. Cohen, Paul A. Hargrave and Grazyna Adamus for their helpful advice and use of their laboratory facilities throughout the course of this work. In addition, I would like to thank Drs. Grazyna Adamus and Adrian M. Timmers for their critical reading of this manuscript. This work was supported by a grant from NIH (F32 EY05975) and a non-restricted Departmental Grant awarded to the Department of Opthalmology from Research to Prevent Blindness, Inc.

References

Adamus G, Romanowska E (1980). Outer membrane proteins of Shigella Sonnei II. Comparative studies on virulent and avirulent strains of phase I. Arch Immunologiae et Therapiae Experimentalis 28:553-558.

Bridges CDB, Alvarez RA, Fong S-L, Liou GI, Ulshafer RJ (1980). Rhodopsin, vitamin A, and interstitial retinol binding protein in the rd chicken. Invest Ophthalmol Vis Sci 28:613-617.

Cheng KM, Shoffer RN, Gelatt KN, Gum GG, Otis JS, Bitgood JJ (1980). An autosomal recessive blind mutant in the chicken. Poultry Sci 59:2179-2182.

Gelsema WJ, De Ligny CL, Van der Veen NG (1979). Isoelectric points of proteins, determined by isoelectric focusing in the presence of urea and ethanol. J Chromatography 171:171-181.

Goldstein, EB (1969). Contribution of cones in the early receptor potential in the rhesus monkey. Nature 222:1273-1274.

Laemmli UK (1970). Cleavage of structural proteins during the assembly of the head of bacteriophage T4. Nature 227:680-685.

Lee NR, Ulshafer RJ, Cohen RJ (1987). cGMP in the rd chicken retina. Invest Ophthalmol Vis Sci (Suppl) 28:344.

Lee NR, Dawson, WW, Ulshafer RJ (1988). Receptor potentials in the maturing dystrophic retina. Invest Ophthalmol Vis Sci (Suppl) 29:243.

O'Farrell PH (1975). High-resolution two-dimensional electrophoresis of proteins. J Biol Chem 250:4007-4021.

Pak, WL (1968). Rapid photoresponses in the retina and their relevance to vision research. Photochem Photobiol 8:495-503.

Ruth, NJ, Ulshafer, RJ, Kelley, KC (1985). Depressed glucose utilization in the rd chicken retina. Invest Ophthalmol Vis Sci (Suppl) 26:62.

Sieving, PA, Fishman, GA (1982). Rod contribution to the early receptor potential (ERP) estimated from monochromat's data. Doc Ophthalmol Proc Ser 31:95-101.

Ulshafer RJ, Allen CB (1985a). Hereditary retinal degeneration in the Rhode Island Red chicken: Ultra-structural analysis. Exp Eye Res 40:865-877.

Ulshafer RJ, Allen CB (1985b). Ultrastructural changes in the retinal pigment epithelium of congenitally blind chickens. Curr Eye Res 4:1009-1021.

Ulshafer RJ, Allen CB, Dawson WW, Wolf ED (1984). Hereditary retinal degeneration in the Rhode Island Red chicken. I. Histology and ERG. Exp Eye Res 39:125-135.
Wray W, Boulikas T, Wray VP, Hancock R (1981). Silver staining of proteins in polyacrylamide gels. Anal Biochem 118:197-203.
Wrigley C (1968). Gel electrofocusing- A technique for analyzing multiple protein samples by isoelectric focusing. Science Tools 15:17-23.

Inherited and Environmentally Induced
Retinal Degenerations, pages 467–489

RETINAL DEGENERATION AND PHOTORECEPTOR MAINTENANCE IN *Drosophila*: *rdgB* AND ITS INTERACTION WITH OTHER MUTANTS

William S. Stark and Randall Sapp

Division of Biological Sciences,
University of Missouri, Columbia,
Missouri 65211 USA

INTRODUCTION

Since its discovery (Hotta and Benzer, 1970) and characterization (Harris et al., 1976), *rdgB* has been one of the most useful mutants in the genetic dissection of vision in *Drosophila*. Harris and Stark (1977) showed that *rdgB*, with its light induced retinal degeneration, has a phototransduction defect and that it interacts with *norpA* (no receptor potential). Since then, *rdgB* has been used in over a score of studies (see Stark et al., 1983; Stark and Sapp, 1987 for references). Most studies took advantage of *rdgB*'s selective loss of R1-6 (sparing R7/8) for electrophysiology and behavior. Receptor demise can be monitored electrophysiologically after one light pulse maximally converts rhodopsin to metarhodopsin (Harris and Stark, 1977). With R1-6 analogous to rods functionally (Miller et al., 1981), *rdgB* conjures up a striking parallel with many vertebrate models of retinal degeneration and with most forms of retinitis pigmentosa in humans. Yet, in spite of its importance, a substantial amount was yet to be learned about degeneration mechanisms and cellular specificities using ultrastructural methods. Stark and Carlson (1982) showed the time course of *w rdgB*'s R1-6 demise in cyclic room lighting and also showed that the degeneration did not cross to the second order neurons in the lamina ganglionaris.

The purpose of this chapter is to further document the ultrastructural aspects of the rdgB mutant. Specifically, receptors in the compound eye and their terminals in the lamina ganglionaris were examined at earlier stages in the light treatment (and with calibrated optical stimuli). Further, the effects of other mutations on the light induced demise of the R1-6 receptors in rdgB's compound eye are demonstrated using ultrastructural and other techniques. Some of our results are presented in the context of recent observations concerning turnover and maintenance of visual receptor membrane (Stark et al., 1988B; c.f. Matsumoto et al., 1988). In the breakdown phase of the turnover process, membrane is internalized for autophagy via coated pits. Coated vesicles merge to multivesicular bodies (MVB's) to be attacked by primary lysosomes (Stark et al., 1988B).

MATERIALS AND METHODS

Flies were reared on a standard medium whose most relevant feature is adequate photopigment precursors supplied through yellow cornmeal and β-carotene supplementation (Miller et al., 1984). Flies were reared at about 23° C on a 12 hr - 12 hr day night cycle of standard fluorescent laboratory room light of about 50 cd/m^2 (Stark et al., 1983). Fixations for EM were performed within the first 2 hours after light onset. For age control, flies were isolated from a cleared vial or bottle and maintained in normal food vials (transferring when necessary).

The $rdgB^{KS222}$ stocks, including w rdgB, y cho rdgB, $norpA^{EE5}$ $rdgB^{KS222}$;cn bw, $norpA^{SUII}$ rdgB, and w rdgB;trp, originated from Seymour Benzer's laboratory at Caltech and have been maintained and studied in our laboratory for over a decade. In general, eye color mutants alter the screening pigments without otherwise affecting retinula cell function. However, in the case of light induced retinal degeneration, it is important to note that eye color pigments have a protective effect. The w (white) mutation and the cn bw (cinnabar brown) mutations eliminate eye color pigments from the eyes. Chocolate (cho) darkens the eye color pigments from the eyes. Cho has clumpy appearance to the secondary pigment cells while intraretinular pigment granules are present but are less dense than normal (unpublished observations). $NorpA^{EE5}$ is an allele of the no receptor potential gene which is effective in eliminating the receptor potential; the SUII allele has a receptor potential but was selected because it is effective in blocking the degeneration in $rdgB^{KS222}$ (Harris and Stark, 1977). The trp mutant is a transduction mutant with a transient receptor potential (see, e.g., Chen and Stark, 1983, for references).

To construct w rdgB;Acph-, (and w;Acph-) white (w) remained linked to rdgB since males have no crossing over. This chromosome was inherited patriclinously by maintaining males across from y attached-X (XX//Y) females. Homozygous w rdgB;ca $Acph\text{-}1^{n13}$ bv was confirmed by screening for enzyme activity (Bell et al., 1972; c.f. Stark et al. 1988B). Briefly, crushed heads in spot wells were compared with homozygous and heterozygous positives; negatives lacked the color from the α-napthyl - Fast red TR complex produced if acid phosphatase had cleaved the substrate, α-naphthyl acid phosphate. Stocks were doubly checked by scoring for the closely linked third chromosomal marker brevis bristles; claret eyes were masked by w. The stock we used to make our stocks acid phosphatase negative was kindly provided by Prof. Ross McIntyre at Cornell University.

Methods of transmission electron microscopy have been refined since early work on the first order synapses (Stark and Clark, 1973) and have recently been outlined thoroughly (Stark and Sapp, 1987). In brief, fly heads were dissected, fixed in a triple aldehyde, rinsed, osmicated, dehydrated and embedded in Spurr. Thin sections were stained with uranyl acetate and lead citrate and photographed in Siemens Elmiskop 101 TEM at 60 or 80 kV.

This laboratory has developed detailed microspectrophotometric (MSP) techniques (Stark and Johnson, 1980). The methodology for applying MSP to quantify photopigment in living flies has been used extensively and was recently presented in detail (Stark et al., 1985). Summarizing, the deep pseudopupil is a virtual image of the magnified tips of the photopigment containing organelles of R1-7 pooled from about 25 ommatidia.

Figs. 1 - 4. Properties of the retina of rdgB newly emerged. **Fig. 1:** Cross section of a distally sectioned ommatidium of y cho $rdgB^{KS222}$ at the level of R2 and R5 nuclei (*). Fly was reared at 23°C under cyclic room light. R7's rhabdomere inserts from the superior central cell located at the 5 o'clock position with R1-6 situated clockwise after R7. Two multivesicular bodies surrounded by numerous primary lysosomes are in R7 (large arrow); MVB's with vs without lysosomes are in R2 and R4 respectively (other large arrows). Numerous intraretinular pigment granules (e.g. small arrows) are present since the strain has eye color pigments. Larger pigment granules (electron dense, bottom and top right) are situated in secondary pigment cells. No obvious signs of degeneration. x 9,300. **Fig. 2:** R4 from w rdgB shows the electron densification and dense reticulum typical of the early stages of degeneration. Fly was reared at 18°C in the dark but fixed at room temperature with room light and dissecting microscope light. Only a scattering of cells show such degeneration. x 15,900. **Fig 3:** R6 in a rdgB fly reared and fixed as in Fig 2. MVB (arrowhead) and primary lysosome (L) are in R6 while Golgi complex (G) is seen in neighboring R7. Note intraretinular pigment granules near the rhabdomere and secondary pigment cell's pigment granule (bottom right) in this red eyed fly. x 34,200. **Fig 4:** High magnification of a rhabdomere showing coated pit (arrowhead). Fly is a newly emerged w rdgB fly reared at room temperature on a day night cycle. x 130,500.

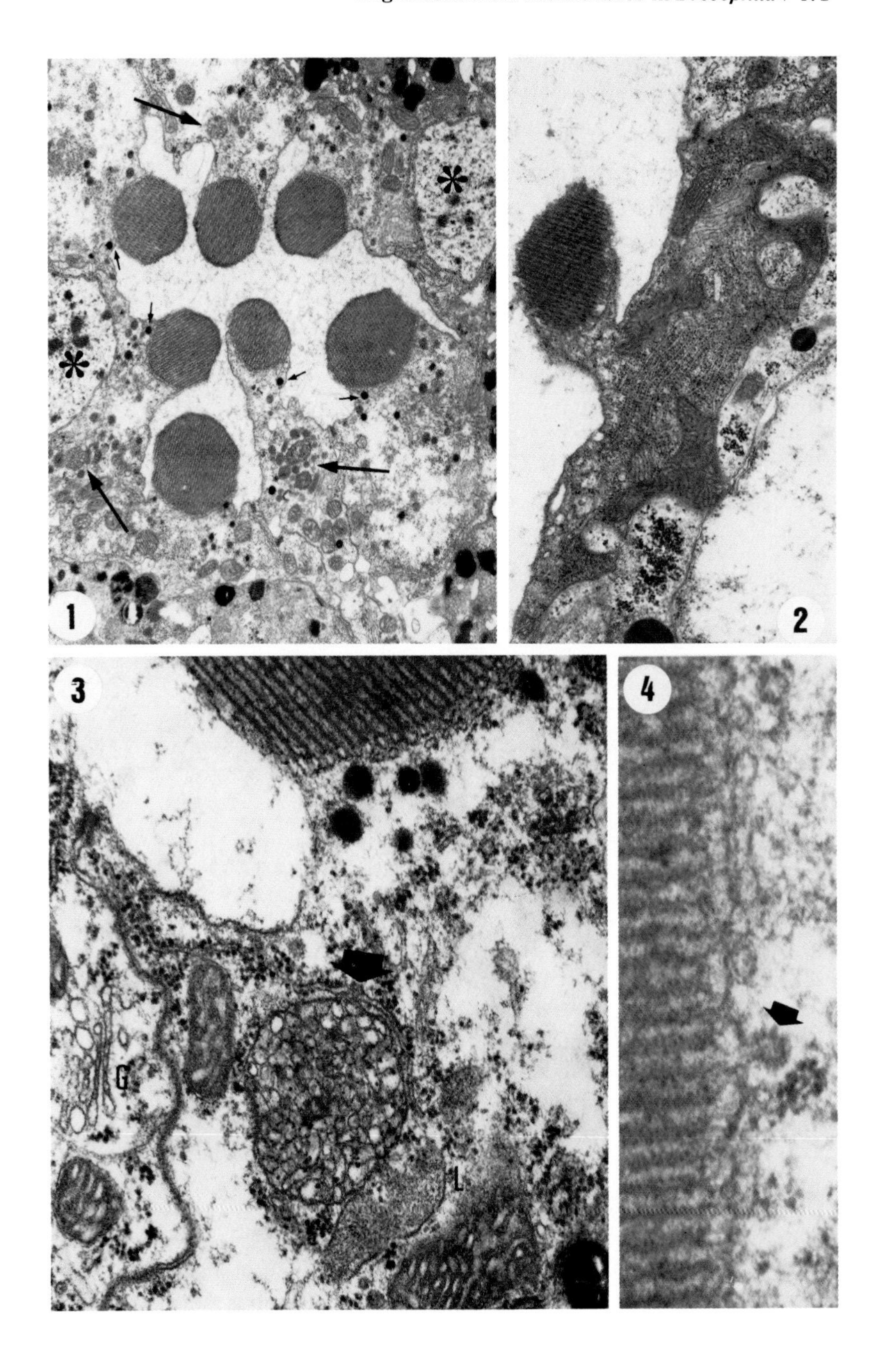
1
2
3
4
G
L

An aperture limits the measuring area to the deep pseudopupil which is dominated by R1-6. When the pseudopupil is indistinct, as when MSP is used to monitor degeneration, the pseudopupil is found as an 80 µm diameter area at 150 µm beneath the most vertical corneal bristles. Photoequilibria are established by 450 nm, which creates maximal metarhodopsin levels, and 620 nm, which nearly maximizes rhodopsin. Light at 579 nm is transmit-

Figs. 5 - 8. The projection from the w rdgB mutant of Drosophila for Figs. 5 - 7 and the projection from the w control for Fig. 8. **Fig. 5:** Pseudocartridges appear normal in this w rdgB specimen which was reared at 18°C in the dark, transferred to room lighting and fixed with normal dissecting microscope illumination shortly thereafter. x 15,900. **Fig. 6:** A cross sectioned optic cartridge from a w rdgB fly reared at 18°C in the dark, and prepared under dim red light (>615 nm). Under light ether anesthesia, the eye was stimulated with blue light (460 nm) at 15.86 log quanta/cm^2, just enough to maximally convert rhodopsin to metarhodopsin. A red beam was used to position the eye for the blue treatment, and the red light was 650 nm at 16.09 log quanta/cm^2·s. Fixation was 30 - 45 min after light treatments. Note that, of the 6 R1-6 terminals surrounding the lucent L1 and L2 profiles (L) some are fairly normal while one is grossly distorted (*). x 17,400. **Fig. 7:** A low magnification of a field of cross sectioned optic cartridges from the "red light control" experiment on w rdgB. All the same rearing conditions, red dissecting light and "sham" red finding light were used as in the Fig. 6 experiment except that the blue pulse was omitted. Note that, even so, most cartridges have one or several swollen R1-6 terminals. x 4,100. **Fig. 8:** The "red light control" experiment of Fig. 6 except that the preparation was a similarly reared w non-rdgB control. The cartridge is normal with 6 R1-6 terminals between the epithelial glia (EG) and the L1 and L2 cells (L). x 14,400.

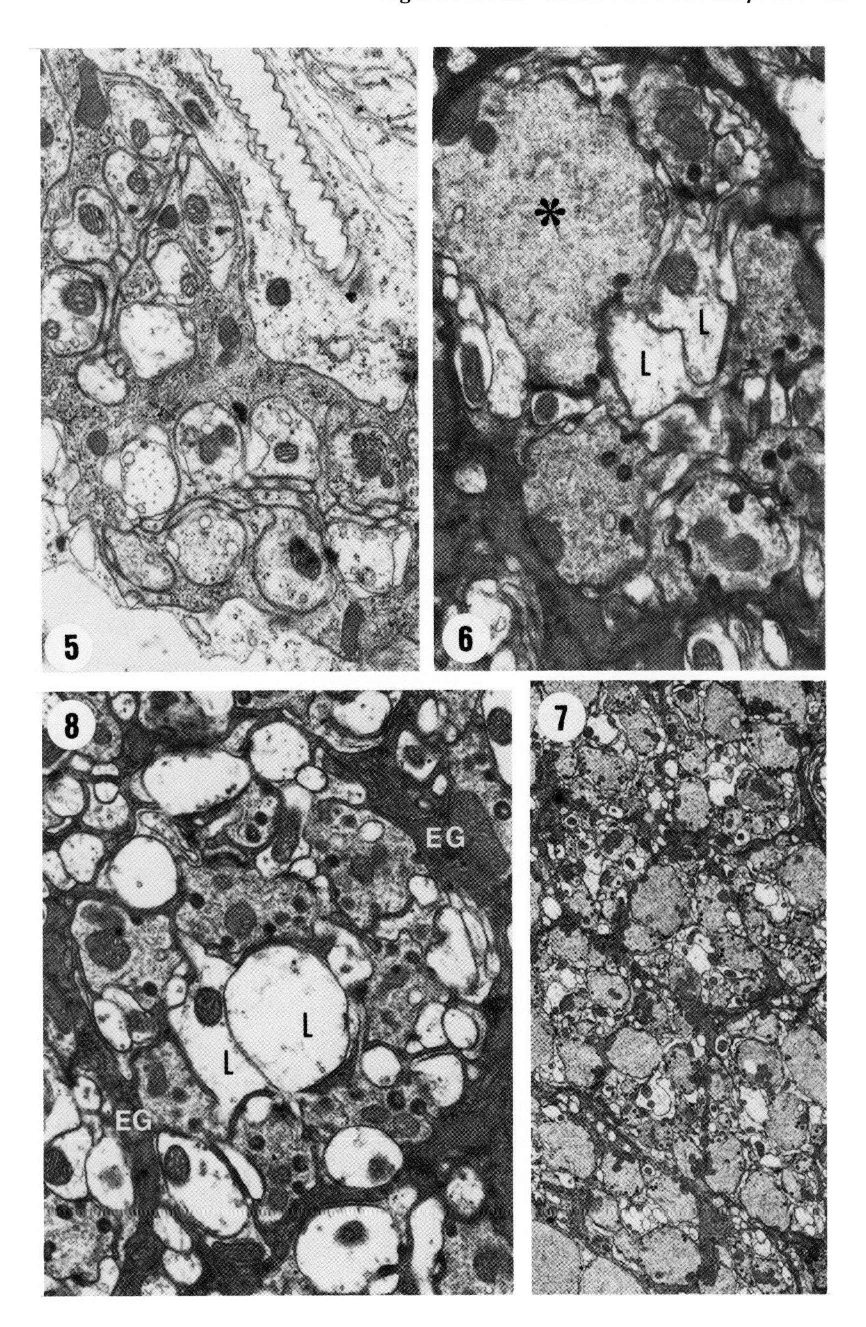
*
L
L
5
6
8
EG
L
L
EG
7

ted the length of the rhabdomeres and is optimal for determining visual pigment levels from absorbance differences since it is near metarhodopsin's absorbance maximum. In several experiments, the compound eye was exposed to calibrated lights. The intensities were measured with a calibrated photodiode as outlined by (Stark et al., 1985). Such quantal calibrations are impossible for the ambient "white" light level for rearing, so in this case, a MacBeth illuminometer was used (Stark et al., 1983). The wavelength accuracy of interference filters was calibrated using a Cary 17 spectrophotometer (Stark and Johnson, 1980).

Figs. 9 - 12. The compound norpA^{EE5} rdgB^{KS222};cn bw mutant aged 3 weeks in cyclic room light. **Fig. 9:** A distal ommatidium of norpA rdgB;cn bw at the level the nuclei of R1, R2, R4 and R7 (*) is very distorted. R1-7 are as labeled. The plasmalemma of R cells facing the intraommatidial cavity has considerable scalloping and extra membrane. x 7,000. **Fig. 10:** A proximally sectioned ommatidium from a norpA rdgB;cn bw fly aged 3 weeks is distorted by the massive elaboration of surplus membrane. R1-8 as labeled. x 9,200. **Fig. 11:** A higher magnification of typical distal retinula cells in norpA rdgB; cn bw aged 3 weeks. R2 fills most of this field with a portion of R1 on the bottom right (on the other side of the belt desmosome, D). Zipper membrane faces the intraommatidial cavity and causes a major infolding in R1's plasmalemma (arrows). Note internalization of membranes with zipper striations (hollow arrowheads). x 28,800. **Fig. 12:** A cross sectioned optic cartridge with R1-6 terminals surrounding laminar monopolars (L) in norpA rdgB;cn bw aged 3 weeks. Note the degenerate body at the 7 o'clock position and the terminal which is somewhat more electron dense than normal (*). T synapses are present in nondegenerate cells (for instance, arrowhead). Section shows a portion of the cartridge to the right which also has a degenerate terminal (arrow). Epithelial glia (eg) surround the R1-6 terminals. x 5,000.

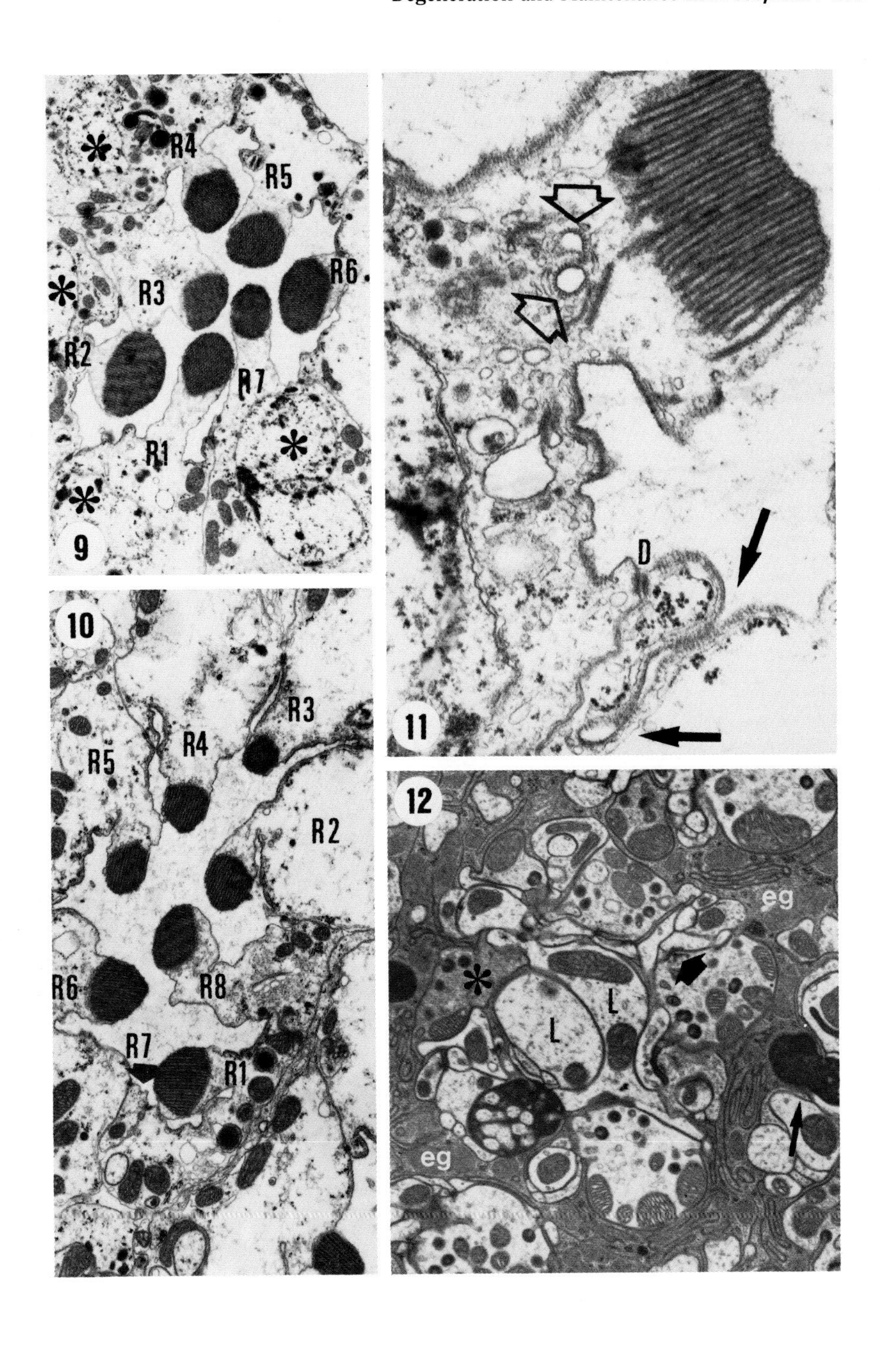
9
R4
R5
R6
R3
R2
R7
R1
10
R3
R4
R5
R2
R6
R8
R7
R1
11
D
12
eg
L
L
eg

RESULTS AND DISCUSSION

We used EM to determine conditions which elicit and inhibit degeneration in *rdgB*. Reared under cyclic light, newly emerged *rdgB* has fairly normal receptors and axons (Figs. 1 & 3 - 5); these retinula cells also exhibit the normal early autophagic steps involved in membrane turnover, namely formation of coated vesicles (Fig. 4) which merge to multivesicular bodies (MVB's) which are attacked by primary lysosomes (Figs. 1 & 3, c.f. Stark et al., 1988B). Only a scattering of cells show the densification and reticulum typical of the early stages of retinal degeneration (Fig. 2, c.f. Stark and Carlson, 1982). We used carefully calibrated stimuli to show that degeneration is initiated in *rdgB* at very low levels of light. Surprisingly, receptor terminals (Figs. 6 & 7, c.f. control, Fig. 8) show abnormalities before somata (Figs, 1, 3 & 4) or axons (Fig 5). The

Figs. 13 - 16. The compound *norpA*^{EE5} *rdgB*^{KS222}*;cn bw* mutant reared at 23°C under cyclic room light. **Fig. 13:** R6 cell in *norpA rdgB;cn bw* newly emerged. Note coated vesicle, presumably of rhabdomeric origin (arrow), and multivesicular body (arrowhead). x 54,000. **Fig. 14:** Retinula cell from *norpA rdgB;cn bw* newly emerged. Note gradation of densities of multivesicular bodies from loosely packed (smallest arrowhead) through very electron dense (three arrowheads of increasing size). x 36,000. **Fig. 15:** A body like a multivesicular body in the intraommatidial cavity (IOC) of *norpA rdgB;cn bw* aged 3 weeks. Coated pits (arrowheads) appear at the plasmalemma of one retinula cell and the MVB-like body. Zippers (Z) are on the plasmalemma of the retinula cells, and one zipper is underlaid with a subsurface cistern (C). x 42,200. **Fig. 16:** Distal section of *norpA rdgB;cn bw* aged 3 weeks. Note multivesicular body in R7 (arrowhead) and coated pit from the plasmalemma (arrow) as well as zippers (Z) and subsurface cisterns (C). R6's rhabdomere is shown as well as a belt desmosome (D). x 27,500.

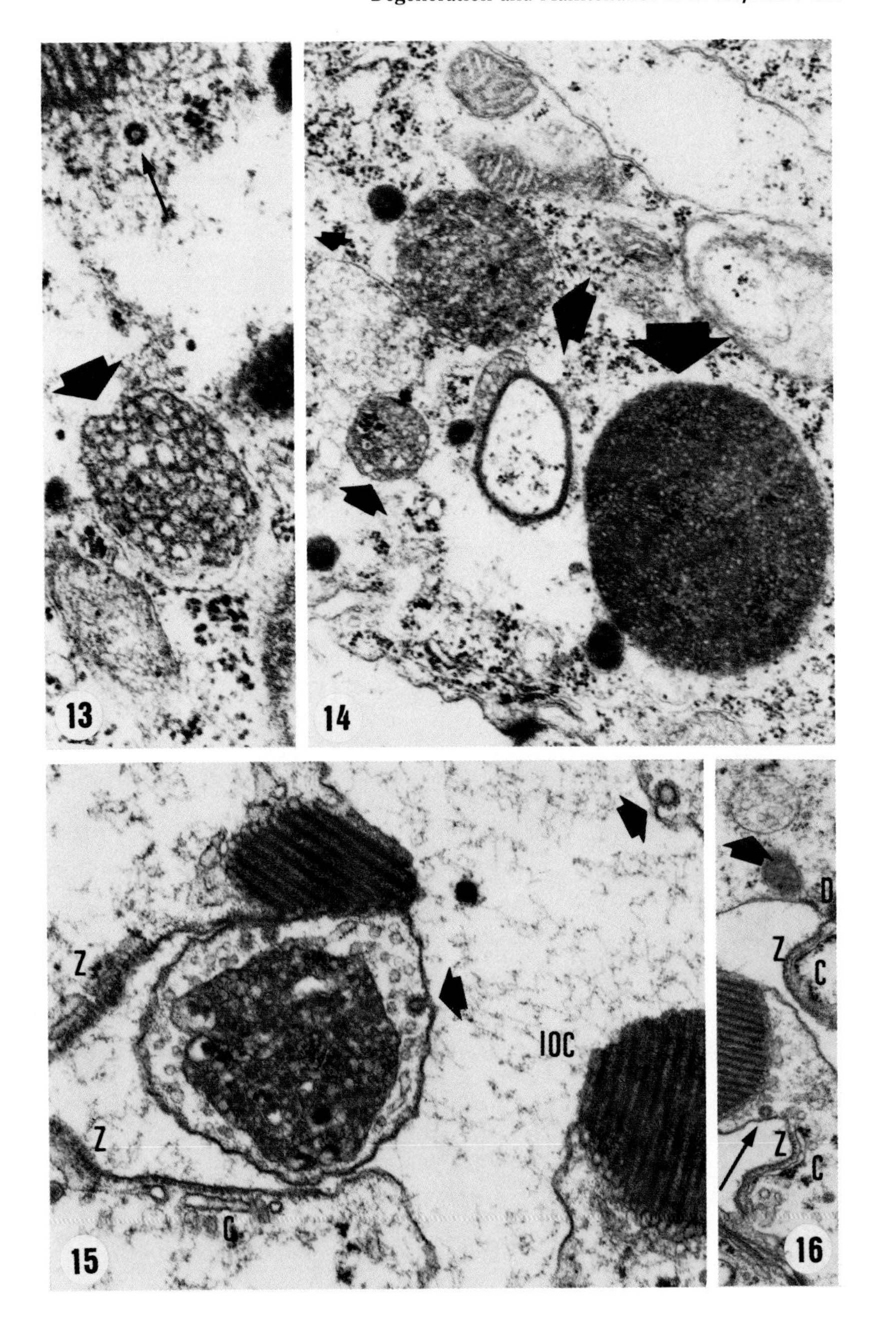
13
14
Z
Z
G
IOC
15
D
Z
C
Z
C
16

terminals become swollen and show abnormalities before somata (Figs. 1, 3 & 4) or axons (Fig. 5). The terminals become swollen and electron dense even with fixation in dim red light after blue and/or red stimulation (see Carlson et al., 1984).

We have been particulary interested in the conditions which inhibit degeneration and have studied *rdgB* compounded with several other mutant stocks. For instance, *ora* (outer rhabdomeres absent) blocks R1-6 rhodopsin synthesis and would be expected to inhibit *rdgB*'s light induced degeneration of R1-6 (Harris and Stark, 1977). The *ora* mutant inhibits degeneration in *rdgB* completely (Carlson and Stark, 1985; Stark and Carlson, 1985, Stark and Sapp, 1987).

NorpA may block degeneration in *rdgB* completely except for the fact that *norpA* shows some degeneration of its own (Stark et al., 1986; Stark and Sapp, 1986; Stark et al., 1988A). The former conclusion is documented in Figs. 9 - 16 for the *EE5* allele of *norpA* which blocks the receptor potential completely. Although the rhabdomeres

Figs. 17 - 20. The *Drosophila* mutant *norpA^{SUII} rdgB* for all figures on this plate. **Fig. 17:** A field dominated by a retinula cell in *norpA^{SUII} rdgB* aged 3 weeks. Note the debris in the intraommatidial cavity. x 33,400. **Fig. 18:** An "island" of debris (arrow) in the intraommatidial cavity associated with a retinula cell at the site of a zipper. *norpA^{SUII} rdgB* animal aged 3 weeks. The zipper as well as the plasmalemma of the adjacent R cell are underlaid with subsurface cisterns (hollow arrowheads). x 56,400. **Fig. 19:** Zippers and subsurface cisterns in *norpA^{SUII} rdgB* aged 3 weeks. x 52,900. **Fig. 20:** An optic cartridge in the lamina ganglionaris of *norpA^{SUII};rdgB* aged 3 weeks. Six R1-6 terminals with round capitate projections are between the epithelial glia (EG) and the central L1 and L2 cells (L). Several T-shaped presynaptic ribbons are clear (arrows). x 11,900.

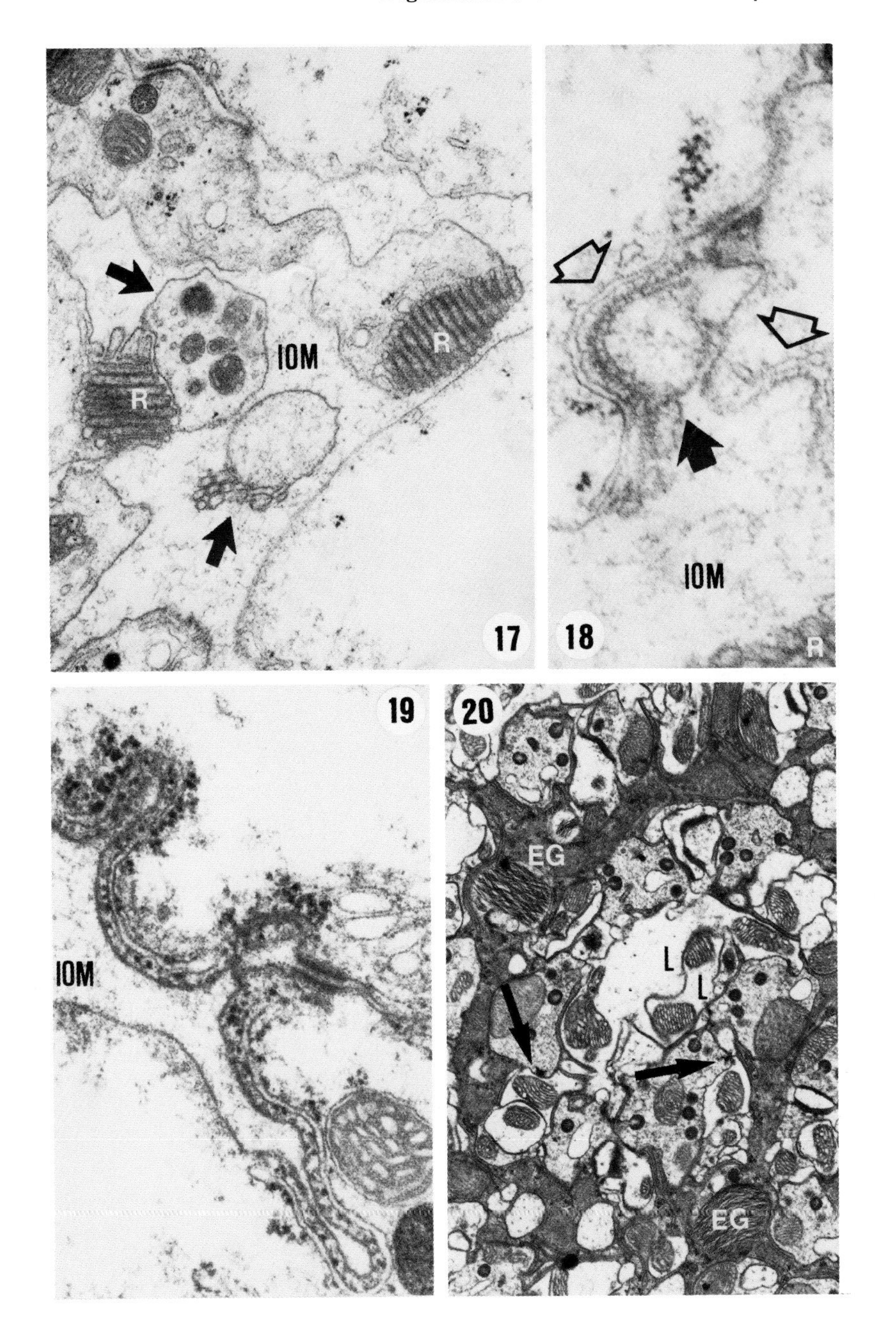
IOM
R
R
17
18
IOM
R
19
IOM
20
EG
L
L
EG

can become smaller with age (Fig. 15, an aspect of norpA gene action, Stark et al., 1988A), somata usually remain nondegenerate and show normal turnover from the formation of coated vesicles to MVB's (Figs. 13 - 16). Very few receptors actually degenerate, though a few do as shown at the level of the receptor terminals (Fig. 12). Thus excitation may be sufficient for degeneration in rdgB. Since norpA blocks degeneration in rdgB, the double mutant looks just like norpA alone and can be used to substantiate the conclusions drawn about the norpA mutant (Stark et al., 1988A). In this regard, "zippers" are a striking feature. Zippers (Figs. 11, 15 & 16) are striated areas of high membrane density sometimes seen apposed to a subsurface cistern (Fig. 15). The amount of zipper membrane facing the intraommatidial cavity increases with age, leaving the ommatidia grotesquely distorted (Figs. 9 & 10). Because norpA is a phospholipase gene (e.g., recently Bloomquist et al.'s 1988 cloning work), it is interesting to speculate that zippers represent

Figs. 21 - 26. The compound mutant w rdgB;trp (newly emerged) for Figs. 21 - 23 and w;trp (aged 2 1/2 weeks) for Figs. 24 - 26. **Fig. 21:** Loosely packed (nascent) MVB's (arrowheads) in distal R2 cell of w rdgB;trp (newly emerged). x 34,500. **Fig. 22:** Loosely packed (nascent) MVB (arrowhead) in R2 cell at intermediate depth of w rdgB;trp. Also shown is an invaginating granule from a secondary pigment cell (hollow arrowhead). x 25,920. **Fig. 23:** A degenerating R1-6 cell in w rdgB;trp (newly emerged) to show that trp does not protect rdgB from the degeneration induced by rearing in cyclic room light. x 9,500. **Fig. 24:** Membrane bound bodies of debris in the intraommatidial cavity of w;trp aged 2 1/2 weeks. Numerous coated vesicles (some labeled, hollow arrowheads) are present. x 36,000. **Fig. 25:** A distally sectioned retinula cell in w;trp aged 2 1/2 weeks to show that the photoreceptor cells in trp can eventually show degeneration. x 17,000. **Fig. 26:** A low magnification field of the lamina of w;trp aged 2 1/2 weeks. Arrowheads show occasional degenerate R1-6 terminals. x 3,700.

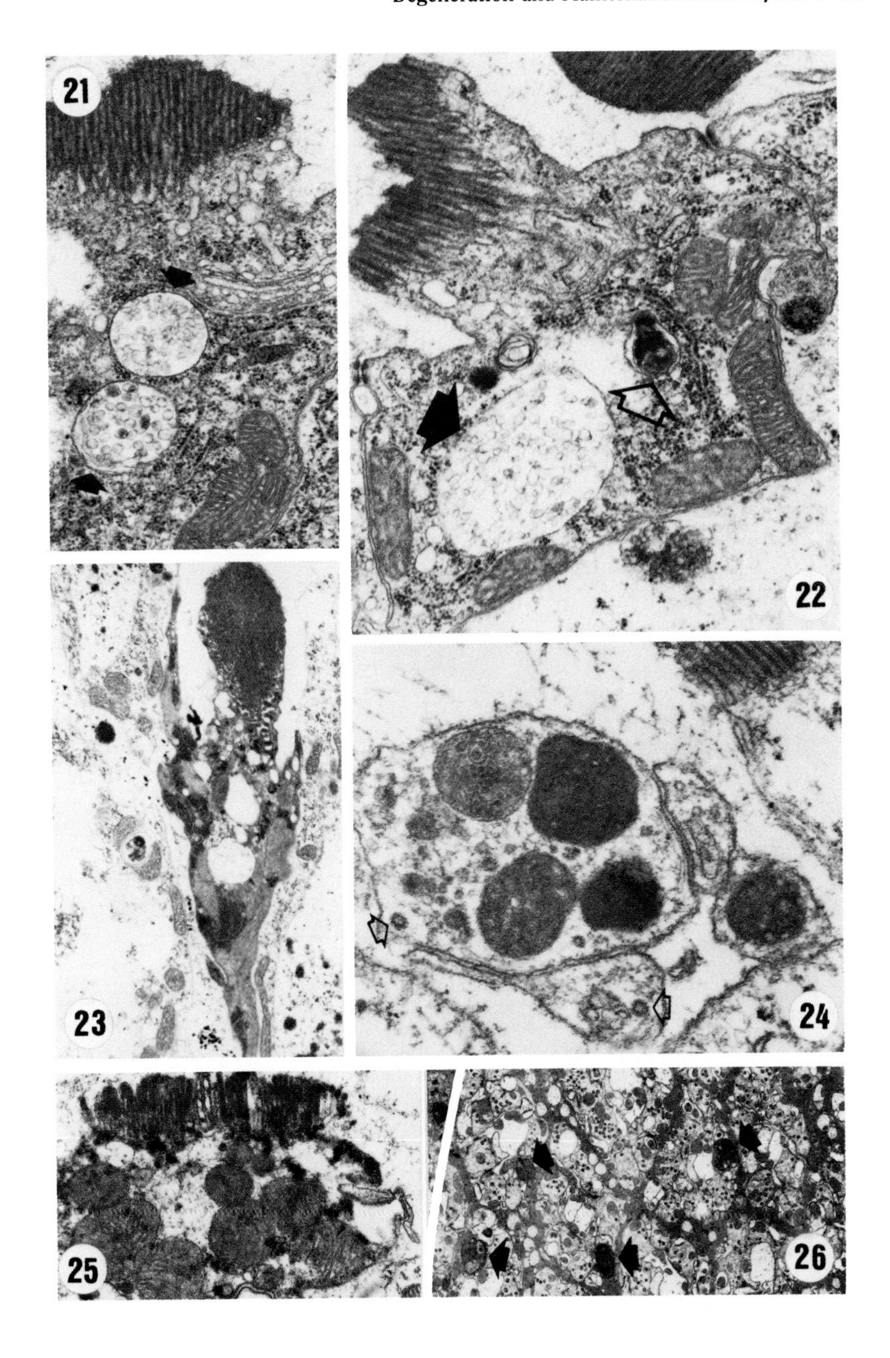
21
22
23
24
25
26

a visualization of a lipid storage defect in the substrate of phospholipase C, namely inositol 4,5-bisphosphate. Sometimes "islands" of "debris" appear to be localized extracellularly (Fig. 15), but these probably represent MVB's which are on peninsulas and would thus be intracellular; these are also present in aged non-norpA flies (Stark et al., 1988B).

Harris and Stark (1977) first showed that norpA inhibits receptor degeneration in rdgB mutants (see also Stark et al., 1983). One allele of norpA (SUII) has a receptor potential but suppresses degeneration in only one allele of rdgB (KS222), suggesting that the rdgB and norpA gene products may interact with each other directly in the process of excitation (Harris & Stark, 1977). Figs. 17 - 20 show the properties of the visual system of the compound mutant $norpA^{SUII}$ $rdgB^{KS222}$ aged 3 weeks in cyclic room light. The plate shows a lack of degeneration in the receptors (Figs. 17 - 19) and the R1-6 terminals (Fig. 20). Other aspects are as in $norpA^{EE5}$ rdgB, including the pre-

Figs. 27 - 31. The strain w;Acph- (aged one week) for Figs. 27 - 28 and the compound mutant w rdgB;Acph- for Figs. 29 - 31. **Fig. 27:** R2 cell in w;Acph- (aged one week) showing large, densely packed MVB (arrowhead) and small, loosely packed (nascent) MVB (hollow arrowhead). x 29,700. **Fig. 28:** Distal R1 cell in w;Acph- showing MVB (arrowhead) merging with a primary lysosome (L). x 36,000. **Fig. 29:** Proximally sectioned ommatidium of w rdgB;Acph- (aged one week in cyclic room lighting) at the level of R8's nucleus (*) showing partial to full degeneration of R1-6. R7's axon (arrowhead) and R8 are normal. x 12,000. **Fig. 30:** Cross sectioned pseudocartridge of w rdgB;Acph- (aged one week in cyclic room light) showing the surviving R7 and R8 axons (hollow arrowheads) and the degenerating R1-6 axons (arrowheads). x 20,300. **Fig. 31:** Cross sectioned optic cartridge of w rdgB;Acph- (aged one week in cyclic room lighting). The 6 R1-6 terminals (R) are degenerating. Surviving laminar monopolars (L) and epithelial glia (EG) are shown. x 8,600.

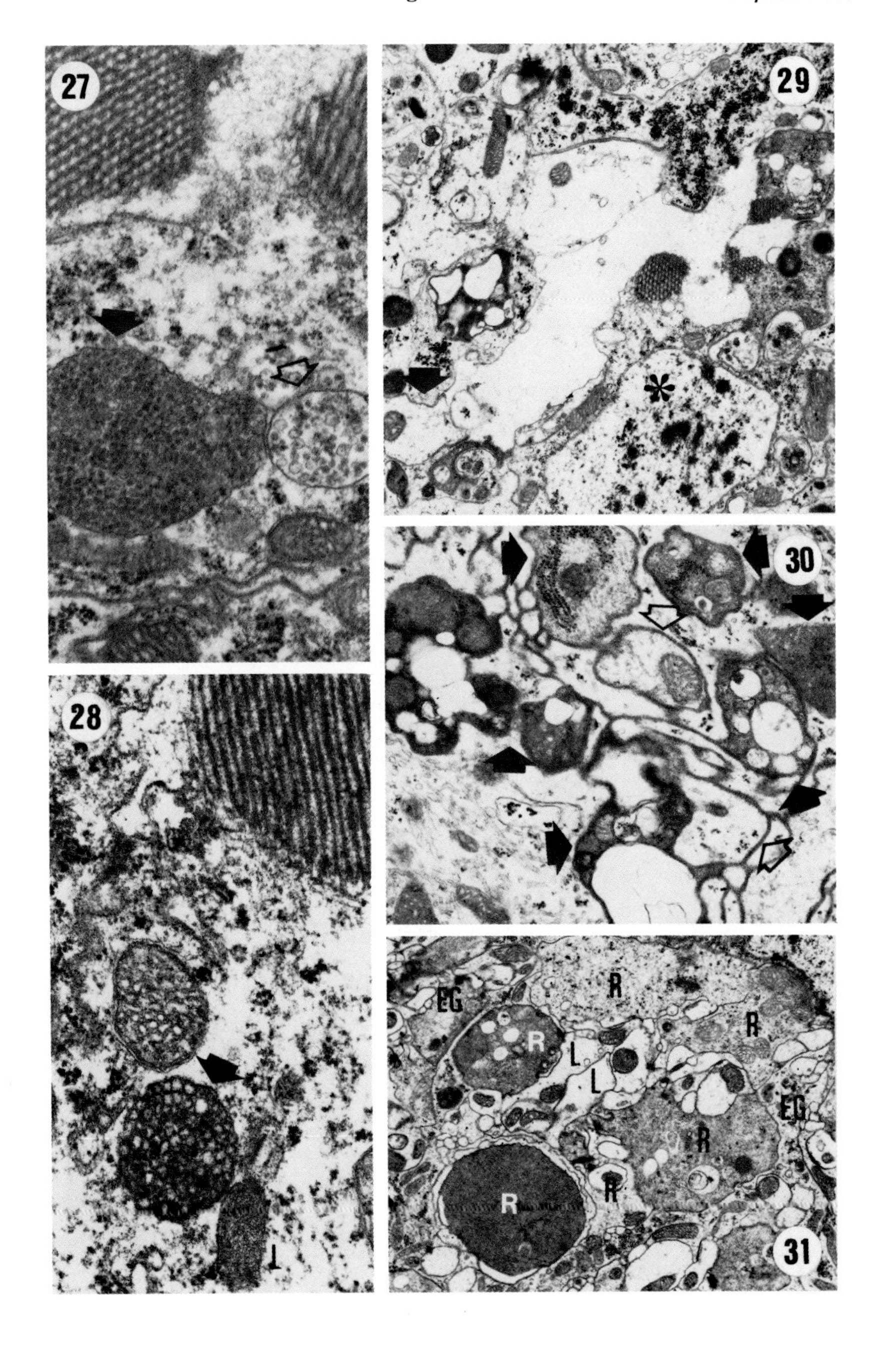
27
28
L
29
*
30
31
EG
R
L
R
EG
R
R
R
R
L

sence of zippers and subsurface cisterns, probable peninsulas with MVB's and the diminution in rhabdomeres.

In contrast to norpA some mutants do not inhibit degeneration in rdgB. For instance, trp does not completely block degeneration in w rdgB in

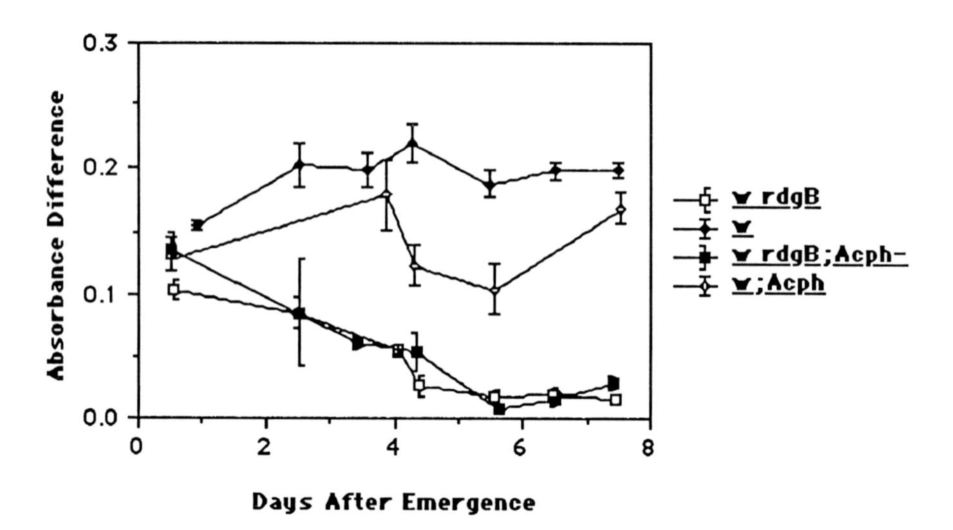

Fig. 32: Microspectrophotometry (MSP) to show that deprivation of the lysosomal enzyme acid phosphatase via the Acph- mutant does not inhibit degeneration in white eyed (w) degeneration mutants (rdgB). The relative amount of visual pigment determined from the deep pseudopupil was used as a quantitative assay of the state of integrity of the visual receptors as a function of time for a week. Absorbance difference at 579 nm after maximal adaptation at 600 vs 450 nm was measured. Each data point is averaged from 4 - 6 animals with the standard error shown. In this experiment, w is the control for w;Acph- to show that Acph-, by itself does not decrease visual pigment. Similarly, the demise shown by w rdgB is paralleled by that of w rdgB;Acph-, demonstrating that Acph- does not inhibit degeneration in w rdgB flies reared on a daily light cycle at room temperature.

flies which are compound mutants (Fig. 23). This EM result substantiates the conclusion made on the basis of MSP data by Chen and Stark (1983). The result shows that diminished depolarization is still sufficient for degeneration. However, cells which are not yet degenerate do show normal MVB's (Figs 21 & 22). The control animals for this study are <u>w;trp</u> (Figs. 24 - 26). Even without <u>rdgB</u>, some receptor somata in <u>w;trp</u> show degeneration (Fig. 25, c.f. receptor terminals, Fig. 26), as Cosens and Perry (1972) had shown earlier. Aged <u>w;trp</u> flies also show some other similarities with the retina of aged flies described earlier for <u>norpA</u> including likely peninsulas with resident MVB's (Fig. 24).

Another mutant which does not inhibit degeneration in <u>rdgB</u> is <u>Acph-</u>. We used electron microscopy (29 - 31) and MSP (Fig. 32) to show that <u>Acph-</u> does not block degernation in <u>w rdgB</u>. We show this at the level of receptor somata (Fig. 29), axons (Fig. 30) and terminals (Fig. 31). The <u>Acph-</u> result contradicts an earlier report (Harris and Stark, 1977) and indicates that this lysosomal enzyme does not accelerate degeneration. Correspondingly, the <u>w;Acph-</u> mutant which we constructed (without <u>rdgB</u>) does not show any obvious alterations in the early stages of receptor cycling (Figs. 27 & 28, c.f. Stark et al., 1988B); specifically, normal MVB's and primary lysosomes are shown.

The advent of molecular approaches applied to visual mutants promises exciting progress in this field and further justifies our efforts to provide a fairly thorough characterization of the visual defects associated with degeneration in mutants. For example, as mentioned previously, <u>norpA</u> has been cloned (Bloomquist et al., 1988) and the nucleotide sequence of this gene confirms the earlier finding that <u>norpA</u> mutants have a phospholipase C defect (Yoshioka et al., 1985). Although <u>rdgB</u> has been impossible to clone to date (C. Zuker, personal communication), another retinal degeneration gene (<u>rdgA</u>) has already provided putative clones (T. R. Venkatesh, personal communication). This molecular work on rdgA, when completed, should relate to the known biochemistry. Earlier, rdgA had been used to demonstrate the retinal localization of the enzyme diglycerol kinase used in the metabolism of phosphatidic acid (Yoshioka et al., 1984). Wong et al. (1985) had used the presence of retinal degeneration in trp to justify cloning this mutant. The most recent molecular studies (C. Montell, personal communication) suggest that trp codes for a 140 kD protein with no known homologies (yet) having 8 transmembrane spans which is localized in the rhabdomeres.

ACKNOWLEDGEMENTS

Supported by NSF grants BNS 8411103 and BNS 8811062 and NIH grant 1 R01 EY07192 and by UMC's Graduate School Research Council. Ms. E Frisch helped with the electron microscopy while Messrs. M. Ferrara, A. Justice and S. Christianson, with work study support, helped in the dark room. Mr. K. Walker helped to construct the Acph- stocks.

REFERENCES

Bell JB, MacIntyre RJ, Olivieri AP (1972). Induction of null-activity mutants for the acid phosphatase-1 gene Drosophila melanogaster. Biochem Gen 6:205-216.

Bloomquist BT, Shortridge RD, Schneuwly S, Perdew M, Montell C, Steller H, Rubin G, Pak WL (1988). Isolation of a putative phospholipase C gene of Drosophila, norpA, and its role in phototransduction. Cell (In press).

Carlson SD, Stark WS (1985). Protection against retinal degeneration in $rdgB^{KS222}$ Drosophila by ora^{JK84} a mutant lacking photoreceptor rhabdomeres. Invest Ophthalmol Vis Sci Suppl 26:131.

Carlson SD, Stark WS, Chi C (1984). Rapid light induced degeneration of photoreceptor terminals in rdgB mutant Drosophila. Invest Ophthalmol Vis Sci Suppl 25:18.

Chen D-M, Stark WS (1983). Sensitivity and

adaptation in the Drosophila phototransduction and photoreceptor degeneration mutants trp and rdgB. J Insect Physiol 29:133-140.

Cosens D, Perry MM (1972). The fine structure of the eye of a visual mutant, A-type, of Drosophila melanogaster. J Insect Physiol 18:1773-1786.

Harris WA, Stark WS (1977). Hereditary retinal degeneration in Drosophila melanogaster: A mutant defect associated with the phototransduction process. J Gen Physiol 69:261-291.

Harris WA, Stark WS, Walker JA (1976). Genetic dissection of the photoreceptor system in the compound eye of Drosophila melanogaster. J Physiol (Lond) 256:415:439.

Hotta Y, Benzer S (1970). Genetic dissection of the Drosophila nervous system by means of mosaics. Proc Nat Acad Sci (USA) 67:1156-1165.

Lindsley DL, Grell EH (1968). Genetic variations of Drosophila melanogaster. Oak Ridge National Laboratory.

Matsumoto E, Hirozawa K, Tagagawak, Hotta Y (1988). Structure of retinular cells in a Drosophila melanogaster visual mutant, rdgA, at early stages of retinal degeneration. Cell Tiss Res 252:293-300.

Miller GV, Hansen KN, Stark WS (1981). Phototaxis in Drosophila: R1-6 input and interaction among ocellar and compound eye receptors. J Insect Physiol 27:813-819.

Miller GV, Itoku KA, Fleischer AB, Stark WS (1984). Studies of fluorescence in Drosophila compound eyes: Changes induced by intense light and vitamin A deprivation. J Comp Physiol 154:297-305.

Stark WS, Carlson SD (1982). Ultrastructural pathology of the compound eye and optic neuropiles of the retinal degeneration mutant (w $rdgB^{KS222}$) Drosophila melanogaster. Cell Tiss Res 225:11-22.

Stark WS, Carlson SD (1985). Retinal degeneration in $rdgB^{KS222}$ is blocked by ora^{JK84} which lacks photoreceptor organelles. Drosoph Inf Serv 61:162-164.

Stark WS, Chen, D-M, Johnson MA, Frayer KL (1983).

The rdgB gene in Drosophila: Retinal degeneration in different mutant alleles and inhibition of degeneration by norpA. J Insect Physiol 29:123-131.
Stark WS, Clark AW (1973). Visual synaptic structure in normal and blind Drosophila. Dros Inf Serv 50:105-106.
Stark WS, Johnson, MA (1980). Microspectrophotometry of Drosophila visual pigments: Determinations of conversion efficiency in R1-6 receptors. J Comp Physiol 140:275-286.
Stark WS, Sapp R (1986). Pathways of maintenance of photoreceptor membrane in normal, vitamin A deprived and mutant Drosophila. Neurosci Abs 12:660.
Stark, WS, Sapp,R (1987). Ultrastructure of the retina of Drosophila melanogaster: the mutant ora (outer rhabdomeres absent) and its inhibition of degeneration in rdgB (retinal degeneration-B). J. Neurogenet. 4:227-240.
Stark WS, Sapp R, Carlson SD (1988). Photoreceptor maintenance and degeneration in the norpA (no receptor potential-A) mutant of Drosophila melanogaster. J Neurogenet (In press).
Stark WS, Sapp R, Hartman CR (1986) Retinal degeneration and membrane cycling in rdgB mutant Drosophila: Interaction with the mutants which block the receptor potential and the lysosomal enzyme acid phosphatase. 7th Internatl Congress of Eye Res Abstract Proc Int Soc Eye Res 4:70.
Stark WS, Sapp R, Schilly D (1988B). Rhabdomere turnover and rhodopsin cycle: Maintenance of retinula cells in Drosophila melanogaster. J Neurocytol 17 (In press).
Stark WS, Walker KD, Eidel JM (1985). Ultraviolet and blue light induced damage to the Drosophila retina: microspectrophotometry and electrophysiology. Curr Eye Res 4:1059-1075.
Wong F, Hokanson KM, Chang L-T (1985). Molecular basis of an inherited retinal defect in Drosophila. Invest Ophthalmol Vis Sci 26:243-246.
Yoshioka T, Inoue H, Hotta Y (1984). Absence of diglyceride kinase activity in the photoreceptor

cells of *Drosophila* mutants. Biochem. Biophys Res Commun 119:384-395.

Yoshioka T, Inoue H, Hotta Y (1985). Absence of phosphatidylinositol phosphodiesterase in the head of a *Drosophila* visual mutant, *norpA* (no receptor potential A). J Biochem 97:1251-1254.

III. INDUCED RETINAL DEGENERATIONS IN LABORATORY ANIMALS

Following the pioneering research begun in the 1950s by Werner Noell, various experimentally induced retinal degenerations have been used to explore photoreceptor and retinal pigment epithelial cell degeneration mechanisms. Of the ten papers in this section, six involve excessive light, the most widely studied means to induce photoreceptor degeneration. These papers include biochemical correlates of light-induced damage, a remarkable protection of photoreceptors from degeneration by hyperthermia, and three studies on the damaging effects of "blue light" and near-UV radiation to photoreceptors and retinal pigment epithelium. Two papers in this section present observations on chemically induced retinal changes. The final two papers deal with animal models of macular degeneration and retinopathy of prematurity, two important human retinal disorders.

Inherited and Environmentally Induced
Retinal Degenerations, pages 493–512

INTENSE-LIGHT MEDIATED CHANGES IN RAT ROD OUTER SEGMENT LIPIDS AND PROTEINS

D.T. Organisciak, H-M. Wang, A. Xie, D.S. Reeves and L.A. Donoso
Dept of Biochemistry, Wright State Univ, Dayton, OH 45435, Dept of Biology (A.X.), Beijing Normal Univ, Beijing, PRC, and Wills Eye Hospital, (L.A.D.), Philadelphia, PA 19107

INTRODUCTION

The absorption of visible light by rhodopsin triggers a complex cascade of molecular events culminating in the transduction of light energy into an electrical response by the photoreceptor cell. Excess light absorption, however, can also result in damage to the visual cell. Since the pioneering work of Noell et al (1966) much has been learned about the process of retinal light damage. It is now known, for example, that light damage in rats is rhodopsin mediated and that both duration of light exposure and retinal irradiance during exposure can affect the extent of damage (Noell et al 1966; Noell, 1979; 1980a,b; Rapp and Williams, 1979; 1980; Williams and Howell, 1983). Age at the time of exposure, genetic and environmental light rearing history are factors that can also influence the extent and type of retinal light damage in rats (Kuwabara, 1970; Noell, 1974; 1979; 1980a,b; O'Steen et al, 1974; Lai et al, 1978; Organisciak et al, 1985; Penn and Anderson, 1987).

The damaging events between the absorption of light by rhodopsin and photoreceptor cell death, however, are still unknown. Light mediated changes in retinal calcium levels (Winkler et al, 1984), opsin packing in rod outer segment (ROS) membranes (Organisciak and Noell, 1977), oxidative processes, and the loss of ROS docosahexaenoic acid (22:6) have been reported (Wiegand et al 1983; Organisciak et al, 1985; Penn and Anderson, 1987). Rats with low levels of ROS 22:6 are also protected against retinal light damage (Organisciak et al, 1987; Noell et al, 1987), but whether the loss of 22:6 is an oxidative process, a phospholipase mediated reaction (Jelsma, 1987), or a combination of both is unclear. Furthermore, little is known about the proteins involved in visual transduction during intense and sustained light exposure.

In this study we have used retinas and ROS isolated from intense light-treated rats to characterize changes in lipids, marker enzymes, DNA and transduction proteins during and after exposure. Our results show that ROS 22:6 decreased during the light period, while retinal peroxide levels are elevated. These events precede the irreversible loss of rhodopsin and visual cell DNA in the light-exposed rats. We also report changes in the levels of S-antigen (S-ag) [Wacker et al, 1977 (48 K protein, arrestin)] and alpha-transducin ($T\alpha$), [Fung and Stryer, 1980 (G-protein)] in ROS during and after light, which suggest that these proteins may also influence the light damage process.

MATERIALS AND METHODS

Animal Maintenance, Light Exposure Protocol:

Weanling albino male Sprague-Dawley rats were obtained from Harlan Inc. Indianapolis, IN and maintained in a weak cyclic-light environment of 4-6 ft cd illumination for 12 hr/day for 40 days (lights on 09:00). During this time the animals were given Purina Rat Chow and water ad libitum. Prior to light exposure all rats were dark adapted 16-18 hrs. Following dark adaptation, they were treated, or not, with green light (490-580 nmeters) for various periods of time. Illumination during exposure was 200-250 ft cd. Light was administered in a series of green Plexiglas chambers designed to maintain environmental and body temperature at 24 °C and 37° C, respectively

(Organisciak et al 1985). All light exposures started at 09:00. Following light treatment, rats were either sacrificed immediately or maintained in a dark environment for up to 2 wks. All animals were sacrificed in halothane saturated chambers.

Retinal Dissection, ROS Isolation, Lipid Extraction:

Retinas were excised from rats within 2 min of death as described by Delmelle et al (1975) and used for the preparation of ROS or for the determination of retinal organic peroxides or DNA. Rod outer segments were isolated from the retinas of 2-3 rats using a discontinuous sucrose gradient technique (Organisciak et al, 1982), modified by the inclusion of small glass beads to reduce or eliminate contamination from retinal pigment epithelial cell (RPE) plasma membranes (Braunagel et al 1988). Purified ROS were obtained from the 32/37% sucrose interface and used directly for enzyme determinations and SDS-PAGE, or precipitated with Krebs-Ringer phosphate buffer and then extracted with chloroform:methanol 2:1. Protein was measured by the technique of Lowry et al (1951).

Analytical Methods:

Organic peroxides were measured by a modification of the enzymatic procedure of Organisciak et al (1983). A crude glutathione peroxidase fraction was isolated from pig liver according to Heath and Tappel (1976) and partially purified by HPLC on a TSK-4000 gel exclusion column (Varian Inc.). The column was maintained at room temperature and eluted with 2.5 mM Tris pH 7.0 at a flow rate of 1 ml/min. Fractions containing proteins of mol wt 25-15 kD were collected, pooled and stored at 4 C for up to 2 wks without loss of activity. According to Ursini et al, (1985), phospholipid hydroperoxide glutathione peroxidase (mol wt 23 kD) is found in high concentration in pig liver and is active with lipophilic hydroperoxides, while glutathione peroxidase (mol. wt. 88 kD) is active with soluble peroxides.

Excised rat retinas were homogenized in 95% ethanol to inhibit endogenous enzymes and then diluted to 25% ethanol with 0.124M Tris buffer pH 7.6, containing 0.2 mM EDTA. Aliquots were incubated for 2 min at room temperature with catalase to reduce H_2O_2 and then for 10 min at 37 C with GSH, NADPH and with, or without, 100 l of the partially purified peroxidase, containing added GSH-reductase (Sigma, Inc.), (Organisciak et al 1983). Following centrifugation at 12,000 x g for 10 min, the supernatants of control and experimental samples were read at 340 nm to measure the conversion of NADPH to NADP. The concentration of organic peroxides was calculated from the loss of NADPH (1:1 mol/mol with peroxide reduction) using its extinction coefficient and the protein concentration of the retinal homogenates.

SDS-PAGE was performed according to Laemmli (1970) on total unwashed ROS fractions isolated from the sucrose gradients. Samples were solubilized with an equal volume of buffer containing 2% SDS and 10% beta-mercaptoethanol at 37 C for 1 hr and then applied to a 4% stacking gel over a 10% acrylamide gel containing 0.1% SDS. Following electrophoresis, the proteins were either stained with 0.2% Coomassie blue or transferred to nitrocellulose.

Transferred proteins were washed, blocked with 3% gelatin (1 hr), and incubated overnight at 4 C with monoclonal antibodies $C_{10}C_{10}$; A_9C_6, against S-antigen (Donoso et al, 1985); TF-15, against alpha-transducin [Navon and Fung, 1988 (kindly supplied by Dr. B. Fung)]; or rabbit polyclonal anti-rhodopsin (provided by Dr. E. Kean). Peroxidase-conjugated anti-mouse or anti-rabbit antibody was used with chromagen for visualization.

The determination of 5'nucleotidase and oubain sensitive NaK-ATPase in ROS has been described (Braunagel et al, 1988). ROS superoxide dismutase (SOD) levels were measured using a RIA by Dr. R. Crouch. Visual cell DNA in the rat retina was estimated as described by Noell et al (1987). Rhodopsin levels were measured in whole eye detergent extracts as previously described (Delmelle et al 1975). In some experiments, rhodopsin was measured in one eye and retinal DNA was measured in the fellow eye. For each, visual cell loss was then estimated by comparison with rhodopsin or retinal DNA levels in unexposed rats maintained in darkness for the same time as the experimental animals.

ROS lipids were extracted and determined as previously described (Organisciak et al 1982; 1987). Glycerophospholipid and fatty acid composition, cholesterol, and organic phosphorus were measured in aliquots from the same samples.

Electron microscopy was kindly performed by Dr. Z.Y. Li on isolated ROS (Li et al 1985).

RESULTS

Characterization of Light Exposed ROS:

The lipid and enzyme profiles of ROS isolated from rats immediately after light treatment were determined as a function of the duration of exposure. As shown in the left panel of Figure 1, ROS 22:6 decreased following 2 hrs of light. After a 4-hr exposure 22:6 was 4 mol % lower than control (50 mol %); it was 8-10 mol % lower after 24-48 hrs of light. The loss of 22:6 was accompanied by decreases in the levels of phosphatidylethanolamine (PE), 4 mol %, and phosphatidylserine (PS), 2 mol %, while the relative level of phosphatidylcholine (PC) in the ROS extracts increased from about 36 mol % to 40 mol %. In these ROS no changes were found in the levels of phosphatidylinositol (2.1 mol %) or sphingomyelin (1.6 mol %). However, an increase in the cholesterol (chol) content of the light exposed ROS did occur. In ROS from unexposed rats, chol was 7 mol %; it was 10 mol % after a 6-hr light exposure and 15 mol % after both 24 and 48 hrs of light.

The levels of organic peroxides in retinal homogenates were also measured as a function of light duration. For these measurements, paired experiments using control and light-exposed rats were performed. As shown in the lower left panel of Figure 1, retinal peroxides were an average of 17% higher than control (9.1 ± 2.4 nmol/mg protein) after 2 hrs of light. Peroxide levels were 19% and 22% higher following 4- and 12-hr light exposure, and 30% higher than control after 24 hrs of light.

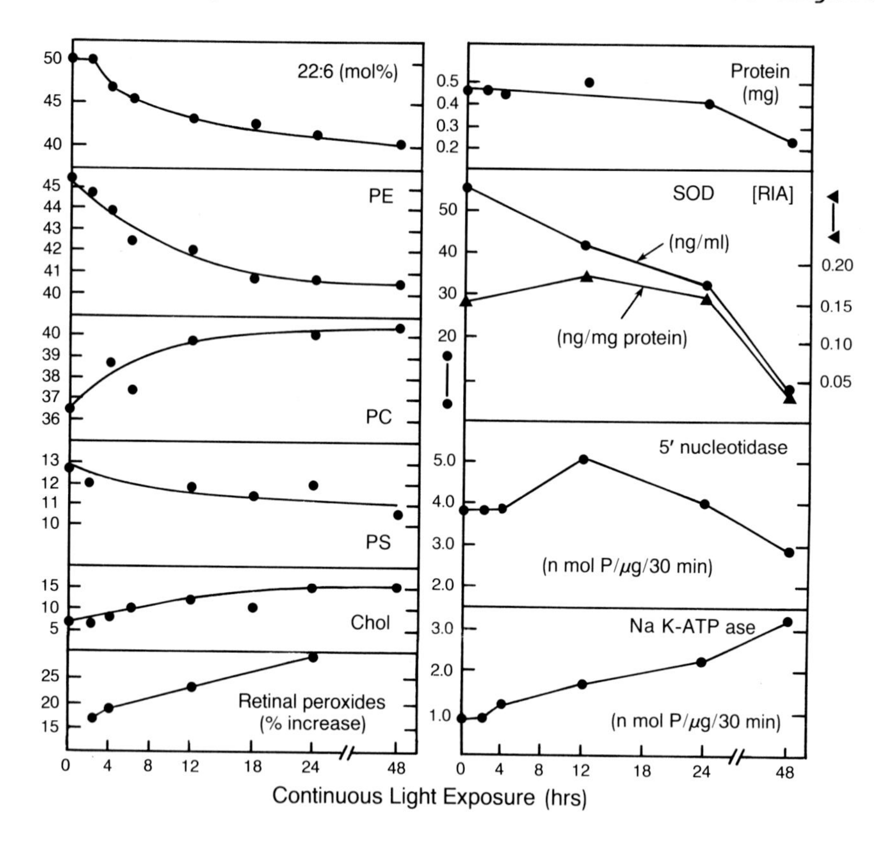

Figure 1. Rod Outer Segment Lipids and Enzymes After Light Damage. Results are the average of two or more separate determinations for ROS isolated immediately after light exposure. Retinal peroxides were measured in retinal tissue extracts. ROS lipids represented as mol %.

The protein content of the isolated ROS and the levels or activities of selected enzymes were measured (right panel Fig. 1). The amount of ROS protein recovered from the sucrose gradients was about 0.5 mg for rats treated with up to 12 hrs of light. ROS protein was 17% lower than control after 24 hrs of light and 50% lower following a 48-hr light exposure.

Superoxide dismutase (SOD), a soluble enzyme of about 40 kD molecular weight, was about 0.15 ng/mg protein in the control and in light-damaged ROS from rats treated for up to 24 hrs with light. In terms of ng/ml, SOD levels decreased by 35-40% during the 24 hr light exposure. After 48 hrs of light, the SOD concentration was 70-80% lower than that found in the control ROS.

The activity of ROS 5'nucleotidase was unchanged during the first 4 hrs of light. 5'nucleotidase was 20% higher after 12 hrs of light but was nearly the same as control following a 24-hr light exposure. During the same 24-hr light exposure, NaK-ATPase activity doubled; it was 3 fold higher following 48 hrs of light. The changes in NaK-ATPase activity may be due to the losses of ROS protein during the longer (e.g, 48 hr) exposure times or to the inclusion of rod inner segment (RIS) membranes in the light-damaged ROS. To confirm the latter possibility we measured the activity of the mitochondrial marker, cytochrome c oxidase. Using ROS isolated from 24-hr light-exposed rats no cytochrome c oxidase activity was detected.

The absence of mitochondria in the light-damaged ROS was confirmed by electron microscopy. As shown in Figure 2, control ROS (A) or 24-hr light-exposed rat ROS (B) were devoid of mitochondria. Other sections were also examined, but in no case were mitochondria observed. In both the control and light-damaged ROS, most of the rods appear to have intact plasma membranes. Some broken rods with exposed discs are present in each as are free disc membranes. Overall, the light-damaged ROS contained more disorganized and vesiculated discs and more broken rods and dissociated discs than control.

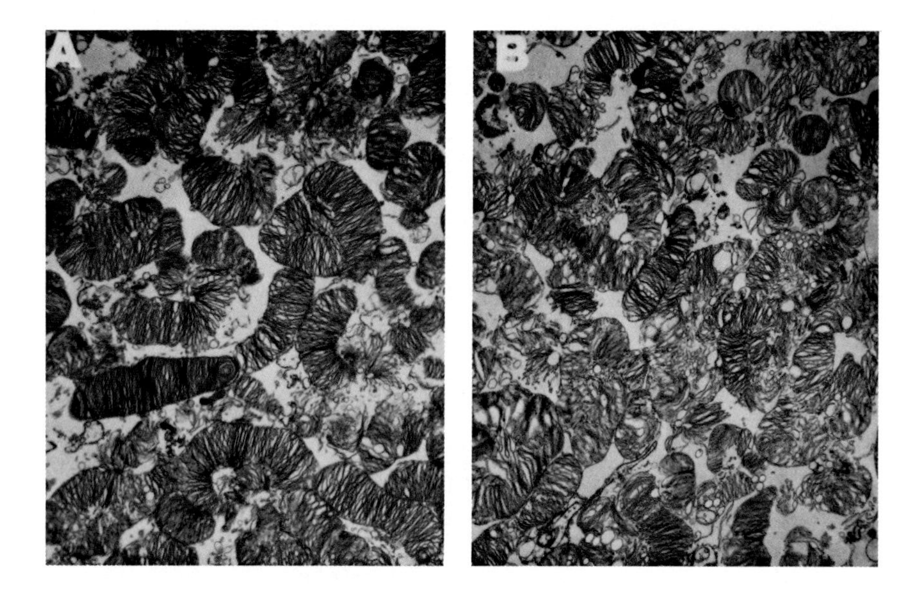

Figure 2. Electronmicroscopy of rat ROS.
ROS were isolated from the 32/37% sucrose interface of discontinuous gradients. A, control; B, 24hr light exposed rats x5000.

The electrophoretic gel profiles of ROS proteins from light-exposed rats were studied to determine if changes occurred during light. Figure 3 contains the results of Coomassie stained SDS-gels and Western transblots probed with various antibodies. Coomassie staining of the ROS proteins from unexposed rats (0 hr) revealed numerous bands with mol wts ranging from > 94 kD down to 30 kD (Fig. 3A). The same proteins were present in ROS from rats treated for 2 hrs with light; however, the staining of retinal S-ag (arrow) was greater, as well as that of a band with a mol wt of about 80-84 kD. The same higher intensity staining of the S-ag and 80-84 kD bands were apparent in ROS of rats treated with light for up to 48 hrs. There were no apparent losses of other high mol wt proteins, although the staining of T-α (upper band of the doublet present between 30 kD and 43 kD) appeared to be less intense in the light-exposed ROS.

A. Coomassie

0 2 4 12 24 48

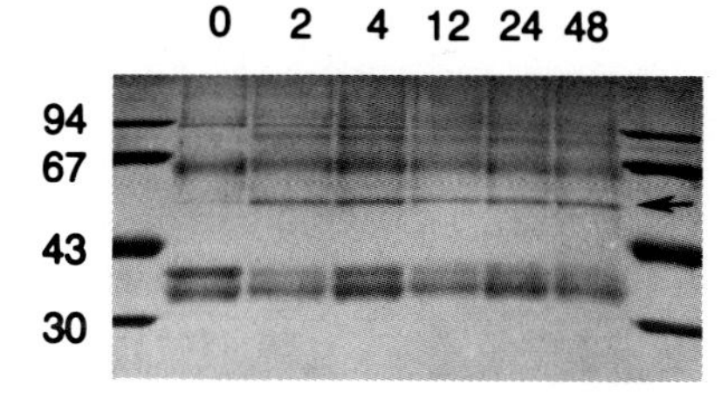

B. Anti S-antigen ($C_{10}C_{10}$)

S-ag 0 2 4 12 24 48 S-ag

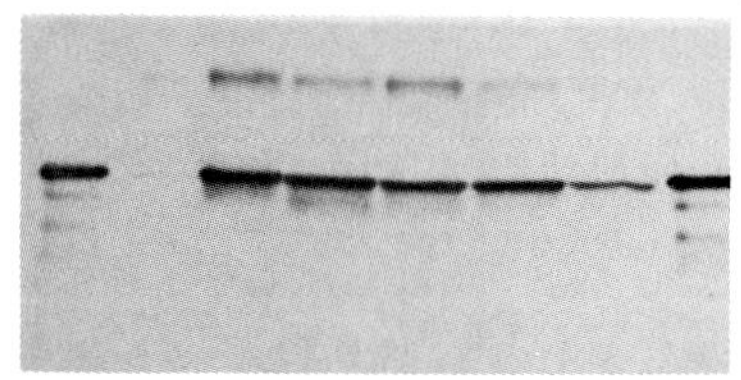

C. Anti Tα (TF-15)

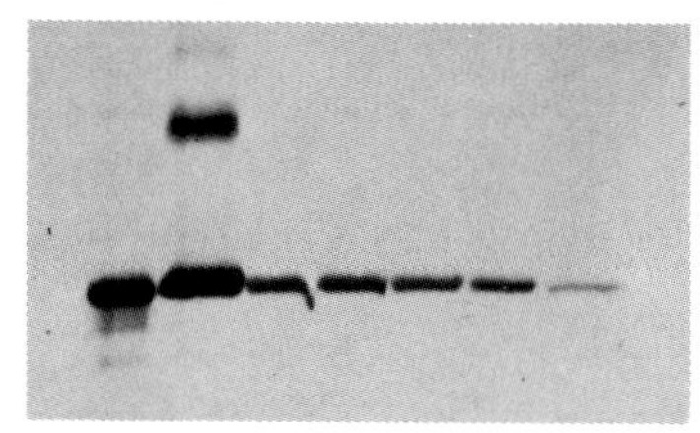

D. Anti Opsin

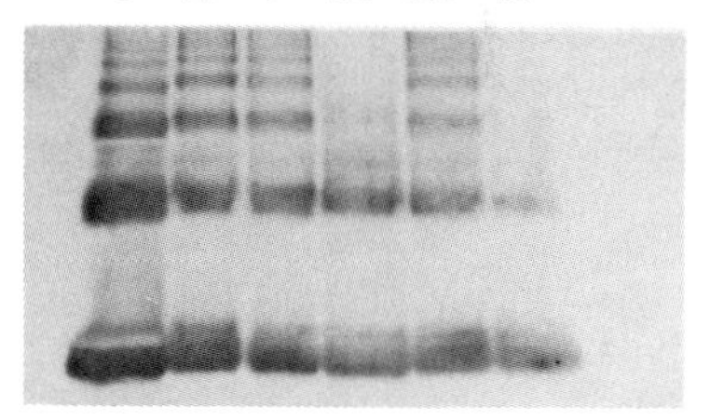

Figure 3. SDS-gel protein profiles of ROS from light exposed rats. ROS were isolated from rats treated for up to 48 hrs with light. A, Coomassie stain; B anti S-antigen monoclonal antibody $C_{10}C_{10}$; C anti transducin-alpha subunit TF-15; D, anti rhodopsin polyclonal. All lanes were loaded with equal protein concentrations (10μg). Arrow, S-antigen; T, transducin; mol wt standards x1000Kd.

When the ROS proteins were probed with monoclonal antibody $C_{10}C_{10}$ against S-ag, the staining of proteins from light-exposed rats (e.g., 2, 4, 12 hrs) was greater than that in the 0-hr control (Fig. 3B). The 80-84 kD band also reacted strongly with $C_{10}C_{10}$. The same results were obtained with A_9C_6, another monoclonal antibody against S-ag, but with a different epitope specificity (data not shown). The staining of both the 48 kD and 80-84 kD bands was similar in the 2-, 4- and 12-hr light-exposed rat ROS but was lower after 24 and 48 hrs of light.

Unlike S-ag, $T\alpha$ was present in high concentration in the 0-hr (unexposed) control ROS. A second high mol wt band was also present (Fig. 3C). Following light exposure for 12, 24 and 48 hrs, $T\alpha$ was lower. The higher mol wt band present in the unexposed-rat ROS was undetectable in the samples from light-treated rats.

Rhodopsin (opsin) staining in the isolated ROS was also greatest in the 0-hr sample and somewhat reduced after 2-, 4- and 24-hr light exposures (Fig. 3D). In the 48-hr light-exposed ROS, a clear loss of opsin was observed.

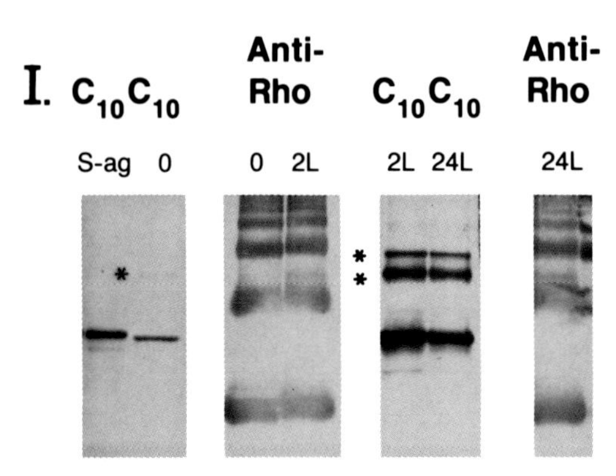

Figure 4. S-antigen and rhodopsin staining of ROS proteins. Unexposed control (0) and ROS from rats exposed for 2 or 24 hr to light (2L, 24L) were run on the same gel and simultaneously transferred for antibody reaction. (*) high mol wt aggregates of S-antigen and other proteins (10 μg).

The 80-84 kD band that stained with anti-S-ag was probed with polyclonal anti-rhodopsin antibodies. As shown in Figure 4 this band (*) was barely detectable in 0-hr ROS, using either antibody. However, the 80-84 kD band was present in both the 2- and 24-hr light treated samples (lower *). Using the anti-rhodopsin antibodies, its migration is identical to the 80-84 kD-$C_{10}C_{10}$ band and present between the dimer and trimer of opsin. A second higher mol wt S-ag-protein complex (upper *) which did not stain with the anti-rhodopsin antibody was also prominent in the 2- and 24-hr lanes.

Subsequent studies showed that this higher mol wt band was soluble, but that the 80-84 kD band precipitated with opsin in washed ROS samples. The 80-84 kD S-ag-opsin aggregate was not sensitive to 10% beta-mercaptoethanol or to hydroxylamine, indicating that it is not a disulfide crosslinked complex and that deacylation of opsin has no effect on its aggregation (O'Brien et al, 1987).

Characterization of ROS After Light Exposure:

ROS proteins were studied as a function of time in darkness after intense light exposure. Figure 5 contains the gel profiles of ROS proteins stained with Coomassie blue and probed with the anti S-ag, Tα , or opsin antibodies. In comparison to the 0-hr unexposed control, ROS from 24-hr light-treated rats had more intense staining of S-ag and less staining of Tα (Fig. 5A). The staining of S-ag remained prominent in the ROS from rats maintained in darkness for 3 hrs after light exposure; however, 1 day later, it was barely detectable. S-ag was undetectable by Coomassie staining following 3, 7 or 14 days of darkness.

Using the more sensitive $C_{10}C_{10}$ antibody probe, the same results were obtained (Fig. 5B). While intense staining of S-ag (and the higher mol wt S-ag aggregates) was apparent in the 24-hr light-exposed ROS and in those after 3 hrs of darkness, S-ag reactivity decreased thereafter. Following 14 days of darkness, S-ag staining by $C_{10}C_{10}$ was nearly the same as in the 0-hr control.

The opposite results were found when ROS proteins were probed with the TF-15 antibody (Fig. 5C). Tα reactivity was high in the 0-hr ROS, but was very low in ROS of rats treated for 24 hrs with light. The intensity of Tα staining increased thereafter and was much greater in the ROS of rats maintained in dark for 14 days after light exposure. Similarly, anti-rhodopsin staining increased as a function of the length of the dark period after light exposure (Fig. 5D). Also shown in Fig. 5D is the 80-84 kD S-ag-opsin complex (see 24L, + 3 hr).

A. Coomassie

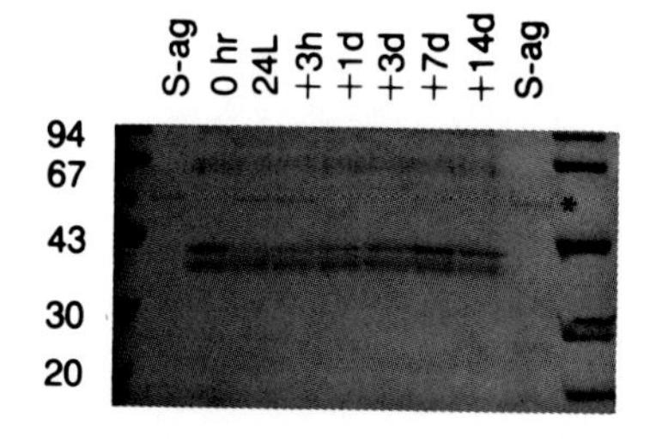

B. Anti S-antigen ($C_{10}C_{10}$)

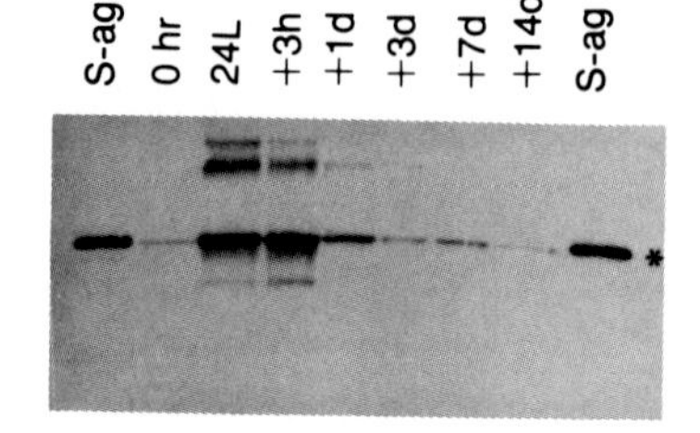

C. Anti Tα (TF-15)

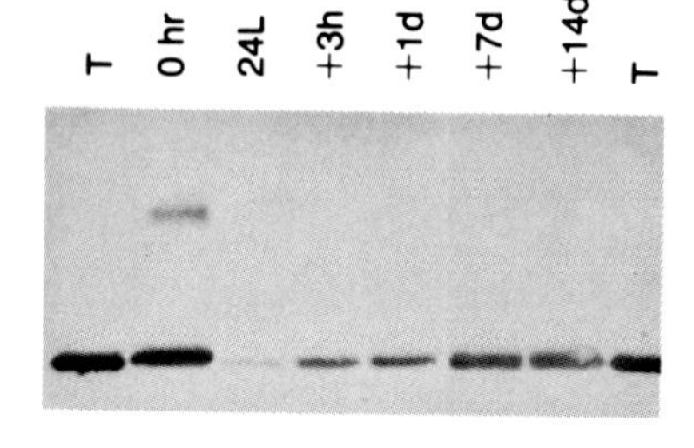

D. Anti Opsin

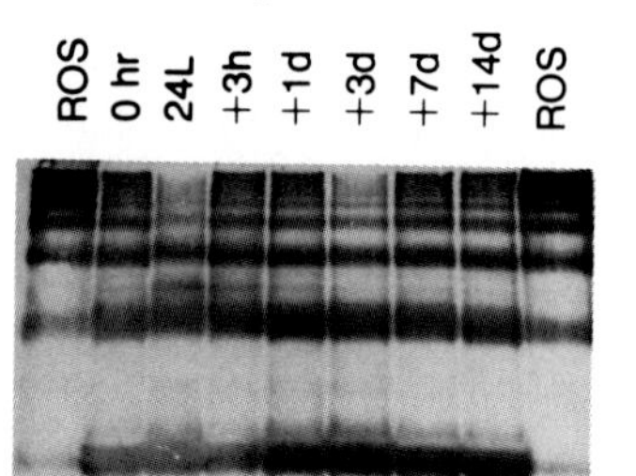

Figure 5. SDS-gel protein profiles of ROS from rats maintained in darkness after 24hr light exposure.

+3hr, 3 hrs dark after 24 hr light exposure; +1 d, +3d,+ 7d + 14d, days of darkness after 24 hr light exposure. A (15μg); B,C,D 7μg protein/lane; * S-antigen; T, transducin; See Figure 3 for details.

Figure 6 illustrates the results for visual cell DNA, whole eye rhodopsin levels and ROS 22:6 measured before light exposure and during the 2-wk dark period following light damage. After a 24-hr light exposure (0 dark) , visual cell DNA was about 10% lower than in the unexposed controls (187 ± 9μg). Rhodopsin was 1.8 nmol/eye before exposure, and 90-95% bleached (less than 0.2 nmol/eye) during the entire 24-hr light period. As previously shown (Fig. 1), ROS 22:6 was reduced by 20% after 24 hrs of light; however, no changes were found in the 22:6 level during the subsequent 2-wk dark period. At all time points ROS 22:6 remained at the same level as that found after 24 hrs of light (80% of control).

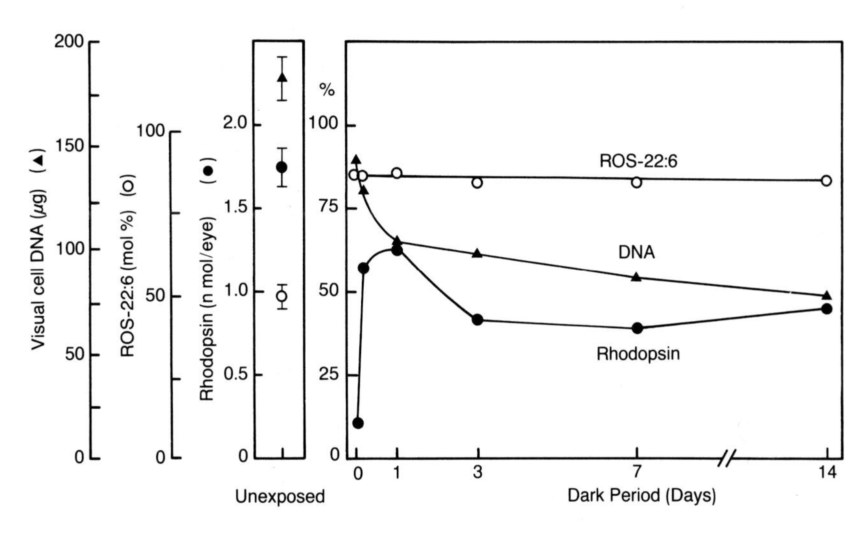

Figure 6. Rhodopsin, DNA and ROS-22:6 Before and After Light Exposure
Values are the average of two-four separate determinations.
Values for the 2wk. dark period are percent of the unexposed control.

Visual cell DNA and rhodopsin levels, however, decreased during the 2-wk dark period. After 1 and 3 days of darkness, the level of DNA was about 65% of control; it was 55% and 50% following 7 and 14 days of dark. Rhodopsin regeneration occurred during the first 24 hrs after light exposure. Following 6 hrs of darkness, it was about 60% of control. One day later rhodopsin was 65% of control. This initial increase was followed by the irreversible loss of rhodopsin from the eye. After 3 days it was 42% control whereas after 7 days of darkness rhodopsin was about 40%. At the end of the 2-wk dark period rhodopsin was slightly higher than at the 7-day time point (45% of control).

Figure 7 shows the time-dependent loss of rhodopsin and DNA from the eyes of light treated rats. Measurements were performed 2 wks after exposure. No losses in rhodopsin were apparent for up to 12 hrs of light exposure. At each time point, the recovery of the visual pigment was the same as

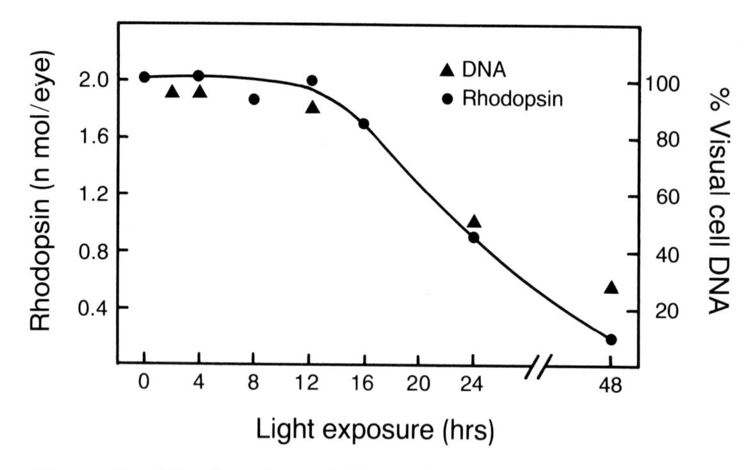

Figure 7. Rhodopsin and Visual Cell DNA
Measurements were performed 2 wks. after exposure for various times. For rhodopsin n = the average of 8 - 10; for DNA, n = 4 - 8.

in the unexposed control (2.0-2.1 nmol/eye). After 16 hrs of continuous light, rhodopsin was 85% of control; it was 45% after 24 hrs of light. Visual cell DNA for the 2-, 4- and 12-hr light exposure periods was 90-95% of the level in control. Nearly equal losses of visual cell DNA and rhodopsin were found for the 24 hr light exposure (50% and 55% respectively). However, after 48 hrs of light visual cell DNA was 30% of control, whereas rhodopsin was only 10% of control. The differences between rhodopsin and visual cell DNA in the surviving visual cells of light-damaged rats may be due to changes in rod length, or ROS opsin content resulting from intense light exposure (Organisciak and Noell, 1977; Penn and Anderson, 1988).

DISCUSSION

Retinal light damage has been studied by characterizing time-dependent changes in the lipids and proteins of ROS isolated from light-exposed rats. Our study shows that the polyunsaturated 22:6 fatty acid and the aminoglycerophospholipids (PE and PS) decrease as a function of light exposure, whereas the relative levels of PC and chol increase in the light-damaged ROS (Fig. 1). The loss of ROS protein was about 17% after 24 hrs of exposure, whereas, depending on the measurement, the level of the soluble enzyme SOD decreased by 35-40%. The loss of SOD correlates with the loss of enzymatic activity (~ 50%) in 24 hr light exposed ROS (data not shown). The activity of the membrane bound 5'nucleotidase did not appear to be particularly affected by light during the shorter exposure periods, however, NaK-ATPase activity was higher in the light-damaged ROS.

The loss of 22:6 from the ROS occurred primarily during light exposure; its levels were not further changed during the 2-wk dark period following light treatment (Fig. 6). Recently, Jelsma (1987) showed that G-protein activation of phospholipase A_2 could lead to the loss of unsaturated fatty acids in isolated ROS. However, the specificity and time dependence of the 22:6 loss from light-exposed rat ROS suggests that other factors may influence its loss in vivo. Certainly, judging by the intensity of light used in these studies, G-protein activation would be expected to be maximal within minutes, however, the loss of 22:6 was not apparent until after 2 hrs of light and was maximal after 24 hrs of exposure. As our study shows, retinal organic peroxides were also elevated during light exposure. These results suggest that peroxidized fatty acids may be particularly susceptible to phospholipase action and that a combination of the enzymatic and non-enzymatic processes may contribute to the overall loss of 22:6 during light exposure.

It is unlikely that the lower 22:6 levels result from contamination of the ROS with RPE plasma membranes, which have low levels of 22:6, because RPE adheres tightly to the glass beads used in the ROS isolation procedure (Braunagel et al, 1988). However, some contamination of our light damaged ROS with RIS membranes probably did occur. RIS plasma membranes are known to contain NaK-ATPase (Bok and Filerman, 1979), and appear to contain higher levels of cholesterol than ROS (Andrews and Cohen, 1983).

Both were elevated in our light-damaged ROS. Electron microscopy of light-damaged ROS revealed that most rods had intact plasma membranes, but an increase in broken rods and swollen or vesiculated discs was observed (Fig. 2). Although mitochondria were not seen and cytochrome c oxidase activity was absent, it is possible that fragments of the RIS membrane remain attached to the isolated-damaged ROS.

The SDS-gel electrophoretic profiles of ROS proteins revealed interesting changes in the levels of S-ag and T_α upon light exposure and during the subsequent 2-wk dark period. Both S-ag and a higher mol wt aggregate of S-ag and opsin were increased during light exposure (Fig. 3, 4). However, the levels of these immunoreactive bands decreased as a function of time in darkness in the ROS of the surviving visual cells (Fig. 5). The opposite effect was seen with T_α. It was high in ROS from unexposed rats and decreased as a function of light duration. In the ROS of rats maintained in darkness after light damage, the level of T_α and rhodopsin increased (Fig. 5).

Recent studies indicate that S-ag may move into the ROS during light exposure (Philip et al, 1987), and that dark-adaptation causes a release of membrane bound S-ag (Broekhuyse et al, 1987). It is also known that S-ag competes with transducin for binding to phosphorylated rhodopsin (Kuhn et al, 1984). Whether the displaced transducin diffuses from the light damaged ROS (as SOD is lost), or moves into the RIS of light exposed rats is not known at present. However, the changes in the levels of T_α and S-ag in ROS from rats maintained in darkness suggest that these are adaptive changes made in response to the animals' light environment. Whether these changes are related to the same mechanisms which alter the opsin content of ROS in dark-maintained rats (Organisciak and Noell, 1977) and the lipid profiles of ROS from rats reared in high light intensities (Penn and Anderson, 1987) will require additional study.

Finally, our study shows that the irreversible loss of rhodopsin and the loss of visual cell DNA from the eyes of light-exposed rats are relatively late-occurring responses to the sequence of damaging reactions that occur during light exposure (Fig. 6, 7). Whether these losses are in response to peroxidative reactions in the retina, to changes in the levels of visual cell transduction proteins, or to a combination of both is still an open question. However, the temporal relationship between the changes in peroxide levels and transduction proteins during light and the subsequent loss of visual cells, suggest that each may be a causative factor in the light damage process.

ACKNOWLEDGMENTS

We thank Drs. B. Fung and E. Kean for their generous gifts of the transducin and rhodopsin antibodies used in this work. Thanks also to Dr. R. Crouch for the SOD determinations, Dr. L. Prochaska for the cytochrome c oxidase measurements and Dr. Z.Y. Li for the electron microscopy. Supported by USPHS EY-01959, the Stanley Petticrew research fund at Wright State and BMRS grant 241327 (DTO); EY-05095 and the Henry Bower Research Laboratory for Macular Disease (LAD).

REFERENCES

Andrews LD, Cohen AI (1983). Freeze-fracture studies of photoreceptor membranes: New observations bearing upon the distribution of cholesterol. J Cell Bio 97:749-755.

Bok D, Filerman B (1979). Autoradiographic localization of Na^+-K^+ATPase in retinal photoreceptors and RPE with ^{3}H-ouabain. Invest Ophthal Vis Sci (Suppl.) 18:224.

Braunagel SC, Organisciak DT, Wang H-M (1988). Characterization of pigment epithelial cell plasma membranes from normal and dystrophic rats. Invest Ophthalmol Vis Sci 29:1066-1075.

Broekhuyse RM, Janssen APM, Tolhuizen EFJ (1987). Effect of light-adaptation on the binding of 48-kDa protein (S-antigen) to photoreceptor cell membranes. Curr Eye Res 6:607-610.

Delmelle M, Noell WK, Organisciak DT (1975). Hereditary retinal dystrophy in the rat: Rhodopsin, retinol and vitamin A deficiency. Exp Eye Res 21:369-380.

Donoso LA, Merryman CF, Edelberg KE, Naids R, Kalsow C (1985). S-antigen in the developing retina and pineal gland: A monoclonal antibody study. Invest Ophthalmol Vis Sci 26:561-567.

Fung BK, Stryer L (1980). Photolyzed rhodopsin catalyzes the exchange of GTP for bound GDP in retinal rod outer segments. Proc Nat Acad Sci 77:2500-2504.

Heath RL, Tappel AL (1976). A new sensitive assay for the measurement of hydroperoxides. Anal Biochem 76:184-191.

Jelsma CL (1987). Light activation of phospholipase A_2 in rod outer segments of bovine retina and its modulation by GTP-binding proteins. J Biol Chem 262:163-168.

Kuhn H, Hall SW, Wilden U (1984). Light-induced binding of 48-kDa protein to photoreceptor membranes is highly enhanced by phosphorylation of rhodopsin. FEBS Letters 176:473-478.

Kuwabara T (1970). Retinal recovery from exposure to light. Amer J Ophthalmol 70:187-190.

Laemmli UK (1970). Cleavage of structural proteins during the assembly of the head of bacterophage T_4. Nature 227:680-685.

Lai Y-L, Jacoby RO, Jones AM (1978). Age-related and light-associated retinal changes in Fischer rats. Invest Ophthalmol and Vis Sci 17:634-638.

Li ZY, Tso MOM, Wang H-M, Organisciak DT (1985). Amelioration of photic injury in rat retina by ascorbic acid. Invest Ophthalmol Vis Sci 26:1589-1598.

Lowry OH, Rosebrough NJ, Farr AL, Randall RJ (1951). Protein measurement with the folin phenol reagent. J Biol Chem 193:265-275.

Navon SE, Fung BK (1988). Characterization of transducin from bovine retinal rod outer segments. J Biol Chem 263:489-496.

Noell WK (1974). "Hereditary retinal degeneration and damage by light". Estratto dagli Atti Simposio di Oftalmologia Pediatrica. Parma, Italy pp. 322-329.

Noell WK (1979). Effects of environmental lighting and dietary vitamin A on the vulnerability of the retina to light damage. Photochem Photobiol 29:717-723.

Noell WK (1980a). "There are different kinds of retinal light damage in rats" In Williams TP, Baker BN (eds): The Effects of Constant Light on Visual Processes. NY: Plenum Press, pp. 3-28.

Noell WK (1980b). Possible mechanisms of photoreceptor damage by light in mammalian eyes. Vision Res 20:1163-1171.

Noell WK, Organisciak DT, Ando H, Braniecki MA, Durlin C (1987). "Ascorbate and dietary protective mechanisms in retinal light damage of rats: Electrophysiological, histological and DNA measurements" In Hollyfield JG, Anderson RE, LaVail MM (eds)): Degenerative Retinal Disorders: Clinical and Laboratory Investigations. NY: Alan R. Liss, pp. 469-483.

Noell WK, Walker VS, Kang BS, Berman S (1966). Retinal damage by light in rats. Invest Ophthalmol 5:450-473.

O'Brien PJ, St. Jules RS, Reddy TS, Bazan NG, Zata M (1987). Acylation of disc membrane rhodopsin may be nonenzymatic. J Biol Chem 262:5210-5215.

Organisciak DT, Favreau P, Wang H-M (1983). The enzymatic estimation of organic hydroperoxides in the rat retina. Exp Eye Res 36:337-349.

Organisciak DT, Noell WK (1977). The rod outer segment phospholipid/opsin ratio of rats maintained in darkness or cyclic light. Invest Ophthalmol Vis Sci 16:188-190.

Organisciak DT, Wang H-M, Kou AK (1982). Rod outer segment lipid opsin ratios in the developing normal and retinal dystrophic rat. Exp Eye Res 34:401-412.

Organisciak DT, Wang H-M, Li ZY, Tso MOM (1985). The protective effect of ascorbate in retinal light damage of rats. Invest Ophthalmol Vis Sci 26:1580-1588.

Organisciak DT, Wang H-M, Noell WK (1987). "Aspects of the ascorbate protective mechanism in retinal light damage of rats with normal and reduced ROS docosahexaenoic acid" In Hollyfield JG, Anderson RE, LaVail MM (eds): Degenerative Retinal Disorders: Clinical and Laboratory Investigations. NY: Alan R. liss, pp. 455-468.
O'Steen WK, Anderson KV, Shear CR (1974). Photoreceptor degeneration in albino rats: dependency on age. Invest Ophthalmol 13:334-349.
Penn JS, Anderson RE (1987). Effect of light history on rod outer-segment membrane composition in the rat. Exp Eye Res 44:767-778.
Philip NJ, Chang W, Long K (1987). Light-stimulated protein movement in rod photoreceptor cells of the rat retina. FEBS Letters 225:127-132.
Rapp LM, Williams TP (1979). Damage to the albino rat retina produced by low intensity light. Photochem Photobiol 29:731-733.
Rapp LM, Williams TP (1980). "A parametric study of retinal light damage in albino and pigmented rats" In Williams TP, Baker BN (eds): The Effects of Constant Light on Visual Processes. NY: Plenum Press, pp. 133-159.
Ursini F, Maiorino M, Gregolin C (1985). The selenoenzyme phospholipid hydroperoxide glutathione peroxidase. Biochim Biophys Acta 839:62-70.
Wacker WB, Donoso LA, Kalsow CM, Yankeelov JA, Organisciak DT (1977). Experimental allergic uveitis: Isolation, characterization, and localization of a soluble uveitopathogenic antigen from bovine retina. J Immuno 119:1949-1958.
Wiegand RD, Giusto NM, Rapp LM, Anderson RE (1983). Evidence for rod outer segment lipid peroxidation following constant illumination of the rat retina. Invest Ophthalmol Vis Sci 24:1433-1435.
Williams TP, Howell WL (1983). Action spectrum of retinal light-damage in albino rats. Invest Ophthalmol Vis Sci 24:285-287.
Winkler BS, Noell WK, Smith JC (1984). A rhodopsin mediated light damage in the isolated rat retina is calcium dependent. Invest Ophthalmol Vis Sci (Suppl.) 25:57.

Inherited and Environmentally Induced
Retinal Degenerations, pages 513–522

FACTORS AFFECTING THE SUSCEPTIBILITY OF THE RETINA TO LIGHT DAMAGE

Muna I. Naash*, Matthew M. LaVail+, and Robert E. Anderson*

*Cullen Eye Institute and Department of Biochemistry, Baylor College of Medicine, Houston, TX
+Departments of Anatomy and Ophthalmology, University of California, San Francisco, CA

INTRODUCTION

The extent of retinal damage due to constant illumination is determined by a number of conditions. Among these are the species (LaVail, Gorrin, Repaci 1987a,c), light history (Penn, Anderson 1987; Penn, Naash, Anderson 1987), age (O'Steen, Anderson, Shear 1974), body temperature (Barbe, Tytell, Gower 1988; Noell, et al 1966), diet (Anderson et al. 1987; Farnsworth, Stone, Dratz 1979; Noell, Albrecht 1971; Organisciak et al. 1986), and genetic makeup (LaVail et al. 1987b) of the animal model employed. It is also known that susceptibility of the retina to light damage is influenced, to some extent, by the dark-adapted and steady-state bleach rhodopsin levels of the animal prior to and during exposure (Noell et al. 1966; Williams, Howell 1983). However, the severity of this damage is also determined by the properties of the incident light, including intensity, duration, and wavelength (Noell, Walker 1966; Rapp, Williams 1980; Williams, Howell 1983).

Since the discovery by Noell et al. (1966) that the albino rat retina is particularly sensitive to the damaging effects of light, most studies on light damage have used this model. Few studies on light damage have been carried out on the mouse, in spite of a number of general advantages the mouse has over the rat as an experimental animal, including several mutations that cause inherited retinal degeneration (LaVail et al. 1987a,c). In addition, LaVail et al. (1987b) have recently found dramatic differences in retinal sensitivity to light between two strains of mice and their F1 heterozygotes. In their study, albino mice of different inbred strains were exposed to constant fluorescent lighting at an illuminance level of 1200 lux for intervals of 1-6 weeks. Under these conditions, the photoreceptors in retinas of albino BALB/cByJ (BALB/c) mice rapidly degenerated, whereas the photoreceptors in retinas of albino c57BL/6^{J}-c^{2J} (B6-c^{2J}) mice were remarkably more resistant to light damage. F1 heterozygotes produced from these two strains displayed an intermediate degree of susceptibility to light-induced degeneration. These findings demonstrate that phenotypically identical populations with different genetic constitutions can show markedly different sensitivities to light, and that single gene or polygenic factors should be included as a determinant of the severity of light damage.

In the present paper, we have sought the biochemical basis for the differences in susceptibility to constant light in the mice with different genetic constitutions.

MATERIALS AND METHODS

Animals and General Methods

Breeding pairs of BALB/c and B6-c^{2J} were obtained from the Jackson Laboratory (Bar Harbor, ME). Offspring of these two strains and their F1 heterozygotes were treated as previously described (LaVail et al, 1987b). Animals were dark-adapted overnight and sacrificed by cervical dislocation, following which their retinas were removed for biochemical analysis. The following biochemical determinations were made on whole retinas: fatty acid composition, rhodopsin, glutathione reductase, glutathione-S-transferase, glutathione peroxidase (total, Se-dependent, and Se-independent), vitamin E, vitamin C, and free amino acids.

In order to confirm that the mice used for biochemical analysis displayed the differences in susceptibility to light damage as shown previously (LaVail et al., 1987b), representative mice from the litters were exposed to constant fluorescent lighting at an illuminance level of 1800 lux for periods of 2 or 3 weeks. Following the light exposure, the mice were sacrificed by cervical dislocation, their eyes removed and prepared for light microscopic examination of 1 μm thick sections as described elsewhere (LaVail and Battelle, 1975). All procedures involving the mice adhered to the ARVO Resolution on the Use of Animals in Research and the guidelines of the UCSF and Baylor College of Medicine Committees on Animal Research.

Antioxidants

Glutathione enzyme activities were measured in retinal homogenates by previously detailed procedures (Naash, Anderson 1984; Naash, Nielsen, Anderson 1988). Glutathione peroxidase (Se-dependent and non-dependent), glutathione-S-transferase, and glutathione reductase were all measured spectrophotometrically by following the oxidation or reduction of NADP(H) at 340 nm.

The retinal levels of vitamins E and C were determined on retina homogenates by liquid chromatography (Nielsen, Naash, Anderson 1988). The values were compared to retinal lipid phosphorus and cytoplasmic protein, respectively.

Rhodopsin, Lipid, and Amino Acid Analysis

Whole retina rhodopsin was measured by adaptation of a method described by Fulton et al (1982). Retinas from dark-adapted mice were excised under dim red light and the visual pigment extracted in 2% Emulphogene in phosphate buffer. Rhodopsin absorbance was obtained from pre- and post-bleach difference spectra scanned from 700-350 nm.

Lipids were extracted from whole retinas using a Bligh and Dyer (1959) chloroform-methanol partition. The phospholipids were isolated and their fatty acids quantified by gas liquid chromatography (GLC) as previously described (Rouser et al 1966; Wiegand, Anderson 1982).

Amino acids were measured in the TCA soluble fraction of retina homogenates using an amino acid analyzer (Rapp et al 1985).

Statistical Analysis

Comparisons of mean values were made using a two-tailed Student's t-test.

RESULTS

Morphological observation of the mouse retinas following constant light exposure confirmed the expected degree of degeneration of each of the strains based on previous studies (LaVail et al 1987a,b,c). After 2 weeks of light exposure, the most severely damaged region of the retina (posterior retina in the superior hemisphere) in the light-

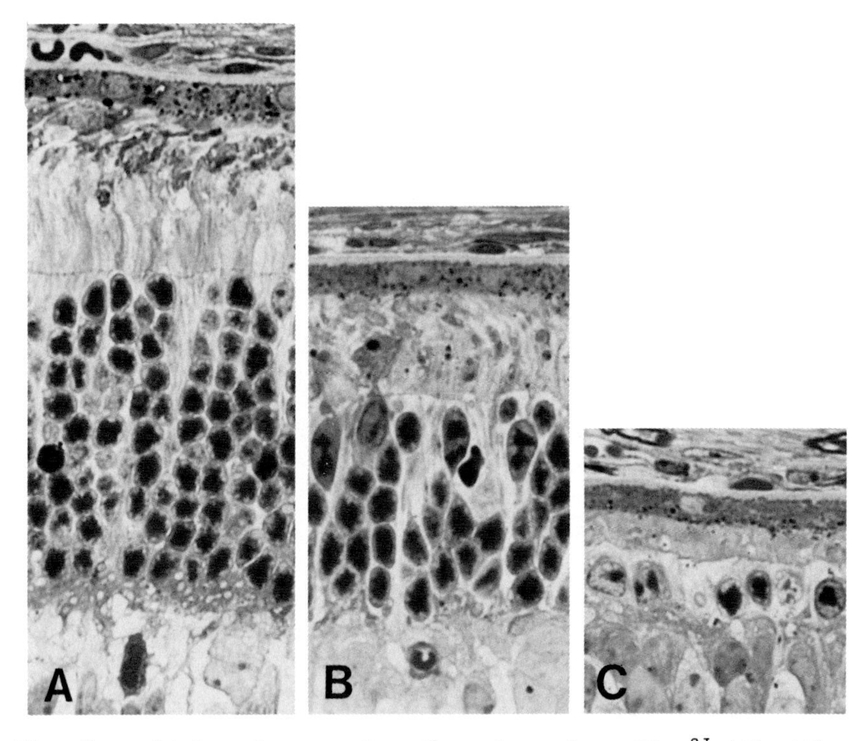

Fig. 1. Light micrographs of retinas from B6-c^{2J} (A), B6-c^{2J} X BALB/c F1 heterozygote (B) and BALB/c (C) albino mice after 2 weeks in constant light. All micrographs taken from the most severely damaged region in the superior hemisphere of the eye. In the B6-c^{2J} retina (A), photoreceptor cell loss is minimal and distinct fragments of outer segments remain. In the F1 retina (B), almost half the photoreceptor nuclei have been lost, the inner segments are shortened and disorganized, and few, if any, outer segment membranes are present. The illustrated region of the BALB/c retina (C) shows loss of most photoreceptor cells, with only a single layer of rod and cone nuclei persisting.

resistant B6-c^{2J} mice showed only a 5-10% loss of photoreceptor nuclei and minimal disruption of inner segments (Fig. 1a). Although the photoreceptor outer segments were clearly disrupted, distinct outer segment fragments were present (Fig. 1a). By contrast, the light-sensitive BALB/c retinas displayed about 90% loss of photoreceptors in the most damaged region (Fig. 1c). The F1 heterozygotes typically showed an intermediate degree of degeneration (Fig. 1b). The F1 mice showed the greatest variability in the degree of degeneration, and following 3 weeks of light exposure some appeared more like BALB/c than the typical F1 appearance shown in Fig. 1b. In all 3 lines of mice, there was less degeneration in the inferior hemisphere and far periphery of the superior hemisphere than in the posterior retina of the superior hemisphere. Further details of spatial distribution and temporal aspects of the light-induced degeneration in these mice have been presented elsewhere (LaVail et al, 1987a,b,c).

TABLE 1. Activities of glutathione dependent enzymes in mouse retinal homogenates.

ENZYME	BALB/c	B6-c^{2J}	F1
Glutathione reductase*	73.0 ± 14.8 n=7	66.4 ± 16.9 n=7	78.3 ± 6.4 n=3
Glutathione-S-transferase	305.3 ± 63.9 n=7	365.2 ± 58.0 n=6	280.4 ± 17.4 n=3
Glutathione peroxidase (tot)	71.5 ± 6.6 n=6	83.6 ± 18.5 n=7	76.1 ± 4.8 n=3
Se-dependent glutathione peroxidase	46.5 ± 12.9 n=4	43.8 ± 8.4 n=6	41.6 ± 6.2 n=3
Se-independent glutathione peroxidase	31.3 ± 6.5 n=4	39.4 ± 9.3 n=4	unmeasured

*Specific activities of glutathione enzymes are expressed in nmole (±SD) GSH oxidized or reduced/min/mg cytoplasmic protein.

TABLE 2. Levels of retinal antioxidants and rhodopsin.

ANTIOXIDANT MOLECULES	BALB/c	B6-c^{2J}	F1
Vitamin E	2.22 ± 0.16 n=6	1.64 ± 0.18 n=6	1.89 ± 0.65 n=4
Vitamin C	21.41 ± 3.20 n=6	18.47 ± 6.50 n=8	27.6 ± 2.6 n=4
Rhodopsin	0.38 ± 0.09 n=3	0.42 ± 0.06 n=3	0.42 ± 0.09 n=3

Vitamin E is expressed in mmole/mole lipid phosphorus (±SD).
Vitamin C is expressed in nmole/mg cytoplasmic protein (±SD).
Rhodopsin is expressed in nmole/retina (±SD).

TABLE 3. Levels of retinal amino acids

AMINO ACID	BALB/c	B6-c^{2J}	F1
Taurine	1078.7 ± 200 n=3	978.9 ± 301 n=3	821.7 ± 220 n=3
Aspartate	35.9 ± 3.2 n=3	63.2 ± 9.4 n=3	53.2 ± 17.6 n=3
Glutamate	138.3 ± 26 n=3	131.5 ± 30.5 n=3	124.8 ± 32.2 n=3
Glutamine	164.7 ± 28 n=3	176.9 ± 70.5 n=3	131.3 ± 23.5 n=3
Glycine	68.3 ± 23 n=3	57.7 ± 15.4 n=3	50.1 ± 18.6 n=3

Amino Acids are expressed in nmole/mg cytoplasmic protein (±SD).

The activities of the glutathione dependent enzymes in whole retinal homogenates of the three groups are shown in Table 1. These levels are comparable to those from adult rats (Naash, Anderson 1988a). The only difference was between glutathione-S-transferase activity in the F1 compared to the B6-c^{2J} ($p < 0.05$). The amount of rhodopsin was the same for all groups (Table 2). However, the levels of vitamins C and E were lower in the B6-c^{2J} group. The amino acid levels shown in Table 3 were the same between groups except for aspartic acid, which was lower in the BALB/c animals. The fatty acid composition of whole retinal total lipids presented in Table 4 showed no differences between groups.

TABLE 4. Fatty acid composition of total lipids from whole mouse retinas (mole percent ±SD, n=4)

FATTY ACID	BALB/c	B6-C^{2J}	F1
16:0	24.4 ± 0.5	25.6 ± 1.8	25.6 ± 1.8
18:0	22.2 ± 0.7	23.4 ± 1.5	21.1 ± 1.5
18:1	10.6 ± 0.4	12.6 ± 3.7	10.8 ± 0.6
20:1	0.5 ± 0.5	0.8 ± 0.7	0.7 ± 0.3
20:4	7.1 ± 0.6	6.7 ± 1.4	6.7 ± 0.4
22:4	0.9 ± 0.5	0.8 ± 0.2	0.7 ± 0.0
22:5	0.6 ± 0.1	0.4 ± 0.2	0.6 ± 0.1
22:6	26.8 ± 0.7	22.3 ± 4.5	25.6 ± 1.5

DISCUSSION

The intermediate susceptibility of the F1 heterozygotes of BALB/c and B6-c^{2J} animals is consistent with the notion that single gene or polygenic factors may control light damage in the mouse retina. Previous studies in albino rats point to a potential role for lipid peroxidation in light damage (Anderson et al 1987; Delmelle 1979; O'Steen, Anderson, Shear 1974; Wiegand et al 1983; Wiegand, Giusto,

Anderson 1982). Accordingly, we examined biochemically the retinas of the three groups of mice which differ in their susceptibility to light damage. Of the parameters tested, only vitamins E and C and aspartic acid levels were different, and these appear to give little information on the question. Indeed, the higher levels of vitamins E and C in the BALB/c mice are consistent with greater protection from light damage, yet BALB/c mice are more sensitive to the damaging effects of light than are B6-c^{2J}. Therefore, it appears that the genetic factors that afford protection to the B6-c^{2J} mice are not any we measured in the present study. Currently, we are comparing soluble and membrane bound proteins of the two strains by two-dimensional electrophoresis to determine if there are any differences. Preliminary results are encouraging and we have found several differences that are passed to F1 progeny in a predictable manner. Whether these different protein patterns are predictors of susceptibility to light damage in these animals is yet to be determined.

ACKNOWLEDGEMENTS

The authors thank Gregg Gorrin and Douglas Yasamura for technical assistance and Janice Jackson for editorial assistance in the preparation of this manuscript. This research was supported by grants from the RP Foundation Fighting Blindness, the National Eye Institute (EY00871, EY04149, EY07001, EY02520, and EY04753), the National Society to Prevent Blindness, Research to Prevent Blindness, Inc., and That Man May See, Inc. Dr. LaVail is an RPB Senior Scientist Research Investigator.

REFERENCES

Anderson RE, Wiegand RD, Rapp LM, Maude MB, Naash MI, Penn JS (1987). Studies on biochemical mechanisms of retinal degeneration. In Sheffield JB, Hilfer SR (eds): "Cell and Developmental Biology of the Eye, Vol. 6: The Microenvironment and Vision," New York: Springer Verlag, pp 159-168.

Barbe MF, Tytell M, Gower DJ (1988). Hyperthermia protects against light damage in the rat retina. Science 241:1817-1819.

Bligh EG and Dyer WJ (1959). A rapid method of total lipid extraction and purification. Can. J. Biochem. Physiol. 37:911-917.

Delmelle M (1979). Possible implication of photo-oxidation reactions in retinal photo-damage. Photochem. Photobiol. 29:713-716.

Farnsworth CC, Stone WL, Dratz EA (1979). Effects of vitamin E and selenium deficiency on the fatty acid composition of rat retinal tissues. Biochim. Biophys. Acta 552:281-293.

Fulton AB, Manning KA, Baker BN, Schukar SE, Bailey CJ (1982). Dark-adapted sensitivity, rhodopsin content, and background adaption in pcd/pcd mice. Invest. Ophthalmol. Vis. Sci. 22:386-393.

LaVail MM, Battelle BA (1975). Influence of eye pigmentation and light deprivation on inherited retinal dystrophy in the rat. Exp. Eye Res. 21:167-192.

LaVail MM, Gorrin GM, Repaci MA (1987a). Strain differences in sensitivity to light-induced photoreceptor degeneration in albino mice. Curr. Eye Res. 6:825-834.

LaVail MM, Gorrin GM, Repaci MA, Thomas LA, Ginsberg, HM (1987b). Genetic regulation of light damage to photoreceptors. Invest. Ophthalmol. Vis. Sci. 28:1043-1048.

LaVail MM, Gorrin GM, Repaci MA, Yasumura D (1987c). Light-induced retinal degeneration in albino mice and rats: Strain and species differences. In LaVail MM, Hollyfield JG, Anderson RE (eds): "Retinal Degeneration: Contemporary Experiments and Clinical Studies," New York: Alan R. Liss, Inc., pp 439-454.

Naash MI, Anderson RE (1984). Characterization of glutathione peroxidase in frog retina. Curr. Eye Res. 3:1299-1304.

Naash MI, Nielsen JC, Anderson RE (1988). Regional distribution of glutathione peroxidase and glutathione-S-transferase in adult and preterm human retinas. Invest. Ophthalmol. Vis. Sci. 29:149-152.

Naash MI, Anderson RE (1989). Glutathione-dependent enzymes in intact rod outer segments of rats. Exp. Eye Res., in press.

Nielsen JC, Naash MI, Anderson RE (1988). The regional distribution of vitamins E and C in human adult and preterm infant retina. Invest. Ophthalmol. Vis. Sci. 29:22-26.

Noell WK, Albrecht R (1971). Irreversible effects of visible light on the retina: Role of vitamin A. Science 172:76-80.

Noell WK, Walker VS, Kang BS, Berman S (1966). Retinal damage by light in rats. Invest. Ophthalmol. 5:450-473.

Organisciak DT, Wang H-M, Li Z-Y, Tso MOM (1986). The protective effect of ascorbate in retinal light damage of rats. Invest. Ophthalmol. Vis. Sci. 26:1580-1588.

O'Steen WK, Anderson KV, Shear CR (1974). Photoreceptor degeneration in albino rats: Dependency on age. Invest. Ophthalmol. 13:334-339.

Penn JS, Anderson RE (1987). Effect of light history on rod outer segment membrane composition in the rat. Exp. Eye Res. 44:767-778.

Penn JS, Naash MI, Anderson RE (1987). Effect of light history on retinal antioxidants and light damage susceptibility in the rat. Exp. Eye Res. 44:779-788.

Rapp LM, Naash MI, Wiegand RD, Joel CD, Nielsen JC, Anderson RE (1985). Morphological and biochemical comparisons between retinal regions having differing susceptibility to photoreceptor degeneration. In LaVail MM, Hollyfield JG, Anderson RE (eds): "Retinal Degeneration: Contemporary Experiments and Clinical Studies," New York: Alan R. Liss, Inc., pp 421-438.

Rapp LM, Wiegand RD, Anderson RE (1982). Ferrous ion-mediated retinal degeneration: Role of rod outer segment lipid peroxidation. In Clayton R, Haywood J, Reading H, Wright A (eds): "Problems of Normal and Genetically Abnormal Retinas," New York: Academic Press, pp 109-119.

Rapp LM, Williams TP (1980). A parametric study of retinal light damage in albino and pigmented rats. In Williams TP, Baker BN (eds): "The Effects of Constant Light on Visual Processes," New York: Plenum Press, pp 135-159.

Rouser G, Siakotos AN, Fleischer S (1966). Quantitative analysis of phospholipids by thin layer chromatography and phosphorus analysis of spots. Lipids 1:85-86.

Wiegand RD, Anderson RE (1982). Determination of molecular species of rod outer segment phospholipids. Methods in Enzymology. 81:297-304.

Wiegand RD, Giusto NM, Anderson RE (1982). Lipid changes in albino rat rod outer segments following constant illumination. In Clayton R, Haywood J, Reading H, Wright A (eds): "Problems of Normal and Genetically Abnormal Retinas," New York: Academic Press, pp 121-128.

Wiegand RD, Giusto NM, Rapp LM, Anderson RE (1983). Evidence for rod outer segment lipid peroxidation following constant illumination of the rat retina. Invest. Ophthal. Vis. Sci. 24:1433-1435.

Williams TP, Howell WL (1983). Action spectrum of retinal light damage in albino rats. Invest. Ophthalmol. Vis. Sci. 24:285-287.

Inherited and Environmentally Induced
Retinal Degenerations, pages 523–538

PHOTORECEPTOR PROTECTION FROM LIGHT DAMAGE BY HYPERTHERMIA

Michael Tytell, Mary F. Barbe and David J. Gower

Dept of Anatomy, Bowman Gray School of Medicine, Winston-Salem, NC 27103 (MT); Dept of Anatomy, Medical College of Pennsylvania, Philadelphia, PA (MFB); Section of Neurosurgery, University of Oklahoma, Oklahoma City, OK 73126 (DJG)

INTRODUCTION

Cells from a wide variety of species all have a common response to metabolic stress. That response consists of a great increase in synthesis of a small number of proteins that have come to be known as the heat shock or stress proteins (HSPs; see, for example, reviews by Schlesinger et al., 1982, and Craig, 1985). Surprisingly, the most prominent HSP produced by stressed cells, a 70,000 Dalton protein (HSP70), is quite similar among widely divergent species. For example, Escherichia coli HSP70 has about 50% of its amino acid sequence in common with that of human cells (Lindquist, 1986). This fact means that the stress response must have developed early in evolution and, more importantly, that the HSPs serve one or more functions that are very basic to cell survival. There are numerous observations showing a strong correlation between HSP synthesis and the acquisition of stress tolerance (Schlesinger et al., 1982; Subjeck and Shyy, 1986) and it has been speculated that the HSPs somehow prevent the irreversible denaturation of vital cell proteins during the abnormal conditions that exist at the time of stress (Minton et al., 1982). Support for that idea has been provided by recent reports that HSP70 can regulate the folding of clathrin (Ungewickell, 1985; Chappell et al., 1986) and of some mitochondrial proteins that are synthesized in the cytoplasm and then must pass through the mitochondrial membranes (Chirico et al., 1988; Deshaies et al., 1988).

Within about the last five years, the potential impact of the stress response on the nervous system has begun to receive more attention. The first report of HSP synthesis subsequent to central nervous system injury was that of White in 1980 (White, 1980). He documented the increased production of HSP70 in brain slices incubated in vitro compared to what was made in the intact brain in vivo and suggested that it played a part in how the brain responded to the injury (Currie and White, 1981). Since then, additional reports of HSP production in the CNS stimulated by stressors like ischemia, hyperthermia, and physical trauma have made it clear that the CNS resembles other tissues with respect to the stress response (Cosgrove and Brown, 1983; Nowak, 1985; Dienel et al., 1986; Gower et al., 1986; Tytell and Barbe, 1987). However, none of the work on the CNS provided any indication of whether the production of the HSPs influenced the extent of cell death after trauma in vivo. The retina is an ideal part of the nervous system in which to investigate this question. It is easy to label newly synthesized retinal proteins via an intra-vitreal injection of radioactive amino acid and exposure of an albino rat to specific schedules of illumination produce relatively easily measured levels of photoreceptor cell death. Summarized below is some of the evidence we have recently obtained suggesting a relationship between elevated HSP levels and resistance of photoreceptors to light damage.

ELEVATION OF RETINAL HSPs BY WHOLE-BODY HYPERTHERMIA

We examined the interaction between the level and duration of hyperthermia and the synthesis and accumulation of HSPs in the retina (Barbe et al., 1988). A minimum heat treatment producing a core (rectal) body temperature of 41°C for 15 minutes was required to significantly stimulate the synthesis of three retinal HSPs four hours after the end of the treatment. These HSPs had approximate relative molecular weights of 64, 74, and 110 kiloDaltons (KDa). Analyses of western blots of retinal samples from control and heat-treated rats using a monoclonal antibody against the stress-inducible 70 KDa HSP (a gift of W.J. Welch; see Welch and Suhan, 1986), showed that the rat retinal HSP64 was immunologically equivalent to the 70 KDa HSP described by others in a variety of cells types and species (see review by

Schlesinger et al., 1982). The lower apparent molecular weight of the rat HSP seems to be a result of the conditions under which the protein is prepared for polyacrylamide gel electrophoresis and not a consequence of a difference in structure (Tytell, unpublished observations). Therefore, the retinal protein will subsequently be referred to as HSP64/70.

Quantification of total retinal HSP64/70 by scanning laser densitometry of the western blots confirmed that, although retinas from untreated rats do contain a basal level of the HSP, the heat treatment did cause a marked increase in retinal HSP64/70 content (Barbe et al., 1988). That heat-induced elevation in HSP64/70 reached a maximum at about 18 hours after the hyperthermia, an observation consistent with earlier results of ours suggesting that the protein has a lifespan of several days in the retina (Tytell and Barbe, 1987).

HYPERTHERMIA REDUCES PHOTORECEPTOR DAMAGE AT TWO LEVELS OF ILLUMINATION

Having determined the heat treatment necessary to significantly increase retinal HSP synthesis and accumulation, we proceeded to test photoreceptor sensitivity to light-induced degeneration in rats heated to 41 or 42°C for 15 minutes. Initially, three levels of illumination were used, 150, 250, or 350 ft-c, to insure that even a weak protective effect of the hyperthermia would be detected. The exposure to the light was begun four hours after the heat treatment, the time at which HSP synthesis was maximal (Barbe, 1987). It was conducted for 24 hours using standard white fluorescent tubes in a white-walled chamber maintained at 30°C as described by Duncan and O'Steen (1985). The illuminance of the tubes was adjusted by partly covering them with a spiral of opaque black tape and it was measured at the level of the rat using a Tektronix photometer.

The effects of the various treatments on retinal photoreceptor survival are illustrated in light micrographs shown in Figure 1. As anticipated, exposure to the two higher levels of illumination, 250 or 350 ft-c, caused considerable loss of photoreceptors. The decrease in the thickness of the outer nuclear layers can be seen by

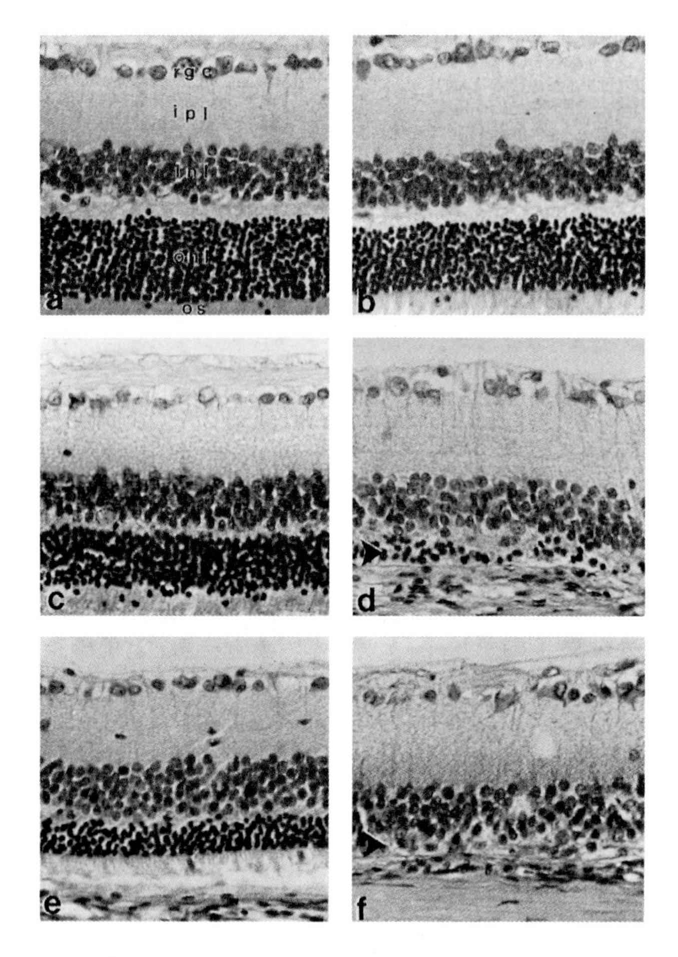

Figure 1. Micrographs showing the effect of prior heat stress (41°C, 15 min.) on light induced photoreceptor damage. Retinal sections were obtained from rats treated as follows and sacrificed two weeks later: (a) control conditions (6 ft-c cyclic light and normal room temperature); (b) heat stress and control illumination; heat stress followed four hours later by a 24 hour exposure to (c) 250 ft-c or (e) 350 ft-c of illumination; no heat stress and a 24 hour exposure to (d) 250 ft-c or (f) 350 ft-c of illumination at the same time of day as the two heat stressed groups. Heat stress alone had little effect on the retina (compare a and b) but greatly reduced the loss of photoreceptor nuclei in retinas from rats later

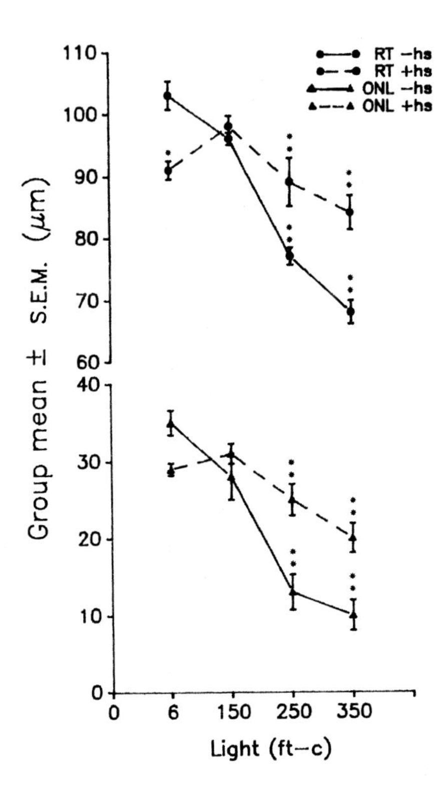

Figure 2. The mean values for the thicknesses of the entire retina (RT) and the ONL for each of the treatment groups described in Figure 1, as well as an additional group of rats exposed to 150 ft-c of illumination in the same way. Thickness is indicated in um along the vertical axis and illumination in ft-c is indicated along the horizontal axis. This graph illustrates the dramatic protection of photoreceptors from light damage across the whole retina in the heat stressed (+hs) rats compared to those exposed only to the light (-hs). There was a slight decrease in RT after the heat stress alone (compare points at 6 ft-c in upper panel), but this did not obscure the protective effect of the heat stress at higher illuminations. At 150 ft-c, no significant difference in either RT or ONL was detected, regardless of the heat treatment. Each point is the mean ± SEM of 7 rats. * and ** indicate significant differences at $0.05 \geq p \leq 0.02$ or $p \leq 0.01$ levels, respectively, between +hs and -hs groups.

---

exposed to the bright illumination (compare the outer nuclear layers (ONL) of panels c and d and e and f. The arrowheads in panels d and f indicate the remnants of the ONL in the retinas from rats exposed only to the light. In panel a, the following layers are labeled: retinal ganglion cell (rgc); inner plexiform layer (ipl); inner nuclear layer (inl); outer nuclear layer (onl); rod outer segments (os). 400X magnification. Hematoxylin and eosin stained.

comparing panels d and f to that of the control rat in panel a of Figure 1. The lowest illumination did not produce a significant loss of photoreceptors (not shown).

Remarkably, the retinas of those rats made hyperthermic prior to the light exposure (Figure 1, panels c and e) had clearly thicker ONLs than those of the unheated rats exposed to the same levels of illumination (Figure 1, panels d and f), indicating that the hyperthermia had made the photoreceptors more resistant to light damage.

To quantitate the protective effect of the hyperthermia, the thicknesses of both the ONL and the entire retina were measured at 12 different sectors as described by O'Steen et al. (1974; 1987). Using those measurements, the mean ONL and total retinal thickness (TRL) were calculated for each treatment group and the results plotted as a function of illumination history in Figure 2. The graphs confirmed that overall, the ONL of retinas from the heat-stressed (HS) animals exposed to 250 or 350 ft-c of illumination remained about twice as thick as those from light-exposed animals that had not been heated.

The upper plot of TRL in Figure 2 showed the same relative pattern as the lower plot of ONL thickness and it appeared that virtually all of the difference between the retinas of the two groups could be accounted for by the differences in the thickness of their ONLs. Thus, the primary effects of the heat treatment and light exposure seemed to be localized to the ONL.

ONSET AND DURATION OF HYPERTHERMIA-INDUCED PROTECTION OF PHOTORECEPTORS

The above results demonstrated hyperthermia-induced protection only at a single time after the heat stress. We wished to determine the time course of the protection; that is, how soon after the heat stress was it first demonstrable, when did it reach its maximum level, and how long did it persist. To answer those questions, groups of rats were heat stressed to 41°C for 15 minutes and exposed to 250 ft-c of illumination for 24 hours at various times thereafter. Two weeks later, the retinas were collected and prepared for standard light microscopic analysis.

Figure 3 shows examples of retinal morphology at sector 4, one of the most light-sensitive regions of the superior retina, in rats with post-heat stress light exposure intervals of 0 - 50 hours. Panel a shows a control rat retina for comparative purposes. A clear time course of post-hyperthermia photoreceptor protection can be seen by comparing the ONLs in panels b-h. Immediately after the heat treatment, little preservation of photoreceptor nuclei is evident (Figure 3b). However, just two hours later, noticeably more photoreceptor nuclei survived the light exposure (Figure 3c).

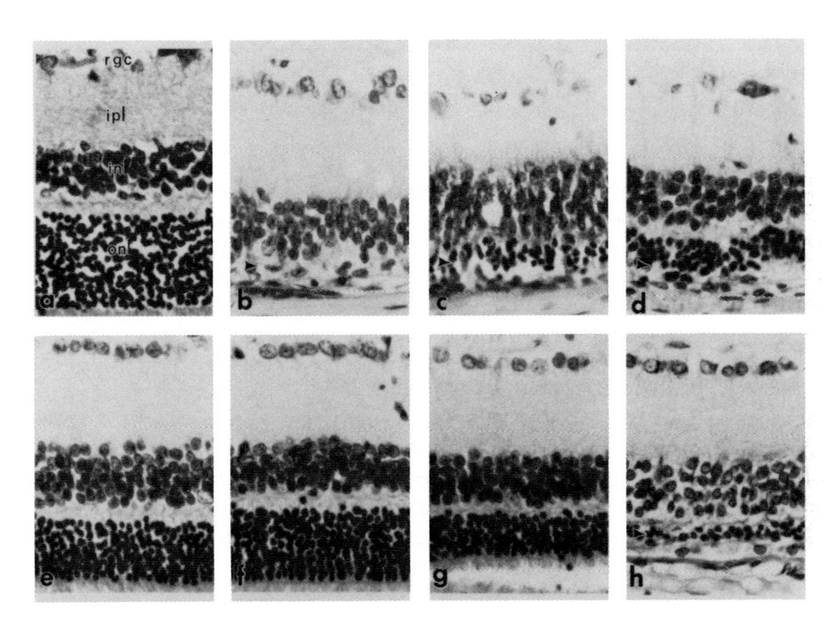

Figure 3. The time course of protection of photoreceptors from light damage. Representative light micrographs from the superior retinas of rats exposed to 250 ft-c of illumination at 0 (b), 2 (c), 4 (d), 10 (e), 18 (f), 25 (g), or 50 (h) hours after heat stress (41°C, 15 min.). Panel a shows a retina from a control rat (abbreviations as in Figure 1). Some preservation of photoreceptor nuclei became evident within two hours after heat stress (arrowhead, panel c) and reached a maximum by 18 hours (panel f). By 50 hours post-heat stress, the protective activity induced by the heat stress had greatly diminished. 400X magnification. Hematoxylin and eosin stained. (From Barbe et al., 1988).

As the interval between the hyperthermia and light exposure increased further, more surviving photoreceptors could be seen in the ONL, and this reached a maximum at 18 hours post-heat stress (Figure 3f). Beyond 18 hours, the protective effect declined, as revealed by the decreasing ONL thickness seen at 25 and 50 hours (Figure 3g and h). Thus, it appeared that the heat-induced increase in photoreceptor resistance to light damage reached its maximum in the first 24 hours after hyperthermia and then declined during the next 24 hours.

The preceding results describe only what happened at a single region of the retina. Was the protective effect similarly expressed in other retinal sectors? Additional rats were treated in the same way as described above and the ONL thickness of each retina was measured at 12 locations as in the previously described study. The results of those measurements, averaged across the corresponding sectors of all the eyes obtained from the rats in each group, are summarized in the set of graphs shown in Figure 4. This regional analysis showed that the protective effect of the heat stress began to be expressed very quickly. Even when the light exposure was started immediately after the end of the heat stress, survival of photoreceptors in the inferior retina was increased relative to that of rat that had not been made hyperthermic. As the interval between the heat stress and light exposure was lengthened, photoreceptors in all parts of the retina became more resistant and, with the exception of the zero hour time point, the effect seemed to be relatively uniform across all regions of the retina. The occurrence of maximal resistance to damage at 18 hours post-heat stress, as suggested by the retinal micrographs shown in Figure 3, was confirmed here (compare 18 hour panel to control). In fact, at this time, the overall mean ONL thickness of the heat-stressed, light exposed rats was not significantly different from that of the control animals. During the next 32 hours, however, that resistance to light damage declined across the whole retina (compare 25 and 50 hours panels with 18 hours panel).

In interpreting the results shown in Figure 4 with respect to the specific time after hyperthermia that the resistance to light damage became detectable, it is important to remember that the period of light exposure lasted 24 hours. Therefore, when the interval between hyperthermia and light exposure was less than 18 hours, the degree of protection was probably a product of two competing processes: (1) the rate of accumulation of the protective activity in the retina (which we suspect to be the HSPs) and (2) the rate at which the photoreceptors became lethally damaged by the light. In other words, the apparent increased resistance of photoreceptors to light damage at zero hours after the heat stress probably does not imply that hyperthermia induces the protective activity within minutes, but rather that some protective activity can accumulate in the retina even while it is being exposed to the damaging level of illumination.

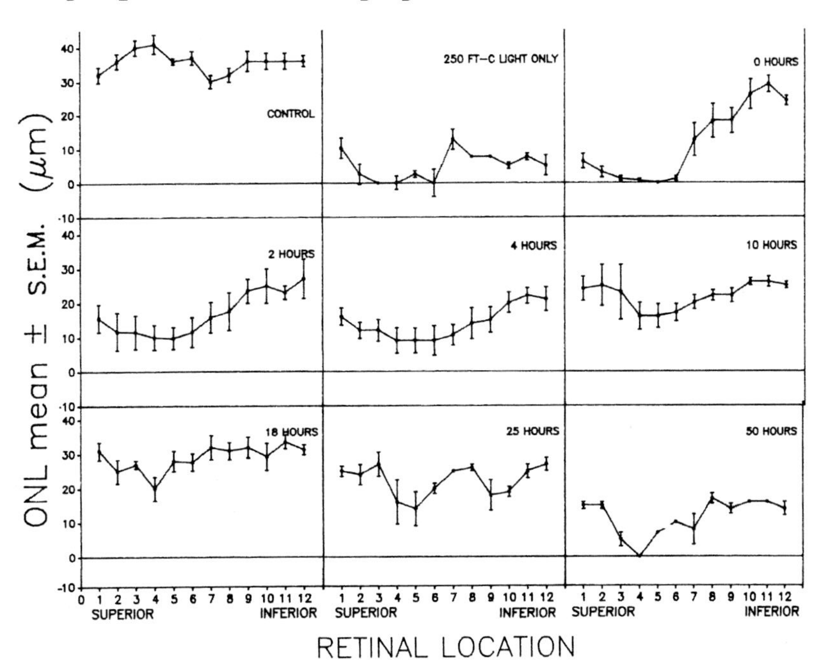

Figure 4. Illustration of the mean changes (±SEM) in ONL thickness at each of 12 retinal sectors in additional rats treated as described in Figure 3. The protective effect of the heat stress appeared to be expressed at all retinal locations, though the greater sensitivity of the superior retina to light damage relative to that of the inferior retina was superimposed on the degree of protection. Four rats per group.

IMMUNOHISTOCHEMICAL DISTRIBUTION

We used the anti-HSP70 monoclonal antibody mentioned earlier to examine the distribution of HSP64/70 in retinas from normal and heat-stressed rats. An example of the results of such an experiment is shown in Figure 5. In the control retina, typically there was some diffuse immunoreactivity throughout all the layers of the retina except the inner and outer segments of the rods. Slightly greater immunostaining was usually found in the cytoplasm of the retinal ganglion cells and in the outer plexiform layer (OPL). After the heat stress, however, the area that includes the rod inner segments (IS) became intensely immunoreactive and strand-like processes of immunoreaction product were seen between the photoreceptor nuclei. Furthermore, a scattering of darkly immuno-stained cells became apparent in the inner nuclear layer and the immunoreactivity of the OPL and retinal ganglion cells became more pronounced than that seen in the control retina. As yet, it is not clear what cellular component or components of the retina could account for the intense immunoreactivity in the OPL and IS. The two most likely candidates are the photoreceptor cell itself and the Muller cell, because both have extensive cytoplasmic projections in those layers. Additional work is in progress to resolve this question.

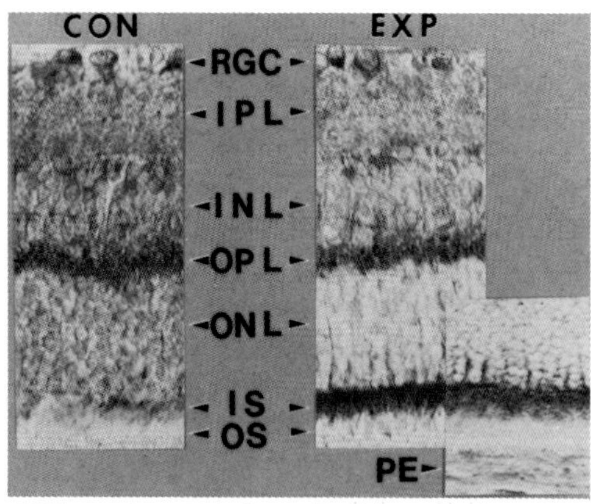

Figure 5. Micrographs of retinal sections observed by differential interference contrast microscopy showing the distribution of immunoreactive HSP64/70 in a typical control rat (CON) and an experimental (EXP) rat heat-stressed to 43°C for 15 minutes and sacrificed 18 hours later. In the CON rat, most of the immunoreactivity is located in retinal ganglion cell (RGC) cytoplasm and in the outer plexiform layer (OPL). There is diffuse lightly stained immunoreactivity in the inner plexiform (IPL) and inner

DISCUSSION

Photoreceptor Protection and Protein Synthesis

The fact that it took almost a full day after the hyperthermic treatment for maximal protection to be expressed argues against the possibility that the protection was a consequence of some direct action upon the photoreceptors of a hormone or other humoral factor whose circulating concentration was altered by heat stress. The time course fits better with the period typically required for a increase in protein synthesis and accumulation to produce a change in cell function. Since our earlier work on retinal protein synthesis after heat stress showed that the most prominent change was in the production of the HSPs (Tytell and Barbe, 1987), and that the time course of increased HSP synthesis and accumulation correlated closely with the acquisition of resistance to light damage by the retina (Barbe, 1987; Barbe et al., 1988), the HSPs remain prime candidates for the protective agents. However, whole body hyperthermia alters other physiological parameters as well, so work in progress is aimed at testing more directly the effects of the HSPs on photoreceptor resistance to light damage.

---

nuclear (INL) layers and very little present in the other retinal layers (ONL and rod inner and outer segments (IS and OS)). The overall darkness of this micrograph was produced optically so that the otherwise unstained layers could be seen. In the heat-stressed retina (EXP), the greatest increase in immunoreactive material is in the IS. Furthermore, strands of darkly stained immunoreactive material can be seen between the photoreceptor nuclei in the part of the ONL closest to the IS, especially in the inset at the lower right. That inset was included to show that the pigment epithelium had relatively little immunoreactivity. Immunoreactive HSP64/70 was detected using an anti-HSP70 monoclonal antibody obtained from W.J. Welch (University of California at San Francisco). Bound primary antibody was detected using a goat-anti-mouse biotin-avidin-horseradish peroxidase system (Vector Co.) and diaminobenzidene as the chromagen. 400X magnification.

The Localization of Induced Retinal HSP64/70 and Photoreceptor Protection

In addition to the correspondence in time of the heat-induced increase in retinal HSPs and photoreceptor protection, there should also be spatial correspondence; that is, the additional HSP produced in the retina from a heat-stressed rat should be located in or around the photoreceptor cells. The dramatic increase in HSP64/70 immunoreactivity in the photoreceptor inner segment layer, as well as a relatively smaller increase in the outer plexiform layer (Figure 5), meets the spatial criterion. Such a distribution of immunoreactivity is most easily explained if it is the photoreceptors themselves that are stimulated by the heat to synthesize more HSP. The additional HSP would then be expected to be most apparent in regions where photoreceptor cytoplasm is most abundant in the rat: at the site of synaptic contacts in the inner plexiform layer and in the inner segment region where the machinery to synthesize new outer segment membrane proteins and phospholipids is located.

How HSPs Might Reduce Photoreceptor Light Damage

Because the HSPs show a striking degree of phylogenetic conservation in their amino acid sequences and changes in their synthesis correlate closely with the acquisition of stress tolerance in many cells, most researchers accept the hypothesis that they are a key aspect of the cell's mechanism of damage control. However, the way in which they might perform such a function remains unclear. Early speculations were that they somehow protected other essential proteins within the cell from being irreversibly denatured by the abnormal conditions that might arise during the period of stress (Minton et al., 1982). Recently, that possibility received considerable support from observations that in normal yeast cells, HSP70 serves to control the folding of proteins destined for mitochondria so that they can pass through the mitochondrial membranes (Chirico et al., 1988; Deshaies et al., 1988).

The above potential mechanism of action of the HSPs could be quite relevant to the photoreceptor exposed to a potentially lethal level of illumination. Under such stress, it is known that within the photoreceptor, ion concentrations may become abnormal and highly reactive lipid peroxides and free radicals are produced (Li et al., 1985; Organisciak et al., 1985; 1987). We suggest that the prior elevation of the retinal HSPs by the heat stress protects many of the photoreceptor proteins from being denatured or oxidized, preserving them in a state so that they can resume or maintain normal cell function after the termination of the abnormal illumination. The increased localization of HSP64/70 in the RIS layer, but not the ROS layer, at first might seem inconsistent with our hypothesis because the site at which light-induced damage is first morphologically visible is the ROS (Kuwabara and Gorn, 1968). However, the primary cause or causes of light-induced photoreceptor cell death are not known, though light-dependent peroxidation of ROS lipids is a well-documented early consequence of light exposure (Weigand et al., 1983). What may be crucial to the survival of a photoreceptor is the extent of damage to the RIS, not just to the ROS, because it is the former that contains much of the biosynthetic and energy-producing machinery of the photoreceptor. The ROS, on the other hand, is primarily a light energy detection structure, portions of which are regularly shed as part of the normal cycle of events taking place in the photoreceptor (reviewed in Lanum, 1978; Besharse, 1986).

Regardless of the specific mechanism by which the hyperthermia leads to enhanced photoreceptor stress tolerance, this observation has several important potential applications. First, in the clinical situation in which surgery must be performed on the retina, it may prove possible to reduce cell loss due to unavoidable surgical trauma to the retina and to speed the recovery rate by first stimulating the production of HSPs using hyperthermia or some other means. Second, even in cases of accidental trauma to the retina, there may be a critical period before the retinal cells actually die during which the enhancement of HSP synthesis or the application of exogenous HSP could reduce the number of cells lost. Third, recent reports suggest that transplantation of healthy pigment epithelial cells into a retina genetically destined to degenerate may reduce the degenerative process (see reports by Turner et al. and Gouras et al., this volume). The survival and effectiveness of such transplants may be significantly enhanced if they are first stimulated to produce stress proteins.

ACKNOWLEDGEMENTS

This work supported in part by NEI grant EY07616-01 to M.T. and a grant to D.J.G. from the N.C. United Way and Bowman Gray School of Medicine. The technical support of Carol R. Hollman is gratefully acknowledged and we thank W. Keith O'Steen for his advice and comments during the course of the work.

REFERENCES

Barbe MF (1987). Heat shock protein production correlates with protection of the retina against light damage. Ph.D. dissertation, Bowman Gray Sch Med.

Barbe MF, Tytell M, Gower DJ, Welch WJ (1988). Hyperthermia protects against light damage in the rat retina. Science 241:1817-1820.

Besharse JC (1986). Photosensitive membrane turnover: Differentiated membrane domains and cell-cell interaction. In Adler, R, Farber, D, "The Retina - A Model for Cell Biology Studies," New York: Academic Press, part 1, pp 297-352.

Chappell TG, Welch WJ, Schlossman DM, Palter KB, Schlesinger MJ, Rothman JE (1986). Uncoating ATPase is a member of the 70 kilodalton family of stress proteins. Cell 45:3-13.

Chirico WJ, Waters MG, Blobel G (1988). 70K heat shock related proteins stimulate protein translocation into microsomes. Nature 332:805-810.

Cosgrove JW and Brown IR (1983). Heat shock protein in mammalian brain and other organs after a physiologically relevant increase in body temperature induced by D-lysergic acid diethylamide. Proc Natl Acad Sci USA 80:569-573.

Craig EA (1985). The heat shock response. CRC Critical Rev Biochem 18:239-280.

Currie RW and White FP (1981). Trauma-induced protein in rat tissues: A physiological role for a "heat shock" protein? Science 214:72-73.

Deshaies RJ, Koch BD, Werner-Washburne M, Craig EA, Schekman R (1988). A subfamily of stress proteins facilitates translocation of secretory and mitochondrial precursor polypeptides. Nature 332:800-805.

Dienel GA, Kiessling M, Jacewicz M, Pulsinelli WA (1986). Synthesis of heat shock proteins in rat brain cortex after transient ischemia. J Cereb Blood Flow Metab 6:505-510.

Duncan TE, O'Steen WK (1985). The diuranl susceptibility of rat retinal photoreceptors to light-induced damage. Exp Eye Res 41:497-507.

Gower DJ, Barbe MF, Tytell M (1986). Stress protein synthesis in response to spinal cord trauma. J Cell Biol 103:78a.

Kuwabara T, Gorn RA (1968). Retinal damage by visible light: An electron microscopic study. Arch Ophthalmol 79:69-78.

Lanum J (1978). The damaging effects of light on the retina. Empirical findings, theoretical and practical implications. Survey Ophthalmol 22:221-249.

Li Z-Y, Tso MOM, Wang H-M, Organisciak DT (1985). Amelioration of photic injury in rat retina by ascorbic acid: A histopathologic study. Invest Ophthalmol Vis Sci 26:1589-1598.

Lindquist S (1986). The heat-shock response. Ann Rev Biochem 55:1151-1191.

Minton KW, Karmin P, Hahn GM, Minton AP (1982). Nonspecific stabilization of stress-susceptible proteins by stress-resistant proteins: A model for the biological role of heat shock proteins. Proc Natl Acad Sci USA 79:7107-7111.

Nowak TS Jr (1985). Synthesis of a stress protein following transient ischemia in the gerbil. J Neurochem 45:1635-1641.

Organisciak DT, Wang H-M, Li Z-Y, Tso MOM (1985). The protective effect of ascorbate in retinal light damage of rats. Invest Ophthalmol Vis Sci 26:1580-1588.

Organisciak DT, Wang H-M, Noell WK (1987). Aspects of the ascorbate protective mechanism in retinal light damage of rats with normal and reduced ROS docoshexaenoic acid. In Hollyfield, JG, Anderson, RE, LaVail, MM, "Degenerative Retinal Disorders - Clinical and Laboratory Investigations," New York: Alan R. Liss, pp 455-468.

O'Steen WK, Anderson KV, Shear CR (1974). Photoreceptor degeneration in albino rats: Dependency on age. Invest Ophthalmol 13:334-339.

O'Steen WK, Sweatt AJ, Eldridge JC, Brodish A (1987). Gender and chronic stress effects on the neural retina of young and mid-aged Fischer-344 rats. Neurobiol Aging 8:449-455.

Schlesinger MJ, Ashburner M, Tissières A (1982). "Heat Shock - From Bacteria to Man." New York: Cold Spring Harbor Laboratory.

Subjeck JR, Shyy T-T (1986). Stress protein systems of mammalian cells. Am J Physiol 250:C1-C17.

Tytell M, Barbe MF (1987). Synthesis and axonal transport of heat shock proteins. In Smith, RS, Bisby, MA (eds): "Axonal Transport," Neurology and Neurobiology, vol 25, New York: Alan R. Liss, pp 473-492.

Ungewickell E (1985). The 70-kd mammalian heat shock proteins are structurally and functionally related to the uncoating protein that releases clathrin triskelia from coated vesicles. EMBO J 4:3385-3391.

Weigand RD, Guisto NM, Rapp LM, Anderson RE (1983). Evidence for rod outer segment lipid peroxidation following constant illumination of the rat retina. Invest Ophthalmol Vis Sci 24:1433-1435.

Welch WJ and Suhan JP (1986). Cellular and biochemical events in mammalian cells during and after recovery from physiological stress. J Cell Biol 103:2035-2052.

White FP (1980). Difference in protein synthesized *in vivo* and *in vitro* by cells associated with cerebral microvasculature. A protein synthesized in response to trauma? Neurosci 5:1793-1799.

Inherited and Environmentally Induced
Retinal Degenerations, pages 539–553

CLEAR PMMA VERSUS YELLOW INTRAOCULAR LENS MATERIAL. AN ELECTROPHYSIOLOGIC STUDY ON PIGMENTED RABBITS REGARDING "THE BLUE LIGHT HAZARD".

Sven Erik G. Nilsson, Ola Textorius, Björn-Erik Andersson and Barbro Swenson

Department of Ophthalmology, University of Linköping, S-581 85 Linköping, Sweden

ABSTRACT

Pigmented rabbits were exposed for 3.5 h to light (retinal irradiance 60-70 mW/cm^2, i.e. below the level of thermal damage) from a Xenon lamp, passing IR filters and a fiber optic system as well as a Perspex CQ (clear PMMA) IOL material in front of one of the eyes (the "PER" eye) and a yellow (blue light absorbing) filter (potential IOL material) in front of the other eye (the "YEL" eye). The difference in spectral distribution of light transmitted by the two filters may be important. Does the yellow filter offer significant protection against "the blue light hazard"? DC ERG recordings were performed before, 1 day after and 4-6 days after exposure. The c- (mainly pigment epithelium (PE)) and b-wave (neuroretina) amplitudes were measured and the c_{PER} / c_{YEL} as well as the b_{PER} / b_{YEL} ratios calculated. Both ratios were found to be reduced after exposure, for the c-wave 30-33% ($p < 0.05 - 0.001$), and for the b-wave 12-20% ($p < 0.01 - 0.02$). This means that both the PE and the neuroretina were injured in the PER eye, the PE more than the neuroretina. At day 4-6 the c- and b-wave ratios were found to have returned to initial levels, indicating that the damage was reversible to a large extent. Thus, the yellow filter offered a better protection than the Perspex material. There were no ophthalmoscopic fundal changes 1-2 h after exposure. After 1 day minimal changes were seen in 3/16 YEL eyes and somewhat more pronounced changes in 8/16 PER eyes, all in the central fundus. The conclusion is that in acute experiments and under the conditions applied the yellow filter protected the PE and the retina against photochemical light injury (mainly "the blue light hazard") significantly better than the Perspex material. Furthermore, the yellow filter was found to protect significantly better than a UV absorbing IOL material.

INTRODUCTION

The neuroretina and the pigment epithelium (PE) may be damaged thermally by near infrared (IR) radiation and long-wavelength visible light. Tissue temperature must be increased more than 20°C above ambient temperature for a pure thermal injury to occur, which requires a fairly high irradiation (Ham et al., 1980a). At an increase of less than 10°C the damage will be purely photochemical, whereas at an increase of 10-20°C the two types will be mixed: thermally enhanced photochemical damage. Even a minimal thermal injury is irreversible and typically involves both PE and photoreceptors (Ham et al. 1980a).

UV radiation, which may reach the retina and PE in aphakes, and short-wavelength visible light ("the blue light hazard") constitute a much greater potential danger to the ocular fundus than long-wavelength radiation (Ham et al., 1976, 1979; Lawwill et al., 1977). Such short-wavelength radiation may give rise to photochemical retinal and PE injury at irradiances that are too low to cause thermal damage (Friedman and Kuwabara, 1968; Ham et al., 1973; Tso, 1973; Harwerth and Sperling, 1974; Lawwill et al., 1977). A temperature increase as modest as 0.1^{o}C above ambient temperature may be sufficient (Ham et al., 1980a). A minimal photochemical injury, which may not become visible until after 48 hours, is characterized by inflammation of the PE, whereas photoreceptor damage is negligible. Healing without any loss of visual acuity will occur with time (Ham et al., 1980a). Direct current electroretinography (DC ERG) has been shown to be a very sensitive method (c-wave reduction) for detecting early PE damage, e.g. by visible light (Skoog and Jarkman, 1985). In contrast to the typical photochemical damage just described, Zigman (1987) found that low-level monochromatic UV-A radiation caused specific photoreceptor damage in aphakic grey squirrels. Solar retinitis is an example of injury which is mainly photochemical in nature and caused by short-wavelength solar irradiation (Noell et al., 1966; Friedman and Kuwabara, 1968; Ham et al., 1976, 1980b).

The sensitivity to light damage is much greater in central than in peripheral parts of the ocular fundus (Kuwabara and Gorn, 1968; Marshall et al., 1972; Lawwill, 1973; LaVail and Gorrin, 1987), although a certain protection is offered by the yellow pigment of the primate fovea (Jaffe and Wood, 1988). It is well known that albino animals are extremely sensitive to light injury and that they do not tolerate what is considered to be "normal" light levels (Noell et al., 1966; Kuwabara and Gorn, 1968). PE melanin was thought to protect the underlying photoreceptors from such damage (LaVail 1980). However, recent studies on chimeras and translocation mice found light injury to be independent of PE melanin (LaVail and Gorrin, 1987). The melanin in the iris may instead be of greater significance, since with pupil dilation light damage occurs much faster than with normal pupil size (Noell et al., 1966; Rapp and Williams, 1980).

UV-B and UV-C radiation is absorbed by the cornea and UV-A by the lens in the phakic human eye (Sliney and Wolbarsht, 1980). When the lens is removed, e.g. in cataract surgery, UV-A may reach the fundus of the eye, however, and it is not blocked by a polymethacrylate (PMMA) intraocular lens (IOL). Mainster (1978) found that the PMMA IOL transmittance of 350 nm UV radiation was 30 times that of the natural lens. Kraff et al. (1985) presented some evidence that UV filtering IOL:s might reduce the incidence of cystoid macular edema, and for such reasons UV absorbing IOL:s are now used quite often.

Against this background one must ask the important question whether long-term exposure to low-level ambient light, nowadays containing fairly large amounts of short-wavelength radiation (e.g. the blue light hazard), may be dangerous to the PE and neuroretina, and more so in aphakic than in phakic eyes. Although not proven, several authors suggest a possible causal relation between accumulated doses of blue light (phakic and aphakic eyes) and UV-A radiation (aphakes) on one hand and age-related macular degeneration on the other hand. Many of these authors therefore also recommend that blue light and/or UV absorbing sunglasses, contact lenses or IOL:s be used (see below).

The present study was undertaken to investigate whether in short-term experiments the retina and PE would be protected against photochemical damage in a better way by a yellow (blue light absorbing) potential IOL material, which leaves colour vision undisturbed, than by clear PMMA or UV absorbing IOL materials. Some of the results were presented earlier in a preliminary form (Nilsson et al., 1989).

MATERIAL AND METHODS

Eyes, with maximally dilated pupils, of young adult pigmented rabbits under general pentobarbital anaesthesia were exposed to light from a Xenon lamp, passing IR filters and fiber optics, for 3.5 hours. A clear PMMA (Perspex CQ) IOL material was placed in front of one eye (the "PER" eye). Fig. 1 shows the spectral distribution of light transmitted. Rel. radiant power was 25%, 50% and 75%, respectively, at about 405, 455 and 485 nm, respectively. This Perspex material was compared to a yellow (blue light absorbing) potential IOL material, placed in front of the second eye (the "YEL" eye). Rel. radiant power was 25%, 50% and 75%, respectively, at about 445, 465 and 490 nm, respectively. The filters (including the UV filter to be mentioned below) differed in transmission mainly below about 500 nm. Psychophysical tests showed that enough blue light was transmitted by the yellow filter not to interfere significantly with colour vision. The yellow filter was in turn compared to a UV absorbing IOL material, placed in front of one of the eyes (the "UV" eye). The transmission curve of the UV filter showed that rel. radiant power was 25%, 50% and 75%, respectively, at about 415, 460 and 485 nm, respectively.

The fiber optic was in close apposition to the filter, which was placed at a distance of about 2 mm from the eye. The cornea was kept moisturized. The irradiance transmitted by the two fiber optics was equalized before reaching the filters. Retinal irradiance was 60-70 mW/cm^2, i.e. below the level of thermal damage. The area of the fundus exposed was about 12 mm in diameter. (Measurements regarding irradiances and transmission curves were performed by the Swedish National Defence Research Institute, Linköping.)

Retinal and PE function was assessed by means of DC ERG, which permits recording of the slow c-wave, representing mainly the PE (Steinberg et al., 1970; Oakley and Green, 1976; Shimazaki and Oakley, 1985), in addition to the faster a- and b-waves, reflecting photoreceptor (Brown and Wiesel,

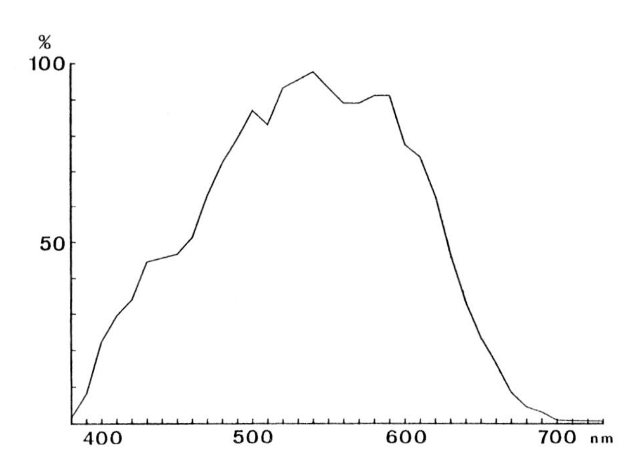

Fig. 1. Transmission curve of Perspex CQ (clear PMMA) intraocular lens material.

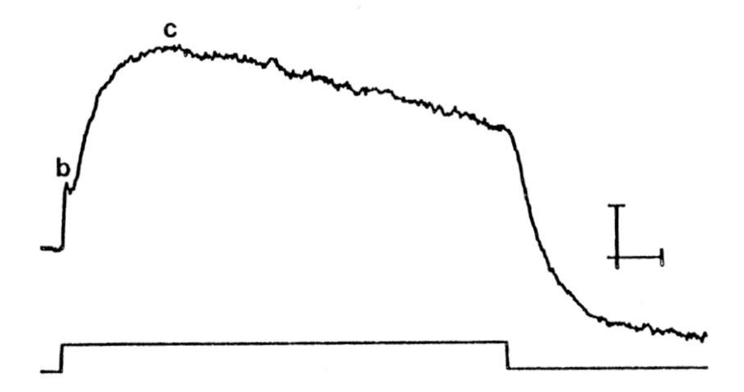

Fig. 2. Rabbit DC ERG with b- and c-waves in response to a 10 sec light stimulus (indicated on lower line), 3.5 log rel. units above b-wave threshold. (The a-wave is not seen at this stimulus intensity level.) Amplitude and time calibrations 500 V and 1 sec, respectively.

1961) and inner retinal (Newman, 1979) activity, respectively (Fig. 2). ERGs were recorded before, 1 day (20-23 hours) after and, for some animals, 4-6 days after exposure. Ten sec light stimuli, 3.5 log rel. units above b-wave threshold were presented in uniocular "ganzfeld" hemispheres. The recording set-up included recording and reference electrodes in the form of matched calomel half-cells connected to a contact lens on the eye and a plastic chamber on the forehead by means of saline-agar bridges, low-drift DC amplifiers and a Hewlett-Packard computer system (Nilsson and Andersson, 1988). B- and c-wave amplitude were measured, and amplitude ratios (c_{PER} / c_{YEL}; b_{PER} / b_{YEL}; c_{UV} / c_{YEL}; b_{UV} / b_{YEL}) were calculated and used to compare PE and retinal functions for the two eyes. (Using such ratios eliminates influence on the results of from day-to-day variations in absolute amplitude values.) The paired t-test was employed for statstical analysis, and p-values < 0.05 were considered to indicate statistically significant differences.

Two and 1 mm thick Perspex and yellow filters were compared in 10 and 6 rabbits, respectively, before and 1 day after exposure. Seven of these rabbits (3 with 2 mm and 4 with 1 mm filters) were studied also 4-6 days after exposure. One mm thick UV and yellow filters were compared in 5 rabbits before exposure as well as 1 day after exposure. The ocular fundi were evaluated regarding visible changes by means of indirect, binocular ophthalmoscopy at 1-2 hours, 1 day and 4-6 days after exposure. After completion of the experiments the rabbits were sacrificed by a lethal dose of pentobarbital.

RESULTS

Comparison Of Clear Perspex And Yellow IOL Materials

There was no significant ($p > 0.1$) change in c-wave amplitude for the YEL eye (2 mm filters) from the pre-exposure value (1472.9 ± 144.7 mV) to the value 1 day after exposure (1732.6 ± 130.2 mV), showing that the PE of the YEL eye was not suffering from photochemical light damage. On the contrary, the c-wave of the PER eye was significantly ($p < 0.05$) reduced from 1517.3 ± 160.1 mV before exposure to 1181.4 ± 101.6 mV 1 day after exposure, indicating significant PE injury in the PER eye. For 2 mm filters the c_{PER} / c_{YEL} ratio was significantly ($p < 0.001$) reduced (33%) from a pre-exposure value of 1.020 ± 0.019 to a value 1 day after exposure of 0.682 ± 0.027 (Fig. 3). The b_{PER} / b_{YEL} ratio was significantly ($p < 0.02$) reduced as well (12%), from 0.952 ± 0.038 to 0.836 ± 0.019. The results show that both the PE and the neuroretina were significantly damaged in the

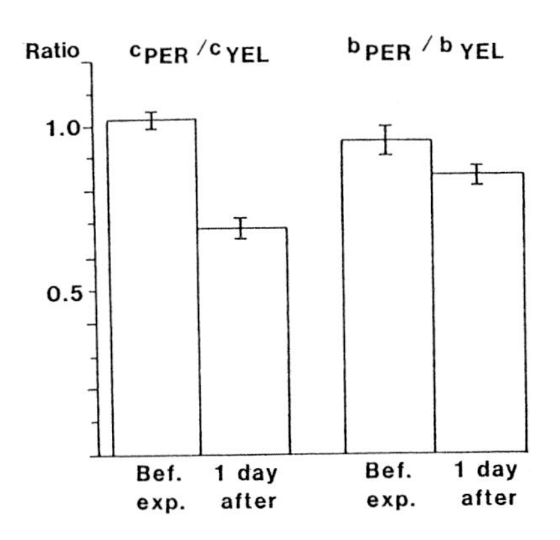

Fig. 3. Comparison of Perspex material and a yellow filter, one in front of each eye, before and 1 day after a 3.5 hour exposure to a retinal irradiance of 60-70 mW/cm^2. The c- and b-wave ratios are both significantly reduced 1 day after exposure. (S.E. indicated on each column.)

PER eye, the PE more than the neuroretina. Thus, the yellow filter protected the ocular fundus against photochemical damage significantly better than the Perspex material.

For 1 mm thick filters, which are more similar in thickness to actual IOL:s both the c-wave ($p < 0.05$) and b-wave ($p < 0.01$) ratios were again significantly reduced 1 day after exposure. The c-wave ratio decreased by 30% from 0.906 ± 0.055 to 0.631 ± 0.112, and the b-wave ratio by 20% from 0.970 ± 0.044 to 0.774 ± 0.069. It may be concluded also in this case that the PE as well as the retina were injured in the PER eye, and that the yellow filter offered a significantly better protection than the Perspex material.

Regarding the group of animals studied also 4-6 days after exposure, the initial c-wave (0.894 ± 0.042) and b-wave (0.919 ± 0.039) ratios were likewise significantly (c: $p < 0.02$; b: $p < 0.05$) reduced to (c: 0.627 ± 0.097; b: 0.760 ± 0.055) 1 day after exposure (Fig. 4). They returned, however, at day 4-6, to a level (c: 0.971 ± 0.053; b: 0.887

± 0.050) not significantly (c: $p > 0.1$; b: $p > 0.1$) different from the initial one. It thus seemed that the damage was largely reversible, a fact that fits with an injury that is mainly photochemical.

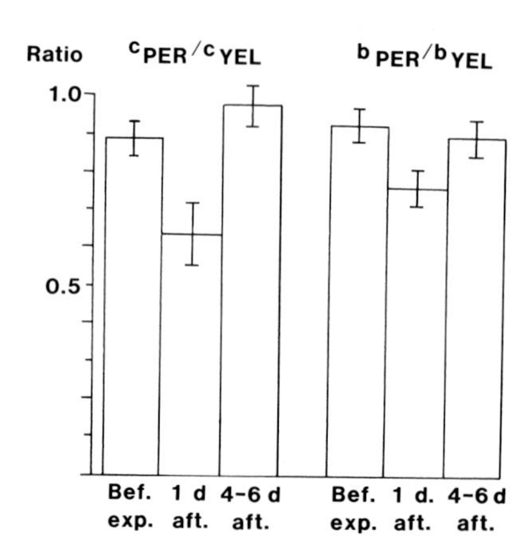

Fig. 4. Comparison of Perspex material and a yellow filter, one in front of each eye, before, 1 day after and 4-6 days after a 3.5 hour exposure to a retinal irradiance of 60-70 mW/cm^2. The c- and b-wave ratios are both significantly reduced 1 day after exposure but return to a level not significantly different from the initial one within 4-6 days after exposure. (S.E. indicated on each column.)

Ophthalmoscopically visible changes were not present at 1-2 hours after exposure. At day 1 a minimal grey to white streak was seen in a few (3/16) of the YEL eyes, whereas some (8/16) PER eyes showed a grey to white area, up to 1.5 disc diameter in size. At day 4-6 most (6/7) YEL eyes exhibited atrophy with pigmentation in an area up to a size of 2.5 disc diameters. All (7/7) PER eyes showed similar changes but in an area up to a size of 4.5 disc diameters. When changes were bilateral, they were always much more prominent in the PER eye, at day 1 as well as at day 4-6. This visible damage was too restricted in extension to influence significantly on the ERG potentials.

Comparison Of UV And Yellow IOL Materials

The initial c_{UV} / c_{YEL} ratio (1.092 ± 0.029) was significantly ($p < 0.01$) reduced (17%) at day 1 after exposure (0.908 ± 0.055), indicating significant PE injury in the UV eye. This means that the yellow filter protected the PE better than the UV filter (Fig. 5). On the other hand, the initial b_{UV} / b_{YEL} ratio (0.997 ± 0.015) had not changed significantly ($p > 0.1$) when checked 1 day after exposure (0.930 ± 0.063), which means that there was no significant neuroretinal injury.

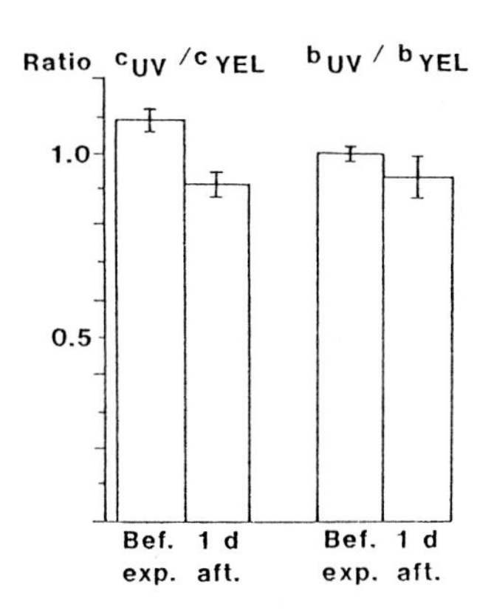

Fig. 5. Comparison of UV and yellow filters, one in front of each eye, before and 1 day after exposure to a retinal irradiance of 60-70 mW/cm^2 for 3.5 hours. The c-wave ratio is significantly reduced, whereas the b-wave ratio is not, 1 day after exposure. (S.E. indicated on each column.)

DISCUSSION

We demonstrated in the present study that, in short-term exposure experiments, a yellow (blue light absorbing) potential IOL material protected the PE and neuroretina against photochemical light damage more effectively than a clear PMMA (Perspex CQ) IOL material and also more effectively than a UV absorbing IOL material, at least regarding the PE. (One must keep in mind, however, that a permanent PE injury will affect photoreceptor function in the long run.) It is essential that a yellow filter intended for use as IOL:s transmits enough blue light so as not to disturb colour vision significantly. The present material was found to be fully acceptable in standard tests.

The ERG recordings reflect the activity of the entire ocular fundus. The ophthalmoscopically visible alterations were too small to significantly influence on the ERG potentials. Thus, the ERG changes seen represented areas that were macroscopically intact - at least 4-6 days after exposure. This in turn indicates that the damage was purely photochemical. Such an assumption is strengthened by the facts that PE injury, demonstrated in ERG, was more pronounced than neuroretinal damage and that the ERG changes were reversible within 4-6 days. Furthermore, it is quite possible that the neuroretinal injury reflected as b-wave reductions is not a primary damage to the neuroretina but secondary to PE injury. The more pronounced visible changes in minor parts of the central fundus appear to be due to thermally enhanced photochemical damage, partly because of a higher sensitivity in central than in peripheral parts of the fundus (Kuwabara and Gorn, 1968; Marshall et al., 1972; Lawwill, 1973; LaVail and Gorrin, 1987), and partly because of a certain focusing effect of the optical apparatus of the eye.

When discussing a possible need for using UV filters or blue light absorbing (yellow) filters it is essential to discuss at the same time whether there is any evidence that the PE and neuroretina may be photochemically damaged by accumulated doses over the years of low-level ambient light, particularly involving the blue light hazard (and also UV-A, regarding aphakes). During ageing a certain loss of cones and of PE cells, which are also nondividing, occurs (Marshall 1978, 1987a). Furthermore, the function of the PE cells declines, resulting in an accumulation of lipofuscin granules (Feeney, 1973). It is not unlikely that the effects of light on such age-compromised structures may be more dangerous than on the cells of a younger eye (Marshall, 1987a). With increasing age the human lens accumulates a yellow pigment, which like the macular pigment is well suited to protect the PE and retina against blue light (Lerman, 1980). This may compensate for a higher vulnerability of the structures of the ocular fundus occurring at the same time. In addition, it seems that lens opacities protect against AMD (Gjessing 1925, 1953) and drusen formation (Guyer et al., 1986). Therefore, it does not seem unreasonable to use IOL:s that filter out not only UV-A but also much of the blue light when in surgery replacing the yellow, cataractous lenses.

A possible relation between long-term exposure to ambient light and development and/or worsening of AMD has been proposed, but nor proven, by many authors (e.g. Gass and Norton, 1966; Young, 1981). More specificly one has pointed to the blue light hazard (short-wavelength visible light) regarding phakic as well as aphakic eyes and/or to UV-A radiation as to aphakic eyes (Ham et al. 1980a, 1986; Sliney and Wolbarsht, 1980; Ham, 1983; Mainster et al., 1983; Kraff et al., 1985; Kirkness and Weale, 1985; Fishman, 1986; Rosen, 1986; Mainster, 1987; Marshall, 1987b; Young, 1988). Most of these authors also recommend the use of filtes, e.g. sunglasses, which absorb short-wavelength visible light (phakic and aphakic eyes) and UV-A (aphakic eyes) when ambient light is strong. The value of IOL:s which filter out UV-A has been emphasized by many investigators within the group referred to above, and such lenses are now used in many instances. Blue light absorbing (yellow) IOL:s (at the same time absorbing UV-A) are not available on the market yet, but several author stress the importance of a development in this direction (Kirkness and Weale, 1985; Ham et al., 1986; Rosen, 1986; Marshall, 1987b). UV-A filtering contact lenses have been designed lately and are sometimes recommended for aphakes (Pitts and Lattimore, 1987), whereas blue light absorbing contact lenses are not available. Operating microscopes are often accompanied by large amounts of short-wavelength radiation, which may be dangerous (Gass and Norton, 1966; Irvine et al., 1971). A yellow filter was found to be of protective value (Parver et al., 1983).

Earlier findings with respect to the danger associated with the blue light hazard are supported by the results of the present investigation, where it was shown that in short-term experiments a blue light absorbing (yellow) filter, which is psychophysically fully acceptable, offered a more effective protection of the neuroretina and the PE than clear PMMA or UV absorbing IOL materials. It thus appears quite reasonable to replace a yellow cataractous human lens by a yellow IOL instead of the currently used IOL:s. We still do not know, however, whether an exposure for a few hours to an irradiance of 60-70 mW/cm^2 is of significance for evaluation of risks involved in daily exposures for 10-20 years to the low-level irradiance of ambient light.

ACKNOWLEDGEMENT

The present investigation was supported by a grant from the Swedish Medical Research Council (Project No. 12X-734).

REFERENCES

Brown KT, Wiesel TN (1961). Localization of origins of electroretinogram components by intraretinal recordings in the intact cat eye. J Physiol (Lond) 158: 257-280.

Feeney L (1973). The phagolysosomal system of the pigment epithelium. A key to retinal disease. Invest Ophthalmol Vis Sci 12: 635-638.

Fishman GA (1986). Ocular phototoxicity: Guidelines for selecting sunglasses. Survey Ophthalmol 31: 119-124.

Friedman E, Kuwabara T (1968). The retinal pigment epithelium. IV. The damaging effects of radiant energy. Arch Ophthalmol 80: 265-280.

Gass JDM, Norton EWD (1966). Cystoid macular edema and papilledema following cataract extraction. Arch Ophthalmol 76: 646-661.

Gjessing HGA (1925). Gibt es einen Antagonismus zwischen Cataracta senilis und Haabscher seniler Makulaveränderungen? Z Augenheilkd 56: 79-90.

Gjessing HGA (1953). Gibt es einen Antagonismus zwischen Cataracta senilis und Haabscher seniler Makulaveränderungen? Acta Ophthalmol 31: 401-421.

Guyer DR, Alexander MF, Auer CL, Hamill MB, Chamberlin JA, Fine SL (1986). A comparison of the frequency and severity of macular drusen in phakic and non-phakic eyes. Invest Ophthalmol Vis Sci 27 (Suppl): 20.

Ham WT Jr (1983). Ocular hazards of light sources: Review of current knowledge. J Occup Med 25: 101-103.

Ham WT Jr, Allen RG, Feeney-Burns L, Marmor MF, Parver LM, Proctor PH, Sliney DH, Wolbarsht ML (1986). The involvement of the retinal pigment epithelium. In Waxler M, Hitchins WM (eds): "Optical radiation and visual health," Boca Raton, FA: CRC Press, pp 43-67.

Ham WT Jr, Mueller HA, Ruffolo JJ Jr (1980a). Retinal effects of blue light exposure. Soc Photo-Optical Instr Eng 229: 46-50.
Ham WT Jr, Mueller HA, Ruffolo JJ Jr, Clarke AM (1979). Sensitivity of the retina to radiation damage as a function of wavelength. Photochem Photobiol 29: 735-743.
Ham WT Jr, Mueller HA, Ruffolo JJ Jr, Guerry D III (1980b). Solar retinopathy as a function of wavelength: its significance for protective eyewear. In Williams TP, Baker BN (eds): "The effects of constant light on visual processes," New York: Plenum Press, pp 319-346.
Ham WT Jr, Mueller HA, Sliney DH (1976). Retinal sensitivity to damage from short wavelength light. Nature 260: 153-155.
Ham WT Jr, Mueller HA, Williams RC, Geeraets WJ (1973). Ocular hazard from viewing the sun unprotected and through various windows and filters. Appl Optics 12: 2122-2129.
Harwerth RS, Sperling HG (1974). Prolonged colour blindness induced by intense spectral lights in rhesus monkeys. Science 174: 520-523.
Irvine AR, Bresky R, Crowder BM, Forster RK, Hunter DM, Kulvin SM (1971). Macular edema after cataract extraction. Ann Opthalmol 3: 1234-1240.
Jaffe GJ, Wood IS (1988). Retinal phototoxicity from the operating microscope: A protective effect by the fovea. Arch Ophthalmol 106: 445-446.
Kirkness CM, Weale RA (1985). Does light pose a hazard to the macula in aphakia? Trans Ophthalmol Soc UK 104: 699-702.
Kraff MC, Sanders DR, Jampol LM, Lieberman HL (1985). Effect of an ultraviolet-filtering intraocular lens on cystoid macular edema. Ophthalmology 92: 366-369.
Kuwabara T, Gorn RA (1968). Retinal damage by visible light. Arch Ophthalmol 79: 69-78.
LaVail MM (1980). Eye pigmentation and constant light damage in the rat retina. In Williams TP, Baker BN (eds): "The effects of constant light on visual processes," New York: Plenum Press, pp 357-387.
LaVail MM, Gorrin GM (1987). Protection from light damage by ocular pigmentation: Analysis using experimental chimeras and translocation mice. Exp Eye Res 44: 877-889.
Lawwill T (1973). Effects of prolonged exposure of rabbit retina to low-intensity light. Invest Ophthalmol 12: 45-51.

Lawwill T, Crockett S, Currier G (1977). Retinal damage secondary to chronic light exposure. Doc Ophthalmol 44: 379-402.
Lerman S (1980). "Radiant energy and the eye." Functional Ophthalmol Ser, Vol 1, New York: Macmillan.
Mainster MA (1978). Spectral transmittance of intraocular lenses and retinal damage from intense light sources. Am J Ophthalmol 85: 167-170.
Mainster MA (1987). Light and macular degeneration: A biophysiocal and clinical perspective. Eye 1: 304-310.
Mainster MA, Ham WT Jr, Delori FC (1983). Potential retinal hazards. Instrument and environmental light sources. Ophthalmology 90: 927-932.
Marshall J (1978). Ageing changes in human cones. XXIII Int Congr Ophthalmol, Kyoto, Japan, Excerpta Medica, pp 375-378.
Marshall J (1987a). The ageing retina: Physiology or pathology. Eye 1: 282-292.
Marshall J (1987b). Ultraviolet radiation and the eye. In Passchier WF, Bosnjakovic BFM (eds): "Human exposure to ultraviolet radiation: Risks and regulations," Amsterdam: Elsevier Science Publishers BV, pp 125-142.
Marshall J, Mellerio J, Palmer DA (1972). Damage to pigeon retinae by moderate illumination from fluorescent lamps. Exp Eye Res 14: 164-169.
Newman EA (1979). B-wave currents in the frog retina. Vision Res 19: 227-234.
Nilsson SEG, Andersson BE (1988). Corneal D.C. recordings of slow ocular potential changes such as the ERG c-wave and the light peak in clinical work. Equipment and examples of results. Doc Ophthalmol 68: 313-325.
Nilsson SEG, Textorius O, Andersson B-E (1980a). Protective effects of a yellow filter against "the blue light hazard". Proc VIIIth Congr Europ Soc Ophthalmol, Lisbon, 1988. Excerpta Med Internatl Congr Ser. (In press.)
Noell WK, Walker VS, Kang BS, Berman S (1966). Retinal damage by light in rats. Invest Ophthalmol 5: 450-473.
Oakley BII, Green DG (1976). Correlation of light-induced changes in retinal extracellular potassium concentration with c-wave of the electroretinogram. J Neurophysiol 39: 1117-1133.
Parver LM, Auker CR, Fine BS (1983). Observations on monkey eyes exposed to light from an operating microscope. Ophthalmology 90: 964-972.

Pitts DG, Lattimore MR Jr (1987). Protection against UVR using the Vistakon UV-bloc soft contact lens. Internatl Contact Lens Clinic 14: 22-29.
Rapp LM, Williams TP (1980). The role of ocular pigmentation in protecting against retinal light damage. Vision Res 20: 1127-1131.
Rosen ES (1986). Pseudophakia and hazards of nonionizing radiation. Seminars Ophthalmol 1: 68-79.
Shimazaki H, Oakley BII (1985). Effects of cesium upon Müller cell membrane responses, K^+_o, and the electroretinogram. Invest Ophthalmol Vis Sci 26 (ARVO suppl): 112.
Skoog K-O, Jarkman S (1985). Photic damage to the eye: selective extinction of the c-wave of the electroretinogram. Doc Ophthalmol 64: 49-53.
Sliney D, Wolbarsht M (1980). "Safety with lasers and other optical sources." New York and London: Plenum press.
Sperduto RD, Hiller R, Siegel D (1981). Lens opacities and senile maculopathy. Arch Ophthalmol 99: 1004-1008.
Steinberg RH, Schmidt R, Brown KT (1970). Intracellular responses to light from cat pigment epithelium: origin of the electroretinogram c-wave. Nature (Lond) 227: 728-730.
Tso MOM (1973). Photic maculopathy in rhesus monkey. A light and electron microscopic study. Invest Ophthalmol 12: 17-25.
Young RW (1981). A theory of central retinal disease. In Sears ML (ed): "New directions in ophthalmic research," New Haven: Yale Univ Press, pp 237-270.
Young RW (1988). Solar radiation and age-related macular degeneration. Survey Ophthalmol 32: 252-269.
Zigman S (1987). Near UV radiation effect on the lens and the retina. In Passchier WF, Bosnjakovic BFM (eds): "Human exposure to ultraviolet radiation: Risks and regulations," Amsterdam: Elsevier Science Publishers B V, pp 143-148.

Inherited and Environmentally Induced
Retinal Degenerations, pages 555–567

REVERSIBLE AND IRREVERSIBLE BLUE LIGHT DAMAGE TO THE ISOLATED, MAMMALIAN PIGMENT EPITHELIUM

E.L. Pautler, M. Morita and D. Beezley

Departments of Physiology (E.L.P., D.B.) and Anatomy and Neurobiology (M.M.), Colorado State University, Fort Collins, Colorado 80523

Most studies have employed cell damage or death as a means of evaluating the photo-toxicity of blue light on the pigment epithelium (Ham et al., 1976; Ham et al., 1984; Crockett and Lawwill, 1984). We chose to investigate the effects of blue light on the functional properties of the isolated bovine pigment epithelium as manifested in the electrical parameters and transport of leucine and chloride.

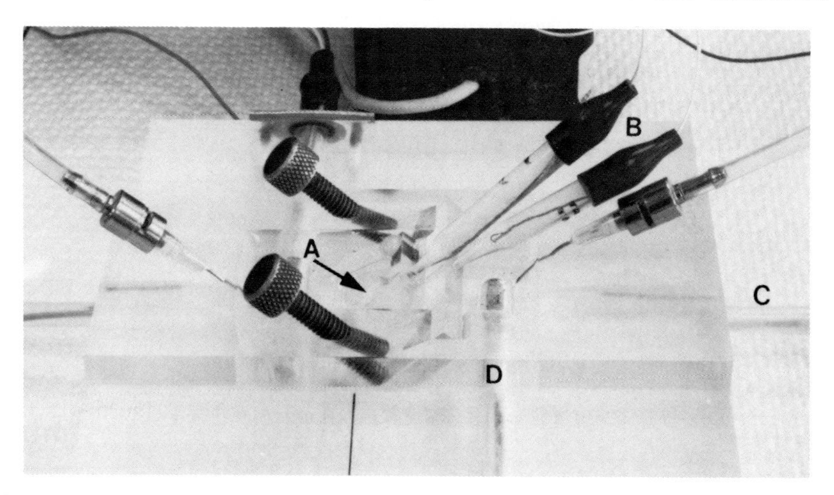

Figure 1. The picture of the Ussing chamber shows where the tissue is placed (A), the position of the voltage (B) and current electrodes (C) as well as the thermostatically controlled heating elements (D) inserted into the sides of the chamber.

The pigment epithelium (PE) with tapetal and choroidal remnants was isolated from perfused cow eyes and mounted in a chip as a membrane separating two compartments of a Ussing Chamber. As shown in Fig. 1, the voltage sensing electrodes are in close apposition to the tissue whereas the current electrodes are located in the distal ends of the chambers. The thermostatically controlled heating elements maintain the bicarbonate buffered saline solution at 34 degrees C. This arrangement permitted monitoring of the transepithelial potential (TEP) and short circuit current (SCC). Radioactive label (5-10 μCi) was added to one side of the chamber and the transport calculated from the slope of the appearance curve in the opposite chamber which was sampled every 10 minutes for 60 minutes. Unless otherwise noted, all blue light exposures were at 20 mW/cm^2 with a peak wave length of 430 nm.

The transport of leucine was studied at a carrier concentration of 10 μM. We have established that at this concentration leucine is transported in the R-->Ch direction by the amino acid L system which is Na independent and has a minimal requirement for ATP, yet exerts a concentrating action (Collarini and Oxender, 1987). Recent work has shown that a pH gradient may contribute to the driving force for the L system of amino acid transport (Mitsumoto et al., 1986, 1988). The transport of leucine in the Ch-->R direction appears to occur via simple diffusion. This polarization of transport results in a net movement in the R-->Ch direction. Because of the discovered susceptibility of the R-->Ch transport to blue light, this unidirectional flux was extensively studied.

The first surprise in our series of experiments was the observation that the TEP and the SCC were reversibly diminished during exposure to blue light. There was substantial variability among the preparations, but all showed this tendency to varying degrees. As shown in Fig. 2 the changes in TEP and SCC were such as to increase the specific resistance (R_t) of the tissue suggesting that transport was being reduced. Similar effects on the TEP and SCC are noted after the application of uncouplers of oxidative phosphorylation (Miller et al., 1982). The appearance curve of leucine being transported in the R-->Ch direction is also presented. Note that after an approximate delay of 10 minutes into the exposure time the leucine transport is markedly inhibited.

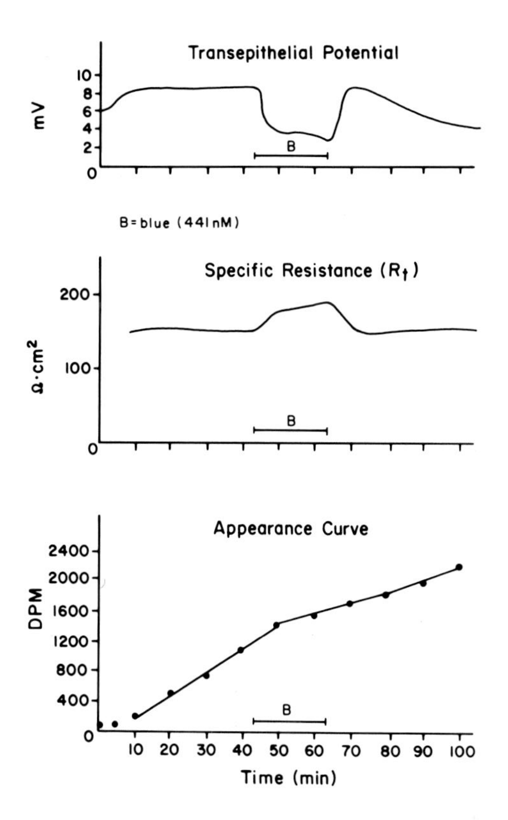

Figure 2. Sample recording illustrating a pronounced effect of blue light on the TEP and SR. The decrease in the SCC was such as to increase the SR. The appearance curve of leucine being transported in the R-->Ch direction in this tissue is also shown.

A similar effect of blue light on leucine transport under standard conditions is shown in Fig. 3. In this particular case a delay period of about 30 minutes after the onset of light is required before the transport rate is diminished. However, there is significant recovery after the cessation of blue light. The companion graph illustrates the effect of .5% ethanol on the inhibitory effect of blue light. Again a significant delay period is noticed but the inhibitory effect on the transport rate is greatly enhanced. A substantial recovery is noted after the blue light is terminated.

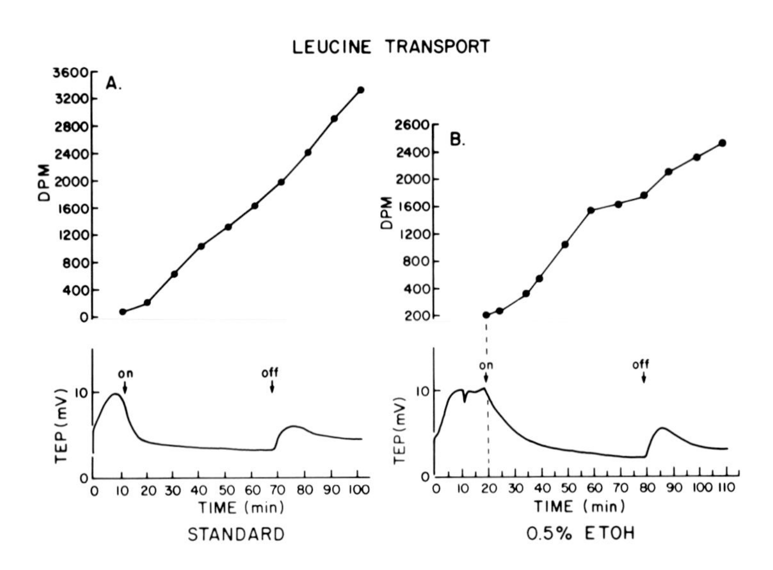

Figure 3. Appearance curves indicating changes in transport of leucine in R-->Ch direction induced by blue light. Fig. 3A was recorded under standard conditions and Fig. 3B illustrates the effect of .5% ethanol on the transport rate. Note the time required for inhibition of transport of leucine.

To determine if blue light, with the accompanying changes in TEP and SCC, altered ionic transport, experiments were performed in which the R-->Ch unidirectional flux of chloride was measured. The transport of chloride by the frog PE is mediated by an active, furosemide sensitive, system (Di Mattio et al., 1983) whereas the dog PE exhibits an additional dependence on the Na/K ATPase (Tsubir et al., 1986). As shown in Fig. 4, there is a significant change in the transport rate as indicated by the slope of the appearance curve during the period of blue light. The inhibition and recovery of transport closely follow the time course of changes in the TEP. Therefore, exposure of the isolated PE to adequate intensities of blue light will inhibit both leucine and chloride transport; however, the time courses are different for the two species. Inhibition of leucine requires considerable time to develop whereas the transport of chloride is closely coupled to changes in the electrical parameters.

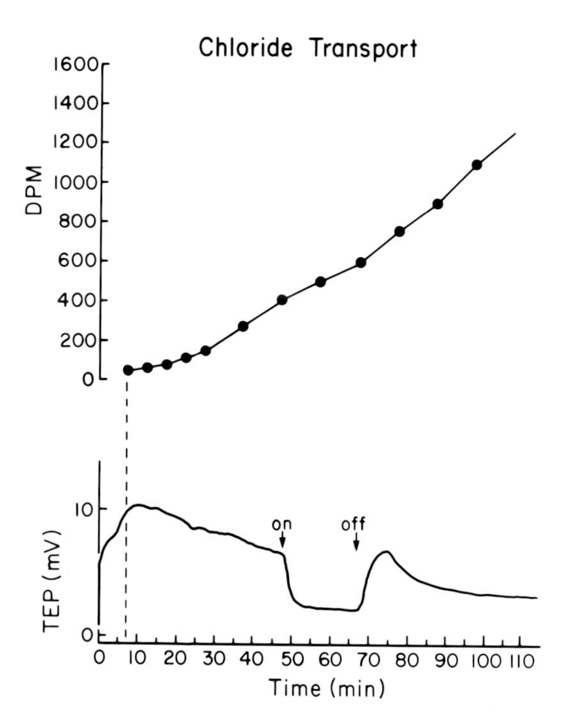

Figure 4. The effect of blue light on chloride transport in the R-->Ch direction. The change in slope of the appearance curve coincides with changes in the TEP whereas there was a substantial delay in leucine transport.

The results of blue and red light on leucine transport are summarized in Fig. 5. The transport of leucine is only inhibited by blue light in the R-->Ch direction which utilizes the L system of amino acid transport. Simple diffusion in the Ch-->R direction is not effected and red light of similar intensity does not alter either transport system.

The next surprise we encountered was the dramatic effect of ethanol in sensitizing changes in transport and the electrical properties of the PE exposed to blue light. As shown in Fig. 4, .5% ethanol significantly enhanced the inhibition of leucine transport by blue light. Even more pronounced effects were noted on the magnitude of the TEP changes elicited by blue light.

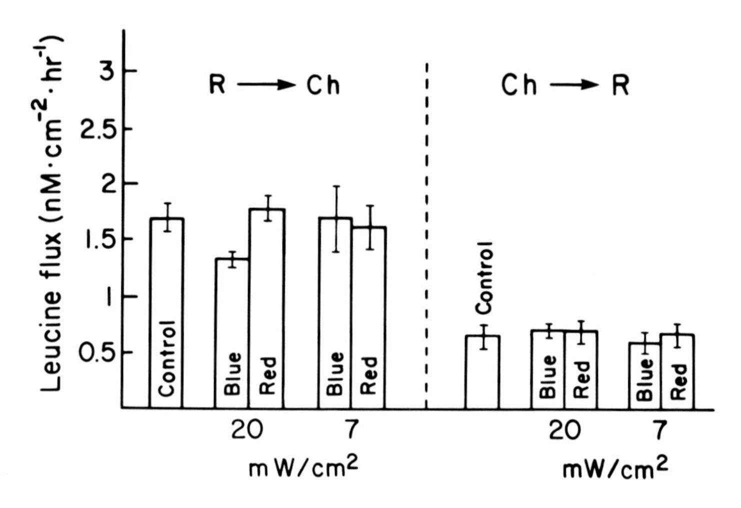

Figure 5. Summary of the effects of blue and red light at two different intensities on leucine transport in the R-->Ch and Ch-->R directions. Leucine transport was only inhibited by the 20 mW/cm^2 blue light in the R-->Ch direction.

Fig. 6 shows the magnitude of response elicited by a 30 sec. irradiation of the PE with 20 mW/cm^2 of blue light 15 minutes after the administration of ethanol. The extreme sensitivity is appreciated by realizing that a concentration of .125% doubles the depolarization of the TEP and that this level of alcohol is just slightly more than the amount which would qualify a person for a D.U.I. Within the range of concentrations covered, there is a linear relationship between the change in TEP elicited by blue light and the level of ethanol. It should be mentioned that 1-butanol is an even more potent sensitizer to blue light than ethanol by a factor of 8-10. The companion figure illustrates that 15-20 minutes is required for alcohol to exert its maximum sensitizing action.

To help identify the molecular species which is mediating the effects of blue light on the TEP, an action spectrum was determined for the elicited potential changes. The action spectrum is a plot of the reciprocal quantal flux required for an equal depolarization of the TEP versus wavelength and is a measure of the photosensitivity. The bovine PE consists of both pigmented and non-pigmented regions so action spectra were determined for non-pigmented, non-pigmented with .5% ethanol and pigmented with .5% ethanol. As shown in Fig. 7 the action spectrum is basically the same under all these conditions and reveals a peak which corresponds to a Soret band in absorbance spectroscopy with a shoulder at about 440 nm. The existence of a Soret band is indicative of a heme protein. Considering the absorption spectra of the heme proteins known to be common in mammalian cells, the peak between 410-420 nm with a shoulder at 440-450 favors cytochrome oxidase as the most likely candidate for the sensitizer. It is well established that blue light can either reversibly or irreversibly damage cytochrome oxidase in yeast and beef heart mitochondria at intensities and durations comparable to those employed in our studies (Epel and Butler, 1969; Ninneman et al., 1970a,b). It should also be emphasized that the cytochrome oxidase in the cited studies was only damaged by blue light if O_2 was present, indicating a photodynamic action.

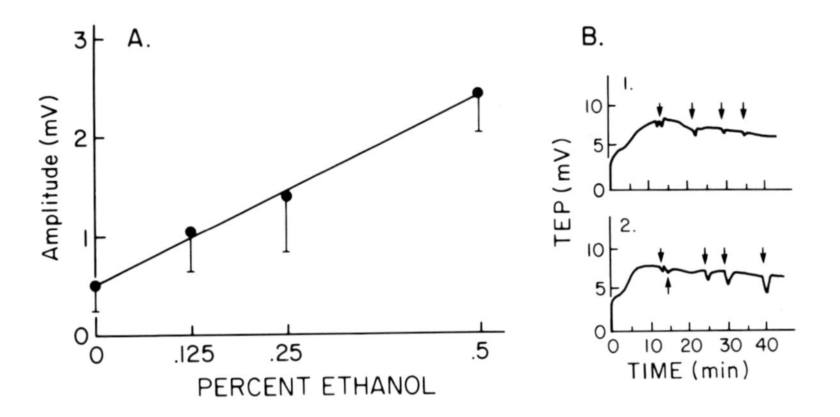

Figure 6. The effects of ethanol on the TEP changes elicited by 30 sec exposures of 20 mW/cm^2 blue light. Fig. 6A shows the amplitude of TEP elicited 15 min after the administration of ethanol. Data is plotted as mean ± SD with n=5. Fig. 6B illustrates the time required for ethanol to exert its full sensitizing action which is generally between 15-20 min.

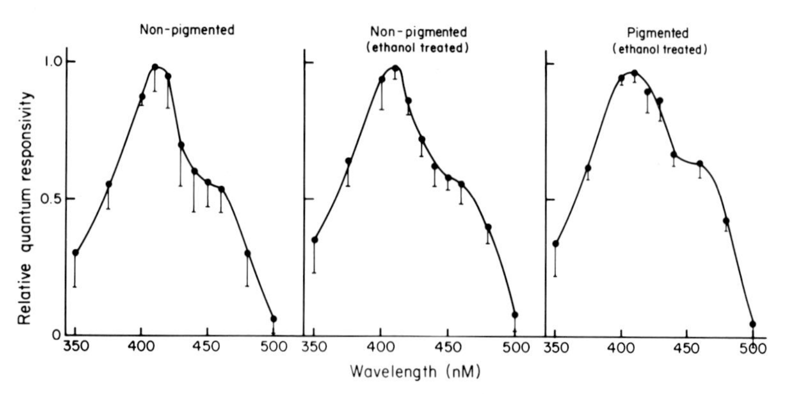

Figure 7. Action spectra of TEP depolarization recorded under three different conditions: non-pigmented PE, non-pigmented PE with .5% ethanol, pigmented PE with .5% ethanol. Plotted as mean ± SD with n=4 for each group. All spectra reveal a peak between 410-420 nm with a prominent shoulder at 440-450 nm.

The reduction of the TEP and SCC, as well as the transport properties of the PE, by blue light is consistent with an inhibition of oxidative phosphorylation through reversible or irreversible damage to cytochrome oxidase. Dr. Caughey has recently shown that ethanol and especially the longer chain alcohols inhibit respiration by a reaction with cytochrome oxidase at a presumed regulatory site (Personal Communication). This may contribute to the sensitizing action of ethanol we have noted.

Since the intensities of blue light required to alter the transport properties of the pigment epithelium are known to produce cell damage in other experimental procedures, we examined some tissues by means of electronmicroscopy to ascertain the site and extent of damage. All tissues for EM were fixed in 2.5% glutaraldehyde-.5% formaldehyde and post-fixed with 1% buffered osmium tetroxide. They were dehydrated and embedded in Poly/Bed 812, sectioned and stained with uranyl acetate and lead citrate, and observed with a Philips EM 200. Fig. 8 shows a section of a tissue exposed to red light for 60 min. with 30 additional min. in the chamber. No significant damage to the mitochondria or other organelles is evident. In all tissues taken at the end of a 60 minute exposure to 20 mW/cm^2 blue light we observed a blistering of the mitochondria as shown in Fig. 9.

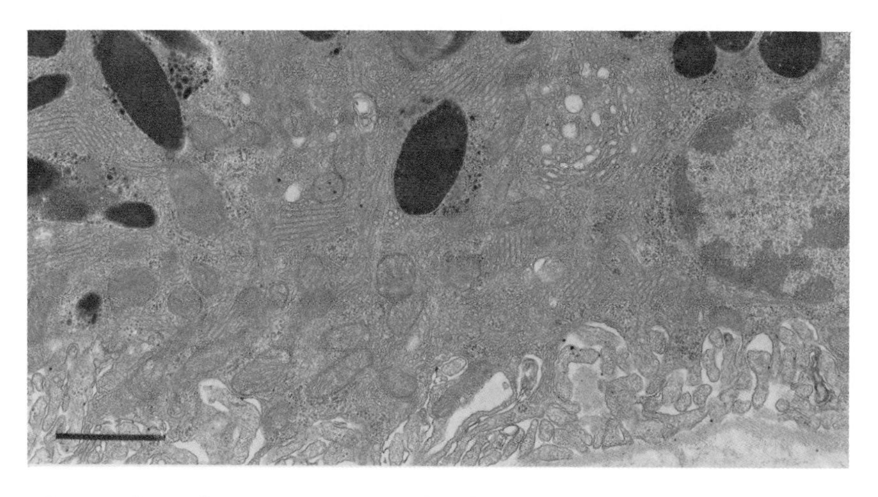

Figure 8. Electronmicrograph of a sample tissue which was exposed to 20 mW/cm^2 of red (600 nm) light for 60 min. with an additional 30 min. recovery period in the chamber. No damage could be detected in the cellular organelles. The line represents 1 μ.

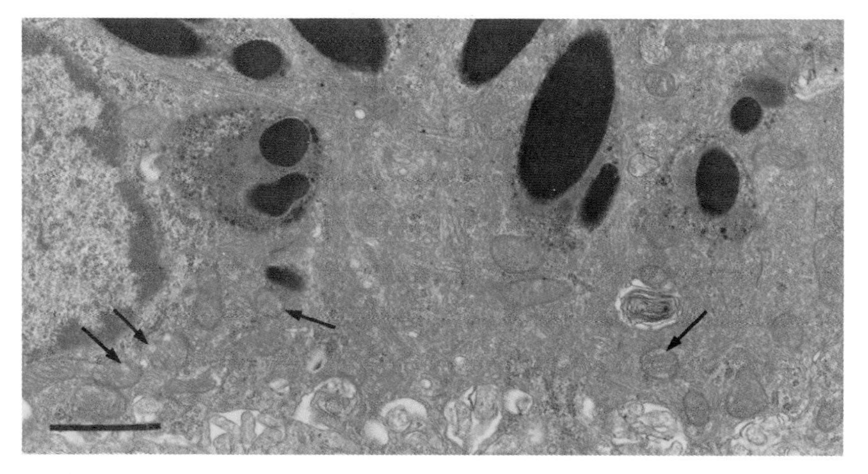

Figure 9. Electronmicrograph of a tissue exposed to 60 min. of blue light with no recovery time. The arrows indicate the appearance of blisters on the mitochondria. This was observed in all cells examined.

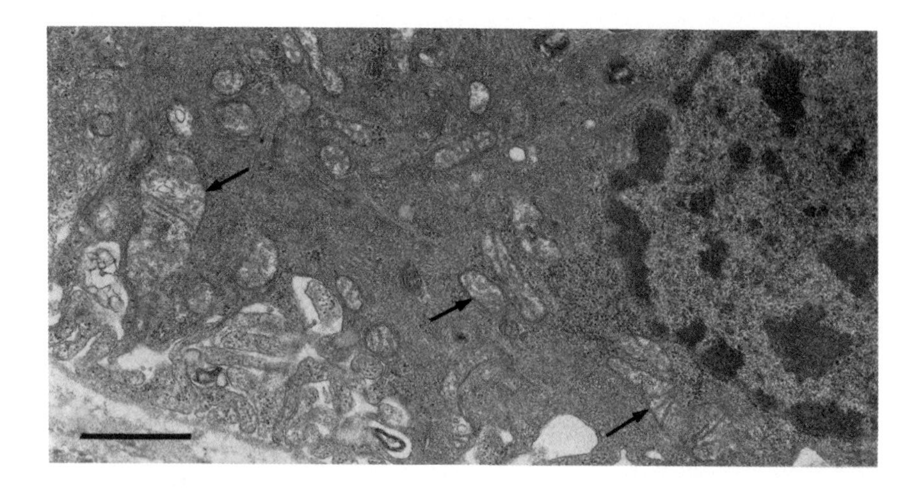

Figure 10. The effects of 60 min. of blue light exposure followed by a 30 min. recovery period of the PE. The arrows indicate that mitochondria have progressed to a swollen state with other signs of deterioration.

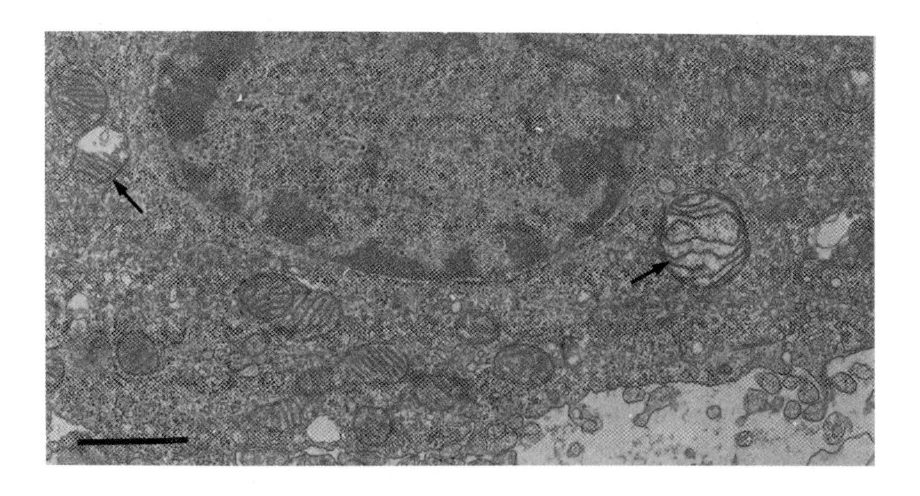

Figure 11. The effects of 60 min. of blue light exposure on a tissue being exposed to .5% ethanol. The blistering and swelling of mitochondria are apparent and seem to be more severe than comparable exposures without ethanol, cf Fig. 9.

If we allowed a 30 min. "recovery" period after the cessation of blue light, the blistering progressed to obvious mitochondrial swelling. (Fig. 10). Tissue exposed to blue light and .5% ethanol is shown in Fig. 11. The mitochondria appear to be in worse shape than usually noted when taken immediately at the end of the blue light exposure.

One of the possible consequences of interfering with the energy supply required by many transport systems would be the accumulation of ions. Calcium has been long suspected to play a key role in cell injury. Indeed, it has been argued that the activation of phospholipases by calcium and deterioration of mitochondrial structure and function appear to be most important in the evolution of cell injury leading to cell death (Humes, 1986). However, it is important to distinguish between the initiating events and the cellular consequences when analyzing the damaging effects of blue light. Also, we must recognize that cells contain a number of potential sensitizers for blue light damage, e.g. riboflavin and melanin, which could be activated depending on the wavelength and energy fluency rate. Accordingly, there might be differences in the emerging pathology depending on the experimental conditions. An excellent discussion of the possible mechanisms of light damage is provided by Noell (1980).

In summary, we have shown:

1. Blue light reversibly reduces the TEP and SCC of the PE.

2. Blue light inhibits the transport of leucine and chloride in the R-->Ch direction with different time courses of action.

3. Ethanol markedly enhances the inhibitory effects of blue light on the transport and electrical properties of the PE.

4. The sensitizing molecule for the TEP changes is a hemo protein with cytochrome oxidase as a likely candidate.

5. Electron microscopy revealed a consistent blistering of mitochondria after blue light exposure which soon progressed to swelling and other signs of damage.

ACKNOWLEDGMENTS

Research sponsored by the Air Force Office of Scientific Research, Air Force Systems Command, USAF, under grant or cooperative agreement number, AFOSR 87-0189. The US Government is authorized to reproduce and distribute reprints for Governmental purposes notwithstanding any copyright notation thereon.

REFERENCES

Collarimi EJ, Oxender DL (1987). Mechanisms of transport of amino acids across membranes. Ann Rev Nutr 7:75-90.

Crockett RS, Lawwill T (1984). Oxygen dependence of damage by 435 nm light in cultured retinal epithelium. Curr Eye Res 3:209-215.

DiMattio J, Degnan KJ, Zadunaisky JA (1983). A model for transepithelial ion transport across the isolated retinal pigment epithelium of the frog. Exp Eye Res 37:409-420.

Epel B, Butler WC (1969). Cytochrome a_3: Destruction by light. Science 166:621-622.

Ham WT Jr, Mueller HA, Sliney DH (1976). Retinal sensitivity to damage from short wavelength light. Nature 260:153-154.

Ham WT Jr, Mueller HA, Ruffalo JJ Jr, Millen JE, Cleary SF, Guerry RK, Guerry D III (1984). Basic mechanisms underlying the production of photochemical lesions in the mammalian retina. Curr Eye Res 3:165-174.

Humes HH (1986). Role of calcium in pathogenesis of acute renal failure. Amer J Physiol 250:F579-F589.

Lawwill T (1982). Three major pathological processes caused by light in the primate retina. A search for mechanisms. In Trans Amer Ophthol Soc, Johnson Press, Rochester, 80:517-579.

Miller S, Hughes B, Machen T (1982). Fluid transport across the retinal pigment epithelium is inhibited by cyclic AMP. Proc Natl Acad Sci 79:2111-2115.

Mitsumoto Y, Sato K, Ohyashiki T, Mohri T (1986). Leucine-proton cotransport system in Chang liver cell. J Biol Chem 261:4549-4554.

Mitsumoto Y, Sato K, Mohri T (1988). Leucine transport-induced activation of the Na^+/H^+ exchanger in human peripheral lymphocytes. Biochim Biophys Acta 939:349-351.

Ninnemann H, Butler WL, Epel WC (1970a). Inhibition of respiration in yeast by light. Biochim Biophys Acta 205:499-506.

Ninnemann H, Butler WL, Epel WC (1970b). Inhibition of respiration and destruction of cytochrome a_3 by light in mitochondria and cytochrome oxidase from beef heart. Biochim Biophys Acta 205:507-512.

Noell WK (1980). Possible mechanisms of photoreceptor damage by light in mammalian eyes. Vis Res 20:1163-1171.

Tsubir S, Manabe R, Tizuba S (1986). Aspects of electrolyte transport across isolated dog retinal pigment epithelium. Amer J Physiol 250:F781-F784.

Inherited and Environmentally Induced
Retinal Degenerations, pages 569–575

COMPARISON OF RETINAL PHOTOCHEMICAL LESIONS AFTER EXPOSURE TO NEAR-UV OR SHORT-WAVELENGTH VISIBLE RADIATION

Robert J. Collier and Seymour Zigman

Department of Ophthalmology Box 314
School of Medicine and Dentistry
University of Rochester
Rochester, New York 14642

INTRODUCTION

Near-UV and short-wavelength visble energy have been shown to be very efficient wavelengths for inducing photochemical lesions to the retina (Ham et al., 1978). Damage to photoreceptors and pigment epithelial cells has been reported after exposure of monkeys (Zuclich & Connolly, 1976; Ham et al., 1978; 1982), rats (Henton & Sykes, 1984), and mice (Zigman & Vaughn, 1974) to radiation of wavelengths shorter than 450 nm. In an attempt to understand the mechanism underlying such lesions, we are studying the damaging effects of exposure to different wavelengths of monochromatic energy. In these studies we assume the cell containing the chromophore that is maximally sensitive to a particular wavelength will be the cell that is initially damaged. To determine which retinal cells are most susceptible to photochemical damage, we have adopted a strategy of assessing the relationship between energy incident on the retina and light-microscopic changes.

In this study, the American gray squirrel was used as it has several retinal properties that we feel are advantageous for studying the actinic effects of light on retinal tissue. Gray squirrel photoreceptors exhibit many morphologic characteristics that are common to both human and monkey, including types of rod and cone membrane infoldings and similar types of synaptic contacts (Cohen, 1964; West & Dowling, 1975). Photoreceptors are easily distinguished. Electroretinographic and single-unit recordings suggest that three cone pigments are present in the gray squirrel retina (Blakeslee, 1983). Finally, the distribution of rods and cones is homogeneous across the retina, a rod-to-cone ratio of 1: 2.7 in central retina that falls to 1: 2.6 in the peripheral retina (West & Dowling, 1975).

MATERIALS AND METHODS

Subjects

Gray squirrels, *Sciurus carolinensis,* were trapped locally under scientific license granted by the State of New York, Department of Environmental Conservation. Squirrels were individually housed in cages with nesting boxes and maintained on a 12L: 12D light cycle before and after light exposure. Food and water were available ad lib. Animal care and treatment in this investigation were in compliance with the ARVO Resolution on the Use of Animals in Research.

Lens Removal

The effects of near-UV exposure were studied in squirrels that had been rendered monocularly aphakic. Squirrels were anesthetized by intraperitoneal (IP) injection of sodium pentobarbital (50 mg/Kg), pupils were dilated (Neosynephrine; 2.5%), corneas were anesthetized (tetracaine-HCl; 0.5%), and eyelids were retracted prior to lens extirpation by phacoemulsification using a Kelman-Cavitron apparatus. Blue-light exposures were to squirrels with intact biological lenses.

Experimental Exposure

Anesthetized squirrels (sodium pentobarbital; 50 mg/Kg; IP) with dilated pupils (Neosynephrine; 2.5%), anesthetized corneas, (tetracaine-HCl; 0.5%), and retracted eyelids were held in a head restraint and aligned to a one-channel Maxwellian optical system that included an ophthalmoscope. The retina was viewed with dim yellow illumination so that the squirrel could be adjusted to bring its retina into sharpest focus and positioned to allow exposure of a retinal region 30° in diameter, located in the superior temporal retina adjacent to the optic nerve.

Exposure was to monochromatic radiation from a 200W super-pressure mercury-vapor arc lamp (Osram HBO) filtered through heat absorbing glass and either a 366-nm (near-UV radiation) or 440-nm (short-wavelength visible radiation) Ditric interference filter (half amplitude bandpass= 10 nm). Near-UV radiant exposures were 0.0, 0.6, 1.8, and 3.6 J/cm^2; while blue-light radiant exposures were 10.2, 12.75, and 15.0 J/cm^2. An intensity-damage relationship was determined by exposing two eyes to each radiant exposure. Eyes were examined by light microscopy 24 hours after exposure.

Light Microscopy

Anesthetized squirrels were sacrificed by transcardiac perfusion of 2.0% paraformaldehyde/ 2.5% glutaraldehyde fixative in phosphate buffer (ph 7.4). Retinas were post fixed in osmium tetroxide, dehydrated in ethanol, and embedded in Epon 812 (Polysciences). One-micron thick sections stained with toluidine blue were studied. Counts of photoreceptor nuclei and thickness of photoreceptor layers were measured to determine the effects of exposure on retinal tissue. For UV-A exposures, data from similar non-exposed retinal locations were analyzed to compare the nonoperated eye with the aphakic eye to control for the effects of lens removal surgery.

RESULTS

Near-UV Radiation Exposure. Retinal lesions were not detected in non-exposed control aphakic eyes (0.0 J/cm^2 exposure group) or non-exposed retinal

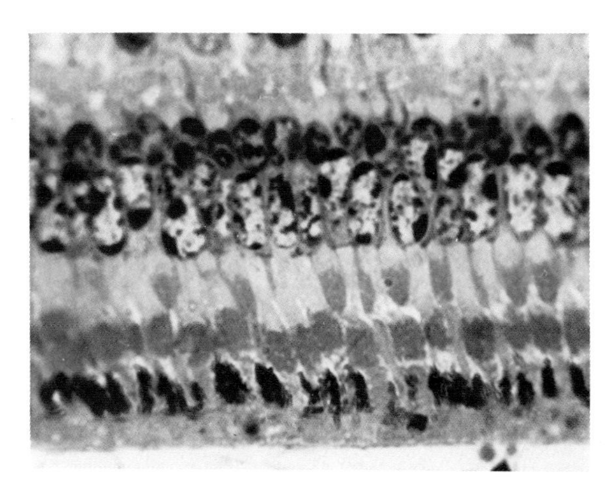

Figure 1. Control retina from eye with lens intact. The inner segments are seen as two bands, rods residing in the vitrad layer and cones in the sclerad layer.

locations in exposed eyes. Figure 1 shows the normal retinal appearance from a non-exposed squirrel eye. The outer nuclear layer (ONL) is composed of two rows of nuclei. An outer row of nuclei contain oval-shaped cone nuclei, while the inner row of nuclei contain small, round, rod nuclei. Rod and cone inner segments are also confined to separate layers. Rod inner segments are seen as granular, dark-staining bodies just below the cone nuclei, while cone inner segments form the second, lower, sclerad tier of inner segments. Rod and cone inner segments exhibit a granular appearance due to the large number of mitochondria found in the ellipsoid region.

Lesions were not observed 24 hours after exposure to 0.6 J/cm^2 (λ= 366 nm). Inner segments of rods and cones had a granular appearance and outer segments were not disrupted. Counts of photoreceptor nuclei and retinal layer thickness were not significantly different than control measurements suggesting that neither viewing of the aphakic retina with the ophthalmoscope nor exposures to low irradiances result in damage to retinal tissues.

The photoreceptor layer, measured from the inner edge of the outer nuclear layer to the tips of the outer segments, was reduced by 18% as a result of exposure to 1.8 J/cm^2. The granular appearance of rod and cone inner segments was not apparent across the lesioned area except at the lesion border where only rod inner segments showed this abnormality. In addition, the "plate like" stacking of rod lamellar disks had been replaced by breaks running down the center of the outer segment (Figure 2). Cone outer segments appeared normal in the exposed retinal area.

Exposure to 3.6 J/cm^2 resulted in a 44% decrease in photoreceptor layer thickness and a reduction in photoreceptor number by 17%. Both rod and cone outer

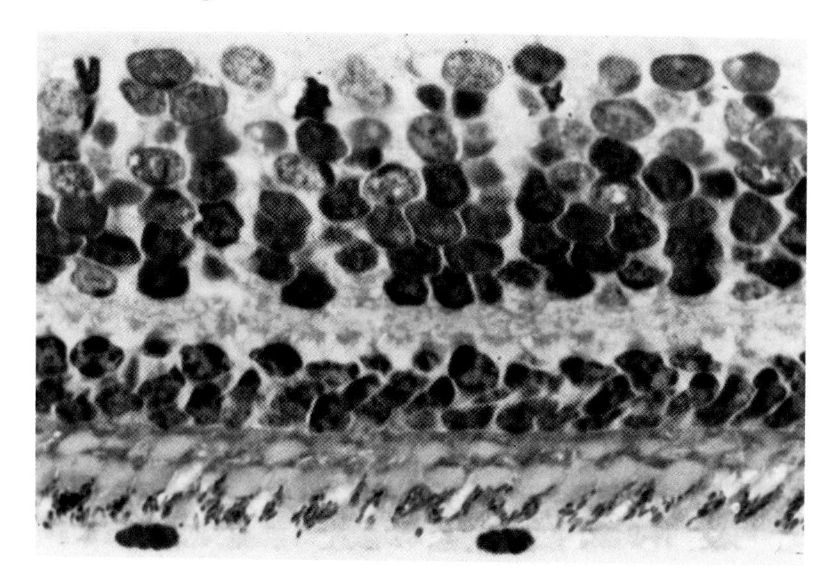

Figure 2. Aphakic retina one day after exposure (radiant exposure= 1.8 J/cm^2). Rod outer segments are degenerative and inner segments are swollen and shortened.

segments were degenerative. No differences in the thickness of the pigment epithelium were measured although regions of dense pigment granule accumulation were evident.

Short-Wavelength Visible Radiation Exposure. No lesions were detected in gray squirrel eyes exposed to 10.2 J/cm^2 when examined 24 hours after exposure to 440-nm radiation. However, lesions were detected in the exposed retinal region of eyes receiving radiant exposures of 12.75 J/cm^2. Rod and cone photoreceptor outer segments were shortened and disorganized, necrotic inner segments were observed, and 33% of outer nuclear layer nuclei were pyknotic (Figure 3). Damage to the pigment epithelium was observed and consisted of flattening of their apical processes. Radiant exposures of 15.0 J/cm^2 resulted in further damage to the pigment epithelium and neuroretina. About 50% of the rod and cone nuclei were pyknotic and the cytosol of the pigment epithelium was vacuolated.

DISCUSSION

Exposure of the retina to ultraviolet radiation can cause lesions to the sensory nerve endings and pigment epithelium. The purpose of this investigation was to determine whether photoreceptors, particularly rods, were more susceptible to near-UV induced retinal damage than the pigment epithelium and to compare these changes with lesions induced by short-wavelength visible radiation exposure.

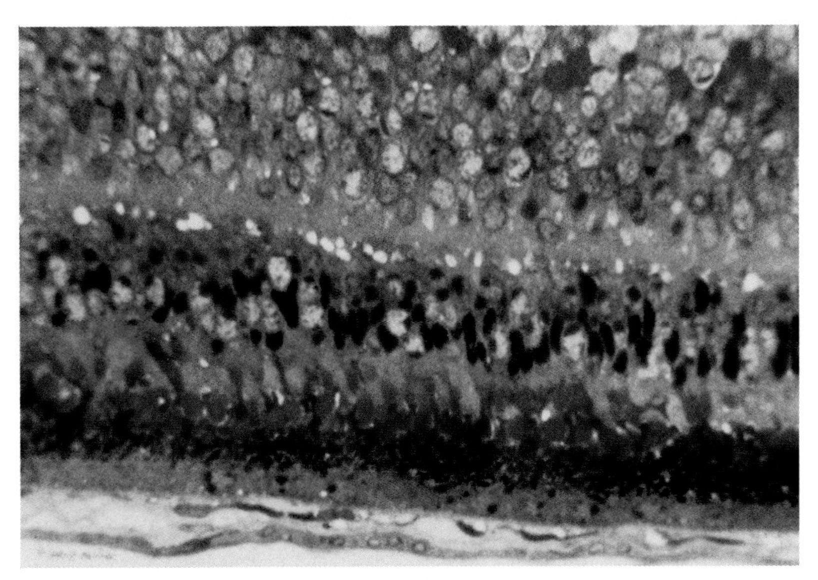

Figure 3. Retina one day after exposure to short-wavelength visible radiation (radiant exposure= 12.75 J/cm^2). Photoreceptor nuclei are pyknotic, inner segments are necrotic, and outer segments are disorganized.

Only photoreceptors were lesioned after exposures just exceeding the damage threshold (radiant exposure= 1.8 J/cm^2). As a result of this exposure, rods appeared to be more severely damaged than did cones. Rod outer segments were disrupted and rod inner segments, which normally exhibit strong, acidophilic, staining properties, stained faintly and lost their granular appearance. Similar changes to cone inner segments were observed after exposure to this irradiance, but cone outer segments were unaffected. Examination of retinal tissue near the border of the lesion showed only rod inner segment damage. Rod outer segments and cone inner and outer segments were devoid of significant lesions. Apical invaginating processes of the pigment epithelium, melanin-pigment distribution within the pigment epithelial cell, and vacuolization of the pigment epithelial cytosol in the exposed aphakic eye were not different from the control eye.

Light microscopic examination of retinal lesions in primate eyes two days after exposure to 350-nm radiation showed thinning of the outer nuclear layer (ONL), loss of rod and cone outer segments, and depigmentation of pigment epithelial cells (Ham et al., 1982). In these experiments, damage thresholds were assessed by determining the power required to produce an ophthalmoscopically visible minimal lesion immediately after exposure. Based on this criterion, Ham concluded that the action spectrum for light damage resembles the absorbance of melanin in the pigment epithelium. This criterion may overestimate threshold as Schmidt and Zuclich (1980) have shown that light-microscopic lesions can be detected at intensities one-half those intensities required to detect lesions with an ophthalmoscope in the monkey.

Following exposure to monochromatic blue light, Ham et al. (1978) and Sperling et al. (1980) reported damage to the pigment epithelium, including loss of pigment granules and vacuolization of the pigment epithelial cytosol. Lesions to the gray squirrel retina from short-wavelength visible radiation exposure were to the pigment epithelium and photoreceptors. Where UV-A lesions resulted in a loss of the granular appearance of inner segments and reduced staining properties, blue-light lesions resulted in darkly stained necrotic inner segments.

The strong, acidophilic staining properties and granular appearance of the inner segments can be attributed to the presence of many large mitochondria in the inner segment. It was these two properties of the inner segment that were affected by threshold UV-A lesions. These data can be interpreted to suggest that near-ultraviolet exposure leads to photic injury to the cell by an interruption of respiration, either by inactivating photoreceptor mitochondrial enzymes or by physically destroying individual mitochondria. These changes would result in a loss of energy to support metabolic processes. It has been shown that near-UV radiation can affect mitochondria by inactivation of cytochrome c (Kashket & Brodie, 1963) or by the production of superoxide anions following irradiation of NADH and NADPH coenzymes (Cunningham, Johnson, Giovanazzi, & Peak, 1985). Other possible mechanisms to explain the results include cell damage from the production of tryptophan oxidation products (Zigman, 1984) or DNA photoproducts (Brash & Haseltine, 1982).

Acknowledgments

This work was supported by Fight for Sight Student Fellowships financed by grants from James Thurber Foundation and Burroughs Wellcome Co., to Fight for Sight, Inc., New York City (R.C.). Partial support was provided by the Lighting Research Institute (S.Z. & R.C.); N.I.H. research grant #EY00459 (S.Z.), Research to Prevent Blindness (S.Z.), and Center Support Grant #EY01319 (Center for Visual Sciences).

REFERENCES

Blakeslee B (1983). Electrophysiological studies of spectral mechanisms in the retinas of ground squirrels and tree squirrels. Unpublished doctoral dissertation, University of California, 1983.

Brash DE & Haseltine WA (1982). UV-induced mutation hotspots occur at DNA damage hotspots. Nature 298:189-192.

Cohen AI (1964). Some observations on the fine structure of the retinal receptors of the American gray squirrel. Investigative Ophthalmology and Visual Science 3(2):198-216.

Cunningham ML, Johnson JS, Giovanazzi SM, & Peak MJ (1985). Photosensitized production of superoxide anion by monochromatic (290 - 405 nm) ultraviolet irradiation of NADH and NADPH coenzymes. Photochemistry and Photobiology 42:125-128.

Ham WT Jr, Mueller HA, Ruffolo JJ Jr, Guerry D III, & Guerry RK (1982). Action spectrum for retinal injury from near-ultraviolet radiation in the aphakic monkey. American Journal of Ophthalmology 93:299-306.

Ham WT Jr, Ruffolo JJ Jr, Mueller HA, Clark AM, & Moon ME (1978). Histologic analysis of photochemical lesions produced in rhesus retina by short-wavelength light. Investigative Ophthalmology & Visual Science 17(10):1029-1035.

Ham WT Jr, Mueller HA, Ruffolo JJ Jr, & Clarke AM (1978). Sensitivity of the retina to radiation damage as a function of wavelength. Photochem Photobiol 29:735-743.

Henton WW & Sykes SM (1984). Recovery of absolute threshold with UVA-induced retinal damage. Physiology and Behavior 32:949-954.

Kashet ER & Brodie AF (1963). Oxidative phosphorylation in fractionated bacterial systems. Journal of Biological Chemistry 238:2564-2570.

Schmidt RE & Zuclich JA (1980). Retinal lesions due to ultraviolet laser exposure. Invest Ophthalmol Vis Sci 19:1166-1175.

Sperling HG, Johnson C, & Harwerth RS (1980). Differential spectral photic damage to primate cones. Vision Research 20:1117-1125.

West RW & Dowling JE (1975). Anatomical evidence for cone and rod-like receptors in the gray squirrel, ground squirrel, and prairie dog retinas. Journal of Comparative Neurology 159:439-460.

Zigman S (1984). The role of tryptophan oxidation in ocular tissue damage. In Schlossberger HG, Kochen W, Linzen B, & Steinhart H (eds): "Progress in tryptophan and serotonin research," Berlin: Walter de Gruyter and Co, pp 449-467.

Zigman S & Vaughan T (1974). Near-ultraviolet light effects on the lenses and retinas of mice. Investigative Ophthalmology and Visual Science 13(6):462-465.

Zuclich JA & Connolly J (1976). Ocular damage induced by near-ultraviolet laser radiation. Investigative Ophthalmology and Visual Science 15(9):760-764.

Inherited and Environmentally Induced
Retinal Degenerations, pages 577–583

NUCLEOTIDE-INDUCED RETINAL CHANGES

Fumiyuki Uehara, Norio Ohba, Munefumi Sameshima, Katsuhide Takumi, Kazuhiko Unoki, Takashi Muramatsu, Douglas Yasumura and Matthew M. LaVail

Departments of Ophthalmology (F.U., N.O., M.S., K.T., K.U.) and Biochemistry (T.M.), Kagoshima University Faculty of Medicine, Kagoshima-shi, Japan 890. Departments of Anatomy and Ophthalmology (F.U., D.Y., M.M.L.), University of California, San Francisco, School of Medicine, San Francisco, CA 94143

INTRODUCTION

Glycoconjugates on the cell surface are thought to be involved in a variety of cell functions. The non-reducing termini of sugar chains of such glycoconjugates are usually occupied by sialic acid. We have previously suggested that abnormalities of sialic acid may lead to retinal degeneration (Uehara et al., 1987). This suggestion was based on the comparison of glycoconjugates in the degenerative retina of the C3H mouse and the normal retina of the C57BL mouse by means of lectin reactions of two-dimensional gel blots. We observed that some spots of glycoconjugates are detected only in the degenerative retina with *ricinus communis* agglutinin-1, probably because of the unmasking of ß-galactose residues normally masked by the terminal sialic acid (Uehara et al., 1987). In addition, trans-scleral administration of wheat germ agglutinin (specific for sialic acid and/or N-acetyl-D-glucosamine) into the subretinal space resulted in a marked destruction of the photoreceptors and the outer nuclear layer of the rat retina (Uehara et al., 1988).

Abnormalities of sialic acid in the termini of sugar chains could possibly be caused by abnormal activities of sialyltransferase, which catalyzes sialylation of oligosaccharides of glycoconjugates. Thus, it is possible that modification of sialyltransferase may cause retinal degeneration. Some nucleotides are reported to be effective inhibitors or activators of sialyltransferase (Bernacki, 1975; Kijima-Suda, 1985). It might be possible, therefore, to modify experimentally the activity of retinal sialyltransferase by the application of these nucleotides. In this paper, we report histological changes of the rat retina induced by intravitreal administration of nucleotides.

MATERIALS AND METHODS

Adult albino rats (Wistar; 200-300 gm) were housed under a lighting cycle of 12 hr:12 hr light and dark. They were anesthetized with intramuscular injection of ketobarbiturate hydrochloride (20 mg/kg) and intraperitoneal injection of sodium pentobarbiturate (25 mg/kg) before intravitreal injection of nucleotides and eye enucleation. Topical anesthesia of benoxinate hydrochloride was also applied to the eye during all the procedures. The nucleotides used in this study were cytidine 5'-diphosphate (CDP), guanosine 5'-diphosphate (GDP) and uridine 5'-diphosphate (UDP). At 2 hr after the onset of the light cycle, 5 µl of nucleotide or sucrose (1 mg/ml in PBS) was injected into the vitreous cavity of the rat eye [8 eyes (4 rats)/reagent] through the pars plana ciliaris by means of a 27 gauge microinjector following corneal incision (1mm) with a razor blade under an operation microscope. Intravitreal injections were done 9 additional times at 30-min intervals following the first injection. At one hr after the last injection, the rat eyes were enucleated, quickly immersed in 2% glutaraldehyde in 0.1M phosphate buffer at pH 7.3 for 24 hr, rinsed in PBS, postfixed with 2% osmium tetroxide in 0.1 M phosphate buffer for 2 hr, dehydrated in a series of ethanols, and embedded in epoxy resin. Semi-thin sections (2 µm thick), stained with toluidine blue, were observed by light microscopy; ultrathin sections, stained with uranyl acetate and/or lead citrate, were examined with a JEM-100B electron microscope.

RESULTS

By light microscopy, CDP induced vacuolar changes in the inner layers of the retina [nerve fiber layer (NF), ganglion cell layer (GCL), inner plexiform layer (IPL), inner nuclear layer (INL), and outer plexiform layer (OPL); Fig. 1a]. GDP induced vacuolar changes in the inner layers (NF, GCL, IPL, INL, and OPL) and a reduction of staining of the apical portions of the photoreceptor outer segments with toluidine blue compared to the basal portions (Fig. 1b). UDP also induced a reduction of staining of the apical portions of the outer segments (Fig. 1c), but did not cause changes in the inner retina. Sucrose induced no morphological changes in the retina (Fig. 1d).

By electron microscopy, no changes were observed in the outer segments of the CDP- or sucrose-injected eyes (Fig. 2), whereas vacuolar changes of the disc membranes were observed in the apical portions of the outer segments of the GDP- or UDP- injected eyes (Fig. 3).

DISCUSSION

CDP and GDP induced similar vacuolar changes of the inner retina, while GDP and UDP induced changes of the apical portions of the outer segments. Sucrose, whose molecular weight is similar to that of CDP, GDP and UDP, induced

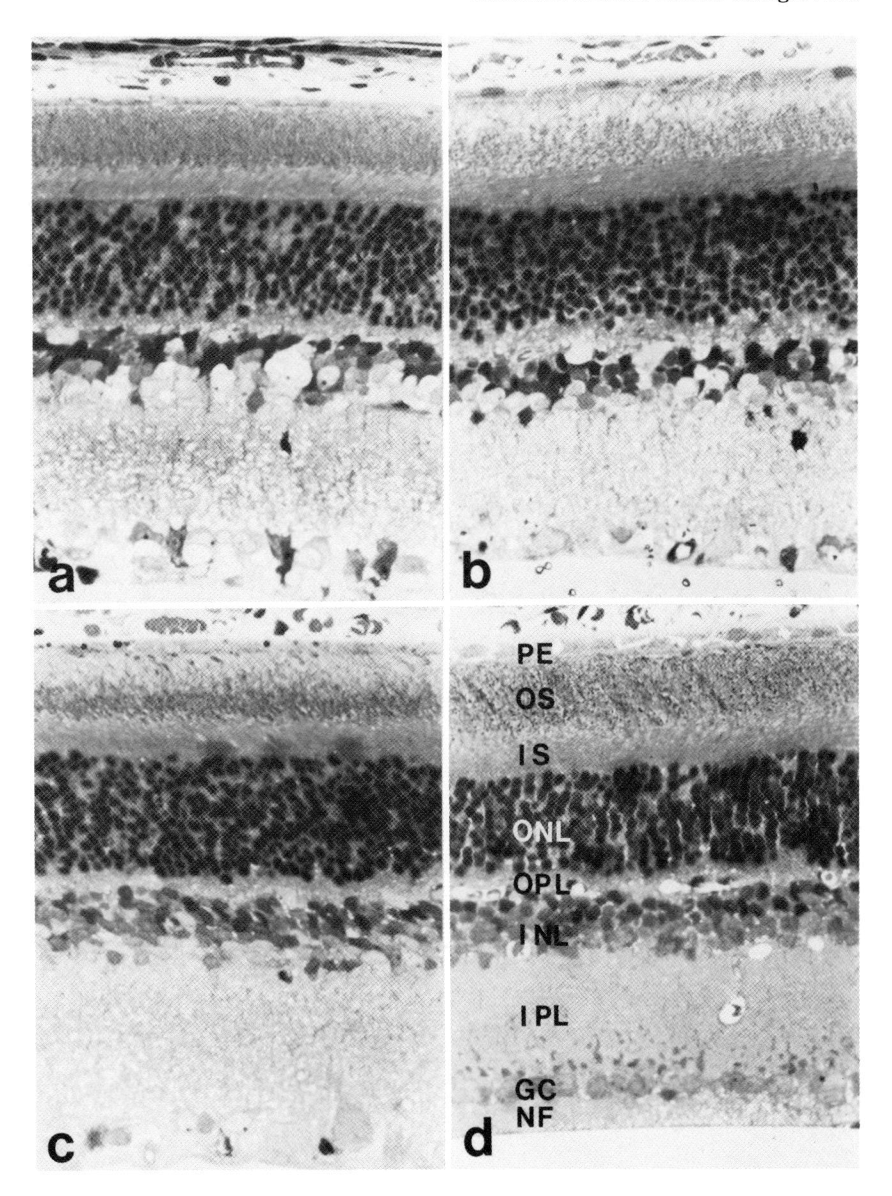

Figure 1. Light micrographs of rat retinas after intravitreal injection of either CDP (a); GDP (b); UDP (c); or sucrose (d). NF, nerve fiber layer; GC, ganglion cell layer; IPL, inner plexiform layer; INL, inner nuclear layer; OPL, outer plexiform layer; ONL, outer nuclear layer; IS, photoreceptor inner segments; OS, photoreceptor outer segments; PE, pigment epithelium. Toluidine blue. x400.

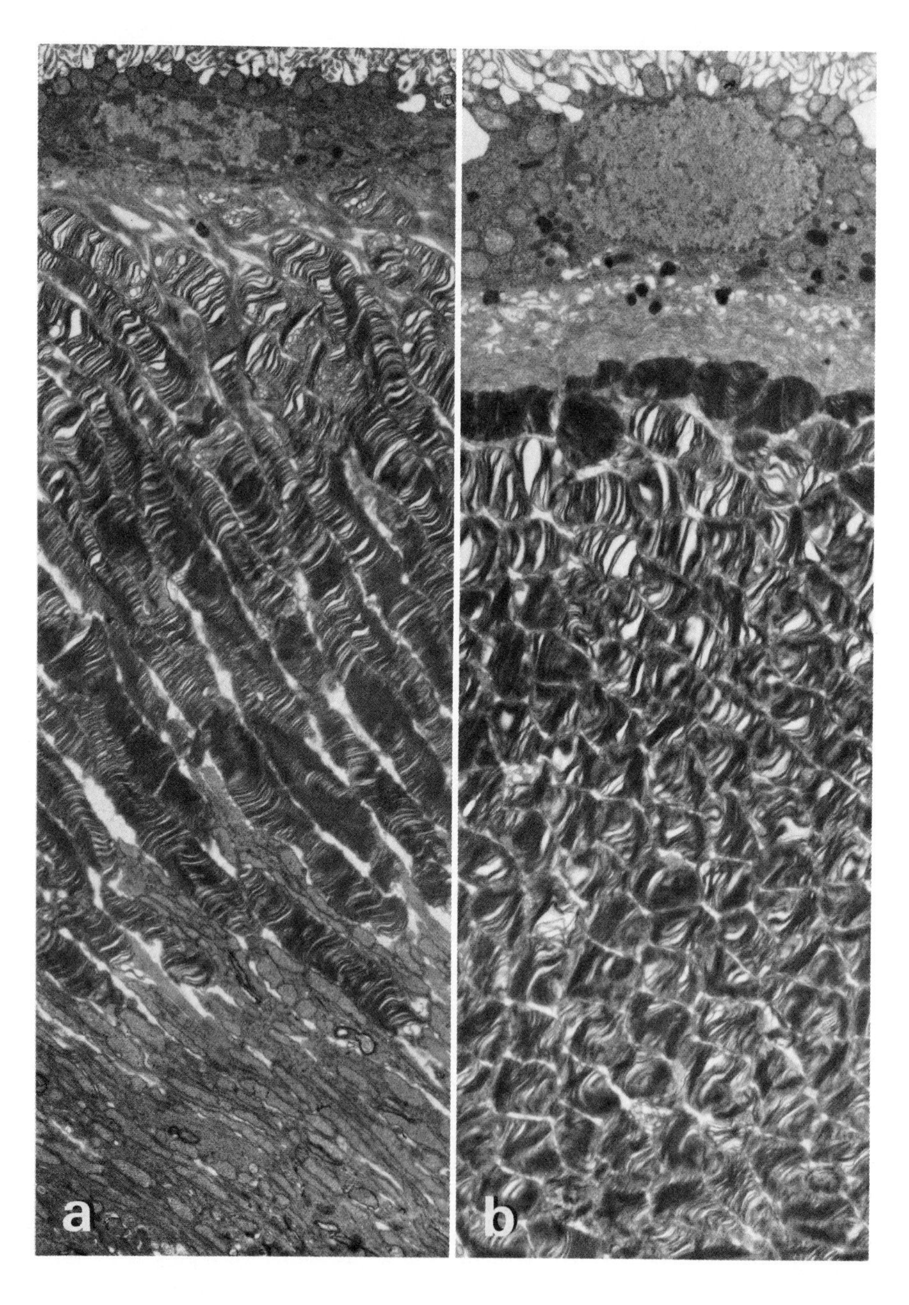

Figure 2. Electron micrographs of photoreceptor outer segments and inner segments with intravitreal injections of either CDP (a); or sucrose (b). x5000.

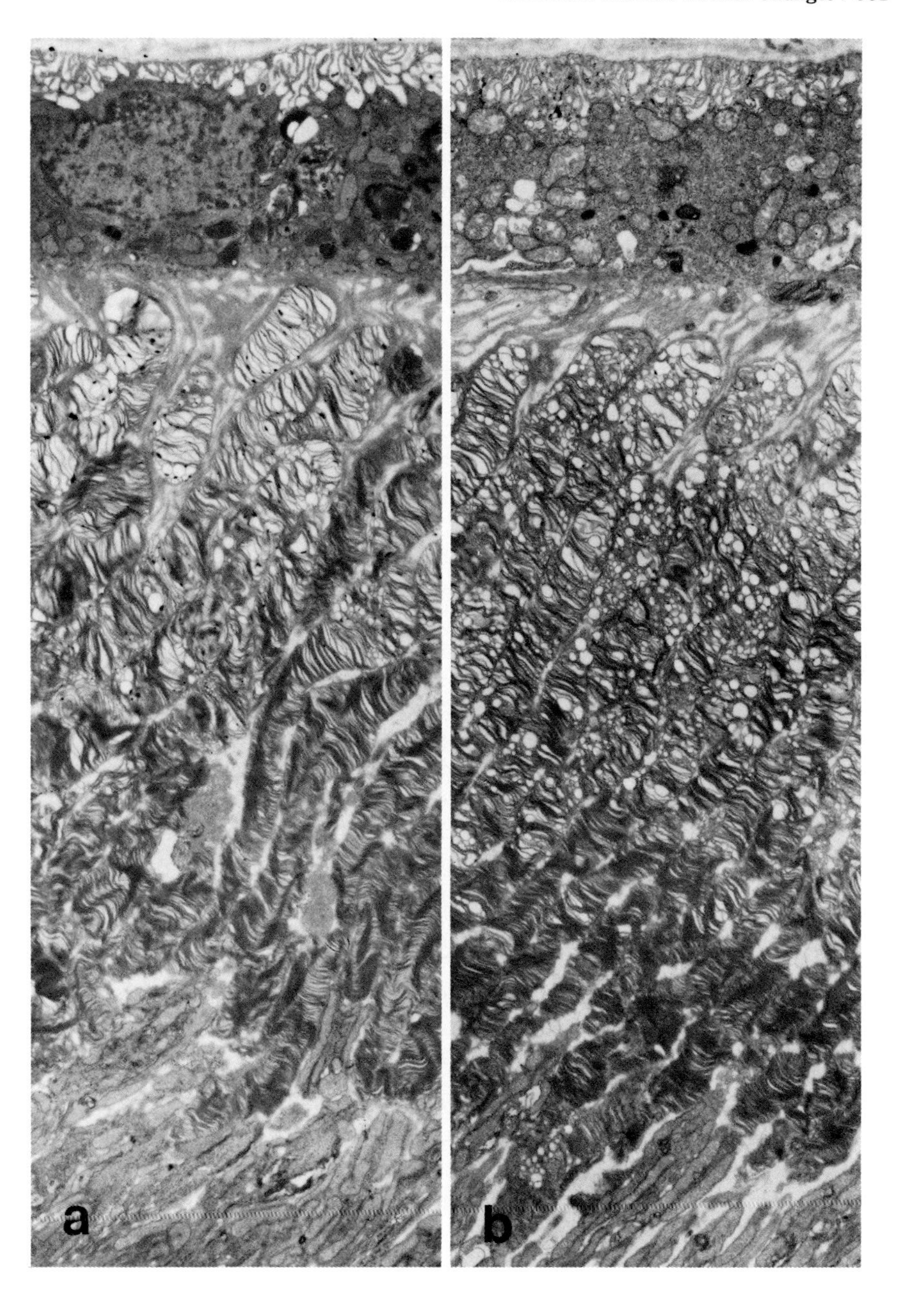

Figure 3. Electron micrographs of photoreceptor outer segments and inner segments with intravitreal injections of either GDP (a) or UDP (b). x5000.

no morphological changes in the retina, which suggests that the retinal changes induced by the nucleotides may be caused by their physiological effects. CDP and UDP have been reported to be effective inhibitors and activators, respectively, of sialyltransferase activity (Bernacki, 1975; Kijima-Suda et al., 1985). We have observed that a low concentration of GDP acts as an activator of the sialyltransferase of the retina, whereas a high concentration of GDP acts as an inhibitor (unpublished data). We cannot conclude that the retinal changes in this study were caused by modification of the sialyltransferase activity without biochemical analysis of the retinas. However, we would like to propose this hypothesis as one possibility — that inhibition (CDP) and activation (UDP) of the activity of the sialyltransferase may induce vacuolar changes of the inner retina and of the apical portions of the photoreceptor outer segments, respectively. GDP may act as sialyltransferase inhibitor in the inner retina, where the concentration of GDP may be high. The inner retina is reported to be rich in sialoglycolipids (especially gangliosides; Uehara et al., 1987), whose sialic acid may play an important role in maintaining the inner layers.

In the rat retina, sialoglycoconjugates are reported to be found predominantly near the apical portions of the outer segments (Porrello and LaVail, 1986). In this region, GDP may act as sialyltransferase activator where the concentration of GDP may be low. We have previously suggested that a decrease of sialic acid occurs on the most apical portions of the outer segments in preparation for recognition and phagocytosis by the pigment epithelium (Uehara et al., 1985). Thus, it is possible that the content of sialic acid in the glycoconjugates of the apical portions of the photoreceptor outer segments may regulate some interactions between the outer segments and the pigment epithelium. When the sialyltransferase activity is activated (e.g., by GDP or UDP) and the sialic acid does not decrease, the interaction may be altered and the apical portion of the outer segment may be adversely affected.

There may be other mechanisms by which the injected nucleotides induced the retinal changes. For example, GDP and UDP have also been reported to be effective inhibitors or activators of other glycosyltransferases (Bernacki, 1975; O'Brien, 1978; Kijima-Suda, 1985). However, the directions of the effects on the other glycosyltransferases (fucosyltransferase, galactosyltransferase, etc.) are different between GDP and UDP, which cannot explain the similar morphological changes of the outer segments induced by GDP and UDP.

Pyrimidine 5'-nucleotidase has been reported to be localized on the apical processes of the pigment epithelium and the distal ends of the outer segments (Irons, 1987). It is possible that the vacuolar changes of the apical portions of the outer segments may be induced by modification of this enzyme by the injected nucleotides.

In any case, further studies are needed to clarify the mechanism of the nucleotide-induced retinal changes described here.

ACKNOWLEDGMENTS

The authors thank Ms. Gloria Riggs for secretarial assistance. This work was supported in part by research grant No. 61480367 from the Japanese Ministry of Education, Science and Culture, a research grant for retinochoroidal atrophy from the Ministry of Health and Welfare of Japan, NIH Research Grant EYO1919, Core Grant EYO2162 and funds from the Retinitis Pigmentosa Foundation Fighting Blindness, Research to Prevent Blindness and That Man May See, Inc. Dr. LaVail is the recipient of a Research to Prevent Blindness Senior Scientific Investigators Award.

REFERENCES

Bernacki RJ (1975). Regulation of rat-liver glycoprotein: N-acetylneuraminic acid transferase activity by pyramidine nucleotides. Eur J Biochem 58: 447-481.

Irons MJ (1987). Redistribution of Mn^{++}-dependent pyrimidine 5'-nucleotidase (MDPNase) activity during shedding and phagocytosis. Invest Ophthalmol Vis Sci 28: 83-91.

Kijima-Suda I, Toyoshima S, Itoh M, Furuhata K, Ogura H, Osawa T (1985). Inhibition of sialyltransferases of murine lymphocytes by disaccharide nucleotides. Chem Pharm Bull 33: 730-739.

O'Brien PJ (1978). Characteristics of galactosyl and fucosyl transfer to bovine rhodopsin. Exp Eye Res 26: 197-206.

Porrello K, LaVail MM (1986). Histochemical demonstration of spatial heterogeneity in the IPM of the rat retina. Invest Ophthalmol Vis Sci 27: 1577-1586.

Uehara F, Muramatsu T, Sameshima M, Kawano K, Koide H, Ohba N (1985). Effects of neuraminidase on lectin binding sites in photoreceptor cells of monkey retina. Jpn J Ophthalmol 29: 54-62.

Uehara F, Muramatsu T, Takumi K, Ohba N (1987a). Two-dimensional gel electrophoretic analysis of lectin receptors in the degenerative retina of C3H mouse. In Hollyfield JG, Anderson RE, LaVail MM (eds): "Degenerative Retinal Disorders: Clinical and Laboratory Investigations", New York: Alan R Liss, pp 219-227.

Uehara F, Takumi K, Unoki K, Sameshima M, Ohba N (1987b). Immunohistochemical localization of gangliosides in the monkey retina. Acta Soc Ophthalmol Jpn 91: 1103-1106.

Uehara F, Sameshima M, Takumi K, Unoki K, Muramatsu T, Ohba N (1988). Sialic acid in the retina and its significance in the retinal degeneration. In Ohyama M, Muramatsu T (eds): "Glycoconjugates in Medicine", Professional Postgraduate Services, pp 310-315.

Inherited and Environmentally Induced
Retinal Degenerations, pages 585–600

GLIAL MARKERS IN IODOACETATE RETINAL DEGENERATION

Yusuf K.Durlu, Sei-ichi Ishiguro and Makoto Tamai

Department of Ophthalmology, Tohoku University School of Medicine, Sendai.

INTRODUCTION

Iodoacetate has been known to produce selective impairment of retinal function experimentally(Noell,1951;Schubert and Bornschein,1951). By light microscopy, Noell(1952) demonstrated that rod cells were most susceptible and irreversibly affected after intravenous injection of this agent. However,the cone cells of the cone-dominant retina of the ground squirrel are also susceptible to intracardiac iodoacetate treatment (Farber et al., 1983).

The morphological changes of glial cells in iodoacetate retinal degeneration including proliferation and mitosis (Karli, 1954), lamellar figures, increase of glycogen granules and ribosomes, phagocytosis of the visual cell debris seen in the Müller cells(Orzalesi et al.,1970),the alterations correlated with endoplasmic reticulum (Babel and Stangos, 1973) and the disappearance of Müller cell processes from the photoreceptor layer(Farber et al., 1983) have been reported. Electrophysiologically, b-wave was firstly affected prior to a-wave elimination after intravenous iodoacetate injection in the rabbit (Noell, 1951).

Glucose 6-phosphatase (G6Pase) (D-Glucose 6-phosphate phosphohydrolase,E.C. 3.1.3.9), is a key glycogenolytic and gluconeogenic enzyme that catalyzes the hydrolysis and synthesis of glucose 6-phosphate(G6P)(Nordlie,1971).G6Pase multicomponent system consists of three translocases(T1,T2 and T3) and a relatively nonspecific phosphohydrolase: phosphotransferase being associated with the endoplasmic reticulum(ER) in liver and kidney which has been postulated by Arion et al. (1980). T1 facilitates the movement of G6P,T2 of inorganic phosphate and pyrophosphate,and T3 of glucose between the cytosol and lumen of the ER. Although, Fishman and Karnovsky(1986) reported that in brain where G6Pase activity is mostly present in neuronal cell bodies and dendrites (Broadwell and Cataldo, 1983),glucose 6-phosphohydrolase is not associated with a glucose 6-phosphotranslocase. However in the rabbit retina ,G6Pase is mostly localized in a modified astrocyte (Polyak,1957) that is Müller cell(Lessell and Kuwabara, 1964). It has been proposed that the biochemically measured G6Pase activity in the microsomal fraction of the rabbit retina might be conveniently evaluated as a glial marker (Durlu et al., 1988).

Glutamine Synthetase(GS)(L-glutamate ammonia ligase,E.C. 6.3.1.2), a key enzyme of nitrogen metabolism which is exclusively localized in Müller cells catalyzes the formation of glutamine from glutamate (Riepe and Norenburg,1977, Moscona and Linser, 1983). GS can also catalyze the exchange of the amide group of glutamine for ammonia known as γ-glutamyl transferase (γ-GT) activity (Pahuja and Reid, 1985).

Glial fibrillary acidic protein(GFAP) is an intermediate filament which was first isolated from the multiple sclerosis plaques (Eng et al., 1971). In the normal rabbit retina,GFAP immunoreactivity is only observed in astrocytes (Schnitzer, 1985). However, some pathological states of the retina and culturing of Müller cells lead Müller cells to express GFAP (Bignami and Dahl, 1979; Shaw and Weber, 1983; Eisenfeld et al., 1984; Erickson et al., 1987; Wakakura and Foulds, 1988).

In this study, enzymological and immunohistochemical methods were used to probe the metabolic response of glial cells to experimentally induced retinal degeneration by iodoacetate.

MATERIALS AND METHODS

Adult domestic albino rabbits weighing 2.5-3.0 kg of both sexes delivered from a local breeder were used. All animals were treated in accordance with the ARVO Resolution on the Use of Animals in Research. Sodium iodoacetate(NaIA) in physiologic saline was given intravenously at a dose of 40 mg/kg after adjusting the pH to 7.0 by 0.01N NaOH. Except 15 min., 1h. and 4 hrs., the total dose was divided into two equal portions 6 hrs. apart. Eyes were enucleated immediately at the indicated time , the anterior portion of the eye and vitreous were removed and retinal tissue was collected using a spatula. All tissues were weighed and homogenized with 5 volumes of ice cold 0.25M sucrose solution containing 1mM EDTA (pH 7.0) by a glass-Teflon homogenizer. Subcellular fractionation was performed as described (Hayasaka, 1974). Crude homogenate and microsomal fraction were used for G6Pase assay, and soluble fraction was used for GS and γ-GT assays. G6Pase assays were performed at pH 6.5, 30°C as described previously (Baginski et al., 1974).The incubation time was not more than 10 min. which was in a linear range. Full disruption of the microsomal preparations was made as described (Lange et al.,1980). In Triton X-100 treated samples, 0.1%(w/v) final concentration of this detergent gave the maximum latency of G6Pase activity. Latency is the percentage of the activity of fully disrupted microsomes that is not expressed in untreated microsomes and calculated as described previously (Arion et al.,1975). Intactness of each microsomal fraction was assessed using 2mM mannose 6-phosphate(M6P) as a substrate (Nordlie et al., 1983). The total and specific activities are expressed as nanomol(nmol) inorganic phosphate(Pi) released/min and nmol Pi released/min/mg protein respectively. Glutamine synthetase (GS) and γ-glutamyl transferase (γ-GT) assays (Meister ,1985) were done at pH 6.8 and pH 6.0 respectively which were found to be the optimum pH in the rabbit retinal soluble fraction.For each assay appropriate blanks (for GS , - glutamate and - ATP and for γ-GT, - glutamine and - ADP) were made and subtracted from the full assay system. The total and specific activities are expressed as

γ-glutamyl hydroxamate(γ-GH) released/min and unit (U)/mg protein respectively. A unit(U) of GS activity is defined as the amount of enzyme that forms 1μmol of γ-GH for 15 min at 37°C (Meister, 1985). Total enzyme activities of G6Pase,GS and γ-GT were measured using two eyes of the same NaIA-treated rabbit as at least two eyes were required for differential centrifugation.The protein concentrations were determined by the method of Lowry et al. (1951) using bovine serum albumin as standard. Statistical analysis of the data was carried out by student t-test.

The glial fibrillary acidic protein(GFAP) immunostaining was performed by the avidin-biotin-peroxidase complex method (Hsu et al.,1981). After enucleation, the anterior portion of the eye ball was sliced off and fixed overnight with periodate-lysine-paraformaldehyde fixative(McLean and Nakane, 1974).The sections were cut at a 3μm thickness and digested with pepsin(0.046 g in 100ml 0.01N HCl) at 0°C for 2 min. Following treatment with 0.3% hydrogen peroxide-methanol for 15min. at room temperature they were incubated overnight at 4°C,with 1:100 diluted mouse anti-human GFAP, clone 6F2 (commercially purchased from Synbio,Uden),then with second antibody(biotin-conjugated horse anti-mouse IgG. 1:200 dilution,Vector Lab. Inc.,Burlingame,CA) overnight at 4°C.The sections were incubated with avidin-biotin-peroxidase complex for 1 hr at room temperature.The first antibody was omitted from control. Antibodies were dissolved in PBS containing 0.05% Tween 20 and 1% bovine serum albumin and sections were rinsed with PBS containing 0.05% Tween 20 between incubations.Color was developed by incubation in 50mM Tris-HCl,pH 7.6, 0.02% 3,3'-diaminobenzidine-HCl, and 0.005% hydrogen peroxide for 10 min. at room temperature.The sections were counterstained with methylgreen,dehydrated and mounted.

RESULTS

Intravenously given sodium iodoacetate(NaIA) in the rabbit selectively destroyed the visual cells.The pyknosis seen at the outer nuclear layer was marked by 24 hrs(Fig.1b) with the disruption of rod outer and inner segments.At 5 days(Fig.1c) the degenerative process was remarkable and by 4 weeks(Fig.1d) the visual cell layer was lost,though the inner layers of the retina seemed to be less affected from the morphological point of view by light microscopy.

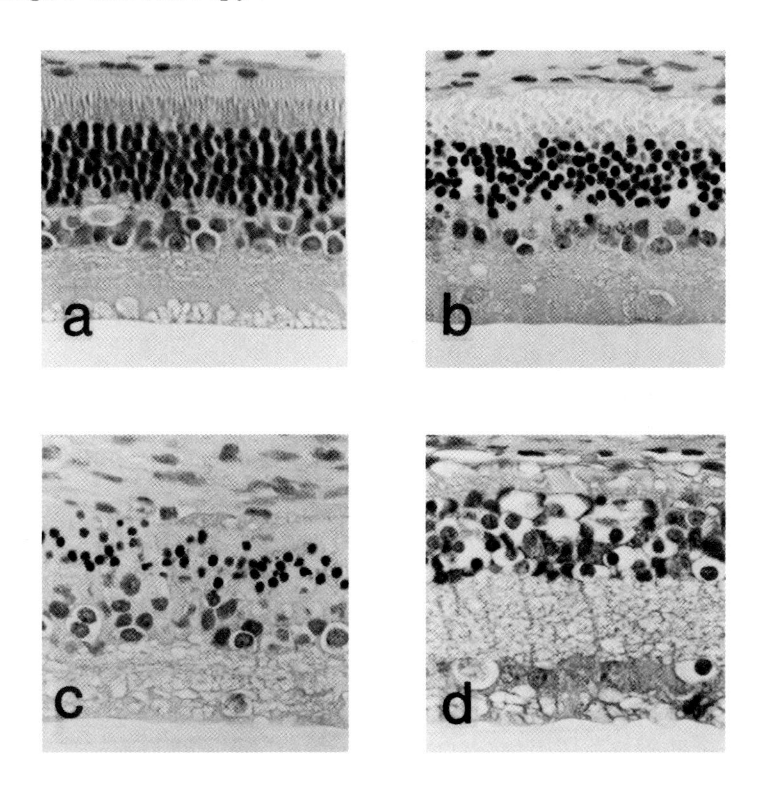

Figure 1. (a)Midperipheral retina of the control adult rabbit (b)One day after NaIA injection.Pyknosis at the outer nuclear layer and disrupted rod inner and outer segments are observed.Midperipheral retina.(c)Five days after NaIA injection. Visual cell degeneration is progressed. Midperipheral retina.(d)Four weeks after NaIA injection.The complete degeneration of the photoreceptor cell layer. Central retina.H.E. staining,x380.

The subcellular distribution study of G6Pase showed that 55.5±4.6%(n=4)(mean ± SD) of the total activity was present in the microsomal fraction of the control rabbit retina. Subcellular distribution of this enzyme in other fractions was as follows: nuclear(17.4±2.9%),mitochondrial(8.6±1.2%),lysosomal(18.5±4.0%), and no activity in soluble fraction.All fractions were supplemented with Triton X-100 as described in materials and methods. 82.6±13.5% activity of the starting homogenate was recovered in total subcellular fractions.It was found that the microsomal preperations of the control rabbit retina expressed 86.5±5.7%(n=4) intactness and 11.5±3.9%(n=4) latency towards to glucose 6-phosphate (G6P). No statistical significance was found between the control and NaIA treated rabbits regarding the intactness and latency to G6P by the method defined in materials and methods.

The total and specific G6Pase activities of the crude retinal homogenate after NaIA treatment was shown in Fig 2.The activities were indicated as the % of the control. The total activity of the control was 332±12 nmol/min(n=4) and the specific activity 21.4±1.9 nmol/min/mg protein(n=4). Slight increase was seen at both the total and specific activities of the retinal homogenate within first hours after NaIA treatment. On 20 days in contrast to increase of specific activity(148% of the control, $p<0.001$),the total activity was decreased to 72% of the control($p<0.001$).In the microsomal fraction of the control rabbit retina, the total and specific activities were found as 129±21 nmol/min (n=4) and 132±19 nmol/min/mg(n=4),respectively.The activity profiles after NaIA treatment were seen in Fig 3. The specific activity of G6Pase was 56% of the control by the 5th day ($p<0.01$).

Concerning the GS, the control retinal soluble fraction showed the total activity of 0.62±0.21 μmol γ-GH/min(n=4),and specific activity of 2.43± 0.55 U/mg. Total and specific activities of γ-GT in the control were measured as 4.1±0.2μmol γ-GH/min(n=3) and 17.2±4.1 U/mg, respectively. Both the total and specific activities of GS and γ-GT enzymes showed prominent decrease by the 20th day after NaIA treatment ($p<0.01$,Figs. 4 and 5).

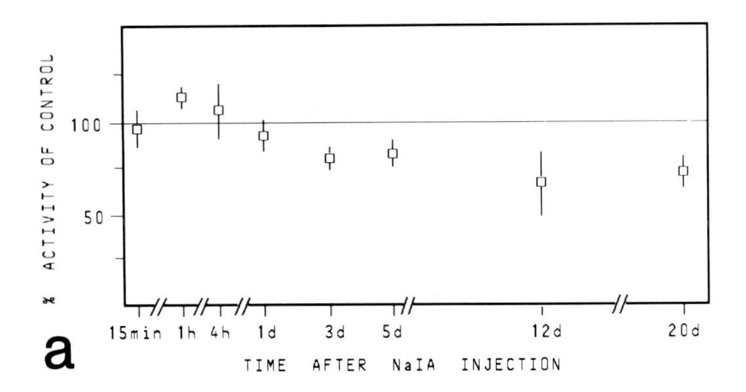

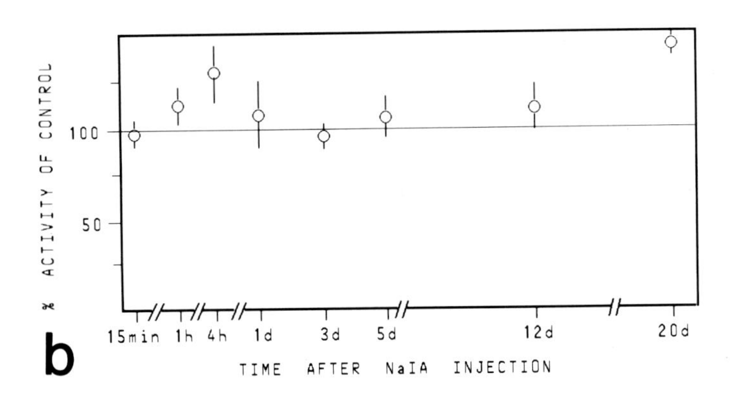

Figure 2. The total(a) and specific(b) activities of G6Pase assayed in crude retinal homogenate after NaIA injection.The activities are expressed as the percentage of the control. Each point represents the mean of three seperate experiments and the bars indicate the standard deviation.The assay conditions are given at materials and methods.

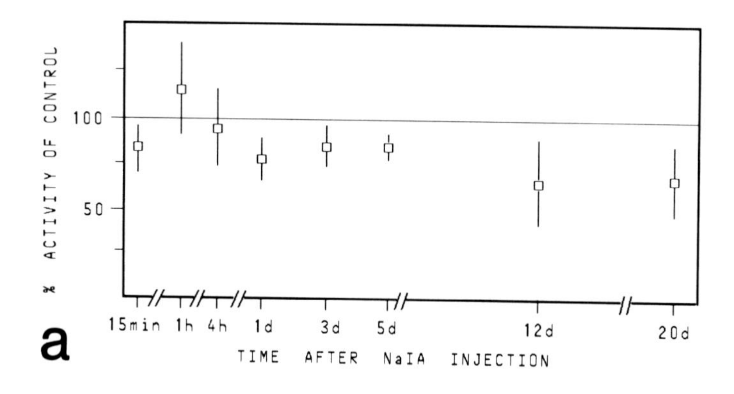

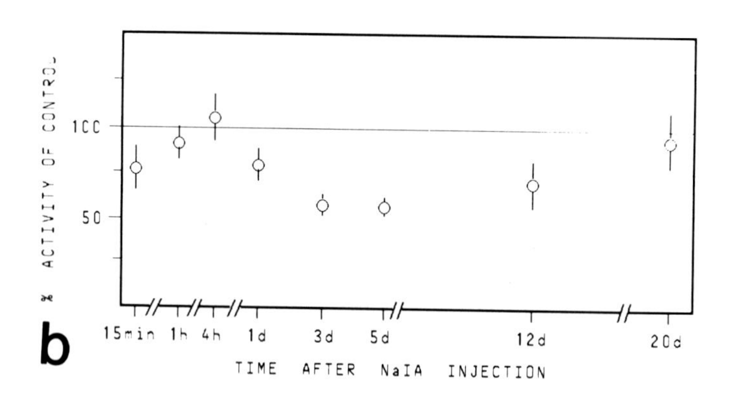

Figure 3.The total(a) and specific(b) activities of G6Pase assayed in microsomal fraction of the retina after NaIA injection.Further details are shown in the legend of Fig 2.

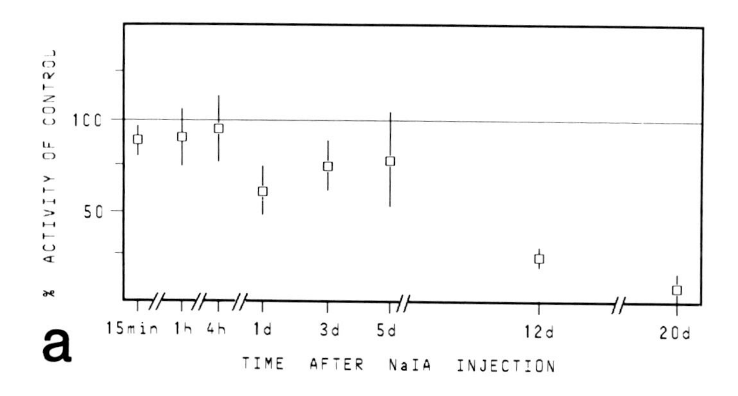

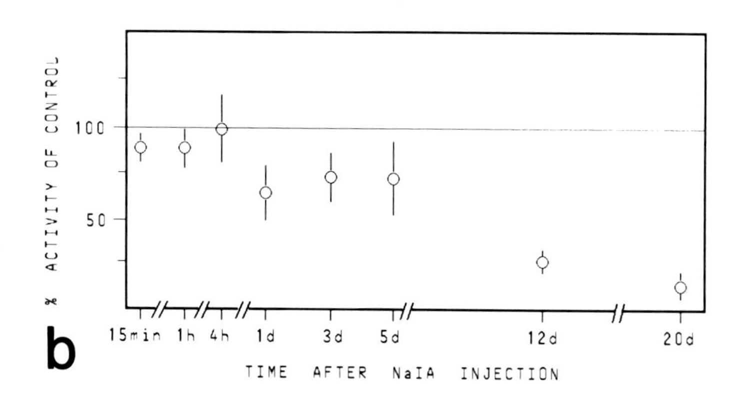

Figure 4.The total(a) and specific(b) activities of glutamine synthetase assayed in retinal soluble fraction after NaIA injection.Further details are shown in the legend of Fig 2.

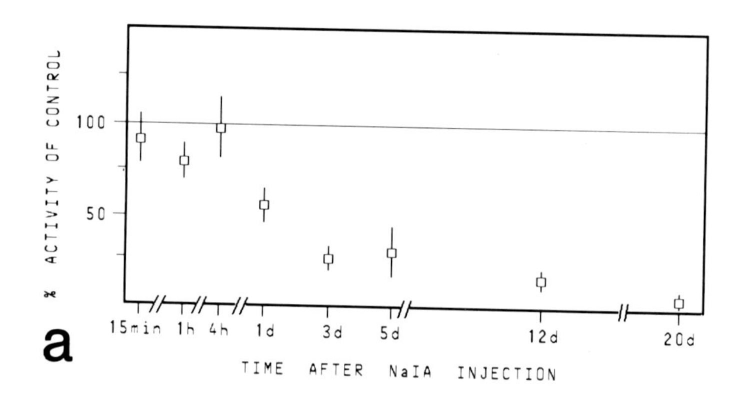

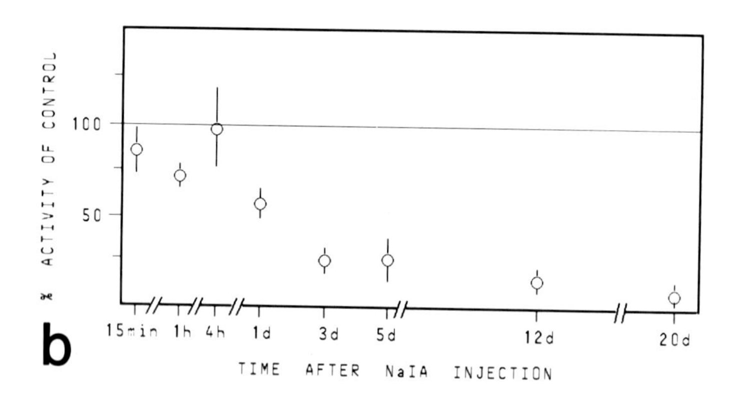

Figure 5.The total(a) and specific(b) activities of γ-glutamyl transferase assayed in the retinal soluble fraction after NaIA injection. Further details are shown in the legend of Fig 2.

GFAP immunoreactivity was expressed in Müller cells by the day 5th of iodoacetate retinal degeneration(Fig 6a).The first antibody-omitted adjacent sections didn't show any immunoreactivity(Fig 6b). By light microscopic immunohistochemistry,GFAP immunoreactivity was observed only in astrocytes and no staining was found in Müller cells of the control rabbit retina.

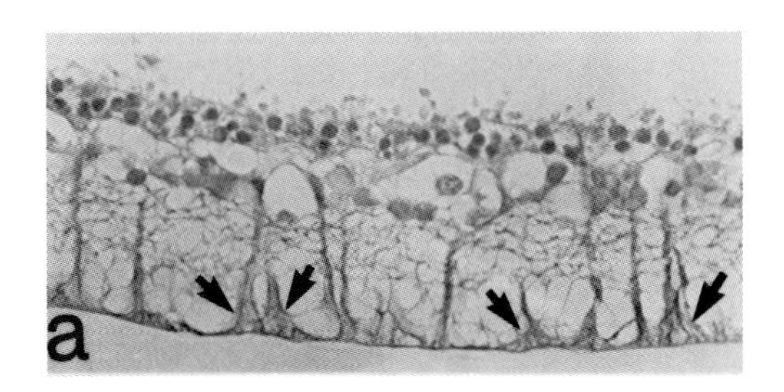

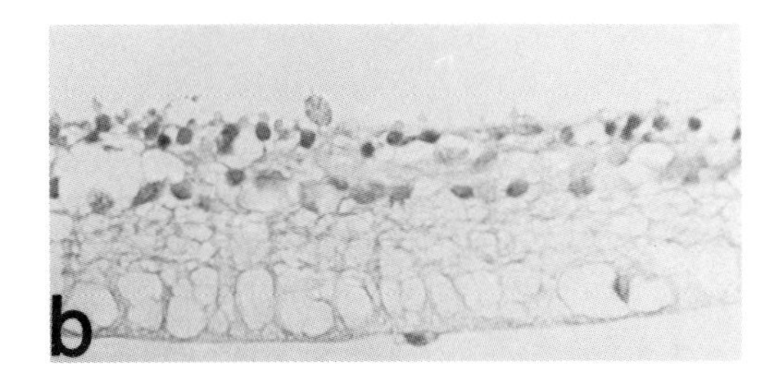

Figure 6. GFAP expression(a) in Müller cells(arrows) of the fifth day iodoacetate-degenerated rabbit retina.Control retina(b) does not show any immunoreactivity.Counterstained with methyl green, x380.

DISCUSSION

We have shown that specific activity of G6Pase in the crude homogenate of NaIA-treated rabbit retina was increased by the 20th day (Fig 2b) ($p<0.001$), whereas the activity was not increased in the microsomal fraction (Fig. 3b).Since retinal homogenates contain high activity of acid phosphatase (Hayasaka, 1974), significant levels of nonspecific phosphatases may contribute to hydrolyze G6P.However Burchell et al.(1988) reported that in microsomal preperations, the level of nonspecific phosphatases is about 3% of the normal adult level of G6Pase activity in liver. In our experiments, it is found that rabbit retinal G6Pase activity is mostly recovered in microsomal fraction (55.5±4.6% of total activity) which is in accordance with the previous results studied at liver and kidney (Nordlie ,1971). In addition, the optimum pH of the rabbit retinal G6Pase assayed at microsomal fraction lies between 5.5-6.0 in undisrupted preperations and in the absence of its substrate(G6P) retinal enzyme is very heat unstable after solubilization with Triton X-100 (Durlu et al., 1988). Pyridoxal 5'-phosphate which has been reported to inhibit liver phosphotranslocase(Gold and Widnell, 1976) also inhibits the rabbit retinal phosphotranslocase specifically at low concentrations (unpublished result).The latency towards to G6P(11.5±3.9%,n=4) and M6P(86.5±5.7%, n=4) was observed. It is thought that rabbit retinal G6Pase differs from nonspecific phosphatases by several distinct properties but it is apparent that microsomal fraction reflects the activity changes of G6Pase more correctly in the degenerated retina.

GS has several functions in the retina including the role in anabolic reactions, ammonia detoxification, pH regulation and possible role in neurotransmission by inactivation of glutamate released by photoreceptors (Pahuja and Reid,1985, Pahuja et al.,1985). Purification of GS from bovine retina has given evidence that the γ-GT activity is the catalytic function of the native enzyme(Pahuja et al.,1985). In the previous study of LaVail and Reif-Lehrer(1971) the GS activity of dystrophic mouse retina was measured as 55.6% of the normal one on postnatal day 18, although at the same time no significant difference of the GS specific activity was found. We observed prominent decline of both the total and specific activities of GS and γ-GT in iodoacetate retinal degeneration of the rabbit when the visual cell degeneration was progressed (Figs 4 and 5).

GFAP expression in Müller cells has been reported after perforating eye injury and optic nerve section (Bignami and Dahl, 1979), in retinal degeneration (Shaw and Weber, 1983; Eisenfeld et al.,1984) and after culturing of Müller cells(Wakakura and Foulds,1988). The amount of GFAP increase was demonstrated by Erickson et al. (1987) at the detached retina of Müller cells. Recently, Sarthy and Fu(1988) reported that in mouse photoreceptor degeneration, GFAP mRNA levels were elevated in Müller cells. Iodoacetate retinal degeneration in the rabbit also caused GFAP expression of Müller cells by the fifth day(Fig 6).We also observed that GFAP immunoreactivity was confined to astrocytes and no staining was found in Müller cells of the control(untreated) rabbit retina by light microscopic immunohistochemistry. These findings are in accordance with the previous results reported by the other authors (Schnitzer,1985; Wakakura and Foulds,1988).

It is concluded that, the glial markers of glycogenolysis/gluconeogenesis, glutamate-glutamine cycle and cytoskeletal protein metabolism of the retina are affected in iodoacetate retinal degeneration which causes photoreceptor cell loss.

ACKNOWLEDGMENTS

This study was supported in part by the Ministry of Education, Science and Culture,Japan and S.Mishima Fund. Y.K.Durlu is grateful to the Japanese Ministry of Education,Science and Culture for research scholarship and Retinitis Pigmentosa Foundation Fighting Blindness for travel fellowship.The authors wish to thank Dr.S.Hara for critical comments and Mrs. Y.Sagara,Miss E.Sato and Mr.Y.Abe for technical assistance.

REFERENCES

Arion WJ, Wallin BK, Lange AJ and Ballas LM (1975). On the involvement of a glucose 6-phosphate transport system in the function of microsomal glucose 6-phosphatase. Mol Cell Biochem 6(2):75-83.

Arion WJ, Lange AJ, Walls HE and Ballas LM (1980). Evidence for the participation of independent translocases for phosphate and glucose 6-phosphate in the microsomal glucose 6-phosphatase system. J Biol Chem 255(21):10396-10406.

Babel J and Stangos N (1973).Essai de correlation entre l'E.R.G. et les ultrastructures de la retine. 1.Action du monoiodoacetate.Arch Opht (Paris) 33(4):297-312.

Baginski ES, Foa PP and Zak B (1974).Glucose 6-phosphatase.In Bergmeyer HU(editor):"Methods of Enzymatic Analysis", New York. Academic Press, vol 1, pp.876-880.

Bignami A and Dahl D (1979).The radial glia of Müller in the rat retina and their response to injury.An immunofluorescence study with antibodies to the glial fibrillary acidic(GFA) protein. Exp Eye Res 28:63-69.

Broadwell RD and Cataldo AM (1983).The neuronal endoplasmic reticulum: its cytochemistry and contribution to the endomembrane system. I. Cell bodies and dentrites. J Histochem Cytochem 31(9):1077-1088.

Burchell A, Hume R and Burchell B (1988).A new microtechnique for the analysis of the human hepatic microsomal glucose-6-phosphatase system. Clinica Chimica Acta,173:183-192.

Durlu YK, Ishiguro S, Tamai M (1988).Some properties of rabbit retinal glucose-6-phosphatase.92nd Japanese Congress of Opthalmology Abstr.,Kyoto.

Eisenfeld AJ, Bunt-Milam AH and Sarthy PV (1984). Müller cell expression of glial fibrillary acidic protein after genetic and experimental photoreceptor degeneration in the rat retina. Invest Opthalmol Vis Sci 25:1321-1328.

Eng LF, Vanderhaeghen JJ,Bignami A and Gerstl B (1971).An acidic protein isolated from fibrous astrocytes.Brain Res 28:351-354.

Erickson PA, Fisher SK, Guerin CJ, Anderson DH and Kaska DD (1987). Glial fibrillary acidic protein increases in Müller cells after retinal detachment. Exp Eye Res 44:37-48.

Farber DB, Souza DW and Chase DG (1983).Cone visual cell degeneration in ground squirrel retina:Disruption of morphology and cyclic nucleotide metabolism by iodoacetic acid. Invest Ophthalmol Vis Sci 24:1236-1249.

Fishman RS and Karnovsky ML (1986).Apparent absence of a translocase in the cerebral glucose 6-phosphatase system.J Neurochem 46:371-378.

Gold G and Widnell CC (1976).Relationship between microsomal membrane permeability and the inhibition of hepatic glucose 6-phosphatase by pyridoxal phosphate. J Biol Chem 251(4):1035-1041.

Hayasaka S (1974).Distribution of lysosomal enzymes in the bovine eye.Jap J Ophthalmol 18:233-239.

Hsu SM, Raine L and Fanger H (1981).Use of avidin-biotin-peroxidase complex (ABC) in immunoperoxidase techniques: a comparision between ABC and unlabeled antibody procedures.J Histochem Cytochem 29:577-580.

Karli P (1954). Contribution experimentale a l'etude de la pathogenie des retinoses pigmentaires. Ophthalmologica 128:137-162.

Lange AJ, Arion WJ and Beaudet AL (1980).Type Ib glycogen storage disease is caused by a defect in the glucose 6-phosphate translocase of the microsomal glucose 6-phosphatase system. J Biol Chem 255(18):8381-8384.

LaVail MM and Reif-Lehrer L (1971).Glutamine synthetase in the normal and dystrophic mouse retina.J Cell Biol 51:348-354.

Lessell S and Kuwabara T (1964).Phosphatase histochemistry of the eye.Arch Ophthalmol 71:851-860.

Lowry OH, Rosenbrough NJ, Farr AL and Randall RJ (1951).Protein measurement with the Folin phenol reagent.J Biol Chem 193:265-275.

McLean IW and Nakane PK (1974).Periodate-lysine-paraformaldehyde: A new fixative for immunoelectron microscopy.J Histochem Cytochem 22:1077-1083.

Meister A (1985).Glutamine synthetase from mammalian tissues.In Meister A(editor):"Methods in Enzymology", New York. Academic Press, vol 113, pp.185-199.

Moscona AA and Linser P (1983).Developmental and experimental changes in retinal glial cells:cell interactions and control of phenotype expression and stability. Current Topics in Developmental Biology,vol 18,pp 155-188.

Noell WK (1951).The effect of iodoacetate on the vertebrate retina.J Cell Comp Phsiol 37:283-307.

Noell WK (1952).The impairment of visual cell structure by iodoacetate.J Cell Comp Phsiol 40:25-45.
Nordlie RC (1971). Glucose-6-phosphatase. In Boyer PD(editor):"The Enzymes",New York.Academic Press, vol 4,pp. 543-609.
Nordlie RC,Sukalski KA, Munoz JM and Baldwin JJ (1983).Type Ic, a novel glycogenosis. Underlying mechanism.J Biol Chem 258(16):9739-9744.
Orzalesi N, Calabria GA and Grignolo A (1970).Experimental degeneration of the rabbit retina induced by iodoacetic acid. Exp Eye Res 9:246-253.
Pahuja SL, Mullins BT and Reid T (1985).Bovine retinal glutamine synthetase.1.Purification, characterization and immunological properties.Exp Eye Res 40:61-74.
Pahuja SL and Reid T (1985). Bovine retinal glutamine synthetase. 2.Regulation and properties on the basis of glutamine synthetase and glutamyl transferase reactions. Exp Eye Res 40:75-83.
Polyak S (1957).The Vertebrate Visual System.University of Chicago Press,London.
Riepe RE and Norenburg MD (1977).Müller cell localisation of glutamine synthetase in rat retina. Nature 268: 654-655.
Sarthy PV and Fu M (1988).Induction of the GFAP gene in mice with retinal degeneration. Proceedings of the International Society for Eye Research,vol 5,p. 151.
Schnitzer J (1985). Distribution and immunoreactivity of glia in the retina of the rabbit. J Comp Neurol 240:128-142.
Schubert G and Bornschein H (1951).Spezifische Schadigung von Netzhautelementen durch Jodazetat. Experientia 7:461.
Shaw G and Weber K (1983).The structure and development of the rat retina: An immunofluorescence microscopical study using antibodies specific for intermediate filament proteins. Eur J Cell Biol 30:219-232.
Wakakura M and Foulds WS (1988).Immunocytochemical characteristics of Müller cells cultured from adult rabbit retina. Invest Ophthalmol Vis Sci 29:892-900.

Inherited and Environmentally Induced
Retinal Degenerations, pages 601–622

MACULAR PATHOLOGY IN MONKEYS FED SEMIPURIFIED DIETS

Lynette Feeney-Burns*, Martha Neuringer†, and Chun-Lan Gao*
*Mason Institute of Ophthalmology, University of Missouri-Columbia, MO and
†Oregon Regional Primate Research Center, Beaverton, OR

INTRODUCTION

We reported previously (Malinow et al 1980) that monkeys fed semipurified diets had maculas that lacked the characteristic yellow carotenoid pigment (xanthophylls) for which the macula lutea is named. The experimental diets in this study contained recommended amounts of calories, vitamins and minerals but no unrefined plant or animal products and therefore no xanthophylls. Fundus photographs of these axanthochromic maculas showed mottling of the retinal pigment epithelium (RPE), and fluorescein angiograms revealed numerous punctate and diffuse areas of hyperfluorescence without leakage; in these respects, they resembled human maculas in the early stages of age-related macular degeneration (AMD). Since an animal model of this disease is, as yet, unknown, we have carried out an investigation of the underlying defects responsible for the clinical picture in these maculas.

The eyes of several monkeys in each dietary category were obtained for histologic examination. We now report on the morphologic changes found in maculas of monkeys maintained for up to 14 years on semipurified diets (SPD) that were either deficient (SPD-D) or sufficient (SPD-S) in protein; both SPD used casein as the sole source of protein and contained no taurine. In addition, the macular findings in animals fed the two experimental dietary regimens are compared to those of monkeys living under the same conditions but maintained for similar periods of time on a control diet, i.e., a standard stock diet sufficient in protein, taurine and xanthophylls.

METHODS AND MATERIALS

Animals

The monkeys were maintained at the Oregon Regional Primate Research Center and treated in a humane manner according to the guidelines described in NIH Publication No 86-23. They were housed indoors in cages with a 12 hour light/dark cycle. The age, sex and other information concerning each monkey are given in Table 1.

The content of the three diets used in this study is given in Table II. The protein used in the SPD was vitamin-free casein which contained no measurable taurine (Hayes et al 1975).

The eyes of the monkeys were examined as described previously (Malinow et al 1980, Feeney-Burns et al 1981). Serial fundus photographs and fluorescein angiograms were obtained on several of the SPD animals, a final set being taken close to the scheduled time of termination of the experiment. A few animals of each SPD group died spontaneously or became ill and were euthanized. In some cases fixation of the histologic specimens was delayed, and the maculas of these animals (see Table 1) were not usable for morphometric analysis.

Tissue Preparation

All eyes were enucleated and then placed in a double-aldehyde fixative (Feeney-Burns et al 1981) prior to dissection; previous experience indicated that the retinas, particularly those of the protein deficient monkeys, were too fragile to withstand manipulation of the fresh eye. A slit was made at the equator to aid penetration of fixative. Eyes in fixative were shipped to Missouri where macular blocks were dissected after about 72 hours of fixation.

Maculas were bisected through the fovea and one half was post-fixed with osmic acid. The other half was processed without osmic acid so that the osmiophilic lipids of lipofuscin granules in the RPE would not obscure other electron dense components (e.g. melanin) of some granules.

Table 1. Animals studied, appearance of the macula in vivo, and comments on tissue specimens

PC#	Sex	Eye	Age	Macular Pigment	Comment*
Chow Diet					
9952	F	OD/OS	10	+	1
8980	F	OD	11	+	1
8712	F	OD/OS	14	+	1
Protein sufficient semipurified diet (SPD-S)					
7830	M	OD	7	-	2
7895	M	OS	8	-	2,4
7910	M	OS	9	-	1
5983	M	OD	10	-	3
7802	F	OD	10	-	1
7848	M	OS	10	-	4
5943	M	OD	13	-	4
5988	F	OS	13	-	1
6012	M	OS	13	-	1
5935	F	OD	13	-	1
Protein deficient semipurified diet (SPD-D)					
7798	M	OS	9	-	2
5834	F	OS	7	-	2,4
5954	M	OD	10	-	3
7800	M	OD	10	-	1
5990	F	OS	12	-	2
5936	M	OD	14	-	1
5939	M	OD	14	-	1

*1 - Animal well at termination, excellent specimen for morphometry.
2 - Animal ill but specimen usable for morphometry.
3 - Animal died. Delay in tissue fixation. Specimen not usable for morphometry.
4 - Neural retina detached from RPE during tissue processing

Table 2. Diets used in this study[A]

	Chow[B]	SPD-S	SPD-D
% of calories:			
Protein	15.6	13.9[C]	3.7[C]
Carbohydrate	73.0	50.4[D]	60.6[D]
Fat	11.7	35.7[E]	35.7[E]
Fiber g/100g	4	7.4	7.4
Vitamin A U/gm	30	7.9	7.9
Other carotenoids[F] mg/gm	22.43[G]	0.5	0
α-tocopherol mg/kg	55	63.5	63.5
Ascorbic acid mg/kg	750	318	318
Other vitamins, salts, etc.[A]			

[A]SPD-S, protein sufficient semipurified diet; SPD-D, protein-deficient semipurified diet (Malinow et al, 1980).
[B]Purina Monkey Chow[R] supplemented with fresh fruit
[C]Vitamin-free casein
[D]Corn starch, sucrose
[E]Corn oil, changed to butter plus corn oil at age 4-7 yr (see Portman et al 1981).
[F]Xanthophylls, see Malinow et al, 1980 for details of analyses
[G]From alfalfa, soy beans, etc.

Specimens were flat embedded in epoxy resin so that the foveola was present in the first sections from each of the two macular blocks. Serial 1 μm thick sections were taken at 10 μm intervals (step serials) through the fovea. Sections were stained with toludine blue, then examined and photographed by light microscopy. Block faces were trimmed by cutting through the center of the foveola and including as much lateral parafoveal retina as possible for thin sectioning. Electron micrographs were made using a JEOL 1200 EX instrument.

Morphometry

Counts of photoreceptor nuclei and RPE cells were done using a Zeiss Photoscope I equipped with a 40X oil immersion planapochromat objective and an eyepiece measuring grid. For counts of foveal photoreceptor nuclei, a 1 μm wide area at the foveal pit was counted for each monkey. For RPE cell counts, three to five slides of 0.5 μm thick sections passing through the foveola at 20 μm progressive depths were used for each specimen.

RESULTS

In Vivo Findings

Clinical examinations revealed that both groups of animals fed semipurified diets lacked yellow luteal pigment in their maculas (axanthrochromia) owing to the absence of vegetable pigments in their diets (Table 1). As reported earlier (Malinow et al 1980), these axanthochromic maculas also showed mottling of the RPE and scattered non-leaking hyperfluorescent areas in the early phase of the fluorescein angiogram. The examples published in that report remain representative of the clinical findings in this larger sample of rhesus monkeys. Two of the monkeys whose color photographs and matching fluorescein angiograms were published in that report (#5988, SPD-S and #5834, SPD-D) were also part of the present study, their eyes being obtained 2 and 5 years respectively after those clinical pictures were taken.

Comparison of the fundus photographs and fluorescein angiogram groups showed several differences among the three dietary groups. In general, the maculas of monkeys on the control diet had less pigmentary mottling and fewer window defects than did the maculas of the SPD monkeys (Figure 1); however, control animals were not entirely free of such abnormalities. The areas of hyperfluorescence and the number of bright spots in the macula were clearly greater among the SPD groups of animals than among those on control diet. Also, the hyperfluorescence in the SPD monkeys was somewhat brighter owing to the lack of macular pigment that normally helps to screen the choroidal flush.

In addition, the maculopathy of the SPD-D animals showed a greater rate of change than that of the SPD-S animals. This difference is shown in serial fundus photographs (Figure 2) and fluorescein angiograms. The number of bright yellow spots visible in the fundus photographs, and seen as punctate hyperfluorescence in the fluorescein angiograms, increased during a two year period in the maculas of SPD-D monkeys (Figure 2C,D) whereas the few spots in the SPD-S monkey's macula remained stable (Figure 2A,B). The basis for the differences in magnitude of punctate and confluent areas of hyperfluorescence was revealed by subsequent histologic examination of the macular tissue.

Histopathology

Tissue from the SPD animals were very fragile compared to that of the chow fed monkeys. Eyes from the deficient animals were particularly difficult to enucleate without damage to the globe. The unusual delicacy of these specimens was one of the most apparent differences between the SPD-S and SPD-D groups of monkey eyes.

Low magnification histologic sections through the foveas of specimens from the three groups did not reveal differences in retinal thickness, cell number, or abnormal structural features such as drusen or absence of RPE cells (Figure 3). However, the maculas of the SPD monkeys had a greater frequency of vacuolated RPE cells than did maculas of monkeys on the chow diet (Figures 3,6,7,8). Moreover, there were greater numbers of these cells in the maculas of those on the SPD-D diet than in the monkeys on the SPD-S diet (Figures 6B and C). As documented in Feeney-Burns et al (1981), these cells, being refractile to incident light and also more transparent, produced the bright yellow spots in the fundus photographs and the punctate hyperfluorescence in the fluorescein angiograms mentioned above. Their morphology is described further below.

Comparison of the morphology of photoreceptors among the three groups revealed little or no differences in the inner or outer segments. Alterations such as vesiculation, segmentation, and dishevelment of the disks were found in

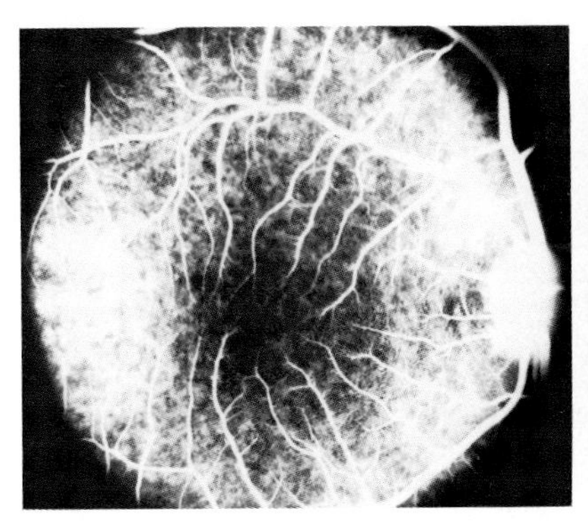

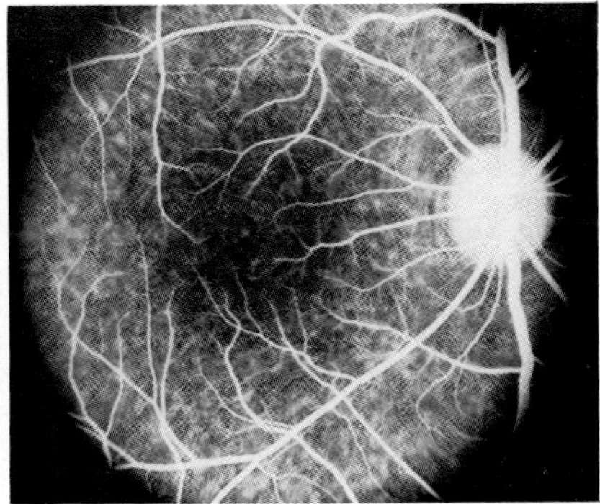

Figure 1. Fluorescein angiograms of monkeys in the 3 dietary groups.
(A) Monkey fed chow diet. The uniformly dark macular region is due to high concentration of RPE melanin and yellow macular pigment. (B) Monkey fed protein sufficient semipurified diet (SPD-S). Fluorescein angiogram shows hyperfluorescence throughout the macular region.

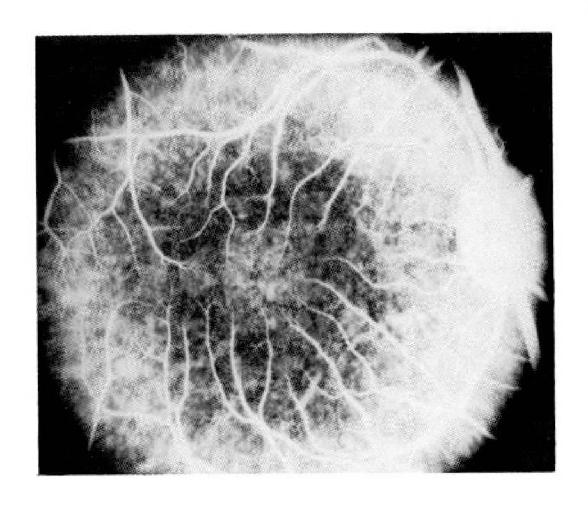

(C) Monkey fed a protein deficient semipurified diet (SPD-D). Fluorescein angiogram shows diffuse hyperfluorescence of macular region and numerous small bright spots.

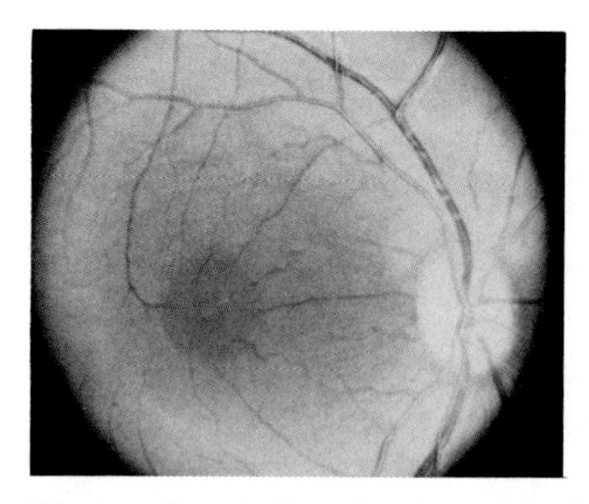
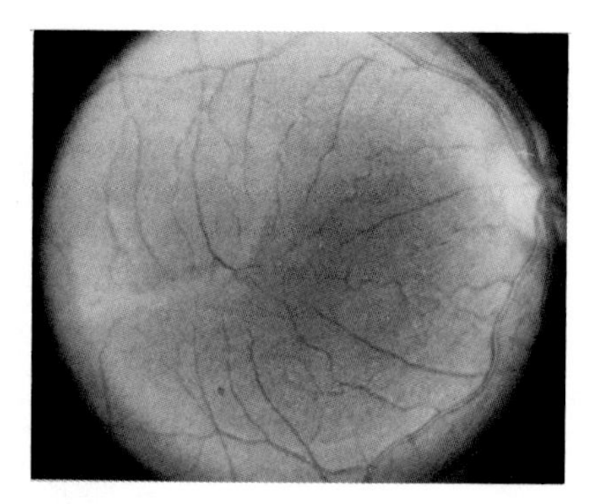

Figure 2. A,B. Fundus photographs of the left eye of an SPD-S monkey taken two years apart (monkey #7830, at 5 and 7 yrs of age respectively). Since macular pigment is lacking, the darkness of the macular region is due to RPE pigments. Note diminution in this pigment in B.

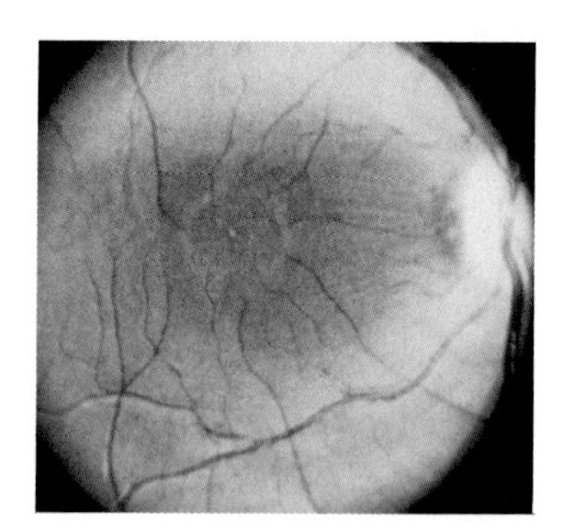
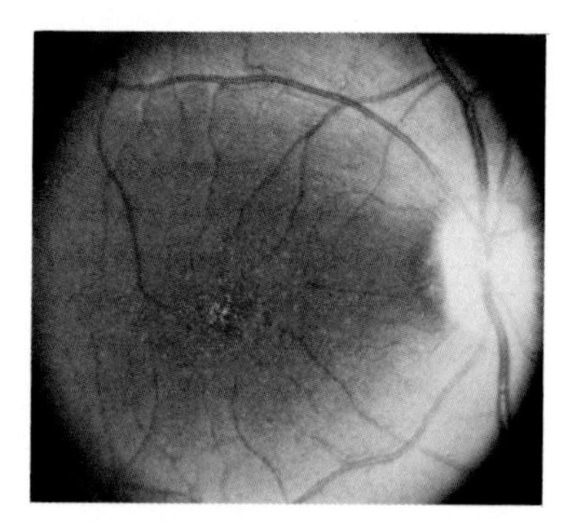

C,D. Similar series to A,B, in SPD-D monkey (monkey #5834, at 8 yr and 10 yr of age respectively). Tiny spots, apparent in the first photograph, have increased in number and confluency two years later.

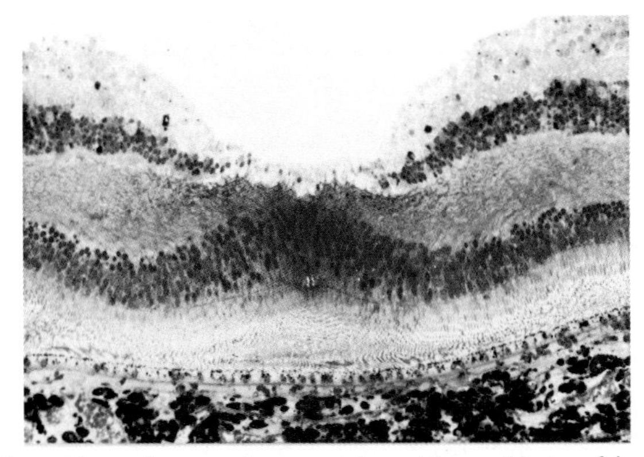

Figure 3. Foveal sections. A. Chow diet, 14 yr.

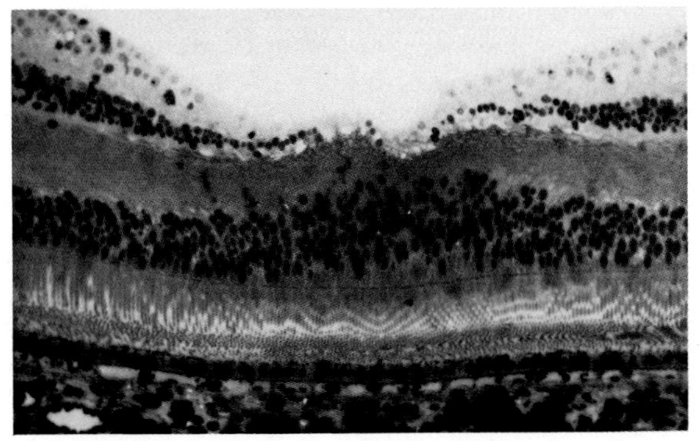

Figure 3B. SPD-S, 10 yr. Monkey 7802

Figure 3C. SPD-D, 14 yr. Monkey 5936. Tissue does not stain well in this group of specimens.

the SPD groups, but these putative abnormalities were also seen in the chow control specimens (Figure 4). Loose fitting outer segment plasma membranes were common in cones of all groups. A comparison of lengths of the outer segments in the various groups could not be accurately carried out owing to the tendency for tissue to sag at this layer of the retina; nonetheless, it was apparent that despite long-standing protein deficiency, the outer segments were present in about normal dimensions (Figures 3, 4 and 7). Also, the interphotoreceptor matrix was present, and apparently normal, even after 14 years of the protein deficient diet (Figure 4).

Differences among the three groups of monkeys were found in the outer plexiform layer, specifically in the cone pedicles (Figure 5). In both the SPD-S and SPD-D monkeys foveal and perifoveal cone synaptic terminals were enlarged and round rather than pyramidal in shape and contained lipid whorls (Figure 5). The lipids melted in the electron beam causing tears in the sections. A lucent zone of cytoplasm was incorporated in the lipid whorl and this assembly often appeared to perforate the plasma membrane of the photoreceptor. Interestingly, these alterations were more pronounced in the SPD-S animals than in the SPD-D animals; the lipid whorls were smaller and less disruptive in monkeys on the deficient diet. The areas of vacant cytoplasm in the cone pedicles in the SPD-D specimens were clearly derived from swollen mitochondria (Figure 5C); the exact origin of the larger vacant areas in the SPD-S specimens could not be determined. In several SPD-D specimens gaps occurred in the row of synaptic terminals (data not shown), suggesting that photoreceptors had been lost.

The photoreceptor axons comprising Henle's layer showed no differences in membrane ultrastructure when the three dietary groups were compared. However, the retina of several SPD-S and SPD-D monkeys split (retinoschisis) near the junction of the axons with the synaptic terminals. This was never seen in the chow fed monkeys. The degree to which the retinoschisis was a mechanical artefact resulting from the fragility of the SPD specimens could not be determined.

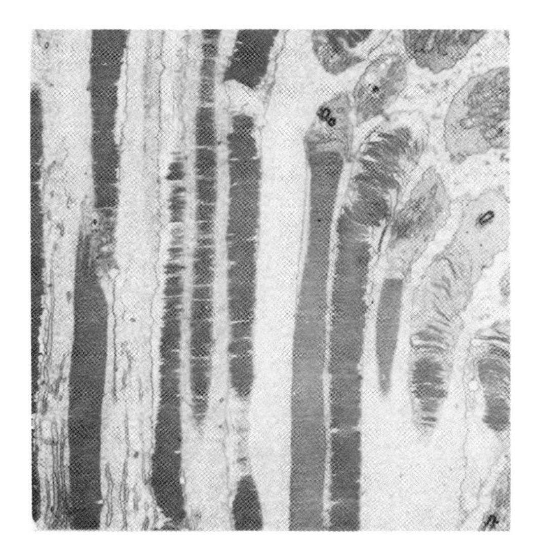

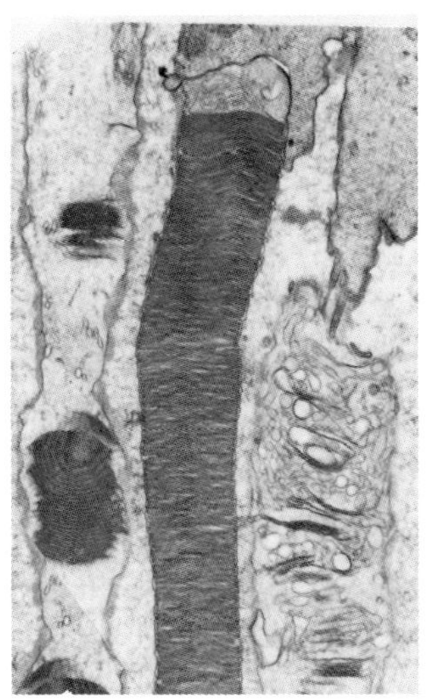

Figure 4. Comparison of foveal cone outer segments in monkeys fed for 14 yrs with chow (A,B) or SPD-D (C). Note similarity of loose plasma membrane sleeves and disk misalignments, and presence of IPM.

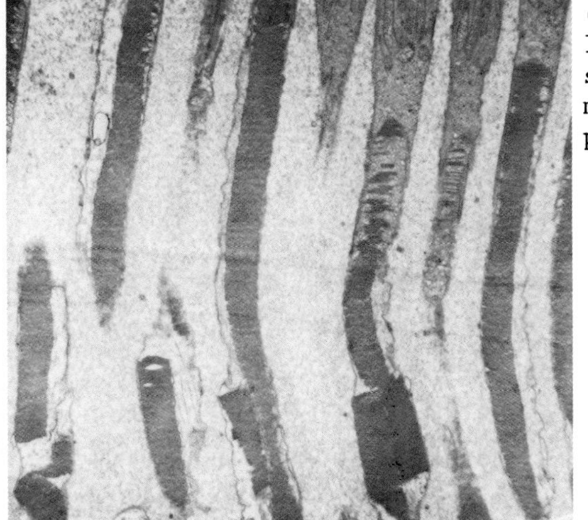

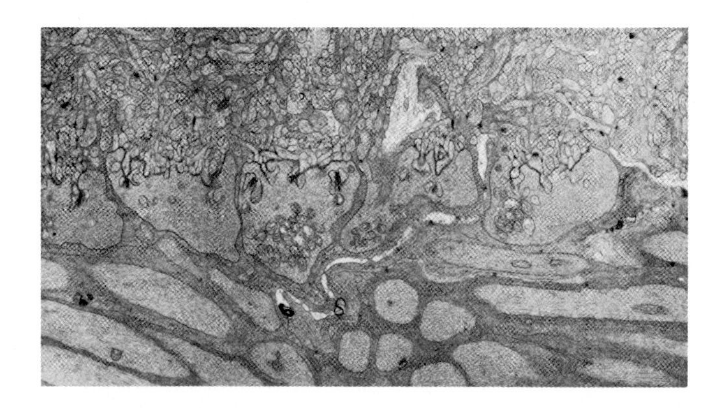

Figure 5. Comparison of foveal cone synapses. A. Chow

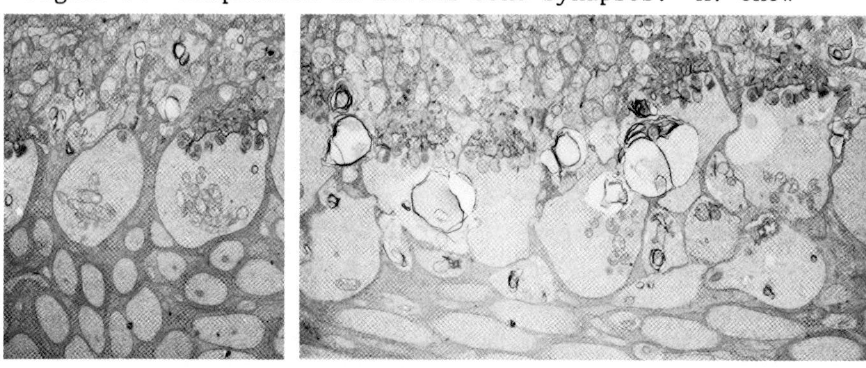

B. SPD-S 14 yr. Foveal (left) parafoveal (right)

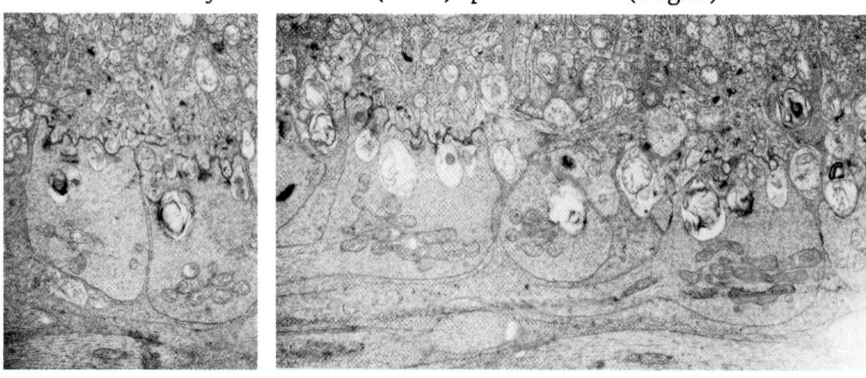

C. SPD-D 10 yr. Foveal and parafoveal synapses.

The most striking finding in the RPE layer was the presence of cells filled with vacuoles of fairly uniform size (Figures 3B,C, 6B,C, 7, 8). These vacuolated RPE cells were found in increased numbers in both groups of the SPD monkeys compared to the chow fed controls, moreover SPD-D specimens had more than SPD-S specimens, as stated above. Oil red O staining of frozen sections of a single specimen (data not shown) confirmed that the contents of the vacuoles were lipids. They were not fixed by osmic acid treatment and therefore they were dissolved by the solvents used during dehydration of the tissue; thus numerous vacuoles were left in their place. In the vacuolated RPE cells the larger vacuoles (1.3-1.4 μm) appear to form by coalescence of smaller (0.2-0.7 μm) vacuoles and vacant spaces in the basal cytoplasm. The vacuoles appeared to be delimited by tangles of smooth endoplasmic reticulum (SER) rather than by a true limiting membrane. The SER had lost its distinct osmiophilic structure and appeared amorphous as vacuoles increasingly consumed the cytoplasmic space. Mitochondria and basal infoldings of these RPE cells seemed unaffected by the initial vacuolization process. Melanosomes were often clustered within autophagic vacuoles; this suggests a basis for the loss of the dark disk of macular pigmentation noted in serial fundus photographs (Figure 2B).

Chow fed monkey specimens showed little or no difference between foveal and perifoveal RPE cells regarding either cell shape or their moderately low content of lipofuscin (Figures 6,9). In contrast, differences between foveal and perifoveal RPE cells were apparent in the SPD animals, particularly with regard to their lipofuscin granules (Figure 6). RPE cells of the fovea were abnormally short or cuboidal, contained fewer lipofuscin granules than non-foveal cells, and the cytoplasm, by light microscopy, had a muddy or amorphous appearance (Figure 6B). In contrast, RPE cells of the perifoveal region had a more normal columnar shape and were engorged with distinctly visible, densely stained lipofuscin granules (Figure 6B, lower). This distinction between the two zones of the macula was more pronounced in the SPD-S group than in those on the protein deficient diet; specifically, the RPE cells of SPD-S animals had extremely sparce cytoplasmic space because of the enormous number of lipofuscin granules (Figure 10). The SPD-D specimens, by comparison, had perifoveal cells with large areas of cytoplasm visible despite the obvious lipofuscin load (Figure 6C).

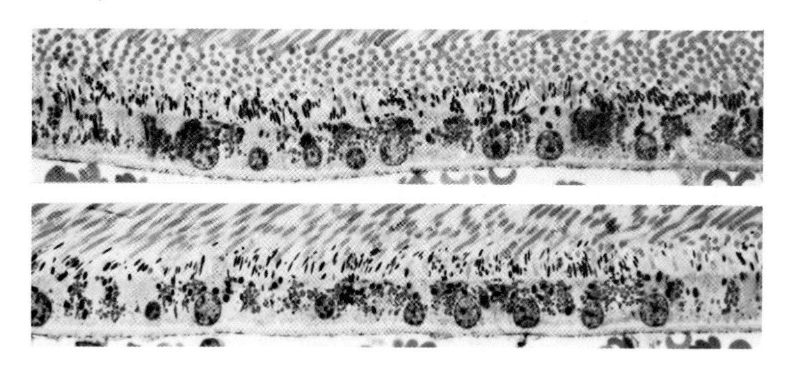

Figure 6. RPE of fovea (upper) and perifovea (lower) in the three dietary groups. All at X600 A. Chow, virtually no difference in appearance of cells in the two locations.

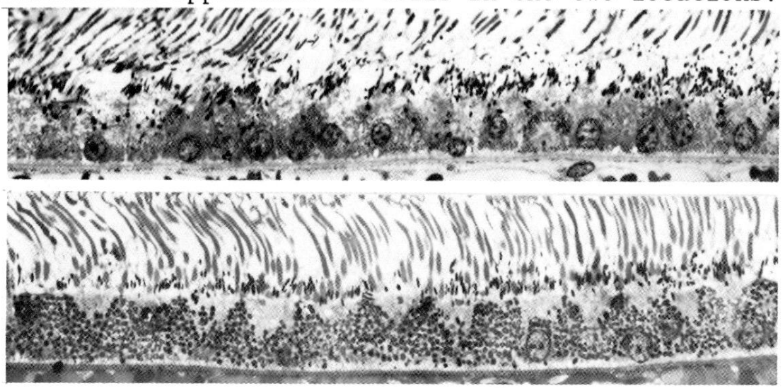

B. SPD-S 14 yr. Foveal RPE cytoplasm is amorphous, perifoveal, engorged with lipofuscin granules.

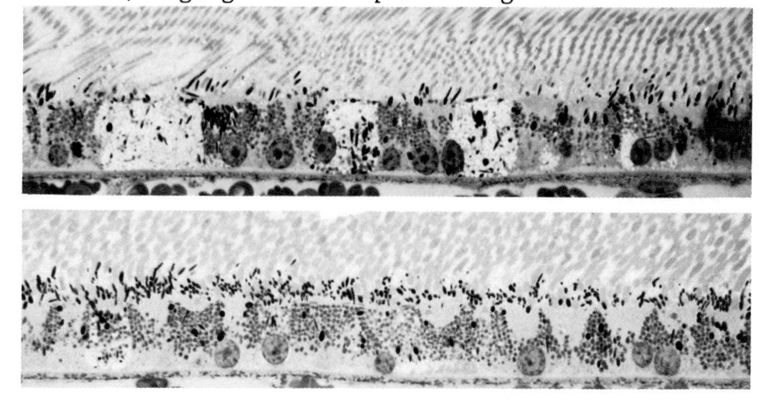

C. SPD-D 14 yr. Fovea, vacuolated RPE; perifoveal lipofuscin abundant.

Figure 7. Vacuolated RPE in foveal region. These cells were found in every specimen where punctate window defects were seen on fluorescein angiography. Note good morphology of neural retina, moderate disturbance of Henle's layer and cone synapses (arrow, inset), and absence of pathology in Bruch's membrane. Light micrograph X 220, X 800. Monkey 5988.

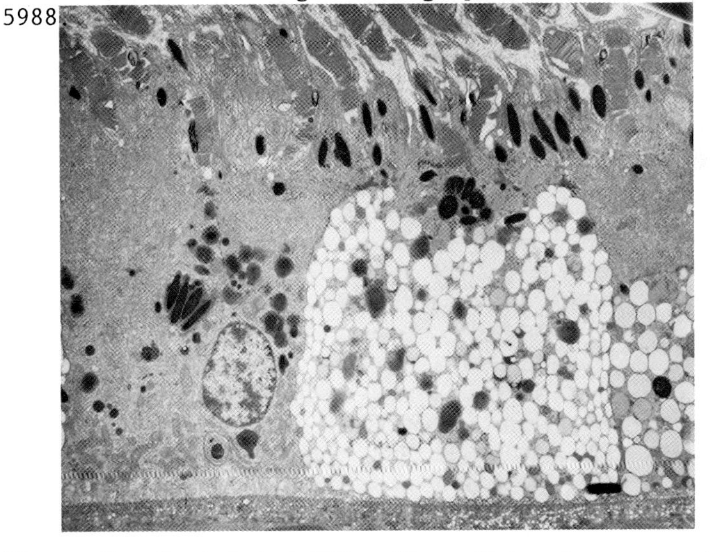

Figure 8. Vacuolated RPE cells. Note also sequestration of melanin granules in autophagic vacuoles. Monkey 5834. X2800.

Morphometry

Figure 11 illustrates differences among the dietary groups in the number of RPE cells per linear millimeter of macular sections through the foveola. The data are plotted according to the age of the animal. Because of technical difficulties detailed in Table 1, only 5 animals comprise each experimental group of data. The number of RPE cells per millimeter was less in the combined SPD groups compared to the age-matched chow fed monkeys with a borderline p value of 0.052. Despite the similarity in the values for the SPD-S and the SPD-D groups, when the two SPD groups were analyzed separately the p value was 0.025 (Kruskal-Wallis test) for the SPD-D monkeys but did not reach statistical significance for the SPD-S group. Additionally, when the data were analyzed for the effect of age, only the SPD-D monkeys showed a significant loss of cells with age ($p=0.0023$).

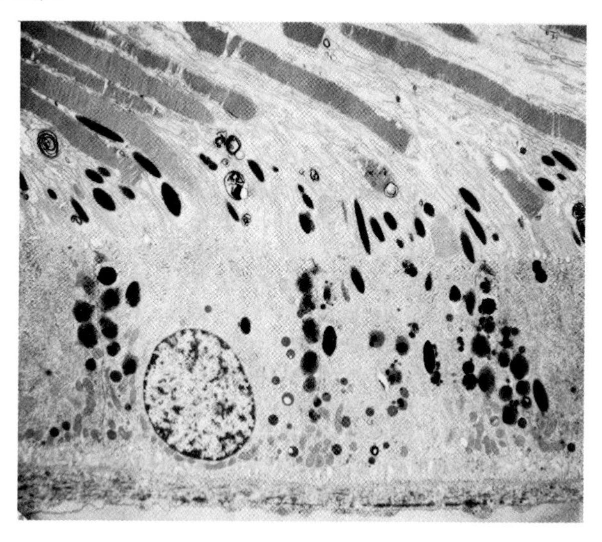

Figure 9. RPE and foveal cone outer segments of chow fed monkey, 14 yr old. Note lipid deposits amid apical microvilli and moderate lipofuscin in RPE. X 2800

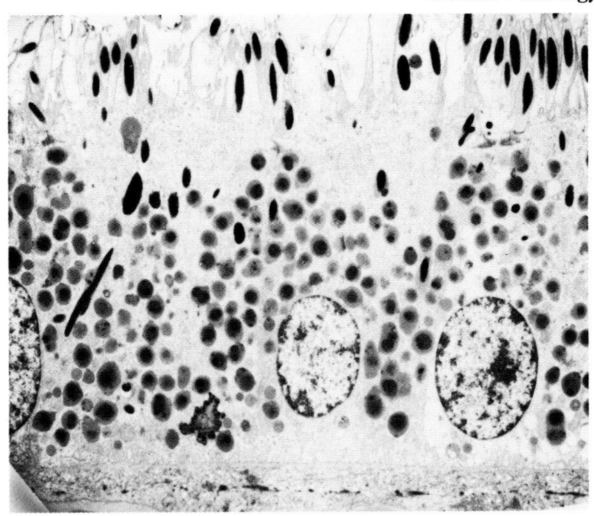

Figure 10. Electron micrograph of perifoveal RPE of a 14 yr old SPD-S monkey (5988); light micrograph shown in Figure 6B. Quantity and quality of lipofuscin granules differ from control, Figure 9. X 2800.

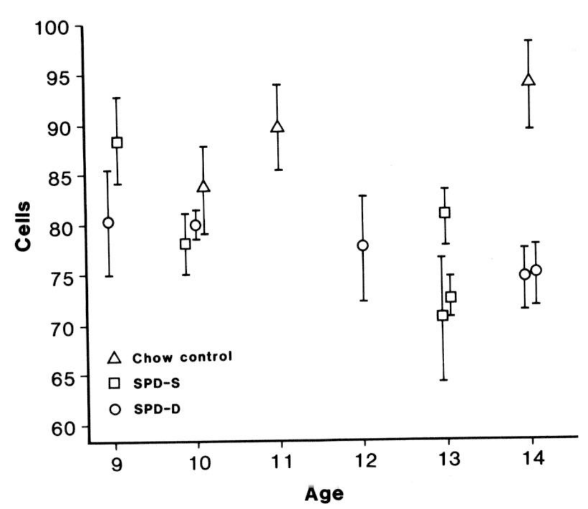

Figure 11. Means and standard deviations of cell counts from specimens of the three dietary groups. Data were obtained by light microscopy from 3-5 step serial sections, 20 um apart, through the foveola.

The number of photoreceptor nuclei in a millimeter of sectioned foveal retina, such as that shown in Figure 3, were 472 +/- 44 for the chow group, 476 +/- 22 for the SPD-S group, and 508 +/- 79 for the SPD-D group (means +/- SD). These differences were not statistically significant.

DISCUSSION

Monkeys fed semipurified diets for up to 14 years showed a number of abnormalities in macular morphology, the most prominent being vacuolization of macular RPE cells. This appears to be a non-specific morphologic change in primate retina. The defect was initially reported in a group of female monkeys that received contraceptive therapy (Fine 1981), but was also found by us in monkeys on atherosclerotic semipurified diets (Feeney-Burns et al 1981) and has also been reported in humans (El Baba et al 1986). The abundance of these cells in monkeys fed SPD diets suggests that this morphology reflects a common, presumably terminal, degenerative pathway in these cells. We examined sites internal and external to the RPE layer, i.e. the interphotoreceptor space and Bruch's membrane, looking for cell remnants, but no evidence of the presumed in situ cell death was found. Neither absence nor atrophy of RPE was observed nor were drusen seen in any macular specimens. The observed cuboidal shape of foveolar RPE cells of SPD animals suggests that the normally columnar cells may have spread out and flattened to cover denuded sites on Bruch's membrane as RPE cells died.

The documented loss of macular RPE cells in monkeys fed semipurified diets, which was preceded by vacuolization (lipoidal degeneration) of these cells and followed by shortening and flattening of foveal RPE, confirms a long suspected biological process underlying the phenomenon of hyperfluorescence in fluorescein angiograms; this process resembles that of the well studied corneal endothelial cell loss and thinning. Thinning and flattening of RPE, rather than total absence of cells, accounted for some of the observed hyperfluorescence; sequestration of melanin granules in autophagic vacuoles and their probable gradual digestion accounted for additional transparency to fluorescein at the RPE level. Whether or not the abnormal amounts of lipofuscin granules in the RPE cells (particularly noticeable in perifoveal RPE that were still columnar in shape and not affected by vacuolization) contribute to the degeneration of these cells remains to be determined. Massive accumulation of lipofuscin granules in RPE cells is virtually the hallmark of aged human RPE and a feature that has been thought to contribute to thickening and other age-related changes in Bruch's membrane (Feeney-Burns and Ellersieck, 1985). The lack of abnormalities in Bruch's membrane of the monkey eyes in this study, particularly the absence of macular drusen, may help us determine causal elements in the complex etiological history of human age-related macular degeneration.

We showed previously (Malinow et al 1980) that lack of vegetable matter in these SPD diets resulted in absence of normal yellow macular pigmentation, together with the virtual absence of xanthophylls in the plasma. Xanthophylls are normally present throughout the primate retina, with cone-rich areas having high concentrations of lutein and rod rich areas having more zeazanthin (Bone et al 1988, Handelman et al, 1988). The xanthophylls are reported to be in the membranes of the photoreceptor axons in Henle's layer (Bone and Landrum, 1984, Snodderly et al 1984). The biological function(s) of the macular pigment is unknown. It is presumed to enhance visual acuity by absorbing the out-of-focus blue image produced by chromatic aberration; however, such enhancement has not been demonstrated. Also, it has been hypothesized that it has an antioxidant role. These carotenoids are tetraterpene hydrocarbons with structural similarity to vitamin A; whether or not they are attached to proteins in the membrane and/or have a membrane stabilizing effect remains to be determined. Maculas that lacked the yellow pigment were very fragile and prone to split at the site of normally highest xanthophyll concentration even in the protein-sufficient specimens. This suggests that these compounds may affect membrane integrity. The consistent inability to achieve good stain density in tissue sections from SPD-D monkeys suggests that the quantity and/or quality of normally stainable molecules was deficient in these retinas. Although no ultrastructural membrane thinning was discernible in Henle's layer in the xanthophyll deficient SPD groups, the cone pedicles were swollen, possibly as a result of as yet unidentified plasma membrane abnormalities.

The absence of dietary taurine may also have contributed to some of the observed morphological abnormalities. More than a decade ago, taurine deficiency was shown to produce retinal and tapetal degeneration in cats (Schmidt et al 1976). The animals in the present study were started on their dietary regimen before taurine deficiency effects on primate retinas were suspected, but recent work has shown that such effects exist, at least in developing monkeys. Infant rhesus monkeys fed formulas with little or no taurine have reduced plasma taurine levels and morphological changes in photoreceptor outer segments, inner segments, synaptic terminals and RPE cells, as well as reduced visual acuity (Imaki et al 1987, Neuringer et al 1987). Adult monkeys, unlike infants, appear not to require dietary taurine because they possess adequate ability to synthesize it from essential sulfur-containing amino acids. Thus, the monkeys fed the taurine-free SPD-S diet showed no reduction in plasma taurine concentrations compared to chow-fed monkeys (Neuringer et al 1985). Monkeys receiving the protein deficient diet, however, lacked an adequate supply of these precursors, and this group showed a 75% reduction in plasma taurine. Thus, specific changes in the SPD-D group may have been related to long-term taurine depletion as well as to protein deficiency per se. Particularly striking are the similarities between taurine-deprived infant monkeys and the SPD-D monkeys in the presence of swollen mitochondria associated with vacuoles in the cone pedicles. On the other hand, disorganization and vesiculation of outer segment disc membranes, the most prominent change noted in taurine deprivation, was not seen frequently or consistently in the SPD-D group; moreover, these "abnormalities" were seen occasionally in all the groups, including chow controls (Figure 4B). Indeed, several other examples of such abnormal cone outer segment morphology can be found in retinas of monkeys on apparently normal, or at least not taurine deficient diets, as for example in our previous study of maculopathy in cynomologus monkeys (Feeney-Burns et al 1981, Figure 10), in aging monkeys (our unpublished data), in other cases in the literature, e.g. in normal rhesus monkeys (Anderson et al 1980), and interestingly, in hibernating ground squirrels (Reme and Young, 1977).

Despite the presence of some abnormalities in the maculas of these monkeys, perhaps the most striking finding is the relative normality of retinal neurons and RPE cells even in long-term combined deficiencies of xanthophylls, protein, and taurine. Visual function was also preserved, as demonstrated by the absence of an impairment in visual acuity in a subset of these animals (Neuringer et al 1984). It has been a consistent finding that the central nervous system is relatively resilient, compared to other organs, in the face of malnutrition, and this generalization appears to apply to the retina as well.

ACKNOWLEDGMENTS

This work was supported by the following grants: EY 03274 and Research to Prevent Blindness (Departmental) to LFB), and HD 07649, HL 09744 and RR 00163 to MN and/or the ORPRC. This is publication no. 1606 of the ORPRC.

REFERENCES

Anderson DH, Fisher SK, Erickson PA, Tabor GA (1980). Rod and cone disc shedding in the rhesus monkey retina: A quantitative study. Exp Eye Res 30:559-574.

Bone RA and Landrum JT (1984). Macular pigment in Henle fiber membranes. Vis Res 24:103-108.

Bone RA, Landrum JT, Fernandez L, Tarsis SL (1988). Analysis of the macular pigment by HPLC: Retinal distribution and age study. Invest Ophthalmol Vis Sci 29:843-849.

El Baba F, Green WR, Fleischmann J, Finkelstein D, de la Cruz ZC (1986). Clinicopathologic correlation of lipidization and detachment of the retinal pigment epithelium. Amer J Ophthalmol 101:576-583.

Feeney-Burns L, Malinow R, Klein ML, Neuringer M (1981). Maculopathy in cynomolgus monkeys. Arch Ophthalmol 99:664-672.

Feeney-Burns L, and MR Ellersieck. Age-related changes in Bruch's membrane ultrastructure. Amer J Ophthalmol 100;686-697.

Fine BS (1981). Lipoidal degeneration of the retinal pigment epithelium. Amer J Ophthalmol 91:469-473.

Imaki H, Moretz R, Wisniewski H, Neuringer M and Sturman J. (1987). Retinal degeneration in three-month-old rhesus monkey infants fed a taurine-free human infant formula. J Neurosci Res 18:602-614.

Handelman GJ, Dratz EA, Reay CC, VanKuijk JGM (1988). Carotenoids in the human macula and whole retina. Invest Ophthalmol Vis Sci 29:850-855.
Malinow MR, Feeney-Burns L, Peterson LH, Klein ML, Neuringer M (1980). Diet-related macular anomalies in monkeys. Invest Ophthalmol Vis Sci 19:857-863.
Neuringer M and Sturman JA (1980). Visual acuity loss in rhesus monkey infants fed a taurine-free human infant formula. J Neurosci Res 18:597-601.
Neuringer M, Malinow MR, Klein ML, Feeney-Burns L (1984). Visual acuity in rhesus monkeys without macular pigment. Invest Ophthalmol Vis Sci 25 (Suppl 3):145
Neuringer M, Sturman JA, Wen GY, Wisniewski HM (1985). Dietary taurine is necessary for normal retinal development in monkeys. In Oja SS, Ahtee L, Kontro P, Paasonan MK (eds), "Taurine: Biological Actions and Clinical Perspectives" New York, Alan R Liss, pp 53-62.
Portman OW, Alexander M, Neuringer M, Illingworth DR, Alam S (1981). The effects of long-term protein deficiency on plasma lipoprotein concentrations and metabolism in rhesus monkeys. J Nutr 111:733-745.
Reme CE, Young RW (1977). The effects of hibernation on cone visual cells in the ground squirrel. Invest Ophthalmol Vis Sci 16:815-840.
Schmidt SY, Berson EL, Hayes KC (1976). Retinal degeneration in cats fed casein. I. Taurine deficiency. Invest Ophthalmol Vis Sci 15:47-52.
Snodderly DM, Brown PK, Delori FC, Auran JD (1984). The macular pigment. I Absorbance spectra, localization, and discrimination from other yellow pigments in primate retinas. Invest Ophthalmol 25:660-673.

Inherited and Environmentally Induced
Retinal Degenerations, pages 623–642

THE RAT AS AN ANIMAL MODEL FOR RETINOPATHY OF PREMATURITY

John S. Penn and Lisa A. Thum

Cullen Eye Institute, Baylor College of Medicine,
Houston, Texas 77030

Although it has been forty-five years since the first case of retinopathy of prematurity (ROP) was reported (Terry, 1942), and thirty years since oxygen was confirmed as causal in the development of ROP (Campbell, 1950), the mechanism of cytotoxicity of oxygen on the retina is still not understood. ROP manifests itself in the vasculature of the developing infant retina, and occurs in neonates, who due to pulmonary complications are maintained under elevated oxygen. Specifically, an interruption of normal vascular development takes place in the retina. This is followed by an abnormal vascular proliferation that is promoted by return to room air. The new vessels become fibrotic and can retract, leading to retinal detachment.

In the 1950's, animal models of oxygen-induced retinopathy (OIR) were instrumental in establishing the causal role of oxygen in ROP. Now, with donations of human retinal tissue becoming rare due to increased survival of premature infants, new attention is being paid to the development of appropriate animal models that mimic the human in retinal ontogeny and in the pathogenesis of OIR. There are additional, intrinsic problems with clinical studies of ROP. Because the retinal cells of premature infants have the added complications of altered perfusion, metabolic acidosis, and respiratory failure, it may be that they are unable to cope with atrial oxygen levels that would be otherwise inoffensive. Since the experimental animal is without intrinsic respiratory and metabolic disturbances, there are fewer extraneous factors to confuse the picture in the laboratory where only one variable, inspired oxygen, is altered.

Past research has shown hyperoxia-induced abnormalities in the retinal blood vessels of several species. The rat (e.g., Patz, 1954; Ricci, 1987), mouse (Michaelson et al., 1954), kitten (Phelps and Rosenbaum, 1977) and rabbit (Ashton et al., 1972) all have demonstrated ROP-like alterations of retinal vasculature when placed in high oxygen environments during development. No model has proven completely reliable, but the newborn rat offers several attractive features that have drawn the attention of investigators for thirty-five years. These features include the following: 1) the ontogeny of the rat retina is similar to that of the human, 2) the vasculature of the rat develops postnatally, facilitating its study, 3) the newborn rat is vitamin E deficient, like the premature infant, and 4) the rat is an inexpensive and easily maintained tissue source for which pertinent techniques have already been perfected. Arguments against the rat as a model of ROP point to the fact that retinal detachment does not occur in this species. This might be due to the large lens of the rat, which occupies nearly half of the total eye volume (Patz, 1955). In any case, the early etiology of OIR in the rat and ROP in the human is identical, rendering the rat a good, inexpensive model for the study of the pathogenesis of endothelial cell damage and vaso-obliteration and its pharmacologic prevention.

Retinal Ontogeny of the Rat

The development of the rat retinal vasculature has been described many times (e.g., Henkind and De Oliveira, 1967). The retinal vessels are first seen as undifferentiated spindle cells derived from the outer wall of the hyaloid vessel at about 14 days gestation (Ashton, 1966). The cells migrate into the retina and proliferate to form a network. They then evolve into luminized vessels in the nerve fiber layer of the retina (Shakib et al., 1968). As it progresses peripherally, the capillary network varies in density, becoming more abundant near the veins. A zone forms around the arteries that is entirely free of capillaries (Michaelson, 1948). This process led Campbell (1951) to suggest that vessel formation occurs in response to the surrounding tissue's oxygen needs, which are reduced in the vicinity of oxygen-rich arterial vessels. By postnatal day 15, the vessels have reached the periphery of the retina, and the capillary sprouting is complete (Ashton and Blach, 1961). Vessel growth from advancing mesenchymal cells is also seen in the human, but not in the kitten or rabbit, where endothelial budding is implicated (Ashton, 1966). The distinction between the vessel proliferation from undifferentiated spindle cells and pre-existing capillaries is emphasized because the mechanism for the development and plasticity of the vessels may be different (Ashton, 1979; Patz, 1984).

Another attractive aspect of the developing rat retina is its relative deficiency of vitamin E (unpublished data) - a condition that also exists in the retina of the premature human infant and which may compromise its ability to withstand the insult of oxygen exposure. Furthermore, the rat is, in fact, the only lower animal that possesses a true central artery that divides radially and symmetrically into six branches extending to the periphery, which is also the case for humans (Ricci, 1987).

Pathogenesis

The effects of supplemental oxygen on incompletely vascularized retinas can be divided into two phases: an initial vasoconstrictive response, which occurs during exposure, and a subsequent phase of vasoproliferation which takes place after removal to room air. There is evidence that the vasoproliferation may depend on local retinal hypoxia produced by previous vaso-obliteration (Johnson, 1981). It has been suggested that, under hyperoxic conditions, the choroid may furnish excessive concentrations of oxygen to the inner retina, producing retinal vaso-obliteration (Patz, 1984). The inner retina may be particularly susceptible to this condition at the developmental stage where photoreceptors are not yet fully formed and are unable to create an adequate oxygen sink (Kretzer and Hittner, 1985). Upon return to room air, the choroidal circulation is inadequate to overcome the relative hypoxia (Ernest and Goldstick, 1984). Others have speculated that retinal vasoconstriction may be a physiological protective response to hyperoxia (Flower and Blake, 1981; Dollery et al., 1964). When aspirin, which produces an inhibition of prostaglandin-induced retinal vasodilation, was administered to beagle puppies prior to oxygen exposure, a persistent vasodilation was observed, with subsequent severe retinopathy (Flower and Blake, 1981). It seems likely that retinal vascular damage results primarily from *cytotoxic effects of oxygen* on endothelial cells, rather than secondarily from the vasoconstriction and diminished blood flow as was originally postulated (Ashton and Pedler, 1962). It is presumed that, if retinal vascular channels are dilated while increased oxygen levels are present, the endothelial vessel walls are in greater contact with toxic oxygen radicals, which are known to damage cell membranes. Indeed, the first event that can be detected by electron microscopy is a selective injury to the endothelial cells of the immature vessels, with no obvious changes in the neuronal elements of the retina (Ashton and Pedler, 1962). These results support a direct injury to the retinal endothelium.

Experimental History

Oxygen-induced retinopathy in rats was first reported in the early 1950's, by Patz and co-workers (Patz et al., 1953: Patz, 1954). Newborn rats, exposed to 80% oxygen for 21 days, revealed "characteristic lesions of human retrolental fibroplasia" as reported by Patz. These included: 1) nodules of endothelial cell proliferation in the retina, 2) budding of capillaries from the retina into the vitreous, 3) retinal hemorrhages, and 4) retinal edema. In 1958, Brands reported exposing 4-day-old rats to 70% oxygen for six days. One group of rats was sacrificed and examined immediately after exposure, and a second group was examined after seven days' survival in room air. In the first group no histological changes were seen in the retina. The group that was maintained in room air after oxygen exposure displayed alterations similar to those seen by Patz (1954).

In 1961, Ashton and Blach conducted a thorough set of experiments and a review of the previous work of Patz (1954) and Brands (1958). These new experiments consisted of exposure of rats to concentrations ranging from 60-80% at ages varying from newborn to 11 days. Some animals were sacrificed immediately, and others were maintained in room air for 10 to 20 days before examination. In none of these cases was abnormal vasoproliferation in the retina seen. Ashton and Blach concluded by conceding that "our findings do not contradict the theories previously advanced to explain vaso-obliteration and vasoproliferation and they further illustrate the specific sensitivity of growing retinal vessels to oxygen." In closing, the authors suggested that one reason for the lack of vasoproliferation in the rat was that the retinal capillary network was plastic and spontaneously reformed after oxygen exposure. This, they reasoned, prevented the accumulation of a vasoformative factor and lessened the hypoxic shock of return to room air with a compromised vessel complement.

Subsequent to these equivocal findings of the 50's and 60's, there was little attention paid to the rat, while other animal models were explored. In the 1980's, with new and varied techniques available, researchers have begun to reexamine oxygen-induced retinopathy in the rat.

Recent Research on the Rat

Up until the 1980's, the major focus of experimental research on retinopathy of prematurity had been on the dose of oxygen delivered - how much and for how long? The methods used to assess vascular damage were limited to age-old histological techniques. Virtually nothing was known about the toxicity of the oxygen dose delivered to the retina in terms of its biochemical and physiological effects on retinal cells and their antioxidant defenses. With the advent of new quantitative analytical techniques, experimental approaches have changed.

The rationale for using the antioxidant vitamin E as treatment for ROP is based on the premise that oxygen radical damage to cell membranes is the initial event and that, as a radical scavenger, vitamin E may prevent cellular damage. It has been shown that, when cells are gradually exposed to hyperoxia, they build up resistance by increased intracellular production of antioxidants (Crapo and McCord, 1976). It is tempting to postulate that *immature* vessels are at higher risk because they have not matured enough to produce protective antioxidants or have not had the necessary gradual exposure to oxygen to induce an adequate complement of these molecules.

Clinically, a controversy has been generated over the supplementation of vitamin E to premature infants as therapy for ROP. Within this decade there has been a nearly equal split between those clinical trials that show protective effects of vitamin E (Hittner et al., 1981; Johnson et al., 1982; and Finer et al., 1983) and those that refute its efficacy (Puklin et al., 1982; Hillis, 1982; and Phelps et al., 1987).

The work of Hittner and Kretzer and their associates includes an elaborate proposal for the cellular mechanism of ROP and the protective role of vitamin E. This proposal is built on the finding that oxygen-treated infants have a higher incidence of spindle cell gap junctions than normoxic infants (Kretzer et al., 1982). It was postulated that these gap junctions alter the normal vasoformative process. Premature infants are born with relatively low levels of retinal vitamin E, particularly in the avascular region. If administered at the appropriate time (ca. 27 wks gestation; Nielsen et al., 1988), vitamin E supplementation results in a rapid increase in retinal levels. Infants supplemented in this manner display a reduction in spindle cell gap junctions and, according to these authors, a reduction in the severity of the retinopathy.

Subsequent to gap junction formation, a proliferation of the volume of rough endoplasmic reticulum occurs in spindle cells of the oxygen-exposed infant. According to Kretzer and Hittner, this event indicates synthesis and secretion of angiogenic factors by spindle cells. The authors have proposed, therefore, that "the postnatal rat offers an excellent animal model from which to tissue-culture spindle cells from the retina" (Kretzer et al., 1986), presumably for testing their production of angiogenic factors and for attempting pharmacologic prevention of the structural alterations induced in them by oxygen.

In 1985, Yabe used the rat to examine the efficacy of vitamin E in oxygen-induced retinopathy. He found that the retinal vessels of vitamin E-deficient newborn rats showed an increased oxygen sensitivity. He attributed this to the observation that retinal vascular development is delayed in vitamin E-deficient rats. Therefore, Yabe concluded, vascular immaturity is the important variable in predisposing a rat to oxygen-induced retinopathy.

The next experiments concerning oxygen-reared newborn rats were conducted by Ricci and coworkers (Ricci, 1987; Ricci et al., 1988). These experiments explored the effects of atmospheric pressure, in association with elevated oxygen, on severity of retinopathy. Results indicated that normobaric oxygen-rearing resulted in severe retinopathy, while hyperbaric oxygen-rearing protected the retinal vasculature. Ricci suggested that pressure-induced vessel constriction in the choriocapillaris was responsible for reducing the oxygen transport to the inner retina, resulting in less severe toxic effects on the immature retinal vessels. Incidentally, Ricci reported the presence of *extensive vasoproliferation* (including extraretinal vessels) in these studies, thereby reopening the question of neovascularization in the rat model.

A second avenue of investigation by these researchers concerned the role of oxygen-rearing in the subsequent occurrence of "functional disturbances in other areas of the visual system" (Ricci et al., 1988). Their idea stemmed from the fact that the incidence of anisometropia, amblyopia, and strabismus are all higher in pre-term babies than in those born at term (Fledelius, 1976). The authors suggested that a reduced axonal transport may account for these functional disturbances of the visual system. Employing intravitreal injections of ^{35}S-taurine, the authors found a reduced axonal transport in oxygen-reared rats when compared to room air controls. They concluded that non-retinal alterations, such as those listed above, may be the result of reduced axonal transport or synaptogenesis in the optic pathway.

Clearly, animal models are necessary for studies like those of Yabe and Ricci, where rather drastic experimental manipulations of the subject take place before its sacrifice.

Current Investigations in Our Laboratory

Studies of oxygen-induced retinopathy in rats and other experimental animals have employed a variety of experimental conditions. Atmospheric oxygen has ranged from 35% (Ashton et al., 1953) to 100% (Patz et al., 1953); atmospheric pressure has been altered (Ricci, 1978; Campbell, 1951); duration and age at onset of exposure have varied (Ashton and Blach, 1961); often animals have been subjected to a combination of room air and oxygen exposure. With respect to rats, it is most often the case that entire litters, mother included, are exposed. For extremely high oxygen levels (>80%), it has been necessary to alternate surrogate mothers from room air to high oxygen, due to the adult rat's susceptibility to respiratory distress. Because one of the foci of our studies is a consideration of retinal antioxidants, this approach has been avoided, since it may complicate the variable of mothers' physiology and the make-up of nutrients they supply the litters. Lower levels of oxygen (40-50%) have shown somewhat less severe effects on retinal vasculature and, likewise, have been avoided. We have selected an exposure of 60% atmospheric oxygen, which empirically has resulted in a highly significant, repeatable effect.

Onset and duration of exposure are equally important, and we have chosen birth to 14 days for our initial experiments. This duration corresponds to the time it takes the developing retinal blood vessels to extend to the periphery in our control rats. Any retardation of vessel growth toward the peripheral retina is, therefore, easily measured. In fact, Ashton and Blach (1961) assumed that only posterior capillaries were affected by oxygen exposure. This is probably because they chose to expose and analyze rats at later ages, thereby missing the earlier stage that shows slowed progression of both capillaries and major vessels. Some of our newborn rats have been subjected to periods of 60% oxygen followed by periods in room air. This was done in order to test the rat's tendency to neovascularize and its ability to recover normal structure and function.

In order to determine the vascular effects of oxygen-rearing, we examine whole mounted retinas from ink-perfused rats. With the aid of a digitometer, we can estimate the portion of retinal surface area that contains capillaries. Our primary treatment (14 days in 60% oxygen) results in a 35% reduction in the area of the retina containing capillaries, as compared to controls raised in room air (Fig. 1). Two things appear responsible for this reduction: 1) retardation of the progression of capillaries to the periphery of the retina in experimental animals, and 2) obliteration of the capillaries that are immediately adjacent to the larger retinal blood vessels in the central retina. We are currently analyzing the possibility of a more subtle relationship between retinal location and extent of vessel abnormality.

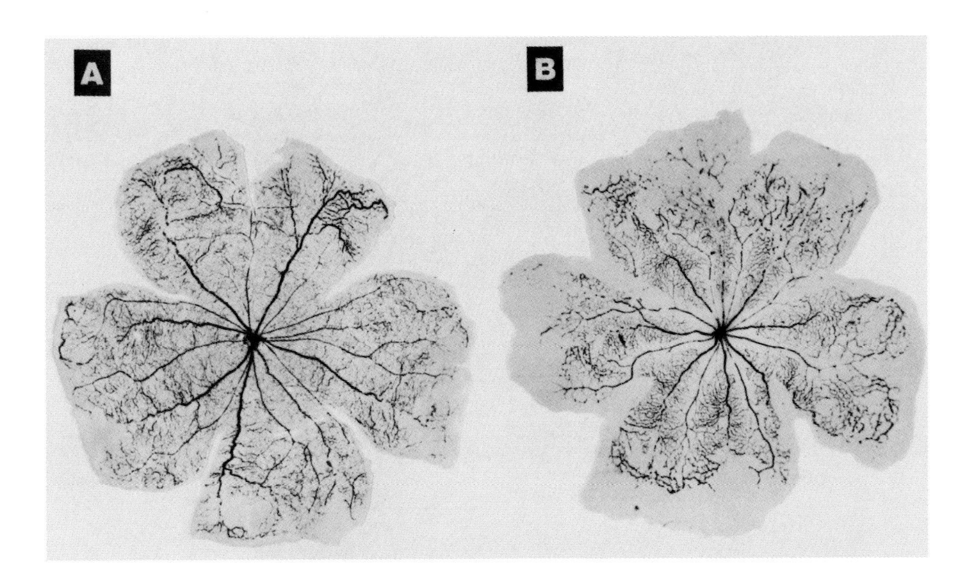

Figure 1. Ink-perfused retinal whole mounts from a room air rat (A) and an oxygen-exposed rat (B) reveal a 35% decrease in the vasculature of the latter.

We have observed arterio-venous shunting in place of the normal bilayered capillary net in the retinas of experimental rats. It is known that under hyperoxic conditions the deep (venous) capillary network of the rat retina will not form, or it will obliterate if it is formed upon subjugation to oxygen (Ashton and Blach, 1961). Our exposure results in the complete loss of the deep net. The shunts that result from this loss are identifiable in perfused whole mounts (Fig. 2) and are easily corroborated in retinal cross-sections. This shunting is emphasized because it is commonly observed in the premature infant retina under hyperoxic conditions (Flynn et al., 1979).

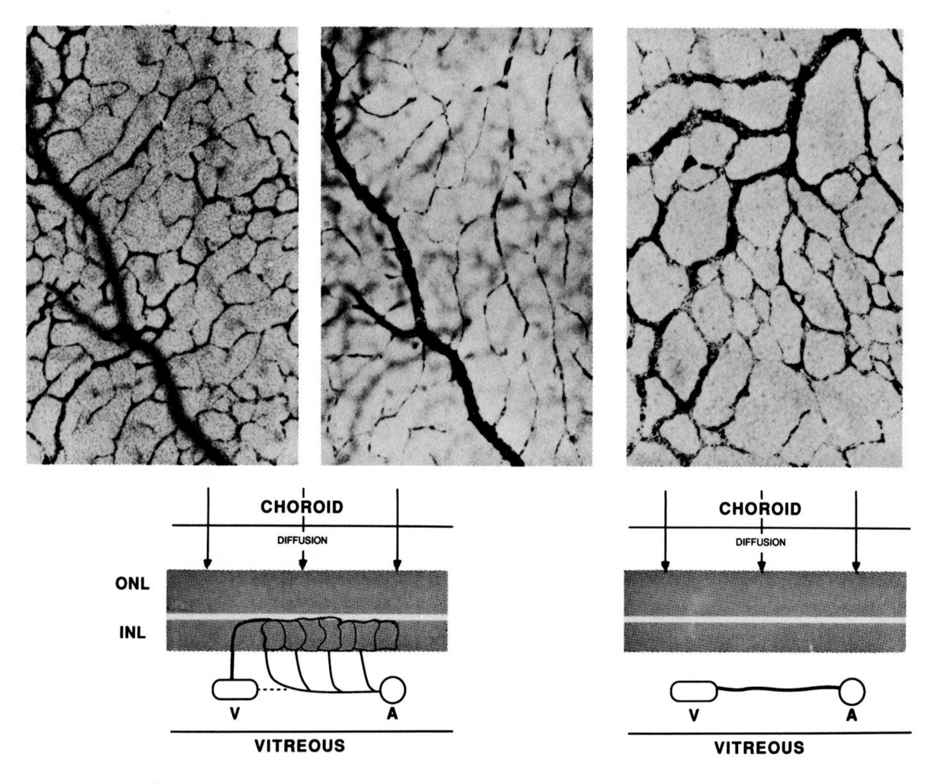

Figure 2. Rats raised in room air (left) develop two layers of capillaries through the depth of the retina. Animals raised in 60% oxygen (right) sustain a complete loss of the deeper capillary net and a partial loss of the superficial one, resulting in arterio-venous shunts.

An attempt has also been made to measure the function of the retina in these two groups of ratlings by electroretinography. Animals of this age and body weight (14 days, 16-24 gms) are very difficult to anesthetize. We presently feel that ERG recordings at this age more likely have reflected the general health and adequate anesthesia of the ratlings, rather than any oxygen-induced functional differences. Although the control responses had slightly higher amplitudes, particularly with respect to the b-wave, variation was great and emphasis is not placed on the difference. However, by one week post-treatment, both control and experimental animals have developed to the point at which they are easily anesthetized and re-awakened. In this case, individual animals can be followed week after week to analyze long-range effects of early oxygen-induced problems.

The a-wave, an ERG parameter attributed to photoreceptor cells, shows maximum amplitudes that are identical for these two groups of 21-day-old rats. The b-wave, a glial cell-initiated response (Miller and Dowling, 1970; 60), however, is reduced by 40% in the oxygen-treated group. By three weeks post-exposure in room air, this reduction has become almost 70% (Fig.3). This is where a functional deficit might be expected to occur - in the inner retina. Photoreceptors receive their oxygen via diffusion from the choroid blood supply, but the inner retina is the responsibility of the retinal vasculature.

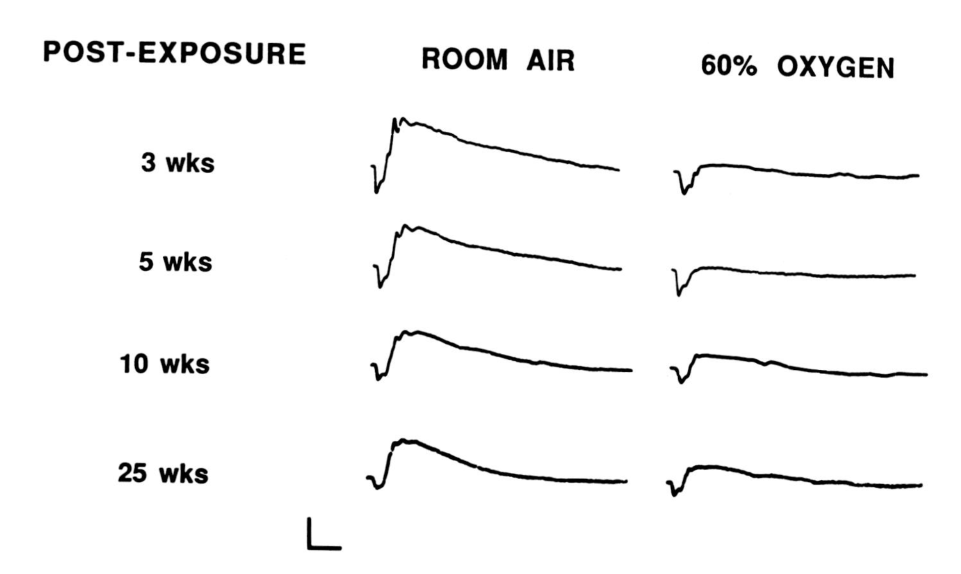

Figure 3. Oxygen-reared rats sustain an irreversible reduction of the b-wave maximum amplitude of the ERG. The a-wave is unaffected by the oxygen exposure.

Therefore, the fact that the b-wave is reduced in hyperoxic animals is not surprising, considering the retinal area receiving normal retinal blood supply was reduced by at least some 30-35%. This vascular reduction is transient. At one week post-exposure, the retinal vasculature of experimental animals appears relatively normal, with only subtle differences in capillary density evident. The ERG, on the other had, has been followed in experimental animals through 35 weeks post-treatment, and no functional recovery has been seen. This represents a permanent functional effect of OIR in an animal model, a possibility that has not been explored in infants with ROP.

In our attempt to determine what biochemical or morphological alterations might correlate with this functional deficit, we targeted the Müller cell, because of its proposed role in the origin of the b-wave. We have found that Müller cells of oxygen-reared rats produce glial fibrillary acidic protein (GFAP) (Fig. 4).

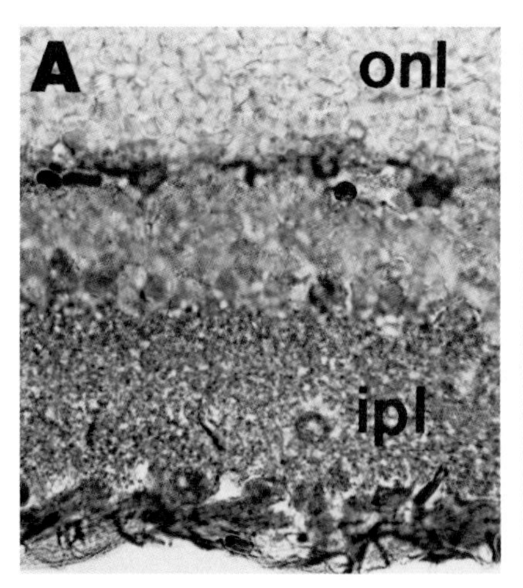

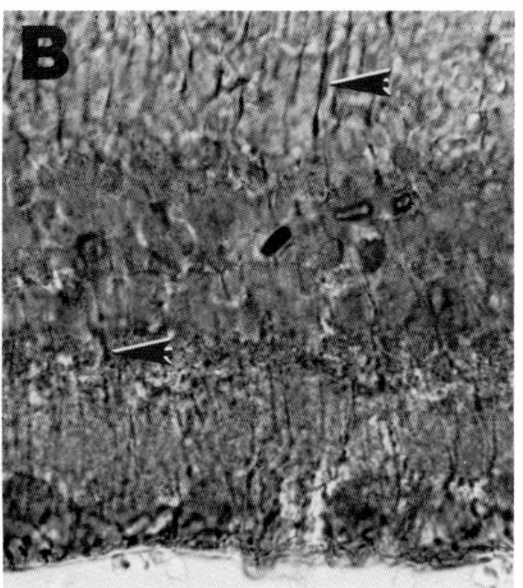

Figure 4. The expression of GFAP in control rats is limited to astrocytes of the nerve fiber layer. In oxygen-exposed rats, the Müller cells (arrows) produce GFAP also. Micrographs are of an 8 wk-old rat raised in room air (A) and an 8 wk-old that spent its first two weeks in 60% oxygen (B).

Production of this protein has been associated with several forms of experimentally induced or inherited retinal degeneration (Bignami and Dahl, 1979; and Eisenfeld et al., 1984). However, we can discern no neuronal cell death in the rat retina as a result of our oxygen exposure. This marks the first indication of a direct effect of oxygen on retinal glia, and it opens a new avenue of investigation toward understanding the pathogenesis of oxygen therapy and realizing its pharmacologic prevention.

The remainder of our experiments have related to the role of peroxidation reactions in oxygen-induced retinopathy. Under normal conditions, the retina is susceptible to damaging oxidative reactions because of its very high levels of substrate molecules for these reactions, its constant bombardment by light, and a relatively high oxygen flux across its membranes. Under high levels of atmospheric oxygen, this last aspect is enhanced. Very little is known of the biochemical mechanisms of retinal oxygen damage, although there has been much recent interest in the biochemistry of oxygen toxicity in general. There has been speculation that endothelial cell damage is due to the action of oxygen-derived free radicals - molecules that contain an odd number of electrons, rendering them chemically reactive (Slater and Riley, 1970). An abundance of excellent substrates for oxidative reactions is present in the retina. Particularly susceptible are polyunsaturated fatty acids, which are found in higher concentrations in retinal tissue than in any other (Anderson and Andrews, 1981). The products of lipid peroxidation include aldehydes, among them malondialdehyde (MDA), which are toxic to the cell. The uncontrolled production of these toxic compounds leads to cell death. In addition to vitamin E, the defense mechanisms against free radicals include various enzyme systems, namely superoxide dismutases, catalases, and peroxidase (Fridovich, 1984). Superoxide dismutase converts superoxide radicals to hydrogen peroxide and oxygen. The hydrogen peroxide produced is metabolized by catalase and glutathione peroxidase. Vitamin E (Dilley and McConnell, 1970), superoxide dismutase (Hall and Hall, 1975), and glutathione peroxidase (Penn et al. 1987) are all present in the retina.

We have found significant differences in the retinal levels of ascorbic acid and α-tocopherol, with control rats displaying 20 and 35% higher levels, respectively, than oxygen-treated rats (Table 1). No differences were found between the two groups in any of the glutathione enzyme activities (glutathione peroxidase, S-transferase, and reductase). Oxygen-reared rats have about twice as much retinal MDA as controls. Further, the levels of peroxidation substrate, docosachexaenoic acid, are significantly reduced in the oxygen-treated group. These two findings indicate that lipid peroxidation has (is) occurred (occurring) during the oxygen exposure. We are currently examining retinal superoxide dismutase and catalase in order to determine if either is compromised.

The reduced levels of retinal vitamin E in oxygen-reared rats encouraged us to follow Yabe's investigation with our own attempt at α-tocopherol supplementation. Manipulation of retinal vitamin E in the newborn rats was accomplished through the diet of the mothers. Dietary supplementation or deprivation began when breeding pairs were combined and lasted through pregnancy and the treatment period (a total of 35-39 days). The two diets were manufactured by Dyets, Inc. (Bethlehem, Pa.), and consisted of 1.0 gm α-tocopherol acetate/kg food or no α-tocopherol acetate. Using this method we were able to cause 3- to 4-fold differences in the retinal vitamin E levels of the newborn animals, regardless of oxygen level. Supplemented rats had retinal α-tocopherol levels of 1.5-2.0 mmol/mol phospholipid, while deprived rats' levels were 0.4 - 0.6 mmol/mol phospholipid. This dietary treatment, in combination with oxygen treatment, affected the levels of other antioxidants as well. Of note was a significant increase in the retinal level of glutathione peroxidase in both the vitamin E-supplemented and vitamin E-deprived rats of the oxygen-exposed group. More important, however, was that vitamin E supplementation lessened the severity of oxygen-induced vaso-obliteration by approximately 10%. Interestingly, both E-supplemented and E-deprived rats raised in oxygen sustained less vaso-obliteration than oxygen-exposed rats with no vitamin E manipulation. We plan to examine the cause of this, beginning with the interrelationship of vitamin E glutathione peroxidase.

In summary, the following characteristics make

oxygen-induced retinopathy in the rat a good model for the study of ROP:

1) Retinal ontogeny is similar in the rat and the human, particularly with respect to spindle cell participation in vascular development. The rat retina is vitamin E-deficient at birth, as is the human retina, and the retinal level of vitamin E can be manipulated.

2) The rat retina has demonstrated a sensitivity to oxygen that parallels the early etiology of the human disorder. The progression to later stages of retinopathy has been questioned in the rat, but recent evidence has renewed the possibility of neovascularization in the model.

3) The rat is an inexpensive and easily maintained tissue source for which pertinent techniques have been developed.

We are confident that we can cause a predictable and reproducible degeneration that morphologically resembles the early stages of retinopathy of prematurity. The retinal function of our rats can be monitored over long periods of time by non-invasive electrophysiological procedures. Further, we can measure and manipulate the levels of retinal antioxidants and this should allow us to test their efficacy in the oxygen-exposed rat. While we await further clinical trials, the pathogenesis of oxygen-induced retinopathy can be studied using the rat model. The identification of any factors in the retina that protect against oxygen damage and the determination of the relationship between their retinal concentrations and the effect of high atmospheric oxygen is an important step. Ultimately, we hope to apply this knowledge to the development of rational therapeutic approaches for the premature infant.

REFERENCES

Anderson RE, Andrews LM (1981) Biochemistry of retinal photoreceptor membranes in vertebrates and invertebrates. In Visual Cells in Evolution (ed Westfall J) Raven Press, New York, pp 1-22.

Ashton N (1966) Oxygen and the growth and development of retinal vessels. Am J Ophthalmol **62**: 412.

Ashton N (1979) The pathogenesis of retrolental fibroplasia, Symposium: RLF. Ophthalmology **86**: 1695.

Ashton N, Blach R (1961) Studies in developing retinal vessels. VIII. Effect of oxygen on the vessels of the ratling. Br J Ophthalmol **45**: 321-340.

Ashton N, Pedler C (1962) Studies on developing retinal vessels. IX. Reaction of endothelial cells to oxygen. Br J Ophthalmol 46: 257-276.

Ashton H, Ward B, Serpell G (1953) Role of oxygen in the genesis of retrolental fibroplasia. Br J Ophthalmol **37**: 513-520.

Ashton N, Tripathi B, Knight G (1972) Effect of oxygen on the developing retinal vessels of the rabbit. I. Anatomy and development of the retinal vessels of the rabbit. Exp Eye Res **14**: 214.

Bignami A, Dahl D (1979) The radial glia of Müller in the rat retina and their response to injury. An immunofluorescence study with antibodies to the glial fibrillary acid (GFA) protein. Exp Eye Res **28**: 63-69.

Brands KH, Hofmann H, Klees E (1958) Geburtsh U Frauenhci K, **18**: 805.

Campbell FW (1951) The influence of low atmospheric pressure on the development of the retinal vessels in the rat. Trans Ophthalmol Soc UK **71**: 287.

Campbell K (1950) Retrolental fibroplasia as a syndrome. Arch Ophthalmol **44**: 245.

Crapo JD, McCord JM (1976) Oxygen-induced changes in pulmonary superoxide dismutase assayed by antibody titrations. Am J Physiol **231**: 1196-1203.

Dilley R, McConnell D (1970) Alpha-tocopherol in the retinal outer segment of bovine eyes. J Membrane Biol **2**: 317-323.

Dollery CT, Hill DW, Mailer CM, Ramalho PS (1964) High oxygen pressure and the retinal blood vessels. Lancet **2**: 291-292.

Eisenfeld AJ, Bunt-Milam AH, Sarthy PV (1984) Müller cell expression of glial fibrillary acidic protein after genetic and experimental photoreceptor degeneration in the rat retina. Invest Ophthalmol Vis Sci **25**: 1321-1328.

Ernest JT, Goldstick TK (1984) Retinal oxygen tension and oxygen reactivity in retinopathy of prematurity in kittens. Invest Ophthalmol Vis Sci **25**: 1129-1134.

Finer NN, Schindler RF, Peters KL et al. (1983) Vitamin E and retrolental fibroplasia: Improved visual outcome with early vitamin E. Ophthalmol **90**: 428-435.

Flower RW, Blake DA (1981) Retrolental fibroplasia: Role of the prostaglandin cascade in the pathogenesis of oxygen induced retinopathy in the newborn beagle. Pediat Res **15**: 1293-1302.

Flynn JT, Cassady J, Essner D et al. (1979) Fluorescein angiography in retrolental fibroplasia: Experience from 1969-1977. Ophthalmol **86**: 1700-1723.

Fridovich I (1984) Oxygen: Aspects of its toxicity and elements of its defense. Curr Eye Res **3**: 1-2.

Hall M, Hall D (1975) Superoxide dismutase of bovine and frog outer segments. Biochem Biophys Res Commun **67**: 1199-1204.

Henkind JD, McCord JM (1976) Oxygen-induced changes in pulmonary superoxide dismutase assayed by antibody titrations. Am J Physiol **231**: 1196-1203.

Hillis AI (1982) Vitamin E in retrolental fibroplasia. New Engl J Med **306**: 867.

Hittner HM, Godio LB, Rudolph AJ et al. (1981) Retrolental fibroplasia: Efficacy of vitamin E in a double-blind clinical study of preterm infants. N Engl J Med **305**: 1365-1371.

Johnson L (1981) Retrolental fibroplasia: A new look at an unsolved problem. Hosp Pract **16**: 109-121.

Johnson L, Schaffer D, Quinn G et al. (1982) Vitamin E supplementation in retinopathy of prematurity. Ann NY Acad Sci **393**: 473-95.

Kretzer FL, Hittner HM (1985) Initiating events in the development of retinopathy of prematurity. In Retinopathy of Prematurity (eds Silverman WA and Flynn JT) Blackwell Scientific Publ, Boston, p 127.

Kretzer FL, Hittner HM, Johnson AT (1982) Vitamin E and retrolenta fibroplasia: Ultrastructural support of clinical efficacy. Ann NY Acad Sci **393**: 145.

Kretzer FL, Mehta RS, Goad D, Hittner HM (1986) Animal models in research on retinopathy of prematurity. In Retinoapthy of Prematurity: Current Concepts and Controversies (eds McPherson A, Hittner HM and Kretzer FL) BC Decker, Philadelphia, PA.

Michaelson IC (1948) The mode of development of the vascular system of the retina: With some observations on the significance for certain retinal diseases. Trans Ophthalmolo Soc UK **68**: 137.

Michaelson IC, Herz H, Lewkowitz E, Kertesz D (1954) Effect of increased oxygen on the development of the retinal vessels. An experimental study. Brit J Ophthalmol **38**: 577.

Miller RF, Dowling JE (1970) Intracellular responses of the Müller (glial) cells of mudpuppy retina: Their relation to b-wave of the electroretinogram. J Neurophysiol **33**: 323-341.

Nielsen JC, Naash MI, Anderson RE (1988) The regional distribution of vitamins E and C in human adult and preterm infant retinas. Invest Ophthalmol Vis Sci **29**: 22-26.

Patz A (1954) Oxygen studies in retrolental fibroplasia. IV. Clinical and experimental observations. Am J Ophthalmol **38**: 291-308.

Patz A (1955) Retrolental fibroplasia: Experimental studies. Am J Ophthalmol **40**: 174-183.

Patz A (1984) Current concepts on the effects of oxygen on the developing retina. Curr Eye Res **3**: 159.

Patz A, Eastham A, Higgenbotham DH, Kleh T (1953) Oxygen studies in retrolental fibroplasia. II. The production of the microscopic changes of retrolental fibroplasia in experimental animals. Am J Ophthalmol **36**: 1511-1522.

Penn JS, Naash MI, Anderson RE (1987) Effect of light history on retinal antioxidants and light damage susceptibility in the rat. Exp Eye Res **44**: 779-788.

Phelps DL, Rosenbaum AL (1977) The role of tocopherol in oxygen-induced retinopathy: Kitten model. Pediatrics **59**: 988.

Phelps DL, Rosenbaum AL, Isenberg SJ, Leake RD, Dorey FJ (1987) Tocopherol efficacy and safety for preventing retinopathy of premaurity: A randomized, controlled, double-masked trial. Pediatrics **79**: 489-500.

Puklin JE, Simon RM, Ehrendranz RA (1982) Influence on retrolental fibroplasia of intramuscular vitamin E administration during respiratory distress syndrome. Ophthalmology **89**: 96.

Ricci B (1987) Effects of hyperbaric, normobaric and hypobaric supplementation on retinal vessels in newborn rats: A preliminary study. Exp Eye Res **44**: 459-464.

Ricci B, Lepore D, Iossa M (1988) Oxygen-induced retinopathy in newborn rats: Orthograde axonal transport changes in optic pathways. Exp Eye Res **47**: 579-586.

Shakib M, De Oliveira LF, Henkind P (1968) Development of retinal vessels. II. Earliest stages of vessel formation. Invest Ophthalmol **7**: 689.

Slater T, Riley PA (1970) Free radical damage in retrolental fibroplasia. Lancet **2**: 467.

Terry TL (1942) Extreme prematurity and fibroplasia overgrowth of persistent vascular sheath behind each crystalline lens: Preliminary report. Am J Ophthalmol **25**: 203.

Yabe H (1985) Effects of vitamin E deficiency on oxygen induced retinopathy in rat. Nippon Ganka Gakkai Zasshi **89**: 624-630.

IV. RETINAL AND PIGMENT EPITHELIAL CELL TRANSPLANTATION

Only four chapters are presented in this section, but these studies offer an exciting new dimension for possible therapy of retinal degenerations: retinal or pigment epithelial cell transplantation. The logic of this approach is straightforward—if a cell carries a defective gene or if photoreceptor cells are lost for any reason, then the retinal disorder might be ameliorated by transplanting cells without a defective gene or by replacement of the degenerated photoreceptors. In the first two studies, normal retinal pigment epithelial cells have been transplanted into RCS rats that have a pigment epithelial cell defect that results in photoreceptor cell death. The transplanted cells dramatically prevented photoreceptor cell death in the adjacent and nearby retina. This marks the first time that neurons have been prevented from dying in any degenerative disorder of the central nervous system. The other two studies deal with the intrinsically more difficult problem of transplantation of neural retinal cells or photoreceptor cells, specifically, into the retinas of animals with inherited or environmentally induced photoreceptor cell loss.

The remarkable success of the pigment epithelial cell transplantation experiments in the RCS rat has already caught the attention of the lay press and many people who are afflicted with various forms of retinal degeneration. It is extremely important to view these experiments in the proper perspective when considering their application to human retinal degenerations. A number of major hurdles must be overcome before transplantation experiments can be considered in human beings. Among these are 1) the determination of the cell(s) carrying the genetic defect(s) in different forms of retinal degeneration, since it makes no sense to transplant normal pigment epithelial cells into retinas with intrinsic photoreceptor cell defects, or to transplant normal photoreceptor cells into retinas with pigment epithelial cell defects; 2) devel opment of surgical procedures to provide maximal distribution of the transplanted cells; and 3) resolution of the problem of transplant rejection. At the same time, however, the area of transplantation should be recognized as an exciting new approach to the study and possible treatment of retinal degenerations.

Inherited and Environmentally Induced
Retinal Degenerations, pages 645–658

PHOTORECEPTOR CELL RESCUE IN THE RCS RAT BY RPE TRANSPLANTATION: A THERAPEUTIC APPROACH IN A MODEL OF INHERITED RETINAL DYSTROPHY

H. J. Sheedlo, L. Li and J. E. Turner

Department of Anatomy, Bowman Gray School of Medicine, Wake Forest University, Winston-Salem, NC 27103

INTRODUCTION

The Royal College of Surgeons (RCS) dystrophic rat, first reported by Bourne et al. (1938), is an autosomal recessive trait and has been extensively studied both biochemically and microscopically (for review, see LaVail, 1981). Retinal pigment epithelial (RPE) cells of RCS dystrophic rats are defective in ingesting shed rod outer segments (ROS), although photoreceptor cells (PRC) appear not to be defective. Thus, membrane debris accumulates in the region between the RPE and outer nuclear layer (ONL), which may directly or indirectly result in the degeneration of photoreceptor cells (PRC) beginning at about postnatal day 12 and by day 60, few PRC's are detectable (LaVail et al., 1975). The accumulation of membranous debris is slowly reduced, such that by six months, little debris remains (Lucas et al., 1955).

In a recent report, RPE cells from normal rats were transplanted through a newly established dorsal scleral/choroidal lesion technique (Li and Turner, 1988a) into the subretinal space of 26 day old RCS dystrophic rats. When examined at 60 days, this retina had an outer nuclear layer about 8-10 cells in thickness in the region under the graft, whereas, normally this layer is 1-2 cells in thickness in these dystrophic rats (Li and Turner, 1988b). Also, this study showed that the surviving PRC's possessed inner and outer segments, an outer plexiform layer persisted and the debris zone was significantly reduced in those areas affected by the RPE graft. At the edge of the graft site, a transition zone was observed exhibiting a diminished PRC thickness and beyond detectable grafted RPE cells, the ONL thickness was typical of 60 day old RCS dystrophic rats. These results revealed the establishment of a successful therapeutic approach in the RCS rat which arrests PRC loss in an inherited retinal dystrophic disease process.

In view of this study, questions concerning the functional capabilities and metabolism of the surviving PRC's and the grafted RPE cells should be considered. Investigating the membrane-bound enzyme (Na^+ + K^+)-ATPase and the visual pigment opsin in the RPE-grafted retina of RCS dystrophic rats would provide evidence to address these questions. Sodium ions, which are essential for the occurrence of photocurrents, enter ROS and diffuse to the rod inner segment (RIS), where they are pumped out by (Na^+ + K^+)-ATPase. This enzyme has been localized immunocytochemically on the plasma membrane of RIS, by not ROS. Further, (Na^+ + K^+)-ATPase is distributed at apical membranes of RPE, within the inner and outer plexiform layers, surrounding cell bodies in the inner and outer nuclear layers and at the periphery of ganglion cell bodies and axons of ganglion cells (Sheedlo and Turner, 1988, McGrail and Sweadner, 1986; Stahl and Baskin, 1984). The light sensitive protein opsin, the most prominent protein of photoreceptor cells, has been localized on the ROS, as well as on RIS and cell bodies in the ONL (Nir et al., 1987; Nir et al., 1984). Therefore, the localization of (Na^+ + K^+)-ATPase and opsin in PRC's would strongly indicate a heightened functional state of these rescued cells.

In this study ultrastructural examination and light microscopic immunocytochemistry for (Na^+ + K^+)-ATPase and opsin distribution were utilized to assess the functional characteristics of grafted RPE cells and surviving photoreceptor cells and other graft affected and nonaffected regions of the retina of RCS dystrophic rats.

MATERIALS AND METHODS

Animals

The pigmented RPE cells were isolated from Long Evans rats at postnatal day 6-8. Pink-eyed RCS dystrophic rats, obtained from NIH, were grafted at postnatal day 26. All rats were maintained on a 12 hour light-dark cycle as described previously (Li and Turner, 1988a). Adult Long Evans rats were used as controls in the immunocytochemical study.

RPE Cell Isolation

Pigmented RPE cells were isolated from rats by a modified procedure of Mayerson and co-workers (1985) as described by Li and Turner (1988a). Eyes of rat pups were removed and rinsed in an eye rinse (garamycin, kanamycin, Hank's balanced salt solution (HBSS)). Two enzyme solutions were utilized to facilitate the isolation of RPE cells. First, the eyes were incubated at 37°C in collagenase and hyaluronidase in HBSS (10 eyes/10 ml) for 40 minutes, then trypsin in HBSS for 50 minutes. After these incubations, the eyes were placed in RPE growth medium (Dulbecco's minimal essential medium, nutrient mixture F12, L-glutamine, fetal calf serum, garamycin, kanamycin, HEPES buffer). The cornea was cut away and the lens and vitreous were removed. The neural retina and attached RPE were peeled away from the eye cup, placed into RPE medium and incubated at 37°C for 1-2 hours. After this incubation the RPE detached from the neural retina as a sheet of cells. The RPE sheets were centrifuged at 1000xg in medium, then washed in calcium magnesium free medium (CMF) in HBSS. The RPE pellet was incubated in trypsin in CMF-HBSS and soybean trypsin inhibitor, fetal calf serum and DNase I were added after this incubation. Dissociation of the RPE sheets into individual cells was facilitated by trituration with small bore Pasteur pipettes. The resulting cell pellet was centrifuged, washed in CMF-HBSS, then concentrated to about 40,000-60,000 cells/µl. Samples of the cell suspension were examined by light and electron microscopy to determine purity and for viability by trypan blue exclusion.

RPE Cell Grafting

The surgical and grafting procedures followed in this study were as described previously (Li and Turner, 1988a; Turner and Blair, 1986). Briefly, albino RCS dystrophic rats, 26 days old, were anaesthetized with sodium pentobarbital, then injected with atropine. The dorsal surface of the eye was exposed by an incision in the superior eye lid and the eye was retracted anteriorly to expose the superior rectus muscle. After this muscle was cut anteriorly, a small lesion was made with a microblade through the sclera and choroid between the two superior vorticose veins (Figures 1A(B), 1B(left)). A 1.0 μl suspension of 40,000-60,000 pigmented RPE cells was injected at the lateral region of the lesion site (Figures 1A(C), 1B(center)). The suspension of pigmented RPE cells was observed as a circular mass lateral to the site of injection. The lesion was closed with a single 10-0 suture (Figures 1A(D), 1B(right)).

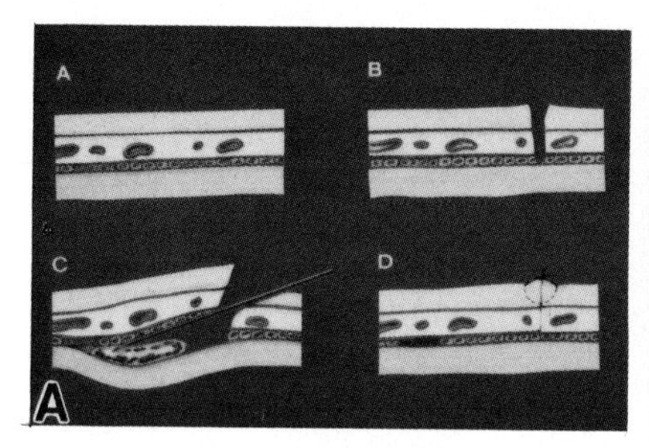

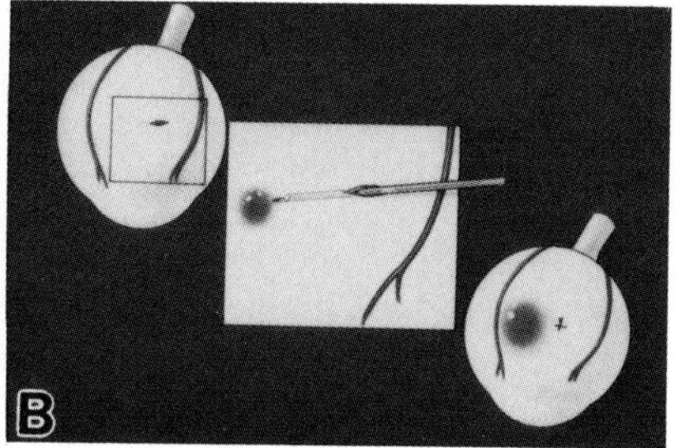

Figure 1. Diagrammatic representation of the surgical and RPE transplantation techniques. A) Observed in cross-section. B) Observed from dorsal surface of eye.

Microscopic Analysis

For light microscopic investigations RPE grafted and sham control RCS dystrophic rats and adult Long Evans rats were anaesthetized with an overdose of sodium pento-barbital and the eyes were enucleated and immersion fixed in Bouin's fixative for five hours at room temperature.

The eyes were dehydrated in an ascending ethanol series and, for grafted retina, trimmed to include the lesion site (observed as a dark circular mass). Trimmed eyes were embedded in paraffin following standard procedures. Paraffin embedded retinas were sectioned at 5 microns and some sections were stained with hematoxylin-eosin, while others were treated immunocytochemically. For electron microscopic examination eyes of anaesthetized RCS dystrophic rats, grafted and control, were removed and fixed in 2.5% glutaraldehyde in sodium phosphate buffer overnight at 4°C. The eyes were trimmed to limits of the lesion site, postfixed in 1% osmium tetroxide, then embedded in Epon following standard procedures. Thick sections (1 µm) were stained with toluidine blue and examined at the light microscopic level, while thin sections, poststained with uranyl acetate and lead citrate, were investigated with a Zeiss EM-10A electron microscope.

Immunocytochemistry

Paraffin sections, 5 µm, were deparaffinized, rehydrated to sodium phosphate-buffered saline (PBS), pH 7.4 and treated immunocytochemically as described previously (Sheedlo and Siegel, 1987; Sheedlo et al., 1987). The antisera used in this study were rabbit antibovine (Na^+ + K^+)-ATPase (31B) antiserum (provided by Dr. G.J. Siegel) and sheep antibovine opsin (Sh-25) antiserum (a kind gift from Dr. D.S. Papermaster). Deparaffinized sections were incubated at room temperature with 20% normal goat ((Na^+ + K^+)-ATPase) or 20% normal rabbit (opsin) serum for 60 minutes. The sections were then incubated in 1:750 (Na^+ + K^+)-ATPase antiserum or 1:2000-1:3000 opsin antiserum, followed by 1:250 diluted goat antirabbit IgG ($F(ab')_2$)-horseradish peroxidase (HRP) conjugate ((Na^+ + K^+)-ATPase) or 1:250 rabbit antisheep IgG ($F(ab')_2$)-HRP (opsin). The primary and secondary conjugates were diluted in 1% NGS or 1% NRS and incubations were overnight at 4°C. For color development, the sections were treated with 0.025% diaminobenzidine, 0.01% hydrogen peroxide in 0.05M Tris-HCl, pH 7.6 for 3-5 minutes. Following this treatment, the sections were washed in distilled water, dehydrated in an ethanol series, then mounted in Permount. The immunostained sections were examined under bright field and Nomarski differential interference optics using a Zeiss Universal light microscope.

RESULTS

Light and Electron Microscopy

As shown in toluidine blue stained-thick Epon sections, retinal regions of RCS dystrophic rats affected by grafted RPE cells exhibited ROS and RIS and an ONL 8-10 cells in thickness. In some grafted areas the debris zone was nonexistant, while in regions lateral to the graft site, this membrane debris was somewhat more pronounced, but not as pronounced as in those regions far removed from the RPE graft (Figure 2). Ultrastructural examination of the RPE-grafted retina revealed melanin-containing cells, putative grafted-RPE cells, attached to Bruch's membrane showing extensive cytoplasmic organization (lysosomes, endoplasmic reticulum, Golgi), which are atypical of RPE cells in the RCS dystrophic rat. Furthermore, apical membrane projections of these cells exhibited a normal structural relationship with ROS (Figure 3). Also, the cells showed ingested shed ROS, which strongly suggests that these melanin-containing cells are functioning RPE cells, which were transplanted from pigmented Long Evans rats (Figure 4).

$(Na^+ + K^+)$-ATPase

Immunostaining for $(Na^+ + K^+)$-ATPase in the retina of control Long Evans rats was most concentrated along the RIS, surrounding cell bodies in the INL and within the IPL and OPL. Less dense $(Na^+ + K^+)$-ATPase immunostain was detected around cell bodies in the ONL (Figure 5A). In 60 day old RCS dystrophic rat, grafted with normal RPE cells at 26 days, those retinal regions affected by the grafted RPE cells exhibited dense $(Na^+ + K^+)$-ATPase immunostaining along the RIS, at the periphery of cell bodies in the INL and IPL and somewhat less dense staining was observed around cell bodies in the ONL. The ONL most directly affected by the RPE graft was about 8-10 cells in thickness (Figure 5B). Even those regions of the retina of 60 day old RCS dystrophic rats not affected by the

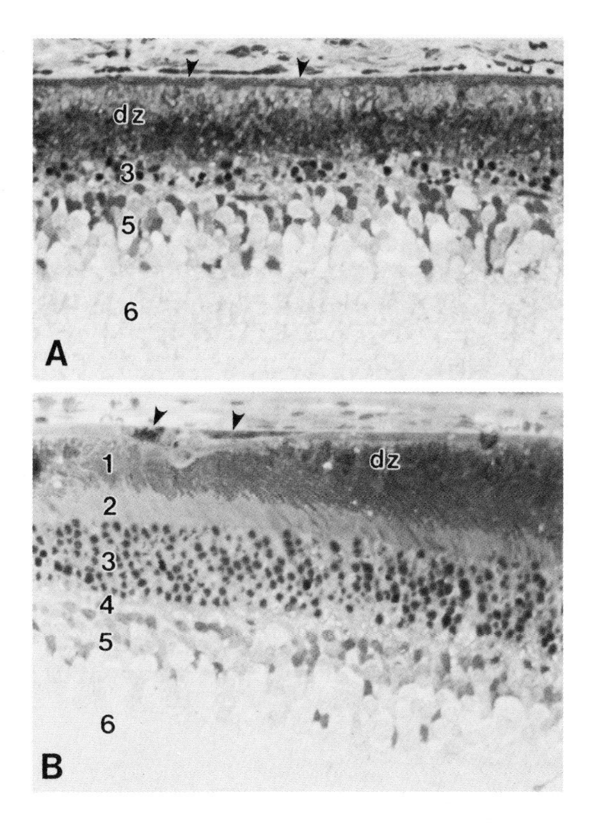

Figure 2. Toluidine blue stained thick-Epon section of sham control and RPE-grafted retina of RCS dystrophic rats. A) Sham control retina. Arrowheads indicate RPE cells. B) Region beneath grafted RPE cells (arrowheads). Magnification: 390x. The following designations are used in Figures 2, 5 and 6: 1-ROS; 2-RIS; 3-ONL; 4-OPL; 5-INL; 6-IPL.

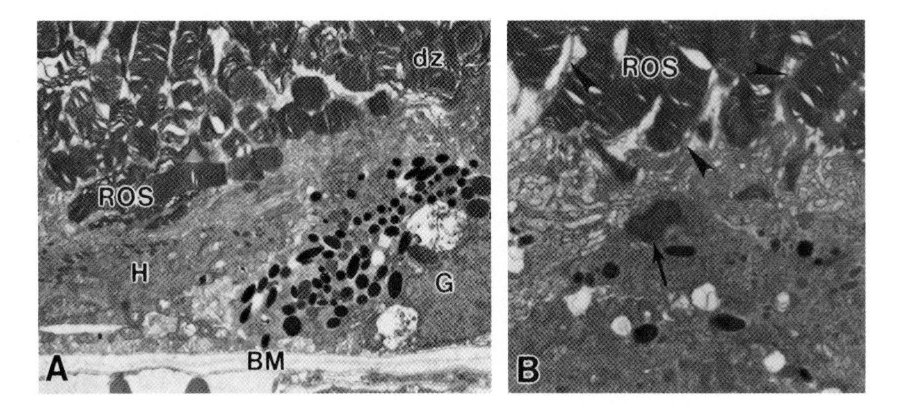

Figure 3. Electron micrograph of RPE-grafted retina of RCS dystrophic rats. A) Grafted (G) and host (H) RPE cells are adjacent and attached to Bruch's membrane (BM). dz, debris zone. B) A grafted RPE cell exhibits microvilli contacting ROS (arrowheads) and an ingested ROS (arrow). Magnification: 2500x (A); 5500x (B).

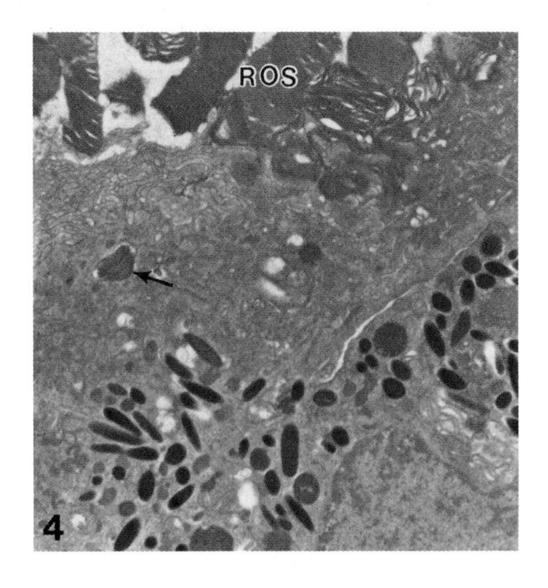

Figure 4. Electron micrograph of grafted RPE cells in retina of RCS dystrophic rats. Arrow indicates ingested ROS. Magnification: 5500x.

graft were immunostained for (Na^+ + K^+)-ATPase, such as around cell bodies in the INL and within the IPL. A few surviving cells in the ONL (1-2 cells thick) were also stained at their periphery; however, no stained RIS were observed due to their complete degeneration. Also, the debris zone did not appear to stain for this enzyme (Figure 5C). In RCS dystrophic rats, the RPE appeared to be stained for (Na^+ + K^+)-ATPase in both graft-affected and -nonaffected retinal regions (Figures 5B,5C).

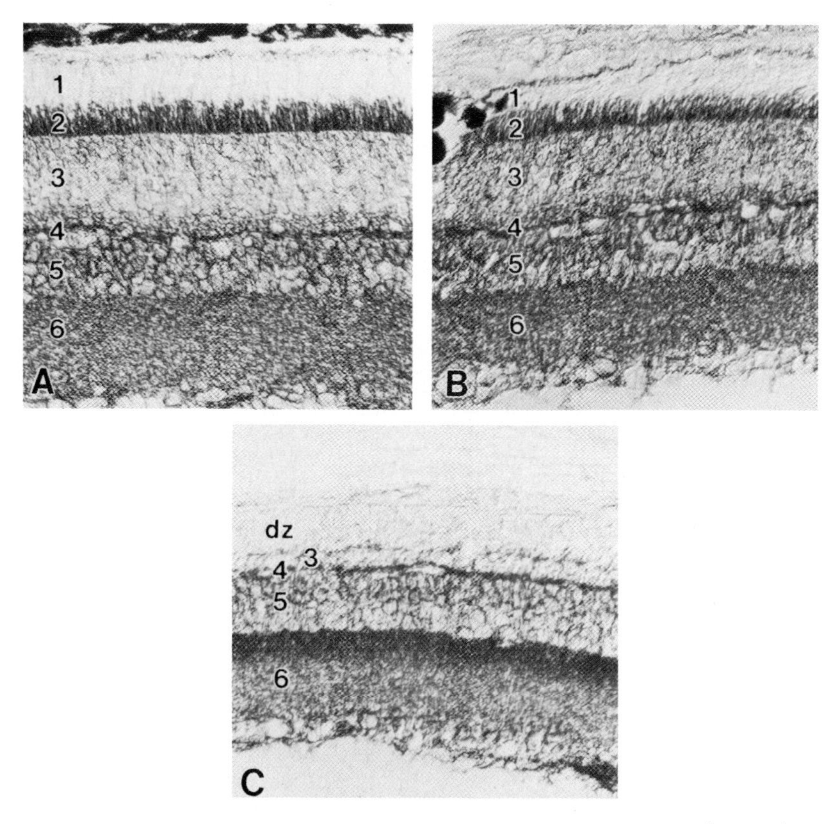

Figure 5. Immunocytochemical localization of (Na^+ + K^+) ATPase in control and RPE-grafted retina. A) Retina of Long Evans rats. B) RPE-grafted retina of RCS dystrophic rats. C) Region of retina not affected by RPE graft in RCS dystrophic rats. dz, debris zone. Nomarski optics. Magnification: 390x.

Opsin

The most dense immunostaining for the photopigment opsin in the retina of Long Evans rats was observed along ROS, while somewhat less dense immunostain was detected along RIS and surrounding cell bodies in the ONL (Figure 6A). In the RPE grafted retina of RCS dystrophic rats, as in the retina of Long Evans rats, rescued PRC's demonstrated dense opsin immunostaining along ROS, while less dense staining was detected along RIS and at the periphery of cell bodies in the ONL. Furthermore, dense immunostaining for opsin was detected in the debris zone (Figure 6B). Degenerated regions of the retina of RCS dystrophic rats which were unaffected by the RPE graft still exhibited opsin immunostain in the debris zone and around a few cell bodies in the ONL (Figure 6C), although ROS and RIS were not detected due to their degeneration. Structural characteristics of the debris zone (dz) were not discernible with opsin immunostaining (Figures 6B,6C).

DISCUSSION

This ultrastructural and immunocytochemical study of RPE-grafted retina of RCS dystrophic rats has clearly shown that the grafted RPE cells and the surviving PRC's exhibit normal functional characteristics. The putative grafted-RPE cells established a normal-appearing structural relationship with ROS of surviving PRC's and, in some cases, have attached to Bruch's membrane (Figures 2,3). Furthermore, the grafted RPE cells have ingested ROS (Figure 4), as well as membrane debris (not shown) which were present in the subretinal space prior to the RPE transplantation. It should be noted that the retinal debris zone in RCS dystrophic rats is maximal at 26 days and still significant at 60 days as shown by opsin immunostaining (Figure 6C), but significantly reduced in graft-affected areas (Figure 6B).

Light microscopic immunocytochemistry revealed that the surviving PRC's in RPE-grafted RCS dystrophic rats are functional as it pertains to two important retinal proteins, the membrane-bound enzyme ($Na^+ + K^+$)-ATPase and the photosensitive protein opsin. Not only are these two proteins synthesized by the rescued PRC's, but the proteins are distributed at their normal respective

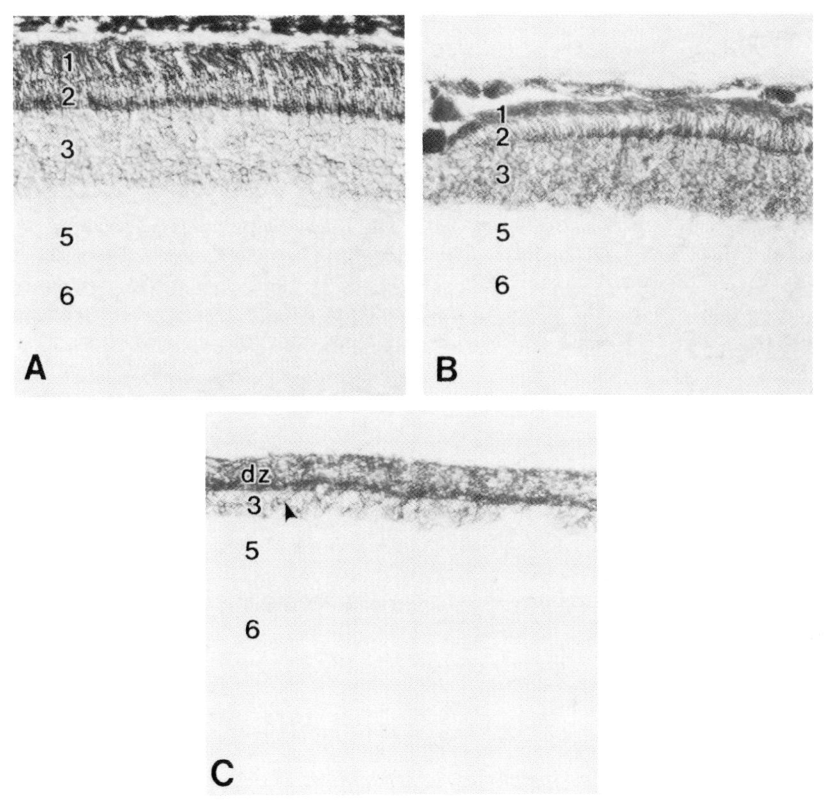

Figure 6. Immunocytochemical localization of opsin in control and RPE-grafted retina. A) Retina of Long Evans rats. B) Retina of RCS dystrophic rats affected by the RPE graft. C) Region of RCS dystrophic rat retina not affected by the graft. Arrowheads indicate an immunostained cell in ONL. dz, debris zone. Nomarski optics. Magnification: 390x.

membrane sites, as determined by contrasting to control retina (Figures 5A,5B,6A,6B). Quite surprisingly, $(Na^+ + K^+)$-ATPase was localized in the IPL and at the periphery of cell bodies in the INL in nongraft affected retinal regions of 60 day RCS dystrophic rats (Figure 5C). It has been theorized that ($Na^+ + K^+$)-ATPase triggers the release of neurotransmitters in synaptic terminals (Stahl and Baskin, 1984). Thus, due to the significantly reduced number of PRC's (ONL 1-2 cells in thickness) and concomitant reduction in PRC originating impulses, it was expected that ($Na^+ + K^+$)-ATPase would be significantly reduced, especially in the dense synaptic IPL region. However, this may indeed occur at later time periods.

In conclusion, surviving PRC's resulting from the grafting of normal RPE cells in RCS dystrophic rats are metabolically functional for ($Na^+ + K^+$)-ATPase and opsin. Thus, it is possible that graft-affected retinal regions with surviving PRC's will respond to light stimuli, which could be determined by ERG recording and other methods.

ACKNOWLEDGEMENTS

The authors wish to acknowledge the expert technical assistance of Ming Lei. This study was supported by grants from the National Eye Institute (EY 04337), Retinitis Pigmentosa International and Retinitis Pigmentosa Foundation Fighting Blindness awarded to J.E.T.

REFERENCES

Bourne MC, Campbell DA, Tansley K (1938). Hereditary degeneration of the rat retina. Br J Ophthalmol 22:613-622.

LaVail MM (1981). Analysis of neurological mutants with inherited retinal degeneration. Invest Ophthalmol Vis Sci 21:638-657.

LaVail MM, Sidman M, Rausin R, Sidman RL (1974). Discrimination of light intensity by rats with inherited retinal degeneration. A behavioral and cytological study. Vision Res 14:693-702.

Li L, Turner JE (1988a). Transplantation of retinal pigment epithelial cells to mature and adult rat hosts: Short and long term survival characteristics. Exp Eye Res in press.

Li L, Turner JE (1988b). Inherited retinal dystrophy in the RCS rat: prevention of photoreceptor degeneration by pigment epithelial cell transplantation. Exp Eye Res in press.

Lucas DR, Attfield M, Davey JB (1955). Retinal dystrophy in the rat. J Pathol Bacteriol 70:469-474.

Mayerson PL, Hall MO, Clark V, Abrams T (1985). An improved method of isolation and culture of rat retinal pigment epithelial cells. Invest Ophthalmol Vis Sci 26:1599-1609.

McGrail KM, Sweadner KJ (1986). Immunofluorescent localization of two different Na,K-ATPases in the rat retina and in identified dissociated retinal cells. J Neurosci 6:1272-1283.

Nir I, Sagie G, Papermaster DS (1987). Opsin accumulation in photoreceptor inner segment plasma membranes in dystrophic RCS rats. Invest Ophthalmol Vis Sci 28:62-69.

Nir I, Cohen D, Papermaster DS (1984). Immunocytochemical localization of opsin in the cell membrane of developing photoreceptors. J Cell Biol 98:1788-1795.

Sheedlo HJ, Turner JE (1988). Immunocytochemical investigation of GFAP and (Na^+ + K^+)-ATPase in the retina of control and RCS dystrophic (rdy/rdy) rats during development. Soc Neurosci (abstr) in press.

Sheedlo HJ, Siegel GJ (1987). Comparison of the distribution of (Na^+ + K^+)-ATPase and myelin-associated glycoprotein (MAG) in the optic nerve, spinal cord and trigeminal ganglion of Shiverer (Shi/Shi) and control (+/+) mice. Brain Res 415:105-114.

Sheedlo HJ, Starosta-Rubenstein S, Desmond TJ, Siegel GJ (1987). (Na^+ + K^+)-ATPase in C_6 glioma and rat cerebrum. J Neuroimmunol 15:173-184.

Stahl WL, Baskin DG (1984). Immunocytochemical localization of Na^+,K^+ adenosine triphosphatase in the rat retina. J Histochem Cytochem 32:248-250.

Turner JE, Blair JR (1986). Newborn rat retinal cells transplanted into a retinal lesion site in adult host eyes. Dev Brain Res 26:91-104.

Inherited and Environmentally Induced
Retinal Degenerations, pages 659–671

TRANSPLANTATION OF RETINAL EPITHELIUM PREVENTS PHOTORECEPTOR DEGENERATION IN THE RCS RAT

P. Gouras, R. Lopez, H. Kjeldbye,
B. Sullivan, M. Brittis
Department of Ophthal. Columbia Univ.
630 W. 168 Street, N.Y., N. Y. 10032

INTRODUCTION

The Royal College of Surgeons strain of rats (RCS) is one of the unique hereditary retinal degenerations in which the cellular nature of the genetic defect is known, although the underlying molecular defect is still being sought. The photoreceptors of this rat degenerate because the retinal epithelium fails to phagocytize the growing outer segments and this leads to total destruction of the receptor cell and subsequent blindness (Bourne et al., 1938; Dowling et al., 1962; Herron et al., 1969; LaVail et al., 1975). The development of techniques to transplant retinal epithelial cells that remain viable and functional (Gouras et al., 1986; Lopez et al., 1987) for relatively long periods of time, (Brittis et al., 1988; Lin-Xi and Turner, 1988) has opened up the possibility of correcting such a genetic defect by transplantation. To this end we have attempted to use a closed eye method to transplant retinal epithelial cells from pigmented normal rats to the subretinal space of the congenic but albinotic, dystrophic rat. These transplanted cells remain viable and phagocytize photoreceptor segments for months after transplantation, locally correcting the defect inherent in the host strain. These changes lead to a much greater survival of photoreceptor nuclei in those areas of the retina where transplant retinal epithelium is found. Therefore this hereditary retinal defect appears to be correctable by the use of cell transplantation. A brief abstract on this approach has been published (Reppucci et al., 1988).

MATERIALS AND METHODS

Infant dystrophic rats from 15 to 25 days old were used for the transplantation experiments because at this time the photoreceptor layer is still intact. Ketamine (2mg/rat/hr) was administered by subcutaneous injection. The surgery was performed under sterile technique. The periocular tissue was cleansed with an iodine solution and the eye was proptosed. The pupil was dilated with 1% tropicamide and 2.5% phenylephrine HCl. A conjunctival flap was formed at the equator and an incision made through the sclera, choroid and neural retina to the vitreal cavity with a stiletto blade. This posterior approach is necessary to avoid damaging the relatively large lens in the rat eye. A microdiathermy probe was introduced through the incision and guided to the retinal surface on the other side of the eye using an operating microscope and a contact lens, the latter designed specifically for the infant rat eye. A small area (0.2-0.5mm) of neural retina was diathermized to prevent any retinal bleeding. Then a glass cannula having a slender shank, a tip diameter of 100-125 microns and flat edge was introduced to produce a bleb detachment with jet stream lift-off of host epithelium and subsequent injection of freshly harvested, dissociated pigmented retinal epithelial cells. In order to facilitate rapid reattachment of the neural retina over the transplant zone, a small bubble of expanding gas (perfluoroethane) was injected into the vitreous just after the incision was tightly sutured with 11-0 vicryl sutures. The entire procedure took approximately 15-30 minutes.

The transplanted cells were obtained from congenic, pigmented rat eyes by trypsin digestion (0.25%) of the eye cup at 37^{0}C for 1 hr from which the neural retina had been removed. The eye was previously placed for 1 hr in Hank's solution containing 50 ug/ml gentamicin and 100 ug kanamycin to insure sterility. The eye cup is then placed in minimal essential media (MEM) with 20% fetal calf serum (FCS) to stop the trypsinization and the RPE cells are gently brushed off. An aliquot of this cell solution is measured in a hemocytometer to determine the cell concentration. Then the entire cell solution is centrifuged at 1000 RPM for 5 minutes; the supernatant is removed and the relatively concentrated cell mixture (10-20ul) pipetted onto a sterile Petri dish. A small air bubble is sucked into the micropipette followed by the cell suspension and a small volume of balanced salt solution, which is used to produce the bleb detachment.

Rats have been sacrificed days and months after transplantation surgery to one eye while the other eye served as a control. The eyes are punctured at the limbus and placed in a solution of buffered 3% glutaraldehyde for 24 hours. After dissection, the tissue was post-fixed in 1% osmic acid in Earle's buffer for 1 hour, dehydrated with ethanol and embedded in Epon. Sections were cut 1 to 2 um thick and stained with toluidine blue for examination by light microscopy. Selected blocks were trimmed and thin sections cut for examination by electron microscopy.

The electroretinogram was recorded with saline wick electrodes and produced by ganzfeld white light flashes. Many responses to the same stimulus were averaged by a Nicolet computer using a flash frequency of 1/sec. The rats were anesthetized with nembutal (5mg/100gm) intraperitoneally and dark adapted for one hour.

RESULTS

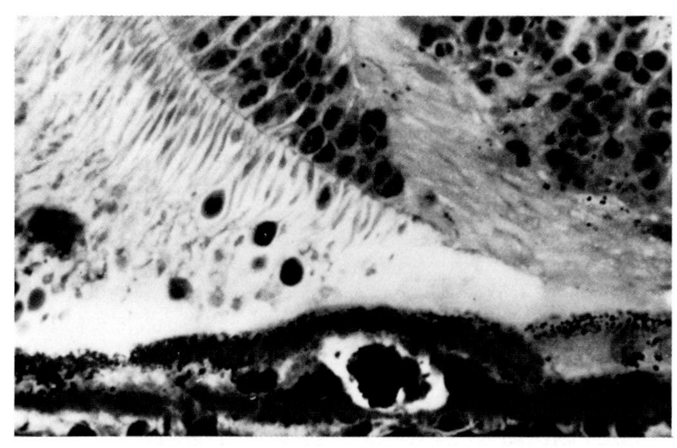

Figure 1 Light micrograph of retina of a 3 month old dystrophic rat that received normal retinal epithelial cell transplants 2.5 months earlier. Photoreceptor outer segments are preserved in this area whereas they are absent in areas without transplant cells or in the contralateral control eye which received no transplants.

Figure 1 shows a light micrograph of transplant cells layered on Bruch's membrane 2.5 months after surgery. Only transplant cells, recognizable by their melanin granules are present occupying the original position of the host epithelium, dislodged by the jet stream. Here slender outer segments of photoreceptors remain on a patch of neural retina, whereas no photoreceptors could be found in the opposite eye which received no transplants.

Figure 2 (top) shows transplant cells within the zone occupied by the outer segments as well as intermingled in a bilayer formed with the host epithelium on Bruch's membrane 2.5 months after surgery. There is considerable preservation of photoreceptor nuclei in this retinal area at 3 months of age when virtually all photoreceptors are lost from these areas of the retina. Figure 2 (bottom) shows retina from a similar area in the control eye where absolutely no photoreceptors are found.

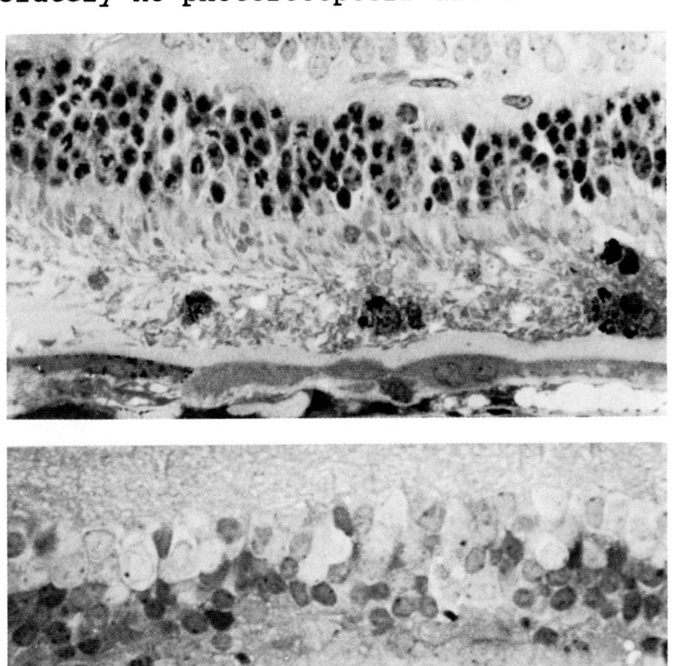

Figure 2. Light micrographs from the retinas of the same 3 month old dystrophic rat. The upper photograph shows the presence of pigmented, transplant retinal epithelial cells from a normal rat in among the outer segments and intermingled with host epithelium bilayered on Bruch's membrane. Many photoreceptor nuclei are present in this area. The lower photomicrograph shows a similar area in the opposite (control) eye. No photoreceptor nuclei are present here or anywhere in this retina.

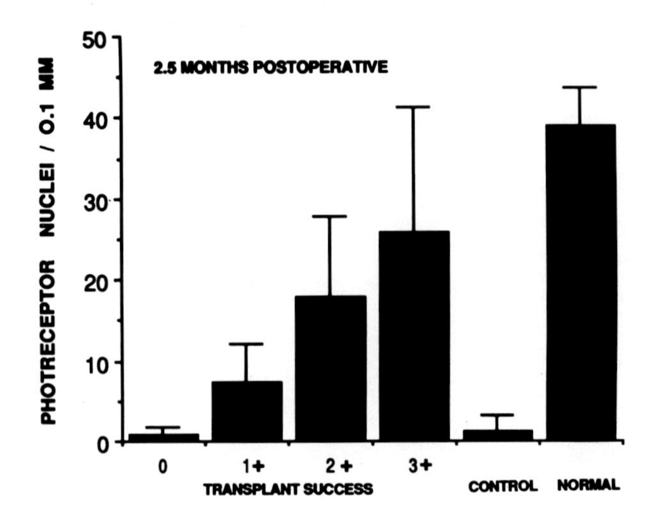

Figure 3 shows the relationship between the amount of receptor nuclei present compared to the number of transplant cells in that area (graded 0 to 3) for eyes receiving transplants (left four columns) compared to the opposite control eye and the normal rat (right two columns). The standard deviation for measurements made on four rats receiving transplants and one normal rat are shown as vertical lines above each bar.

Figure 3 shows a close relationship between the amount of transplant cells found in a particular area of the retina and the number of neighboring nuclei. The more transplant cells found in an area, graded by observation, the greater are the number of receptor nuclei present in that area. In areas where there are no transplant cells (graded 0) there are virtually no photoreceptor nuclei. In the opposite eye which receives no transplant cells and serves as a control, there are also virtually no receptor nuclei. Where the most transplant cells are seen (GRADED 3+), there are the maximum number of photoreceptor nuclei. Transplantation, therefore, acts to preserve the viability of the photoreceptor cell.

Figure 4 shows electron micrographs of transplant cells 48 hours after transplantation into the dystrophic rat retina. A pigmented transplant cell can be seen in Figure 4 (top) with phagosomes in its cytoplasm and wedged in between two host retinal epithelial cells without phagosomes. The transplant cell is in close contact with the outer segments of the photoreceptors, which are still present at this stage (20 days after birth) in the dystrophic rat retina. Figure 4 (bottom) shows a transplant cell containing both melanin granules and numerous phagosomes, much more than we find in the intact retinal epithelium of normal rats. This super abundance of phagosome and phagolysomal material is even more striking in transplant cells which are present more than two months in the dystrophic retina.

Figure 5 illustrates with electron micrographs, transplant cells present in the retina of the dystrophic rats 2.5 months after transplantation. Several transplant cells resting on top of an unpigmented host retinal epithelial cell can be seen containing numerous phagolysosomes as well as fresh phagosomes. The latter can be seen on the upper aspect of the central transplant cell in Figure 5 (top). Another transplant cell is shown in Figure 5 (bottom) which is engorged with phagolysosomes and can be seen to be actively phagocytizing more outer segment discs. We have considered this phenomenon to be a type of "hyperphagocytosis", undoubtedly induced by the super abundance of effete outer segment material.

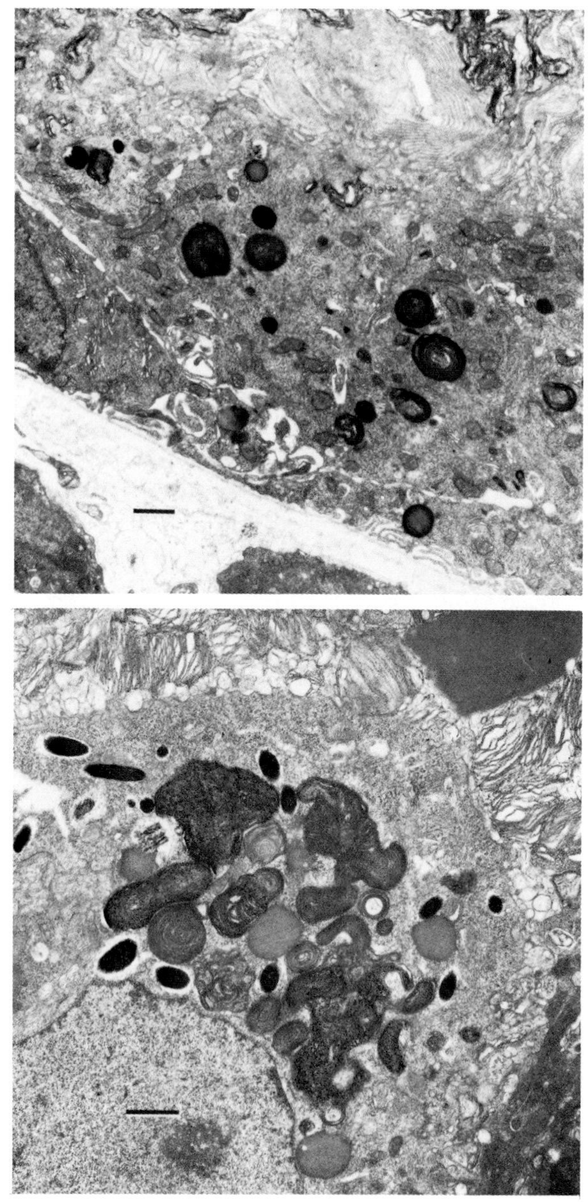

Figure 4. Electron micrographs of transplanted cells in the retina of a twenty day old dystrophic rat 48 hours after transplantation. 1 micron indicated by bar.

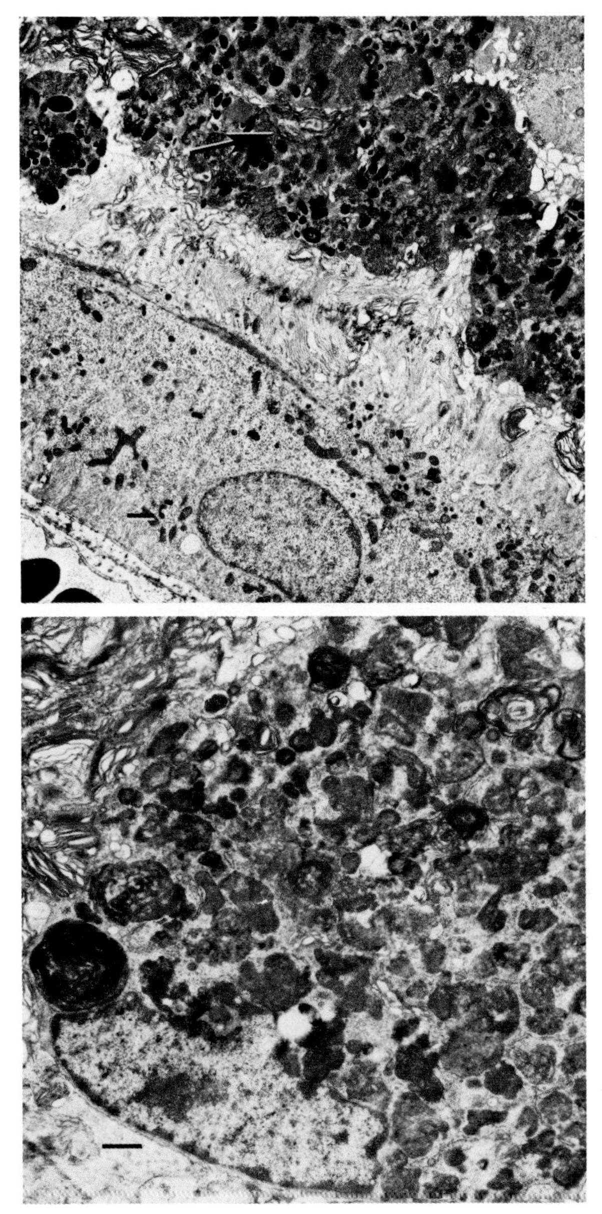

Figure 5. Electron micrographs of transplant cells in the retina of a 3 month old dystrophic rat, 2½ months after transplantation. Arrow (above) indicates a relatively fresh phagosome.

Figure 6 illustrates the degree of this hyperphagocytosis. We believe these amounts will be even higher in transplant cells which are present for longer periods of time in the dystrophic retina.

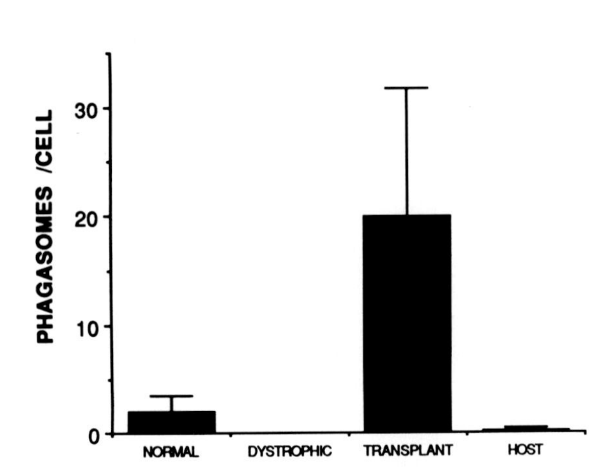

Figure 6. The relationship between the number of phagosomes found in the retinal epithelium of normal rats, dystrophic rats, normal rats transplanted to dystrophic retina 24-48 hours earlier and dystrophic rats, which are hosts to transplants. The standard deviations are shown above each bar.

Electroretinograms have been obtained from five rats which underwent transplantation at 15-20 days after birth and were followed for 2.5 months postoperatively. Four of these rats have been sacrificed and all show some preservation of photoreceptors in those areas of the retina where transplant cells are found. Figure 7 illustrates how the amplitude of the electroretinogram changes with time during the course of the disease (●) and after transplantation (0). The electroretinogram continues to diminish, although at a slightly slower rate in the eye receiving the transplants. At 70 days of age, the amplitude of the electroretinograms are only a few microvolts, near the noise level for its detection. The average response in the transplant eyes is slightly larger than that of the controls but this is not significant. The response obtained from normal rats is between 200-400 microvolts using our current protocol, which requires minimum periods of dark adaptation. The responses from the retinas of the rats receiving transplants are about 1% of this value therefore, it is reasonable to conclude that our current transplant technique can influence only 1%, probably less, of all the photoreceptors in the dystrophic retinas at birth. This is consistent with what we observe histologically.

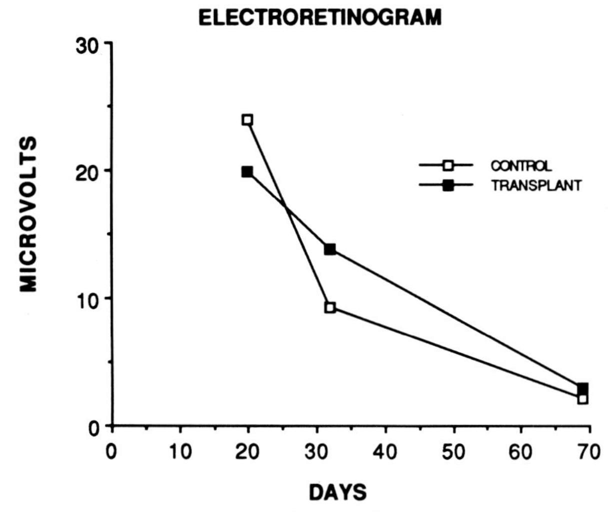

Figure 7. The relationship between the response amplitude of the electroretinogram (ordinate in microvolts) and time after birth for the transplanted (●) and control (0) eye of fine dystrophic rats. Transplantation occurred several days before the initial electroretinogram was obtained.

DISCUSSION

This paper shows it is possible to transplant normal rat retinal epithelial cells into the subretinal space of the dystrophic rat retina, have the host retina reattach over these transplants and restore phagocytosis of the host photoreceptor outer segments, thereby correcting the genetic defect that leads this retina to degenerate. The longer term experiments reveal that this leads to a survival of photoreceptor nuclei in those areas of the retina where transplanted cells are found. There appears to be a strong correlation between the number of transplant cells present and the number of photoreceptors that survive. It is our impression that normal retinal epithelial cells present merely among the outer segments of the photoreceptors are as effective in arresting the degeneration as those either on top of host retinal epithelium or on Bruch's membrane, itself.

The fraction of photoreceptors that can be influenced by our current method, which introduces 5,000 to 20,000 transplant cells into the subretinal space through an anterior approach, is relatively small, less than 1% of all the photoreceptors in this retina. This is too small a number to generate a sufficiently large electroretinogram which would allow us to follow the progress of the degeneration non-invasively. It would be valuable to be able to increase the number of transplant cells delivered and consequently the numbers of photoreceptors that survive. To this end we have been using the transchoroidal approach that has been introduced by Xin-Li and Turner (1988). We find that this can deliver a larger number of transplant cells although it has the disadvantage of making jet stream removal of the host epithelium somewhat more difficult and is not an ideal approach for diseases involving the macular area in primates, which may also be amenable to transplantation as a potential treatment.

REFERENCES

Bourne MC, Campbell DA, Pyke M (1938) Cataract associated with a hereditary retinal lesion the rat. Brit J Ophthal 22: 608.

Brittis M Gouras P, Lopez R, Sullivan B, Kjeldbye H (1988) Retinal epithelial allografts facilitated by cyclosporine. Invest Ophthal Vis Sci 29 (suppl):241.

Dowing JE, Sidman RL (1962) Inherited retinal dystrophy in the rat. J Cell Biol 14:73.

Gouras P, Lopez R, Brittis M, Kjeldbye H, Fasano MK (1986) Transplantation of retinal epithe-ium. IN: Signal Systems in the Normal and Degenerating Retina. B. Ehinger ed, Amsterdam, Elsevier pp. 271-286.

Herron WL, Riegel BW, Myers OE, Rubin ML (1969) Retinal dystrophy in the rat - a pigment epithelial disease. Invest Ophthal 8:595.

LaVail MM, Sidman RL, Gerhardt CO (1975) Congenic strains of RCS rats with inherited retinal dystrophy. J Hered 66:242.

Lin-XI L, Turner JE, (1988) Transplantation of retinal pigment epithelium to immature and adult rat host. Invest Ophthal Vis Sci 29 (suppl):241.

Lopez R, Gouras P, Brittis M, Kjeldbye H (1987) Transplantation of cultured rabbit epithelium to rabbit retina using a closed eye method. Invest Ophthal Vis Sci 28:1131.

Reppucci V, Goluboff E, Wapner F, Syniuta L, Brittis M, Sullivan B, Gouras P (1988) Retinal pigment epithelium transplantation in the RCS rat. Invest Ophthal Vis Sci 29 (suppl):144.

Inherited and Environmentally Induced
Retinal Degenerations, pages 673–686

RETINAL TRANSPLANTS FOR CELL REPLACEMENT IN PHOTOTOXIC RETINAL DEGENERATION.

Manuel del Cerro, Mary F. Notter, Donald A. Grover, John Olchowka, Luke Qi Jiang, Stanley J. Wiegand, Eliot Lazar, and Coca del Cerro.

Departments of Neurobiology and Anatomy, (M.d.C., M.F.N., J.O., L.Q.J., S.J.W., E.L., C.d.C.) and Ophthalmology, (D.A.G.) University of Rochester, Medical School, Rochester, New York, 14642.

INTRODUCTION

The tremendous progress made during the last decade in the field of neural transplantation raises the hope that transplants into adult retinas may be able to repopulate cells destroyed by degenerative disorder. To test this possibility we grafted photoreceptor precursor cells into adult retinas damaged extensively by phototoxic injury. In this experiment, adult Fisher 344 rats, which had been chronically exposed to continuous light irradiation served as hosts. The donor tissue consisted of a suspension of perinatal neural retinal cells (600×10^3 cells/ μl) transplanted with a survival time of ten to one hundred days. The host retinas showed massive destruction of the outer retinal layers, while the transplants appeared as clusters of photoreceptors cells growing within the host retina. Physical continuity, areas of synaptic contacts, and common vascularization existed between the clusters of transplanted cells and hosts. The results show the feasibility of repopulating, at least on a limited bases, photoreceptor cells in an extensively damaged adult eye.

MATERIAL AND METHODS

Zero to two day-old (PN 0-2) rats of the Fisher 344 strain served as donors. The rats were decapitated and their eyes collected in a dish containing ice-cold calcium-magnesium-free balanced salt solution (CMF). The eyes were dissected while still in cold CMF, and the neural retinas were gently separated from the pigment epithelium and cut away from the optic nerve head.

Sixty day old male albino Fisher 344 rats were used as hosts for this study. The rats were kept in a darkened room for three days.Just before the end of the dark adaptation period the animals were anesthetized and mydriatic drops were placed in their eyes. The rats were then housed in individual plastic cages illuminated by fluorescent light tubes. Light intensity was measured at the bottom of the cage to be 3500 lux (0.0929 lux = 1 footcandle). The animals were kept in this environment for 4 weeks, being given water and food ad libitum. Following this exposure period the rats spent one week in a 12 hr light/ 12 hr dark regimen, with the light intensity being fourteen foot candles. After this period, they were used as hosts for transplantation.

Neural retinas from 18 to 22 eyes, PN 0-2, were placed in cold CMF buffer with 0.1% glucose, 100 µg/ml Streptomycin, 2.5 µg/ml Fungizone (Notter et al., 1984). After 2-3 changes of fresh buffer, the tissue was cut into small pieces and transferred to a centrifuge tube containing CMF with 0.02% EDTA. After decanting the buffer, 1 ml of 0.1% trypsin (Sigma, type III) and 0.02% EDTA in CMF was added to the tube containing the retinal tissue. The tube was then incubated for 15 minutes at 37°C and the tissue was triturated into a single cell suspension by aspirating it through a fine-pulled pipette in the presence of 50 µg/ml deoxyribonuclease, which was added to inhibit cell clumping. Fetal calf serum (final concentration, 10%) was added to quench the trypsin activity. Cells were centrifuged and the resulting pellet suspended in 1 ml of Hanks balanced salt solution. Aliquots were used to make cell counts and determine viability. Cell viability as estimated by trypan blue exclusion ranged between 85 and 95 %.

The donor cells were labelled with the fluorescent stain Fast Blue, to allow pos-transplantation identification. Detailed technical description of the procedure is given in a separate publication (del Cerro et al., in press). The basic outline is as follows. A

0.025 - 0.035 % (w/v) solution of Fast Blue (Sigma Co. St.Louis, MO) in normal saline or calcium-magnesium free (CMF) medium,containing 0.6% glucose, was made fresh from a 5 % stock solution. Cell pellets were resuspended in the solution and incubated for 30 min at 4° C. The dye was removed by centrifugation, and the labeled cells were rinsed two times in CMF. A sample of suspended cells was observed under a microscope using fluorescence optics to verify label incorporation and absence of Fast Blue in the medium. Viable transplants were obtained with cell suspensions kept on ice for three hours, the longest time tested. Dissociated cells, labeled and adjusted to the final concentration of 600.000/ µl were injected into the eye of anesthetized hosts. The injection point was located at the 12 o'clock position, behind the eye equator. Needle penetration was limited to 1 mm in order to deliver as many cells as possible within the thickness of the host retina. Survival times ranged from 3 to 100 days pos-transplantation, with the animals receiving repeated ophthalmologic examinations during that period. At the end of the survival period, the animals were perfused through the heart with a mixture of 2% glutaraldehyde and 2% paraformaldehyde, in a 0.1M cacodylate buffer, at a pH of 7.4. The eyes were enucleated, hemisected, and kept overnight in fresh fixative. They were post-fixed in a chromate-osmium tetroxide solution (Dalton, 1955) and embedded in Epoxy resins. One µm thick sections of the tissue were cut and stained with Stevenel Blue (del Cerro et al., 1980) for light microscopy study. Ultrathin sections were cut,"stained" with lead acetate and used for electron microscopical studies. Quantitation of the number of photoreceptor nuclear profiles per one hundred micrometers of retina length was performed by means of an IBAS image analyzer working on line with a Leitz Ortoplan microscope.

RESULTS

The normal fundus of the young Lewis rat presents a centrally located, relatively small round optic nerve head. Six to eight arteries and veins radiate out from the head of the optic nerve in a spoke like fashion. The veins alternate with the arteries and are one and a half to two times as wide as the arteries. As the vessels leave the optic nerve head they tend to protrude directly forward, then gradually slope down towards the periphery. These vessels tend to have a straight course as they extend towards the retinal periphery, but send out multiple arborizations. The far periphery is difficult to observe because of the astigmatism induced by the spheroidal configuration of the lens in the rat eye. The optic nerve varies in color from reddish to orange or pink. In the absence of pigment the large choroidal vessels, which take long sinuous courses, can be easily observed. Fundoscopic observations of the same animals after a period of continuous light exposure shows thinning of the retina and concomitantly a severe attenuation of the vascular bed in both retina and choroid. In some cases the survival and growth of the graft could be followed through the transparent media of the recipient eye.

Histologically the retina of animals exposed to continuous fluorescent illumination, differs dramatically from that of sex and age-matched controls (Figs. 1-2). Light and electron microscopical observations showed the decimation of elements forming the outer retinal layers. The irradiated hosts showed complete absence of photoreceptor inner and outer segments throughout the retina. A few dystrophic rod cells consisting of a small spherical soma and a rudimentary pedicle were found in the retinal periphery. Not only were these cells highly abnormal, but as shown in table II, their numbers represent only a very small percentage of the normal population.

Fig. 1. Normal Lewis rat retina, from a control animal. The inner and outer nuclear layers are indicated. Magnification 375 x.

Fig. 2. The retina of a Lewis rat exposed to continuous light irradiation as described in the text. All of the outer retinal layers have degenerated, as a result the inner nuclear layer (INL) now contacts the retinal pigment epithelium, indicated by arrows. (V) Vessels within the pigment epithelium. Magnification 375 x.

Fig. 3. Photoreceptor cells (R) within an irradiatad retina that have received a transplant of dissociated retinal cells 90 days before sacrifice. Magnification 375 x.

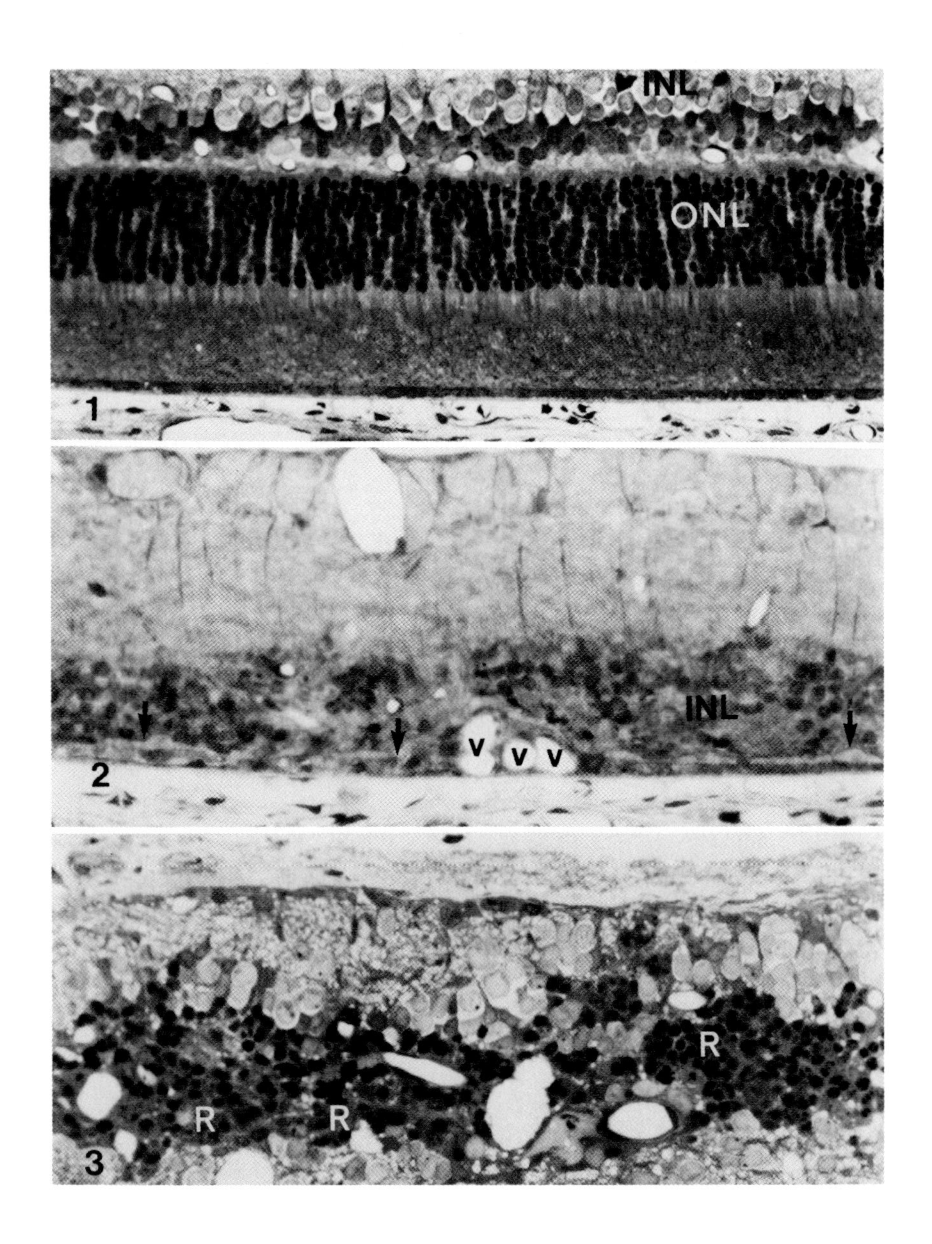
INL
ONL
1
2
INL
V
V
V
3
R
R
R

The apical cytoplasmic processes of the Muller cells, linked to each other by extensive junctional specializations, now formed the outer surface of the neural retina. The Muller cell microvilli, which normally suround the outer segments, appeared hypertrophic, filling to some extent the void left by the degenerated photoreceptors. The retinal pigment epithelium exhibited extensive pathological changes which ranged from focal absence to focal doubling and even tripling of its normal monolayer configuration.

Quantitative observations better delineate the severity and extent of the damage. They indicated that the total retinal thickness was greatly reduced in the peripheral, equatorial, and central retinal sectors, in both the superior and inferior hemispheres (Table 1). Additionally, the number of rod cell nuclei seen in 1 µm-thick sections was reduced to the point of near extintion, particularly in the equatorial and central retina. Data obtained by computer assisted morphometry of saggital sections of the retina, cut along a vertical meridian, are shown in Table 2. The vascular atrophy clinically observed in vivo, had its histological correlate in the form of a few tortuous capillaries crossing the retina in a perpendicular fashion, with greately reduced branching. Additionally, there was an abnormal vascularization of the RPE (Fig. 2) by fenestrated capillaries which upon leaving the apical portion of the RPE cells, extended into the retina itself. In the transplanted eyes, microscopic sections showed that, depending on the experimental group, up to 80% of them had grafted cells, located usually near the injection point. The transplants grew welll, in spite the fact that the host eyes invariably exhibited the profound degenerative changes related to the effects of continuous light exposure. In all cases there was development of clusters of rod cells, their somata forming an irregular outer nuclear layer (Fig.3), with the cells often grouped in rosettes. Fluorescence microscopy showed that the cells forming these clusters carried the Fast Blue label. Rod cell outer segments developed consistently within these rosettes; they had as a constant feature a basal body with a cilium emerging from it. Outer segments also formed, but they were invariably defective, containing collections of irregular cisternae.The outer expansions of the Muller cells contributed to the formation of an outer limiting membrane and to the formation of baskets of filiform projections around the rods inner segments . Synapses occurred in large numbers within the patches of plexiform layer which developed around the transplants. Both conventional and ribbon synapses were present in large numbers within the plexiform layer formed between the inner nuclear layer and the transplanted photoreceptors (Fig. 4).

TABLE 1. Average Retinal Thickness in Micrometers

A) Normal Retinae (N = 6)

	Ps	Es	Cs	Ci	Ei	Pi
Av	121	180	193	202	182	136
SD	3	4	6	8	5	4

B) Light-Damaged Retinae (N = 6)

	Ps	Es	Cs	Ci	Ei	Pi
Av	80	85	111	112	102	91
SD	7	7	4	2	2	8

Ps, Es, and Cs indicate superior peripheral, equatorial and central retinal locations. Ci, Ei,and Pi indicate Inferior central, equatorial,and peripheral retinal locations. Average values and Standard Deviations rounded up to the first integer.

TABLE 2. Photoreceptor-cell nuclear profiles per 100 micrometers of retina.

A) Normal (N = 6)

	Ps	Es	Cs	Ci	Ei	Pi
Av	115	166	174	159	139	100
SD	3	4	2	2	2	9

B) Phototoxic Degeneration (N = 6)

	Ps	Es	Cs	Ci	Ei	Pi
Av	14	1	0	1	1	12
SD	3	3	1	0	0	8

Ps, Es, and Cs indicate Superior peripheral, equatorial and central retinal locations. Ci, Ei, and Pi indicate Inferior central, equatorial, and peripheral retinal locations. Average values and Standard Deviations rounded up to the first integer.

Fig. 4. Synaptic contacts can be seen in the neigborhood of transplanted rod cells, either within small patches of plexiform layer or intermingled with rod cell somas (R), like in this field. The arrows indicate the ribbons present within the photoreceptor terminals. Mag. 25,000 x.

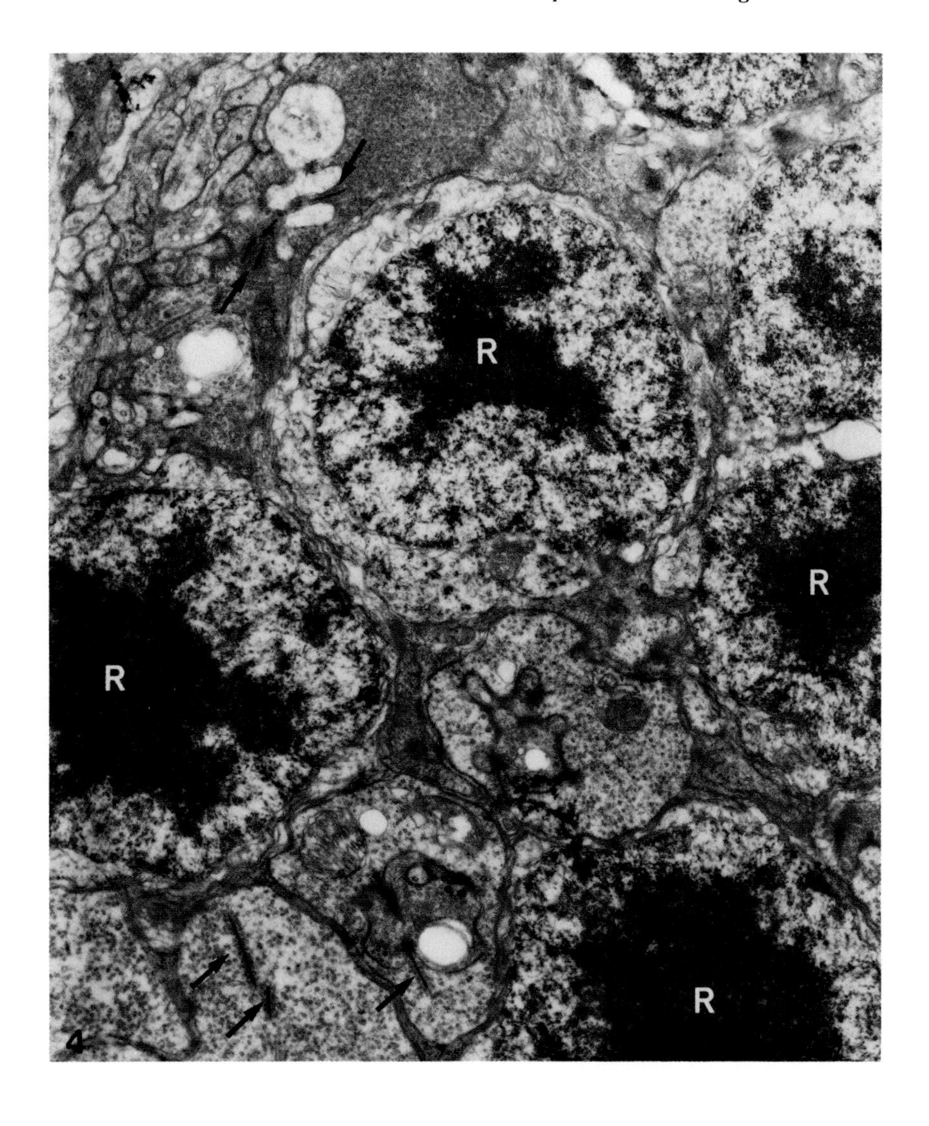
R
R
R
R
4

DISCUSSION

There are two items of particular relevance to the results described. The first point is that of light-induced retinal damage, while the second pertains to the issues concerning retinal tissue transplantation.

The power of visible light to induce irreversible retinal damage under certain conditions, has been well established since the classic experiments of Noell et al. (1966). Our experimental protocol, involving long-term exposure to moderate light intensities, leads to induction of type II light damage (Noell, 1980). This type of damage is characterized by diffuse photoreceptor loss with general preservation of the retinal pigment epithelial cells.

Marked retino-choroidal vascular damage was observed in our studies in conjunction with development of the phototoxic retinopathy. These vascular correlates of photoreceptor decimation have been the object of detailed studies (Bellhorn, 1980). The vascular atrophy may be induced by the lowered metabolic requirements of a retina greatly reduced in total mass, rather than be a direct actinic action, since a comparable vascular attenuation occurs in inherited retinal degenerations (Blanks and Johnson, 1986).

Regarding the transplantation of retinal tissue we have shown the receptivity of the adult rat eye to transplants of embryonic and perinatal retinas, even in transplants made across strains (del Cerro et al., 1984,1985, 1986, 1987). Building upon this foundation we wished to determine if eyes with extensively damaged retinas and severely compromised vascular supply, would also be able to incorporate transplanted cells and support their growth. The foregoing observations give answers to these questions. They demonstrate that normal perinatal retinal cells, even if transplanted to a diseased retinal locus are still able to differentiate and to supply the host with photoreceptor cells. The total mass of the transplants did not reach in any instance that of retinal transplants growing in the anterior chamber of normal eyes (del Cerro et al., 1988), and the laminar organization of the attained by the transplanted cells was less than that seen in the best organized anterior chamber transplants. One possible explanation for these findings is that the differentiation of immature cells into photoreceptors may be follow internal clues and be minimally affected by environmental factors. On the other hand, extrinsic factors may play a large role in guiding the laminar arrangement of the cells. Finally, the extent of the growth may be dependent on the appropriateness of the vascular supply. Thus transplants growing on the highly vascularized iris should fare better than those growing on the poorly vascularized environment of the irradiated retinas. It is also possible that the observed differences in growth and lamination be related to the fact that the transplants to the anterior chamber were made using tissue fragments while transplants into the retina involved cell suspensions.

Two standard questions are posed by the outcome of almost any experiment involving neural grafting, these are: How can the transplant be identified within the host tissue, and to what extent and in what fashion do the transplanted cells interact with the host neurons. In the case of the retinal transplants into light damaged retinas the first question is easily answered. Under our experimental conditions the central and equatorial regions of the hosts loose virtually all receptor cells. Only a few photoreceptor cells remain in the far retinal periphery. The photic insult to the periphery exacerbates the spontaneous tendency to degenerate expressed by the peripheral retina of the Fisher 344 rat (Lai et al., 1978; Shinowara et al., 1982). The combined effects of these two deleterious forces is that our hosts appear to have retinas deprived of photoreceptors to a greater extent than light irradiated animals previously reported in the literature. Thus the presence of clusters of several hundred photoreceptor cells unmistakably identifies the graft. In fact such clusters of photoreceptor cells have never been observed in any of the approximately one hundred phototoxic retinitis-affected eyes we have examined in conjunction with this and other studies. Further the observation of fluorescent labeled cells in the transplanted eyes confirms that these clusters are indeed of cells originated from the donor (See also del Cerro et al., in press). Finally, electron microscopical observations show an important difference between the ultrastructure of the few surviving rod cells seen in the control, non-transplanted phototoxic retinitis affected, animals and those present within the clusters. The latter have a tendency to develop inner and outer segments. Although these structures do not develop normally, their mere presence sets these cells apart from those seen in areas outside the transplant region, where the few remaining rod cells are consistently deprived of both inner and outer segments. Actually the loss of these structures is an early event in the development of phototoxic retinitis (Kuwabara and Gorn, 1968).

The second, and more involved question is that of the synaptic connectivity of the transplanted photoreceptors. Numerous synaptic contacts, including ribbon synapses are found within areas of plexiform layers formed around the clusters of transplanted photoreceptors. Some of the contacts are abnormal, with some occuring as invaginations within the photoreceptor cell soma, instead of being on a pedicle projecting from the cell body; but still retaining enough of the typical morphology to be recognizable as typical outer plexiform layer synaptic contacts. This evidence indicates that the transplanted photoreceptors are capable of synaptically linking to other cells. Two possibilities, which are not mutually exclusive do exist regarding the origin of the synaptic partners to the transplanted photoreceptors. The photoreceptors may be either in synaptic contact with the host inner nuclear layer neurons, or with co-transplanted cells, or both. An unequivocal answer to this question will require differential labeling the host cells with an electron microscopically identifiable tag. Efforts in this direction are currently underway.

Blindness resulting from retinal disease is, in many instances, the consequence of extensive damage to the photoreceptor cell population, while the other cell types forming the neural retina are relatively spared. In this situation transplantation of photoreceptor cells could offer hope for the restoration of some degree of visual function. The results open further possibilities for the study of retinal transplantation.

REFERENCES

Bjorklund A, Stenevi U, Schmidt RH,Dunnet SB, Gage FH (1983) Intracerebral grafting of neuronal cell suspensions . I. Introduction and General methods of preparation. Acta Physiol Scan [Suppl] 522: 1-7.

Dalton, AJ (1955). A chrome-osmium fixative for electron microscopy. Anat Rec 121: 281.

Bellhorn, RW Burns MS Benjamin JV (1980). Retinal vessel abnormalities of phototoxic retinopathy in rats. Invest Ophthalmol Vis Sc 19: 584-595.

Blanks JC, Johnson LV (1986). Vascular atrophy in the retinal degenerative *rd* mouse. J Comp Neurol 254:543-553.

del Cerro M Cogen J del Cerro C (1980 a). Stevenel Blue, an excellent stain for optical microscopy study of plastic embedded tissues. Microscop Acta 83: 117-121.

del Cerro, M Standler N, del Cerro, C (1980b). High resolution optical microscopy of animal tissues by the use of sub-micrometer thick sections and a new stain. Microscop Acta 83: 217-220.

del Cerro M, Gash DM, Rao GN, Notter MF, Wiegand SJ, Gupta M (1984). Intraocular retinal transplants. Invest Ophthalmol Vis Sc 25: 62 (abstr.).

del Cerro M, Gash DM, Rao GN, Notter MF, Wiegand SJ, Gupta M (1985a). Intraocular retinal transplants. Invest Ophthalmol Vis Sc 26: 1182-1185.

del Cerro M, Gash DM, Rao GN, Notter MF, Wiegand SJ, del Cerro C (1985b). Retinal transplants into normal and damaged adult retinas. Soc Neurosci Abstr 11: Part 1:15.

Kwabara T, Gorn RA (1968). Retinal damage by visible light. Arch Ophthalmol 79: 69-78

Kwabara T, Wiedman T.(1974). Development of the prenatal rat retina. Invest Ophthalmol Vis Sc 13: 725-739.

Lai, Y-L, Jacoby RO, Jonas AM (1978). Age-related and light-associated retinal changes in Fischer rats. Invest Opthalmol Visual Sc: 17: 634-638.

Larramendi PCH (1985). Method of retrieval of one to two micron sections from glass for ultrathin sectioning. J Microsc Tech 2: 645-646.

Malik S, Cohen D, Mayer E, Pearlman I (1986). Light damage in the developing retina of the albino rat: An electroretinographic study. Invest Ophthalmol Vis Sc 27:164-167.

Noell WK, Walker VS, Kang BS, Berman S (1966). Retinal damage by light in rats. Invest Ophthalmol Vis Sc, 5: 450- 473.

Noell WK (1980). Possible mechanisms of photoreceptor damage by light in mammalian eyes. Vision Res 20: 1163.

Notter MF, Gash DM, Sladek CD, Scharoun SL (1984). Vasopressin in reaggregated cell cultures of the developing hypothalamus. Brain Res Bull, 12, 307-313 .

Parnavelas JG, Globus A (1976). The damaging effects of continuous illumination on the morphology of the rat retina. Exp Neurol 51: 171-187.

Royo PE, Quay WB (1959). Retinal transplantation from fetal to maternal mammalian eye. Growth 23:313-336.

Shinowara NL, London ED, Rapoport SE (1982). Changes in retinal morphology and glucose utilization in aging albino rats. Exp Eye Res 34: 517 -530.

Turner JE, Blair JR (1986) Newborn Rat Retinal Cells Transplanted Into a Retinal Lesion Site in Adult Host Eyes. Developm Brain Res 26: 91-104.

Turner JE, Blair JE, Chappell T (1986). Peripheral Nerve Implantation into a penetrating Lesion of the Eye: Stimulation of the Damaged Retina. Brain Res 376: 246-254.

Venable JH, Cogeshall R (1965). A simplified lead citrate stain for use in electron microscopy. J Cell Biol 25: 407-408.

ACKNOWLEDGEMENTS: This research has been supported by NEI grant 05262, by the Rochester Eye Bank, and by private contributions. We thank Kim Gesell for expert editorial assistance, and Dorothy Herrera and Nancy Dimmick for their excellent help with the microphotografic illustrations.

Inherited and Environmentally Induced
Retinal Degenerations, pages 687–704

PHOTORECEPTOR TRANSPLANTATION IN INHERITED AND ENVIRONMENTALLY INDUCED RETINAL DEGENERATION: ANATOMY, IMMUNOHISTOCHEMISTRY AND FUNCTION

Martin S. Silverman and Stephen E. Hughes

Central Institute for the Deaf, 818 S. Euclid Ave, (MSS, SEH) and the Dept. of Anatomy and Neurobiology, Washington University (MSS), St.Louis, MO 63110

INTRODUCTION

Several forms of blindness (retinitis pigmentosa, retinal detachment, macular degeneration, and intense light exposure-related blindness) are primarily related to the loss of the retinal photoreceptors (Leibowitz et al., 1980; Gartner and Henkind, 1981). However, destruction of the photoreceptors does not necessarily lead to loss of the remaining retina or the axons that connect the retina to the brain. In such cases, if the photoreceptors could be replaced and if they then could innervate the retina appropriately, some degree of vision might be restored. We therefore investigated the possibility of reconstructing retina in model systems where photoreceptor degeneration is inherited or environmentally induced through transplantation of immature and mature rodent photoreceptors as well as mature human photoreceptors.

While major difficulties with technical aspects of photoreceptor transplantation were evident, several factors made such transplantation potentially promising. The photoreceptor layer of the retina is nonvascularized. The necessity for prompt revascularization, which limits the transplantability of most neural tissue, is therefore not a limitation with photoreceptors. Furthermore, nonvascularized tissue has shown the least amount of tissue rejection. The limited vulnerability of the photoreceptor layer to transplant rejection opens up the potential for the transplantation of tissue that is genetically dissimilar, which could have considerable clinical utility.

The retina does not necessarily undergo glial scar formation when damaged as does the adult central nervous system (Bignami and Dahl, 1979; McConnel and Berry, 1982). This characteristic may contribute to the potential of retinal cells to regrow severed axons within the eye (McConnel and Berry, 1982). Regrowth of photoreceptor axons also might be facilitated by the proximity of the photoreceptors' postsynaptic targets within the adjacent outer plexiform layer. Thus in order for the transplanted photoreceptors to make appropriate connections with the host's retina, growth across substantial intervening neural (or glial scar) tissue would not be necessary.

Finally, successful transplantation of other neural tissue, particularly of embryonic and early postnatal retina to the tecta (Lund and McLoon, 1983; McLoon and McLoon, 1984) and to host retina (Del Cerro et al., 1985; Turner and Blair, 1986) led us to consider these experiments to reconstruct dystrophic retina by transplanting photoreceptors. The results described here indicate that photoreceptors can be transplanted and that they are capable of reestablishing light dependent activation of host retina with photoreceptor dystrophy.

GENERAL METHODS

As models for human blindness resulting from the loss of retinal photoreceptors we used animals in which photoreceptor degeneration is environmentally induced (constant illumination; albino rat) or inherited (the *rd* mouse and the RCS rat) through transplantation of immature (7-8 day old rat or mouse), or mature rat photoreceptors as well as mature human photoreceptors. In these animals almost all photoreceptors are eliminated while the remaining retina is preserved. Animals were handled according to the USPHS Policy for Humane Care and Use of Laboratory Animals and the ARVO Resolution on the Use of Animals in Research.

In order to maintain the organization of the donor photoreceptor layer, we chose not to harvest and purify them by standard techniques (Sarthy and Lam, 1979), since these techniques require dissociating the cells, thereby disrupting the organization and cellular polarity of the photoreceptor layer. Instead, we have developed a technique to isolate the intact photoreceptor layer from the retina (Silverman and Hughes, 1987a; submitted). For labeling of transplanted tissue the isolated outer nuclear layer was cultured overnight in the fluorescent dye DiI (Honig and Hume, 1986).

To reduce bleeding and surgical trauma we devised a surgical approach to the subretinal space through the cornea (transcorneal approach) (Silverman and Hughes, 1987a, submitted). Following appropriate survival times, animals were overdosed and perfused. Cryostat sections of both the control eye and the eye receiving the photoreceptor transplant were then cut (20 μm). Antibody labeling for opsin was performed according to Hicks and Barnstable (1987). Elimination of the primary antibody eliminated specific labeling for opsin.

LIGHT DAMAGED RETINA

By using the transcorneal approach to the subretinal space of the eye, we found that the positioning of the photoreceptor layer between the host's retina and the adjacent epithelial and choroidal tissue layers of the eye could be accomplished while minimizing vascular damage. In addition, we have found that this approach does not appear to disrupt the integrity of the retina, which reattaches to the back of the eye in the immediate area of the transplant with the transplanted photoreceptors interposed between the retina and the retinal pigment epithelium (RPE) (Fig. 1). Using this insertion method it is possible to position the photoreceptors at the posterior pole of the retina (Fig. 1A). Since high visual acuity requires the central retina, the ability to make transplants to various portions of the retina and most importantly to the posterior pole are important if this procedure is to be of eventual clinical utility.

To determine the viability of the transplanted photoreceptors, eyes that received immature photoreceptor transplants were examined at one, two, four, or six weeks after transplantation. We found that the photoreceptors survived transplantation at all times tested (36 out of 54 transplants). More importantly, there was no apparent reduction in viability with longer survival times, indicating that the transplants were stable.

The transplanted cells are easily distinguished from the residual photoreceptors by a number of parameters. First, they are found in discrete patches and have the characteristic columnar stacking arrangement of up to about 12 cell bodies that is particular to photoreceptor cells in the outer nuclear layer of the normal retina. They do not have the flattened appearance of the residual native photoreceptors, but instead have the round, nonpyknotic cell body that is characteristic of normal photoreceptor cells. Furthermore, the transplanted photoreceptors can form rosette configurations (a characteristic of transplanted and cultured photoreceptors, LaVail and Hild, 1971; Lund and McLoon, 1983), while residual photoreceptors are not found in these configurations. To investigate the possibility that our surgical procedure in some manner induced the regeneration of patches of native photoreceptors, we performed sham operations. In these cases all procedures were performed as with the photoreceptor transplants except that no photoreceptors were inserted. While the retina reattached to the back of the eye, in no instance

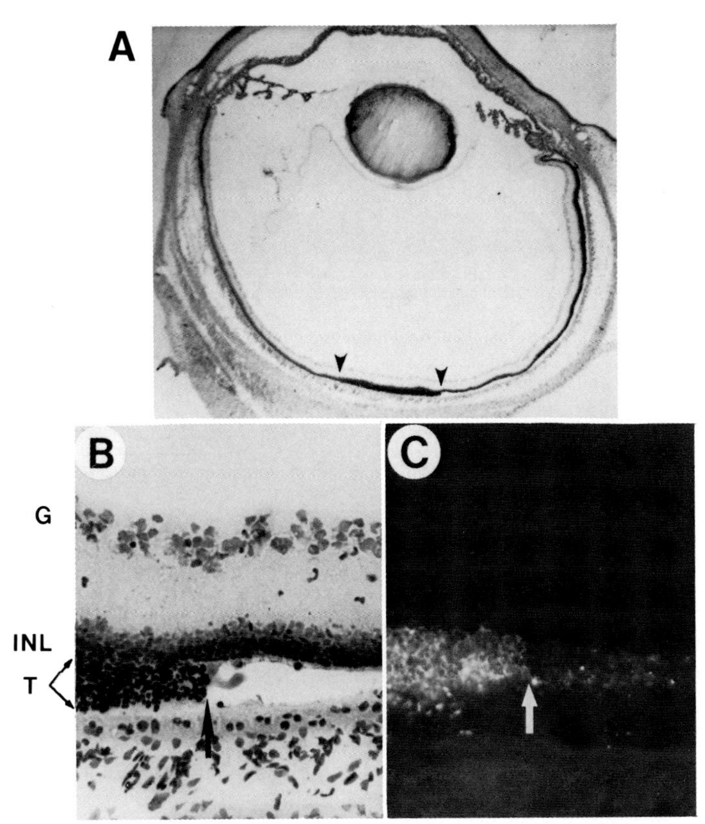

Fig. 1. 4 weeks post-transplantation showing (A) Photoreceptor transplant (between arrowheads) at the posterior pole of the light-damaged host eye. 14X (B) Interface between the transplant and the adjacent retina devoid of outer nuclear layer. Arrows indicate the extent of the transplant (T). G, ganglion cell layer; INL, inner nuclear layer. (C) Antibody RET-P1 specific for opsin. 92X

did we find patches of photoreceptors in these cases.

To positively identify the photoreceptor patches in experimental animals as transplanted tissue, in several cases we labeled the donor outer nuclear layer with the fluorescent marker DiI. As shown in Fig. 2B the photoreceptor patches were fluorescently labeled while the host retina did not show DiI fluorescence. To confirm that the transplants consisted of photoreceptors we used a monoclonal antibody specific for opsin, RET-P1 (provided by C. Barnstable). As opsin is only found in photoreceptors, any cell showing labeling for opsin is therefore identified as a photoreceptor. As can be seen in Figs. 1C and 2C, the transplanted cells stain intensely for opsin whereas other retinal cells, as expected, do not stain. Positive staining for opsin not only identifies these cells as photoreceptors but indicates that these cells are still capable of producing the protein moiety of visual pigment. Retina adjacent to the transplanted photoreceptors shows only a few isolated cells staining for opsin (Fig. 1C). By their location and appearance in H&E stained material, they are identified as the host's residual photoreceptors.

Harvesting the photoreceptor layer from the neonatal retina does not appear to disrupt tissue organization. Once transplanted, the photoreceptor layer was capable of maintaining its characteristic columnar arrangement of photoreceptor cell bodies for all survival times examined, thus forming a new

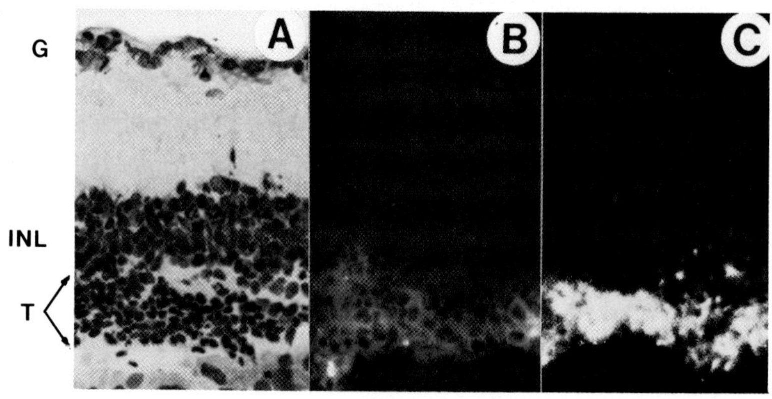

Fig. 2. Three weeks post-transplantation. (A) Transplant and host retina. (B) Transplanted photoreceptors show DiI fluorescence. (C) Antibody RET-P1. 195X

outer nuclear layer within the host's retina. The new layer appears to be attached to the host's outer plexiform layer (Figs. 1B and 2A). This layer normally is the site of synaptic contact between the photoreceptors and the retina.

While the photoreceptors survive, produce opsin and apparently integrate with the host's retina, they do not appear completely normal in that the number of outer segments (OS) is reduced. Photoreceptors lacking OS are still capable of phototransduction (Pu and Masland, 1984). The relative scarcity of OS has also been noted in retina transplanted to the tectum which have also been shown to be functional (Simon and Lund, 1984). This deficiency in OS was thought to be the possible consequence of the lack of appropriate apposition of the RPE to the photoreceptors (LaVail and Hild, 1971). Since in our preparations RPE is present and in apparently normal apposition to the photoreceptors, the scarcity of OS here would not appear to be related to inadequate photoreceptor/RPE contact. This failure of outer segment growth in the presence of photoreceptor apposition to the RPE has also been seen following retinal reattachment (Anderson et al., 1983).

PHOTORECEPTOR TRANSPLANTATION IN INHERITED RETINAL DYSTROPHY

Inherited retinal degeneration afflicts a variety of animals, including humans. Several established animal models exist for inherited retinal dystrophy and retinitis pigmentosa, including the RCS rat and the *rd* mouse. These strains provide model systems in which it is thought that the deficit resides in the photoreceptor (the *rd* mouse) or is related to the pigment epithelium (the RCS rat) (Farber and Lolley, 1974; Mullen and LaVail, 1976).

rd mouse

We have adapted our photoreceptor

transplantation technique to the smaller size of the mouse eye. This modification allows sheets of intact outer nuclear layer to be transplanted to the subretinal space of the mouse eye. We transplanted neonatal (8 day old) photoreceptors from *rd* control mice to the subretinal space of adult *rd* mice. Survival times were for 2 weeks to 3 months. At all survival times we found that the transplanted photoreceptors can survive, becoming physically attached to the outer portion of the host retina and stain positive for opsin. In addition the host retina becomes reattached to the pigment epithelium (Fig. 3A).

In the *rd* mouse almost all photoreceptors are eliminated by day 21. We have found that photoreceptors from a nondystrophic congenic control mouse can be transplanted to their appropriate site within the adult *rd* mouse eye that lacks photoreceptors. These transplanted photoreceptors have been found to survive for as long as so far tested (3 months). This length of time is significant since photoreceptors of the *rd* mouse show signs of degeneration after about 2 weeks and are almost completely eliminated after 3 weeks. The survival of transplanted photoreceptors from congenic normal donors to the adult *rd* mouse supports findings that indicate that the deficit within the *rd* mouse which causes the degeneration of photoreceptors is endogenous to the *rd* photoreceptors themselves.

RCS rat

In our investigation into the feasibility of repopulating the dystrophic retina of the RCS rat we transplanted photoreceptors from 7 to 8 day old RCS controls (normal) to the subretinal space in the eye of adult (3 month old) RCS rats (dystrophic). In these experiments we allowed a two month survival period since in this time period almost all host photoreceptors degenerate in the RCS rat. We found that the grafted photoreceptors do in fact survive transplantation to the subretinal space of the RCS rat (Fig. 3B) and show

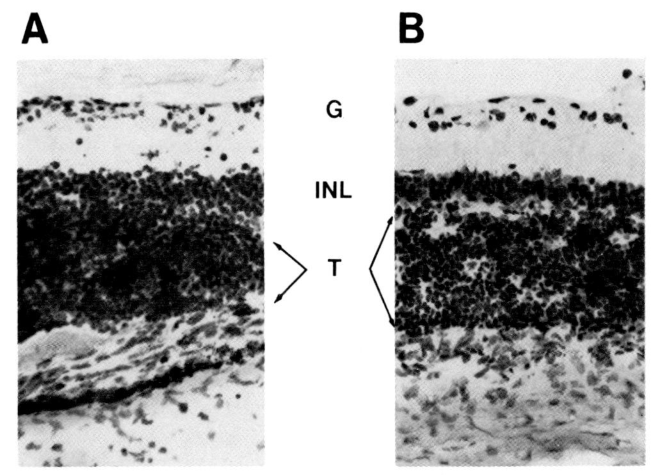

Fig. 3. A. Adult *rd* mouse retina 2 months following transplantation of photoreceptors from a 7 day old *rd* control. B. Adult RCS rat retina 2 months following transplantation of photoreceptors from a 7 day old RCS control. 95X

histotypic as well as immunological characteristics of normal photoreceptors as described above.

We have found that transplanted photoreceptors survive within their homotopic location in the RCS rat whereas the RCS's own photoreceptors do not. It is generally thought that it is not the RCS photoreceptors that are defective, but rather the RCS retinal pigment epithelium that is deficient and thus causes the death of the RCS photoreceptors. Logically, one therefore might expect that the transplanted photoreceptors would be rapidly eliminated from the RCS retina. How then can we explain our results that clearly show that the transplanted normal photoreceptors do in fact survive?

Transplanted photoreceptors might survive in the RCS rat whereas the endogenous RCS photoreceptors do not because transplanted photoreceptors do not produce extensive outer segment growth. Their failure to produce outer segments may protect them from destruction since the accumulation of shed outer segment material within the subretinal space may ultimately lead to photoreceptor elimination.

CRITICAL PERIOD

In our original studies we harvested donor photoreceptors at the earliest ontogenetic time in which the photoreceptors could be isolated from other portions of the retina (7-8 days old), since it is generally thought that more embryonic and undifferentiated neural tissue survives transplantation far better than more mature and differentiated tissue (Das, 1983).

However, we wished to define the "critical period" for photoreceptor transplantation by determining the effect of developmental age on their survival and ability to integrate with the host retina. We subsequently transplanted photoreceptors from 8, 9, 12, 15 and 30 day old rats into light damaged adults. These show progressive development and maturation of the cytological characteristics of mature photoreceptors including mature outer segments (at 15 and 30 days). Using the same criteria as described above we found that for all ages tested the transplants survived for as long as examined (2 months) and integrated with the host retina (Fig. 4A). These observations suggest that photoreceptors have characteristics that differ from other neural tissue that permits them to be transplanted when they are essentially mature while other neural tissue must be at a very immature stage for successful transplantation to occur (Silverman and Hughes, 1987b).

HUMAN PHOTORECEPTOR TRANSPLANTATION

If mature rodent photoreceptors can be transplanted, then the next logical question to us was whether adult human photoreceptors obtained from donors could be transplanted.

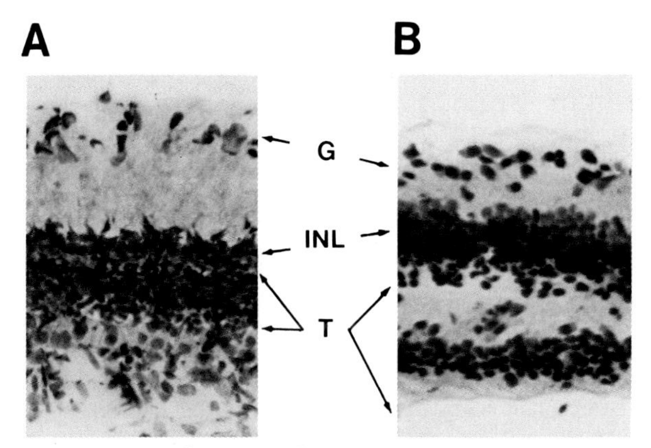

Fig. 4. A. Transplant of mature rat photoreceptors (30 day old donor) to adult light damaged host. B. Transplant of human photoreceptors from adult donor to adult light damaged rat host. T, transplant. 120X

Photoreceptors were taken from the retina of donated human eyes (obtained from the MO Lions and St. Louis Eye Banks) following corneal removal. A portion of the retinas were tested for viability by dye exclusion with trypan blue and didansyl cystine staining. The photoreceptors excluded dye and appeared to be in good condition. Hosts were adult albino rats (immune-suppressed with cyclosporin A or immune-competent) exposed to constant illumination. With immune-suppression successful transplants were seen at all survival times so far examined (1 and 2 weeks; 5 of 9 cases), showing apparent physical integration with the host retina and maintaining morphological features of the outer nuclear layer (Fig. 4B). The transplants stained positive for antiopsin antibody RET-P1, identifying the transplanted cells as photoreceptors and further indicating that they are still capable of producing visual pigment. In contrast, transplants to immune-competent hosts showed signs of rejection within one week of transplantation. Sham operated animals showed no repopulation of the host retina with photoreceptors.

FUNCTION

While it is clear that developing as well as mature photoreceptors can be transplanted and produce opsin, the question concerning their functional characteristics is of paramount importance. First, do transplanted photoreceptors transduce light and second, are they able to activate the host's retina?

Testing for recovery of function following photoreceptor transplantation is complicated by the fact that light-damaged retina is still capable of some light transduction. Therefore recording of ERG signals might not give a conclusive answer - especially in light of the relatively small size of most transplants (covering at most about 30% of the retina). Similarly, recording ganglion cell activity might not yield conclusive evidence of activation of the light-damaged retina by the photoreceptor transplant since the recorded activity might be driven by the residual host photoreceptors that are present following light damage.

If the photoreceptor transplant is capable of activating the host retina above the activity generated by the host's residual photoreceptors, then a method capable of simultaneously sampling the relative activity across damaged retina and adjacent retina receiving a photoreceptor graft might be used to assess the transplants ability to activate light damaged retina. The 2DG functional mapping technique developed by Sokoloff and coworkers (1977) allows the measurement of the relative levels of neural activity for a given stimulus condition. For this reason, the 2DG technique appeared to be an appropriate method of assessing the functional characteristics of the transplant and its ability to activate the light damaged retina.

We therefore compared the patterns of 2DG uptake in the normal retina to that seen in the light-damaged retina, with and without a photoreceptor transplant. We made these comparisons

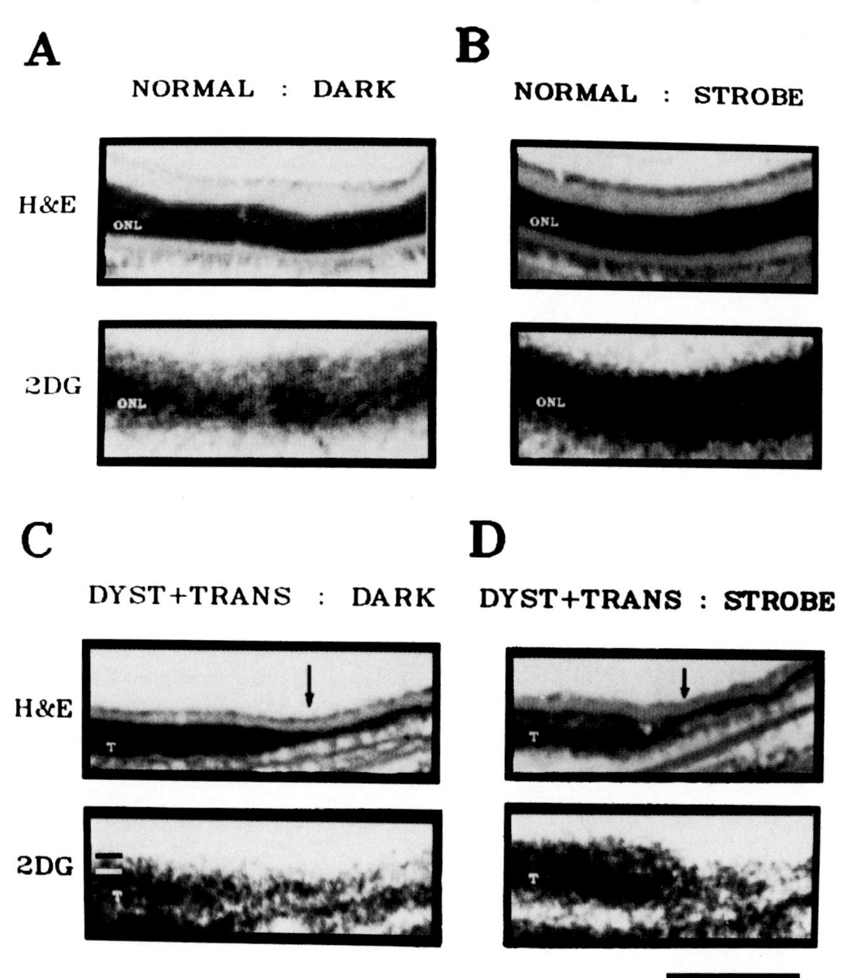

Fig.5. H&E stained retina with corresponding 2-deoxyglucose autoradiographs. A and B normal retina. Sections cut slightly tangentially to expand retinal layers. C. Dystrophic (light-damaged) retina plus photoreceptor transplant (T) left of arrow. Black and white lines at left on 2DG autoradiograph bracket lower 2DG uptake in inner plexiform and ganglion cell layers. D. Dystrophic retina plus photoreceptor transplant left of arrow. ONL; outer nuclear layer, T; transplant, DYST; dystrophic, Bar = 0.5 mm See text for description.

under two different visual stimulus conditions: 1) darkness and 2) strobe flicker at 10Hz. Figure 5 shows the results of these comparisons. As shown in panel A, in darkness 2DG was preferentially taken up in the outer portion of normal retina (photoreceptors and possibly the inner nuclear layer). As shown in panel B, with strobe flicker stimulation 2DG uptake extends through the thickness of the normal retina. These patterns of 2DG uptake are in good agreement with the known physiological characteristics of the retina.

The outer retina might be expected to show high 2DG uptake in the dark since photoreceptors, horizontal and some bipolar cells are maximally depolarized in this situation. As strobe flicker is a strong stimulus for the retina including the amacrine and retinal ganglion cells, 2DG uptake across the entire retina is also to be expected. It therefore appears that the 2DG uptake pattern in normal retina reflects relative degrees of neural activity or neural depolarization, and therefore is a useful indicator of neural activity in the retina as it is in other areas of the nervous system.

In the light-damaged retina which received a photoreceptor transplant, the pattern of 2DG uptake was also dependent on the stimulus conditions. In the dark, preferential uptake of 2DG was limited to the photoreceptor transplant and the adjacent host inner nuclear layer while relatively lower uptake was present in the host's inner plexiform and ganglion cell layers. However, in the strobe flicker condition, high 2DG uptake is present in the transplant and, in addition, extended through the thickness of the host's retina - but only in the area of the photoreceptor graft (Fig. 5D). Adjacent host retina which did not receive the photoreceptor transplant shows relatively low 2DG uptake.

In darkness, both the normal retina and the light-damaged retina receiving the photoreceptor transplant show relatively high uptake of 2DG in the photoreceptor and inner nuclear layers. The similarity in the relative uptake patterns between these cases suggests that the transplanted photoreceptors may have similar functional characteristics as normal photoreceptors (i.e., they depolarize in the dark and are capable of inducing a sustained depolarization of some cells in the host's inner nuclear layer).

In strobe flicker, the light-damaged retina receiving the photoreceptor transplant showed high 2DG uptake through the entire thickness of the retina much like that seen in the normal retina under the same stimulus condition. Adjacent light-damaged retina that did not receive a photoreceptor transplant shows relatively low 2DG uptake. These comparisons show that the pattern of 2DG uptake in the light-damaged retina approximates that seen in the normal retina only in areas of the host retina that received photoreceptor grafts. Adjacent areas of the host retina show relatively low levels of 2DG uptake in both stimulus conditions. The similarity in the 2DG uptake patterns between the light-damaged retina following photoreceptor grafting and the normal retina in both stimulus conditions suggests that the photoreceptor transplant is capable of light-dependent activation of the light-damaged retina.

SUMMARY

In conclusion, we have shown that photoreceptors can be transplanted to retina in which the host's photoreceptors are lost by environmental (constant light) or inherited deficits. Furthermore transplanted photoreceptor cells maintain basic characteristics of normal photoreceptor cells by producing opsin and maintaining an intercellular organization and apposition to the host retina that is similar to that seen in the normal outer nuclear layer.

To accomplish this we have devised a method to isolate the intact photoreceptor layer. This is significant because it will be necessary to maintain tight matrix organization if coherent vision is to be restored to the retina compromised by the loss of photoreceptors. We have further developed a surgical approach which minimizes trauma to the eye and allows controlled positioning of sheets of transplanted photoreceptors to their homotopic location within the eye. In addition these methods for transplantation and isolation of photoreceptors could be utilized to prepare and transplant other retinal layers so that selected populations of retinal cells can be used in other neurobiological investigations.

Photoreceptors can be transplanted when developing or when mature. Not only can mature rat photoreceptors can be transplanted, but we have shown that mature photoreceptors from human donors can be transplanted as well. This is significantly different from neurons which must be immature in order to be transplanted. At present the reason for this difference is not known but has obvious importance for retinal and neural transplantation research in general.

Finally, we have shown that transplanted photoreceptors activate the host's dystrophic retina in a light dependent manner that closely resembles the activation pattern seen in normal retina. This finding taken together with our results showing that human photoreceptors can be transplanted presents the possibility that some forms of human blindness might eventually be ameliorated by photoreceptor transplantation.

The authors are grateful to A. I. Cohen, and N. W. Daw for helpful discussions; to C. Barnstable for gifts of antibody; and to J. Lett for excellent technical assistance. This work was supported by grants from NIH, National Retinitis Pigmentosa Foundation, the Monsanto Company; and an Alfred P. Sloan Fellowship to MSS.

REFERENCES

Anderson, DH, Stern, WH, Fisher, SK, Erickson, PA and Borgula, GA (1983). Retinal detachment in the cat: The pigment epithelial-photoreceptor interface. Invest. Ophthalmol. Vis. Sci. 24:906-926

Bigami, A and Dahl, D (1979). The radial glia of Mueller in the rat retina and their response to injury. An immunofluorescence study with antibodies to glial fibrillary acidic (GFA) protein. Exp. Eye Res. 28:63-69.

Das, GD (1983). Neural transplantation in mammalian brain. In Wallace, RB and Das, GD (eds): "Neural Tissue Transplantation Research," New York: Springer-Verlag, pp. 1-64.

Del Cerro, M, Gash, DM, Rao, GN, Notter, MF, Wiegand, SJ, and Gupta, M (1985). Intraocular retinal transplants. Invest. Opthalmol. Vis. Sci. 26:1182-1185.

Farber, DB and Lolley, RN (1974). Cyclic guanosine monophosphate: Elevation in degenerating photoreceptor cells of the C3H mouse retina. Science 186:449-451.

Gartner, S and Henkind, P (1981). Aging and degeneration of the human macula. I. Outer nuclear layer and photoreceptors. Br. J. Ophthalmol. 65:23-28.

Hicks,D and Barnstable, CJ (1987). A phosphorylation-sensitive antirhodopsin monoclonal antibody reveals light-induced phosphorylation of rhodopsin in the photoreceptor cell body. J. Histochem. Cytochem. 35:1317-1328.

Honig, MG and Hume, RI (1986). Fluorescent carbocyanine dyes allow living neurons of identified origin to be studied in long-term cultures. J. Cell Biol. 103:17-187.

LaVail, MM and Hild, W (1971). Histotypic organization of the rat retina in vitro. Z. Zellforsch. 114:557-579.

Leibowitz, HM, Kruegger, DE, Mounder, LR, Milton, RC, Kini, MM, Kahn, HA, Nickerson, RJ, Pool J, Colton, TL, Ganley JP, Loewenstein JI, and Dawber TR (1980). The Framingham Eye Study Monograph. Surv. Ophthalmol. 24 (Suppl):335-610.

Lolley, RN and Farber, DB (1976). A proposed link between debris accumulation, guanosine 3',5'-cyclic monophosphate changes and photoreceptor cell degeneration in retina of RCS rats. Exp. Eye Res. 22:477-486.

Lund, RD and McLoon, SC (1983). Retinal

transplants. In Wallace, RB and Das, GD, (eds): "Neural Tissue Transplantation Research," New York: Springer-Verlag,, pp. 165-173.

McConnel, P and Berry, M (1982). Regeneration of ganglion cell axons in the adult mouse retina. Brain Res. 241:362-365.

McLoon, LK and McLoon, SC (1984). Early development of projections from embryonic retina transplanted into the host brain of rats. Soc. Neurosci. Abstr. 10: 1035.

Mullen RJ and LaVail, MM (1976). Inherited retinal dystrophy: Primary defect in pigment epithelium determined with experimental rat chimeras. Science 192:799-801.

Pu, GA and Masland, RH (1984). Biochemical interruption of membrane phospholipid renewal in retinal photoreceptor cells. J. Neurosci. 4:1559-1576.

Sarthy, PV and Lam, DMK (1979). Isolated cells from a mammalian retina. Brain Res. 176: 208-212.

Silverman, MS and Hughes, SE (1987a). Transplantation of retinal photoreceptors to light damaged retina. Suppl., Invest. Ophthalmol. Visual Sci. 28:288.

Silverman, MS and Hughes, SE (1987b). Transplantation of retinal photoreceptors to light damaged retina: Survival and integration of receptors from a range of postnatal ages. Soc. Neurosci. Abstr. 17:1301.

Simons, D J and Lund, R D (1984). Physiological responses in retinal transplants and host tecta evoked by electrical or photic stimulation of transplanted embryonic retinae. Soc. Neurosci. Abstr. 10:668.

Sokoloff, L, Reivich, M Kennedy, C Des Rosiers, MH Patlak, CS Pettigrew, KD Sakurada, O and Shinohara, M (1977). The ^{14}C deoxyglucose method for the measurement of local cerebral glucose utilization: Theory, procedure, and normal values in the conscious and anesthetized albino rat. J. Neurochem. 28:897-916.

Turner, JE and Blair, JR (1986). Newborn rat retinal cells transplanted into a retinal lesion site in adult host eyes. Dev. Brain Res. 26:91-104.

Index